Lecture Notes in Computer Science   10520

Commenced Publication in 1973
Founding and Former Series Editors:
Gerhard Goos, Juris Hartmanis, and Jan van Leeuwen

Advanced Research in Computing and Software Science
Subline of Lecture Notes in Computer Science

More information about this series at http://www.springer.com/series/7407

Hans L. Bodlaender · Gerhard J. Woeginger (Eds.)

Graph-Theoretic Concepts in Computer Science

43rd International Workshop, WG 2017
Eindhoven, The Netherlands, June 21–23, 2017
Revised Selected Papers

 Springer

Editors
Hans L. Bodlaender
Utrecht University
Utrecht
The Netherlands

Gerhard J. Woeginger
Rheinisch-Westfälische Technische
    Hochschule Aachen
Aachen
Germany

ISSN 0302-9743   ISSN 1611-3349 (electronic)
Lecture Notes in Computer Science
ISBN 978-3-319-68704-9   ISBN 978-3-319-68705-6 (eBook)
https://doi.org/10.1007/978-3-319-68705-6

Library of Congress Control Number: 2017956719

LNCS Sublibrary: SL1 – Theoretical Computer Science and General Issues

Printed on acid-free paper

This Springer imprint is published by Springer Nature
The registered company is Springer International Publishing AG
The registered company address is: Gewerbestrasse 11, 6330 Cham, Switzerland

Preface

This volume contains the 31 papers that were presented at WG 2017, the 43rd International Workshop on Graph-Theoretic Concepts in Computer Science. The workshop was held during June 21–23, 2017, at the Kapellerput Conference Hotel in Heeze, The Netherlands.

WG conferences cover a wide range of areas, and they connect theory and applications by demonstrating how graph-theoretic concepts can be applied in various areas of computer science. One of the main goals is to present recent results and to identify and explore promising directions of future research. WG has a long tradition that goes back the 1970s. The first three WGs were organized in 1975 by Uwe Pape at TU Berlin, in 1976 by Hartmut Noltemeier in Göttingen, and in 1977 by Jörg Mühlbacher in Linz.

WG usually takes place in Europe. WG 2017 was the fifth time the conference was held in The Netherlands; earlier, WG was held in Amsterdam (1988), Castle Rolduc (near Heerlen, The Netherlands, and Aachen, Germany, 1989), Arnhem (1993), and Elspeet (2003). In addition, it has been organized many times in Germany, three times in France, twice in Austria and twice in Czech Republic, and once in each of England, Greece, Israel, Italy, Norway, Slovakia, Switzerland, and Turkey.

Three excellent invited lectures at WG 2017 were given by Petra Mutzel (Dortmund), Remco van der Hofstad (Eindhoven), and Fedor Fomin (Bergen).

We received 71 submissions in total, and out of these the Program Committee selected 31 papers for presentation at the symposium and for publication in the proceedings. Each submission was reviewed by at least four Program Committee members. We expect the full versions of the papers contained in this volume to be submitted for publication in refereed journals. The Program Committee also selected the winners of the Best Paper Award and the Best Student Paper Award:

Best Paper Award
Yijia Chen, Martin Grohe, and Bingkai Lin:
"The Hardness of Embedding Grids and Walls"

The Program Committee decided to split the Best Student Paper Award between the following two papers:

Best Student Paper Award
Dušan Knop, Martin Koutecky, Tomáš Masařík, and Tomáš Toufar:
"Simplified Algorithmic Metatheorems Beyond MSO: Treewidth and Neighborhood Diversity"

Best Student Paper Award
Steven Chaplick, Martin Töpfer, Jan Voborník, and Peter Zeman:
"On H-Topological Intersection Graphs"

Many individuals and organizations contributed to the smooth running and the success of WG 2017. In particular our thanks go to:

- All authors who submitted their newest research to WG
- Our reviewers and additional referees whose expertise flowed into the decision process
- The members of the Program Committee, who graciously gave their time and energy
- The members of the local Organizing Committee, who made the conference possible
- The EasyChair conference management system for hosting the evaluation process
- The NWO Gravitation program NETWORKS for financial support
- The Mathematics and Computer Science Department at TU Eindhoven
- The members of the local Organizing Committee
- Springer for supporting the Best Paper Awards
- The invited speakers, the other speakers, and the participants for making WG 2017 an inspiring event

August 2017   Hans L. Bodlaender
                                                                    Gerhard J. Woeginger

Organization

Program Committee

Hans L. Bodlaender	Eindhoven and Utrecht, The Netherlands (Chair)
René van Bevern	Novosibirsk, Russia
Andreas Brandstädt	Rostock, Germany
Bart M.P. Jansen	Eindhoven, The Netherlands
Iyad Kanj	Chicago, USA
Mamadou M. Kanté	Aubiere, France
Michael Kaufmann	Tübingen, Germany
Eun Jung Kim	Paris, France
Christian Komusiewicz	Jena, Germany
Stefan Kratsch	Bonn, Germany
Asaf Levin	Haifa, Israel
Haiko Müller	Leeds, UK
Sang-il Oum	Daejeon, South Korea
Felix Reidl	Raleigh, USA
Saket Saurabh	Chennai, India and Bergen, Norway
Pascal Schweitzer	Aachen, Germany
Jan Arne Telle	Bergen, Norway
Ioan Todinca	Orleans, France
Dimitrios M. Thilikos	Athens, Greece and Montpellier, France
Gerhard J. Woeginger	Aachen, Germany (Chair)

Organizing Committee

Mark de Berg
Hans Bodlaender (Chair)
Meivan Cheng
Federico D'Ambrosio
Bart M.P. Jansen
Sándor Kisfaludi-Bak
Jesper Nederlof
Marieke van der Wegen
Tom van der Zanden

Additional Reviewers

Abu-Khzam, Faisal
Adjiashvili, David
Adler, Isolde
Agrawal, Akanksha
Alt, Helmut
Angelini, Patrizio
Baiou, Mourad
Baste, Julien
Bazgan, Cristina
Bekos, Michael
Bergougnoux, Benjamin
Bonamy, Marthe
Bonnet, Edouard
Bonomo, Flavia
Bousquet, Nicolas
Bulteau, Laurent
Bärtschi, Andreas
Cabello, Sergio
Calamoneri, Tiziana
Cameron, Kathleen
Casel, Katrin
Chalopin, Jérémie
Chaplick, Steven
Chiarelli, Nina
Chikhi, Rayan
Chitnis, Rajesh
Cohen, Nathann
Cornelsen, Sabine
Curticapean, Radu
Da Silva, Murilo
Dabrowski, Konrad Kazimierz
Das, Sandip
Dragan, Feodor
Dreier, Jan
Ducoffe, Guillaume
Dürr, Christoph
Eiben, Eduard
Eppstein, David
Escoffier, Bruno
Felsner, Stefan
Ferrara, Michael
Fiala, Jiri
Fleszar, Krzysztof

Fluschnik, Till
Fomin, Fedor
Fulek, Radoslav
Förster, Henry
Gajarský, Jakub
Ganian, Robert
Giannopoulou, Archontia
Golovach, Petr
Gonçalves, Daniel
Guo, Jiong
Gupta, Sushmita
Habib, Michel
Hanauer, Kathrin
Heinsohn, Niklas
Hermelin, Danny
Hong, Seok-Hee
Huang, Chien-Chung
Ito, Takehiro
Jean, Geraldine
Jelínek, Vít
Johnson, Matthew
Joos, Felix
Kaibel, Volker
Kiraly, Tamas
Klavzar, Sandi
Korhonen, Janne H.
Kostitsyna, Irina
Kowalik, Lukasz
Kratochvil, Jan
Kratsch, Dieter
Krithika, R.
Kuinke, Philipp
Kwon, O-joung
Köhler, Ekkehard
Lampis, Michael
Lapinskas, John
Lauri, Juho
Leveque, Benjamin
Liedloff, Mathieu
Limouzy, Vincent
Lokshtanov, Daniel
Löffler, Maarten
Maffray, Frederic

Manlove, David
Markovic, Aleksandar
Mary, Arnaud
Maus, Yannic
Mazoit, Frédéric
McConnell, Ross
Mchedlidze, Tamara
Mertzios, George
Milanic, Martin
Mishra, Sounaka
Mnich, Matthias
Molter, Hendrik
Mondal, Debajyoti
Montealegre, Pedro
Mouawad, Amer
Müller-Hannemann, Matthias
Nederlof, Jesper
Nichterlein, André
Niedermann, Benjamin
Nisse, Nicolas
Nöllenburg, Martin
O'Brien, Michael P.
Obdrzalek, Jan
Ochem, Pascal
Ordyniak, Sebastian
Otachi, Yota
Panolan, Fahad
Pastor, Lucas
Paulusma, Daniël
Pieterse, Astrid
Pilipczuk, Michał
Ponomaryov, Denis
Pyatkin, Artem

Rahman, Md. Saidur
Rao, Michael
Ries, Bernard
Roeloffzen, Marcel
Rutter, Ignaz
Saeidinvar, Reza
Sau, Ignasi
Schaudt, Oliver
Schiermeyer, Ingo
Schlipf, Lena
Segev, Danny
Sinaimeri, Blerina
Sorge, Manuel
Spillner, Andreas
Spoerhase, Joachim
Stavropoulos, Konstantinos
Stewart, Lorna
Strozecki, Yann
Tale, Prafullkumar
Talmon, Nimrod
Thiyagarajan, Karthick
Trenk, Ann
Tsidulko, Oxana
van Leeuwen, Erik Jan
Vaxes, Yann
Vuskovic, Kristina
Walczak, Bartosz
Wasa, Kunihiro
Weller, Mathias
Weltge, Stefan
Wood, David R.
Zhang, Peng
Zoros, Dimitris

Contents

Counting Graphs and Null Models of Complex Networks: Configuration Model and Extensions

Remco van der Hofstad[(✉)]

Department of Mathematics and Computer Science,
Eindhoven University of Technology, P.O. Box 513,
5600 MB Eindhoven, The Netherlands
rhofstad@win.tue.nl
http://www.win.tue.nl/~rhofstad

Abstract. Due to its ease of use, as well as its enormous flexibility in its degree structure, the configuration model has become the network model of choice in many disciplines. It has the wonderful property, that, conditioned on being simple, it is a uniform random graph with the prescribed degrees. This is a beautiful example of a general technique called the *probabilistic method* that was pioneered by Erdős. It allows us to count rather precisely how many graphs there are with various degree structures. As a result, the configuration model is often used as a *null model* in network theory, so as to compare real-world network data to. When the degrees are sufficiently light-tailed, the asymptotic probability of simplicity for the configuration model can be explicitly computed. Unfortunately, when the degrees vary rather extensively and vertices with very high degrees are present, this method fails. Since such degree sequences are frequently reported in empirical work, this is a major caveat in network theory.

In this survey, we discuss recent results for the configuration model, including asymptotic results for typical distances in the graph, asymptotics for the number of self-loops and multiple edges in the finite-variance case. We also discuss a possible fix to the problem of non-simplicity, and what the effect of this fix is on several graph statistics. Further, we discuss a generalization of the configuration model that allows for the inclusion of community structures. This model removes the flaw of the locally tree-like nature of the configuration model, and gives a much improved fit to real-world networks.

1 Complex Networks and Random Graphs: A Motivation

In this survey, we discuss random graph models for complex networks, which are large and highly heterogeneous real-world graphs such as the Internet, the World-Wide Web, social networks, collaboration networks, citation networks, the neural network of the brain, etc. Such networks have received enormous attention in the past decades, partly because they appear in virtually all domains in science. This is also due to the fact that such networks, even though they arise in highly different fields in science and society, share some fundamental properties. Let us describe the two most important ones now.

© Springer International Publishing AG 2017
H.L. Bodlaender and G.J. Woeginger (Eds.): WG 2017, LNCS 10520, pp. 1–17, 2017.
https://doi.org/10.1007/978-3-319-68705-6_1

Scale-Free Phenomenon. The first, maybe quite surprising, fundamental property of many real-world networks is that the number of vertices with degree at least k decays slowly for large k. This implies that degrees are highly variable, and that, even though the average degree is not so large, there exist vertices with extremely high degree. Often, the tail of the empirical degree distribution seems to fall off as an inverse power of k. This is called a 'power-law degree sequence', and resulting graphs often go under the name 'scale-free graphs'. It is visualized for the AS graph in Fig. 1, where the degree distribution of the Autonomous System (AS) graph is plotted on a log-log scale. The vertices of the AS graph correspond to groups of routers controlled by the same operator. Thus, we see a plot of $\log k \mapsto \log n_k$, where n_k is the number of vertices with degree k. When n_k is proportional to an inverse power of k, i.e., when, for some normalizing constant c_n and some exponent τ,

$$n_k \approx c_n k^{-\tau}, \tag{1}$$

then

$$\log n_k \approx \log c_n - \tau \log k, \tag{2}$$

so that the plot of $\log k \mapsto \log n_k$ is close to a straight line. This is the reason why degree sequences in networks are often depicted in a log-log fashion, rather than in the more customary form of $k \mapsto n_k$. Here, and in the remainder of this text, we write $\approx$ to denote an uncontrolled approximation. The power-law exponent τ can be estimated by the slope of the line in the log-log plot. Vertices with extremely high degrees are often called *hubs*, as the hubs in airport networks, or *super-spreaders*, indicating their importance in spreading information, or diseases. For Internet, log-log plots of degree sequences first appeared in a

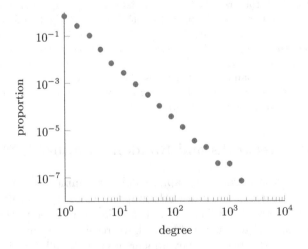

Fig. 1. (a) Log-log plot of the probability mass function of the degree sequence of Autonomous Systems (AS) on April 2014 on a log-log scale from [27] (data courtesy of Dmitri Krioukov).

paper by the Faloutsos brothers [15] (see Fig. 1 for the degree sequence in the Autonomous Systems graph). Here the power-law exponent is estimated as $\tau \approx$ 2.15–2.20. Figure 2 displays the degree-sequence for both the in- as well as the out-degrees in various World-Wide Web data bases.

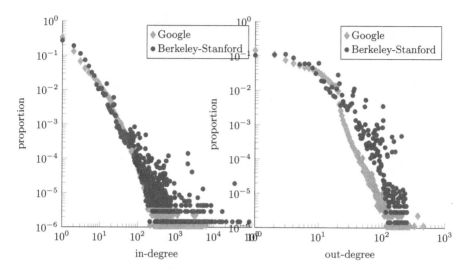

Fig. 2. The probability mass function of the in- and out- degree sequences in the Berkeley-Stanford and Google competition graph data sets of the WWW in [28]. (a) in-degree; (b) out-degree.

Small-World Phenomenon. The second fundamental network property observed in many real-world networks is that typical distances between vertices are small. This is called the 'small-world' phenomenon (see e.g. the book by Watts [35]). In particular, such networks are highly connected: their largest connected component contains a significant proportion of the vertices. Many networks, such as the Internet, even consist of *one* connected component, since otherwise e-mail messages could not be delivered. For example, in the Internet, IP-packets cannot use more than a threshold of physical links, and if distances in the Internet would be larger than this threshold, then e-mail service would simply break down. As seen in Fig. 3(a), the number of Autonomous Systems (AS) traversed by an e-mail data set, sometimes referred to as the AS-count, is typically at most 7. In Fig. 3(b), the proportion of routers traversed by an e-mail message between two uniformly chosen routers, referred to as the *hopcount*, is shown to be at most 27.

For pairs of vertices $u, v \in [n]$ and a graph $G = ([n], E)$, we let the graph distance $\text{dist}_G(u, v)$ between u and v be equal to the minimal number of edges in a path linking u and v. When u and v are not in the same connected component, we set $\text{dist}_G(u, v) = \infty$. We draw U_1 and U_2 uniformly at random from

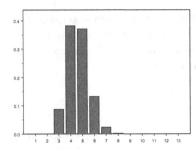 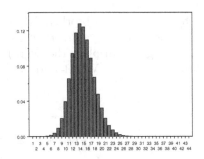

Fig. 3. (a) Proportion of AS traversed in hopcount data. (b) Internet hopcount data. Courtesy of Hongsuda Tangmunarunkit.

$[n]$, and investigate $\mathrm{dist}_G(U_1, U_2)$, which is a random variable even for *determin-istic* graphs due to the occurrence of the two, uniformly at randomly chosen, vertices $U_1, U_2 \in [n]$. Figures 3 and 4 display the probability mass functions of this random variable for some real-world networks.

The nice property of $\mathrm{dist}_G(U_1, U_2)$ is that its distribution tells us something about *all possible* distances in the graph. An alternative and frequently used measure of distances in a graph is the *diameter* $\mathrm{diam}(G)$, defined as

$$\mathrm{diam}(G) = \max_{u,v \in [n]} \mathrm{dist}_G(u, v). \tag{3}$$

[Often, the maximum in (3) is restricted to pairs of vertices u, v that are con-nected, i.e., for which $\mathrm{dist}_G(u, v) < \infty$.] The diameter has several disadvantages, as it is algorithmically more difficult to compute than the typical distances (since one has to measure the distances between all pairs of vertices and maximize over them). Further, it contains far less information than the distribution of $\mathrm{dist}_G(U_1, U_2)$. Finally, the diameter is highly sensitive to small changes of the graph, in that adding a string of connected vertices to a graph may change the diameter dramatically, while it hardly influences the typical distances.

2 Random Graphs and Real-World Networks

In this section, we discuss how *random graph sequences* can be used to model real-world networks. We start by discussing graph sequences:

Graph Sequences. Since many networks are quite large, mathematically, we model real-world networks by *graph sequences* $(G_n)_{n \geq 1}$, where G_n has size n and we take the limit $n \to \infty$. Since most real-word networks are such that the average degree remains bounded, we will focus on the *sparse* regime, where it is assumed that the average degree $\frac{1}{n} \sum_{i \in [n]} d_i$ is uniformly bounded.

Other features that many networks share, or rather form a way to distinguish them, are their *degree correlations*, measuring the extent to which high-degree vertices tend to be connected to high-degree vertices, or rather to low-degree vertices (and vice versa), their *clustering*, measuring the extent to which pairs of neighbors of vertices are neighbors themselves as well, and their *community structure*, measuring the extent to which the networks have more dense connected subparts. See e.g., the book by Newman [31] for an extensive discussion of such features, as well as the algorithmic problems that arise from them, or [19, Sect. 1.4].

Random Graphs as Models for Real-World Networks. Real-world networks tend to be quite complex and unpredictable. Connections often arise rather irregularly. We model such irregular behaviors through a *random process*, thus leading us to study random graphs. By the previous discussion, our graphs will be large and their size will tend to infinity. In such a setting, we can either model the graphs by *fixing their size* to be large, or rather by letting the graphs *grow* large. We refer to these two settings as *static* and *dynamic* random graphs. Both viewpoints are useful. Indeed, a static graph is a model for a snapshot of a network at a fixed time, where we do not know how the connections arose in time. Many network data sets are of this form. A dynamic setting, however, may be more appropriate when we know how the network came to be as it is. In the static setting, we can make assumptions on the degrees so that they are scale free. In the dynamic setting, we can let the evolution of the graph be such that they give rise to power-law degree sequences, thus providing possible explanations for the frequent occurrence of power-laws in real-world networks.

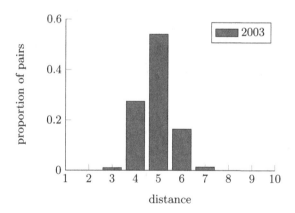

Fig. 4. Typical distances in the Internet Movie Data base in 2003.

3   Random Graph Models as Null Models

Here, we study random graph models as *null models* for network data. In such models, we take certain aspects of real-world networks into account, while ignoring others. This gives a qualitative way of investigating the importance of such empirical features in the real world. Often, real-world networks are compared to *uniform random graphs* with certain specified properties, such as their number of edges or even their degree sequence.

3.1 Null Model 1: Uniform Random Graph

The simplest null model is when we take no properties into account at all except for the size of the network. Then, each of the $\binom{n}{2} = n(n-1)/2$ edges can be there or not, giving $2^{\binom{n}{2}}$ possible graphs. Since we choose the null model uniformly at random from a specific collection of graphs, each of these graphs G of n vertices has probability

$$\mathbb{P}(G) = \frac{1}{2^{\binom{n}{2}}} \tag{4}$$

to be chosen. This probability distribution is the same as choosing each edge independently with probability $p = \frac{1}{2}$. The model where edges are present independently is sometimes called the Erdős-Rényi random graph. It is the simplest possible random graph. In it, we make every possible edge between a collection of n vertices open or closed with equal probability. It has vertex set $[n] = \{1, \ldots, n\}$, and, denoting the edge between vertices $s, t \in [n]$ by st, st is occupied or present with probability p, and vacant or absent otherwise, independently of all the other edges. The parameter p is called the *edge probability*. This above random graph is denoted by $\mathrm{ER}_n(p)$, and the uniform random graph corresponds to $p = \frac{1}{2}$. This model has expected degree $(n-1)/2$, which is quite large. As a result, this model is not sparse at all. We conclude that this is not a good model for complex networks. Erdős [9] used the uniform random graph to show that most graphs have a large complete graph of occupied as well as of vacant edges. This was the first example of the *probabilistic method* of using probabilistic tools to prove deterministic properties.

3.2 Null Model 2: Erdős-Rényi Random Graph with Fixed Number of Edges

In the uniform random graph defined above, we see that there are in expectation $n(n-1)/4$ edges, so that the model is not sparse at all. In this section, we investigate the second attempt at finding a null model, which arises by fixing the total number of edges to be equal to m, and choosing any set of m edges uniformly at random. Particularly, the sparse setting arises when $m = 2\lambda n$ grows linearly with n for some $\lambda > 0$ denoting the asymptotic average degree, fixed.

This gives rise to the *combinatorial Erdős-Rényi random graph*. It is named after Erdős and Rényi, who made profound contributions to its study in [11–14]. The so-called *binomial model* defined in the previous section was first introduced by Gilbert [18]. Informally, when $m = p\binom{n}{2}$, the two models behave very similarly, so that $m = 2\lambda n$ corresponds to $p = \lambda/(n-1) \approx \lambda/n$. The combinatorial model has the nice feature that it produces a uniform graph from the collection of all graphs with m edges, and thus could serve as a *null model* for a real-world network in which the number of vertices and edges is fixed. For fixed m, there are $\binom{\binom{n}{2}}{m}$ graphs that have n vertices and m edges. Since we choose the null model uniformly at random, each of these graphs G of n vertices and m edges has probability

$$\mathbb{P}(G) = \frac{1}{\binom{\binom{n}{2}}{m}}. \tag{5}$$

We take $m = 2\lambda n$, and study the graph as λ is fixed while $n \to \infty$. In this regime, it turns out that the proportion of vertices with degree k converges to the probability mass function of a Poisson random variable (see [19, Sect. 5.4]), i.e., for every $k \geq 0$,

$$P_k^{(n)} = \frac{1}{n} \sum_{i \in [n]} \mathbb{1}_{\{d_i = k\}} \xrightarrow{\ \mathbb{P}\ } p_k \equiv \mathrm{e}^{-\lambda} \frac{\lambda^k}{k!}, \tag{6}$$

where d_i denotes the degree of vertex $i \in [n]$ in the model with $m = 2\lambda n$. It is well known that the Poison distribution has very thin tails, even thinner than any exponential, so that we conclude that the Erdős-Rényi random graph is not a good model for real-world networks with their highly-variable degree distributions.

3.3   Null Model 3: Fixing All Degrees and the Configuration Model

In the third null model, the degrees of *all* vertices are fixed beforehand. This way, we can be sure that the degrees are what we want them to be. Such a model is more flexible than the Erdős-Rényi random graph, which for example always has a positive proportion of vertices of degree 0, 1, 2, etc., as easily follows from (6). One of the difficulties is that it is quite unclear *how many* such graphs there are, as well as how to produce one uniformly at random. For example, it is even non-trivial to figure out which degree sequences are possible. Naturally, the sum of the degrees needs to be even, which is called the *handshake lemma*. Erdős and Gallai [10] gave a precise criterion which degree sequences allow for a simple graph with these degrees. We will not give more details, as the sparse settings that we study typically satisfy this criterion when the sum of the degrees is even.

The work-around to the difficulty in studying simple graphs with prescribed degrees goes under the name of the *configuration model*, which is a beautiful and extremely powerful example of the probabilistic method. Rather than aiming for a simple graph, we construct a *multigraph* and show that, conditioned on simplicity (meaning, no multiple edges and self-loops), the realization is uniform over the set of graphs with those degrees. Fix an integer n that denotes the number of vertices in the random graph. Consider a sequence of degrees $\boldsymbol{d} = (d_i)_{i \in [n]}$. The aim is to construct an undirected graph with n vertices, where vertex j has degree d_j, chosen uniformly from the collection of all simple graphs with these degrees. Without loss of generality, we assume that $d_j \geq 1$ for all $j \in [n]$, since when $d_j = 0$, vertex j is isolated and can be removed from the graph. For there to exist one such graphs, we must assume that the total degree

$$\ell_n = \sum_{j \in [n]} d_j \tag{7}$$

is even. To construct the multigraph where vertex j has degree d_j for all $j \in [n]$, we have n separate vertices and incident to vertex j, we have d_j half-edges. Every half-edge needs to be connected to another half-edge to form an edge, and by forming all edges we build the graph. For this, the half-edges are numbered in an arbitrary order from 1 to ℓ_n. We start by randomly connecting the first half-edge with one of the $\ell_n - 1$ remaining half-edges. Once paired, two half-edges form a single edge of the multigraph, and the half-edges are removed from the list of half-edges that need to be paired. Hence, a half-edge can be seen as the left or the right half of an edge. We continue the procedure of randomly choosing and pairing the half-edges until all half-edges are connected, and call the resulting graph the *configuration model with degree sequence* $\boldsymbol{d}$, abbreviated as $\mathrm{CM}_n(\boldsymbol{d})$.

A careful reader may worry about the order in which the half-edges are being paired. In fact, this ordering turns out to be completely irrelevant since the random pairing of half-edges is *exchangeable*. See e.g., [19, Definition 7.5 and Lemma 7.6] for more details on this exchangeability. Interestingly, one can compute explicitly what the distribution of $\mathrm{CM}_n(\boldsymbol{d})$ is. To do so, note that $\mathrm{CM}_n(\boldsymbol{d})$ is characterized by the random vector $(X_{ij})_{1 \leq i \leq j \leq n}$, where X_{ij} is the number of edges between vertices i and j. Here X_{ii} is the number of self-loops incident to vertex i, and

$$d_i = X_{ii} + \sum_{j \in [n]} X_{ij} \tag{8}$$

In terms of this notation, and writing $G = (x_{ij})_{i,j \in [n]}$ to denote a multigraph on the vertices $[n]$,

$$\mathbb{P}(\mathrm{CM}_n(\boldsymbol{d}) = G) = \frac{1}{(\ell_n - 1)!!} \frac{\prod_{i \in [n]} d_i!}{\prod_{i \in [n]} 2^{x_{ii}} \prod_{1 \leq i \leq j \leq n} x_{ij}!}. \tag{9}$$

This can be proved by carefully checking how many matchings correspond to the same graph G. See e.g., [19, Proposition 7.7] for this result. In particular, $\mathbb{P}(\mathrm{CM}_n(\boldsymbol{d}) = G)$ is the *same* for each *simple* G, where G is simple when $x_{ii} = 0$ for every $i \in [n]$ and $x_{ij} \in \{0, 1\}$ for every $1 \leq i < j \leq n$. Thus, the configuration model conditioned on simplicity is a *uniform* random graph with the prescribed degree distribution, and is thus equal to the null model that we were after in the first place. This is quite relevant, as it gives a convenient way to *obtain* such a uniform graph, which is a highly non-trivial fact and another beautiful example of the probabilistic method.

Interestingly, the configuration model was invented by Bollobás in [4] to study uniform random regular graphs (see also [5, Sect. 2.4]). The introduction was inspired by, and generalized the results in, the work of Bender and Canfield [3]. The configuration model with varying degrees, as well as uniform random graphs with a prescribed degree sequence, were studied in greater generality by Molloy and Reed in [29, 30]. This extension is quite relevant to us, as the scale-free nature of many real-world applications encourages us to investigate configuration models with power-law degree sequences.

We impose some *regularity conditions* on the degree sequence d. In order to state these assumptions, we introduce some notation. Let n_k denote the number of vertices with degree k. We assume that the vertex degrees satisfy the following *regularity conditions:*

Condition 1 (Regularity conditions for vertex degrees). *As $n \to \infty$, there exists a probability mass function $(p_k)_{k \geq 0}$ such that*

$$\frac{n_k}{n} \to p_k. \tag{10}$$

Further, we assume that $n_0 = n_1 = 0$.
(b) Convergence of average vertex degrees. *As $n \to \infty$,*

$$\sum_{k \geq 0} k \frac{n_k}{n} \to \sum_{k \geq 0} k p_k. \tag{11}$$

(c) Convergence of second moment vertex degrees. *As $n \to \infty$,*

$$\sum_{k \geq 0} k^2 \frac{n_k}{n} \to \sum_{k \geq 0} k^2 p_k. \tag{12}$$

The possibility of obtaining a non-simple graph is a major disadvantage of the configuration model. There are two ways of dealing with this complication:

(a) Configuration model conditioned on simplicity. The first solution to the multigraph problem of the configuration model is to throw away the result when it is not simple, and to try again. This construction is sometimes called the *repeated configuration model* (see [7]). It turns out that, when Conditions 1(a)–(c) hold, then (see [19, Theorem 7.12])

$$\lim_{n \to \infty} \mathbb{P}(\mathrm{CM}_n(d) \text{ is a simple graph}) = \mathrm{e}^{-\nu/2 - \nu^2/4}, \tag{13}$$

where

$$\nu = \frac{\mathbb{E}[D(D-1)]}{\mathbb{E}[D]} \tag{14}$$

is the expected forward degree and $\mathbb{P}(D = k) = p_k$ is the random variable with probability mass function $(p_k)_{k \geq 1}$ that appears in Condition 1(a). Thus, this is a realistic option when $\mathbb{E}[D^2] < \infty$. Unfortunately, this is not an option when the degrees obey an asymptotic power law with $\tau \in (2, 3)$, since then $\mathbb{E}[D^2] = \infty$. This is a pity, as values of τ satisfying $\tau \in (2, 3)$ occur rather frequently in real-world networks.

Note that, by (9), $\mathrm{CM}_n(d)$ conditioned on simplicity is a *uniform random graph* with the prescribed degree sequence. Thus, the number of simple graphs $N_n(d)$ with given degree sequence d is equal to

$$N_n(d) = \frac{\mathbb{P}(\mathrm{CM}_n(d) \text{ is a simple graph})}{\mathbb{P}(\mathrm{CM}_n(d) = G)}$$

$$= \mathbb{P}(\mathrm{CM}_n(d) \text{ is a simple graph}) \frac{(\ell_n - 1)!!}{\prod_{i \in [n]} d_i!}, \tag{15}$$

which, with (13), allows us to compute how many there are, as well as to simulate one efficiently. This is the *probabilistic method* at its best!

(b) Erased configuration model. When $\mathbb{E}[D^2] = \infty$, or, when $\tau \in (2,3)$, we see from (13) that the probability of obtaining a simple graph vanishes in the large graph limit. This poses a huge problem. We now propose a highly practical fix for this problem, which, however, is slightly different from a uniform random graph with the prescribed degrees.

Indeed, one way of dealing with multiple edges is to *erase* the problems. This means that we replace $\mathrm{CM}_n(\boldsymbol{d}) = (X_{ij})_{1 \leq i \leq j \leq n}$ by its erased version $\mathrm{ECM}_n(\boldsymbol{d}) = (X_{ij}^{(\mathrm{er})})_{1 \leq i \leq j \leq n}$, where $X_{ii}^{(\mathrm{er})} \equiv 0$, while $X_{ij}^{(\mathrm{er})} = 1$ precisely when $X_{ij} \geq 1$. In words, we remove the self-loops and merge all multiple edges to a single edge. Of course, this changes the precise degree distribution. However, [19, Theorem 7.10] shows that only a small proportion of the edges is erased, so that the erasing does not substantially change the degree distribution. See [19, Sect. 7.3] for more details. Of course, the downside of this approach is that the degrees are changed by the procedure, while we would like to keep the degrees *precisely* as specified.

Let us describe the degree distribution in the erased configuration model in more detail, to study the effect of the erasure of self-loops and multiple edges. We denote the degrees in the erased configuration model by $\boldsymbol{D}^{(\mathrm{er})} = (D_i^{(\mathrm{er})})_{i \in [n]}$, and denote the related degree sequence in the erased configuration model $(P_k^{(\mathrm{er})})_{k \geq 1}$ by

$$P_k^{(\mathrm{er})} = \frac{1}{n} \sum_{i \in [n]} \mathbb{1}_{\{D_i^{(\mathrm{er})} = k\}}. \tag{16}$$

From the notation it is clear that $(P_k^{(\mathrm{er})})_{k \geq 1}$ is a *random* sequence, since the erased degrees $(D_i^{(\mathrm{er})})_{i \in [n]}$ form a random vector even when $\boldsymbol{d} = (d_i)_{i \in [n]}$ is deterministic. Then $P_k^{(\mathrm{er})} \rightarrow p_k$ when Conditions 1(a)–(b) hold, which implies that we remove only a vanishing proportion of the edges. Thus, the erased configuration model provides a good proxy for a uniform random graph with those degrees.

Beyond the Finite-Variance Constraint. Interestingly, there are some results concerning the number of graphs with prescribed degrees even in settings where $\tau \in (2,3)$. For example, Gao and Wormald [17] prove that the number of simple graphs with degree sequence $\boldsymbol{d}$ equals

$$(1 + o(1)) \frac{(\ell_n - 1)!!}{\prod_{i \in [n]} d_i!} \exp\left\{ -n\nu_n/2 - \beta_n/(3n) + 3/4 + \sum_{1 \leq i < j \leq n} \log\left(1 + d_i d_j/\ell_n\right) \right\}, \tag{17}$$

where

$$\nu_n = \mathbb{E}[D_n(D_n - 1)]/\mathbb{E}[D_n], \qquad \beta_n = \mathbb{E}[D_n(D_n - 1)(D_n - 2)]/\mathbb{E}[D_n]. \tag{18}$$

Here, we write D_n for the random variable with probability mass function $\mathbb{P}(D_n = k) = n_k/n$, we assume that $\boldsymbol{d} = (d_i)_{i \in [n]}$ satisfies Conditions 1(a)–(b) and that $d_{\min} \geq 1$ and $\mathbb{E}[D_n^2] = o(n^{1/8})$. They also prove several related results under slightly altered assumptions on the degree distribution.

This result is proved using *rewiring methods*, by carefully studying how many graphs remain on being simple upon rewiring a few of their edges. This gives control on how many simple graphs there are in the first place. While (17) partially answers the question *how many* graphs there are with degree sequence $d = (d_i)_{i \in [n]}$, it does not yet answer the question how to simulate one. Here, we still do not know the answer, and this is a major open problem.

A related result concerns the number of self-loops, as proved with Angel et al. [1]. There, it is proved that the number of self-loops in the configuration model, for rather general degree distributions such that $\sum_k k^2 n_k / n$ tends to infinity, is close to a normal distribution with mean and variance equal to $\mathbb{E}[D_n(D_n - 1)]/(2\mathbb{E}[D_n])$.

3.4   Small-World Properties of the Configuration Model

Having introduced the configuration model as a convenient and flexible model for real-world networks with given degree structures, we now investigate some of its properties. Of course, the scale-free property can be built-in into the model, so that this no longer requires any extra effort. Thus, we look at the small-world nature of the model instead, the key question being to which extent these distances depend on the underlying degree structure. It turns out that the configuration model generally is a *small world*, in the sense that distances are at most logarithmic in the size of the graph, and in some cases it is even an *ultra-small world*.

The first result is with Hooghiemstra and Van Mieghem [21], in the setting where Conditions 1(a)–(c) hold, and we assume that $\nu = \mathbb{E}[D(D-1)]/\mathbb{E}[D] > 1$ (while $\nu < \infty$ by Condition 1(c)). The restriction $\nu > 1$ turns out to be equivalent to the existence of a so-called *giant component*, a connected component in the graph that contains a positive proportion of the vertices. See Molloy and Reed [29, 30] or Janson and Luczak [24] for results concerning the size of the giant component. When the minimal degree $d_{\min}$ satisfies that $d_{\min} \geq 2$, the graph is *almost* connected, in the sense that all but a small number of vertices is in the giant component, while the graph is with high probability connected when $d_{\min} \geq 3$ (see the recent work with Federico and van der Hofstad [16] for these results). When $\nu \in (1, \infty)$, and conditionally on the two random vertices U_1, U_2 being connected, [21] shows that

$$\text{dist}_{\text{CM}_n(d)}(U_1, U_2) \approx \log_\nu(n), \tag{19}$$

and the random variable is tightly distributed around this asymptotic mean. Thus, typical distances grow asymptotically as a *logarithm* of the graph size, which certainly does not grow too fast. When the degrees are uniformly bounded by some constant, it is not hard to see that distances cannot grow more slowly than logarithmically. On the contrary, when considering vertices to be part of a large nearest-neighbor hypercube (such as a square of width $\sqrt{n}$ by $\sqrt{n}$ in $\mathbb{Z}^2$), distances grow much more quickly as $n^{1/d}$, where d is the dimension of the cube. Thus, distances in the configuration model grow much more slowly than that.

The *small-world effect* becomes even more pronounced when considering degree sequences that have infinite variance. In this case, $\nu = \infty$, and, under some extra regularity conditions on the degrees (that in particular guarantee that the power-law exponent of the degrees equals τ), with Hooghiemstra and Znamenski [22], we have shown that, for $\tau \in (2,3)$,

$$\text{dist}_{\text{CM}_n(d)}(U_1, U_2) \approx \frac{2 \log \log n}{|\log(\tau - 2)|}, \tag{20}$$

so that distances are *ultra-small*. Anyone who has done numerical experiments will recognize that it is hard, yet possible, to see that $\log n \to \infty$ as $n \to \infty$. It is nearly impossible, however, to see that $\log \log n \to \infty$ as $n \to \infty$, since $\log \log n$ is still at most 5 when $n \approx 10^{10}$. Thus, this might explain *why* many real-world networks have such small distances in them, as popularized under the name of *six-degrees-of-separation*.

4   Extensions: Other Models

Most of the random graph models that have been investigated in the (extensive) literature are *caricatures of reality*, in the sense that one cannot with dry eyes argue that they describe any real-world network quantitatively correctly. However, these random graph models *do* provide insight into how any of the above features can influence the global behavior of networks, and thus provide for possible explanations of the empirical properties of real-world networks that are observed. Let us discuss three related models.

Small-World Models. The small-world model was introduced by Watts and Strogatz [36] as an attempt to show how spurious long-range connections can shorten distances in complex networks, see also Newman and Watts [32]. In the simplest version, we start with a cycle containing n vertices. Every vertex is connected to a certain fixed number of vertices within a given range. After this, one starts rewiring some edges. The edge (u, v) is rewired and is replaced by the edge (u, v'), where v' is chosen uniformly at random from all the vertices. Thus, the small-world model interpolates between a one-dimensional cycle and an Erdős-Rényi random graph. It turns out that after rewiring an of the edges, for all $a > 0$, the average distances shrink from being of order n to being of order $\log n$. Thus, the small-world phenomenon can be interpreted as arising through highly rare long-distance connections. Kleinberg [25, 26] investigates how one can easily *find* the short paths, which is called the *navigability* of the graph.

Preferential Attachment Models. Most networks grow in time. Preferential attachment models describe growing networks, where the numbers of edges and vertices grow linearly with time. Preferential attachment models were first introduced by Barabási and Albert [2], whose model we will generalize. Bollobás et al. [6] studied the model by Barabási and Albert [2], and later many other papers followed on this, and related, models. See [19, Chap. 8] for details. Here we give a brief introduction.

The model that we investigate produces a *graph sequence* that we denote by $(\mathrm{PA}_n^{(m,\delta)})_{n\geq 1}$ and which, for every time n, yields a graph of n vertices and mn edges for some $m = 1, 2, \ldots$ We start by defining the model for $m = 1$ when the graph consists of a collection of trees. In this case, $\mathrm{PA}_1^{(1,\delta)}$ consists of a single vertex with a single self-loop. We denote the vertices of $\mathrm{PA}_n^{(1,\delta)}$ by $v_1^{(1)}, \ldots, v_n^{(1)}$. We denote the degree of vertex $v_i^{(1)}$ in $\mathrm{PA}_n^{(1,\delta)}$ by $D_i(n)$, where, by convention, a self-loop increases the degree by 2.

We next describe the evolution of the graph. Conditionally on $\mathrm{PA}_n^{(1,\delta)}$, the growth rule to obtain $\mathrm{PA}_{n+1}^{(1,\delta)}$ is as follows. We add a single vertex $v_{n+1}^{(1)}$ having a single edge. This edge is connected to a second end point, which is equal to $v_{n+1}^{(1)}$ with probability $(1 + \delta)/(n(2 + \delta) + (1 + \delta))$, and to vertex $v_i^{(1)} \in \mathrm{PA}_n^{(1,\delta)}$ with probability $(D_i(n) + \delta)/(n(2 + \delta) + (1 + \delta))$ for each $i \in [n]$, where $\delta \geq -1$ is a parameter of the model.

The above preferential attachment mechanism is called *affine*, since the above attachment probabilities depend in an affine way on the degrees of the random graph $\mathrm{PA}_n^{(1,\delta)}$. The model $\mathrm{PA}_n^{(1,\delta)}$ produces a *forest* (i.e., a collection of trees). The general model with $m > 1$ is defined in terms of the model for $m = 1$ as follows. Fix $\delta \geq -m$. We start with $\mathrm{PA}_{mn}^{(1,\delta/m)}$, and denote the vertices in $\mathrm{PA}_{mn}^{(1,\delta/m)}$ by $v_1^{(1)}, \ldots, v_{mn}^{(1)}$. Then we identify or collapse the m vertices $v_1^{(1)}, \ldots, v_m^{(1)}$ in $\mathrm{PA}_{mn}^{(1,\delta/m)}$ to become vertex $v_1^{(m)}$ in $\mathrm{PA}_n^{(m,\delta)}$. In doing so, we let all the edges that are incident to any of the vertices in $v_1^{(1)}, \ldots, v_m^{(1)}$ be incident to the new vertex $v_1^{(m)}$ in $\mathrm{PA}_n^{(m,\delta)}$. Then, we collapse the m vertices $v_{m+1}^{(1)}, \ldots, v_{2m}^{(1)}$ in $\mathrm{PA}_{mn}^{(1,\delta/m)}$ to become vertex $v_2^{(m)}$ in $\mathrm{PA}_n^{(m,\delta)}$, etc. More generally, we collapse the m vertices $v_{(j-1)m+1}^{(1)}, \ldots, v_{jm}^{(1)}$ in $\mathrm{PA}_{mn}^{(1,\delta/m)}$ to become vertex $v_j^{(m)}$ in $\mathrm{PA}_n^{(m,\delta)}$. The resulting graph $\mathrm{PA}_n^{(m,\delta)}$ is a multigraph with precisely n vertices and mn edges, so that the total degree is equal to $2mn$. The original model by Barabási and Albert [2] focused on the case $\delta = 0$ only, which is sometimes called the *proportional* model. The inclusion of the extra parameter $\delta > -1$ is relevant though.

We write

$$P_k(n) = \frac{1}{n} \sum_{i=1}^{n} \mathbb{1}_{\{D_i(n)=k\}} \tag{21}$$

for the (random) proportion of vertices with degree k at time n. For $m \geq 1$ and $\delta > -m$, we define $(p_k)_{k\geq 0}$ by $p_k = 0$ for $k = 0, \ldots, m-1$ and, for $k \geq m$,

$$p_k = (2 + \delta/m)\frac{\Gamma(k + \delta)\Gamma(m + 2 + \delta + \delta/m)}{\Gamma(m + \delta)\Gamma(k + 3 + \delta + \delta/m)} \tag{22}$$

It turns out that $(p_k)_{k\geq 0}$ is a probability mass function (see [19, Sect. 8.3]). The probability mass function $(p_k)_{k\geq 0}$ arises as the limiting degree distribution for $\mathrm{PA}_n^{(m,\delta)}$, i.e., for every $k \geq 1$ (see again [19, Sect. 8.3]),

$$P_k(n) \to p_k. \tag{23}$$

By (22) and Stirling's formula, as $k \to \infty$,

$$p_k = c_{m,\delta} k^{-\tau}(1 + O(1/k)), \tag{24}$$

where

$$\tau = 3 + \delta/m > 2, \qquad \text{and} \qquad c_{m,\delta} = (2 + \delta/m)\frac{\Gamma(m + 2 + \delta + \delta/m)}{\Gamma(m + \delta)}. \qquad (25)$$

Thus, the preferential attachment model, from a very simple microscopic growth model, produces graphs with power-law degree sequences. As such, the preferential attachment mechanism is a *possible* explanation for the omnipresence of power-law degree sequences in real-world networks. Many properties of preferential attachment models have been investigated, such as their distance structure. See [20] for a summary. Due to its dynamic origin, the preferential attachment model is much harder to analyze than the configuration model, which has the simpler representation in terms of uniform matchings.

The Hierarchical Configuration Model. The configuration model has low clustering in the sense that it has few triangles and cliques, which often makes it inappropriate in applied contexts. Indeed, many real-world networks, in particular social networks, have a high amount of clustering instead. A possible solution to overcome this low clustering is by introducing a *community* structure. Consider the configuration model $\mathrm{CM}_n(\boldsymbol{d})$ with a degree sequence $\boldsymbol{d} = (d_i)_{i \in [n]}$ satisfying Condition 1(a)–(b). Now we replace each of the vertices by a small graph. Thus, vertex i is replaced by a local graph G_i. We assign each of the d_i half-edges incident to vertex i to a vertex in G_i in an arbitrary way. Thus, vertex i is replaced by the pair of the community graph $G_i = (V_i, E_i)$ and the inter-community degrees $\boldsymbol{d}^{(b)} = (d_u^{(b)})_{u \in V_i}$ satisfying that $\sum_{u \in V_i} d_u^{(b)} = d_i$. The size of the graph then becomes $N = \sum_{i \in [n]} |V_i|$, where n denotes the number of communities.

This yields a graph with two levels of hierarchy, whose local structure is described by the local graphs $(G_i)_{i \in [n]}$, whereas its global structure is described by the configuration model $\mathrm{CM}_n(\boldsymbol{d})$. This model is called the *hierarchical configuration model*. A natural assumption is that the degree sequence $\boldsymbol{d} = (d_i)_{i \in [n]}$ satisfies Condition 1(a)–(b), while the empirical distribution of the community graphs

$$\mu_n(H, \boldsymbol{d}) = \frac{1}{n} \sum_{i \in [n]} \mathbb{1}_{\{G_i = H, \, (d_u^{(b)})_{u \in V_i} = \boldsymbol{d}\}} \qquad (26)$$

converges as $n \to \infty$ to some probability distribution on graphs with integer marks associated to the vertices. See the works with Stegehuis and van Leeuwaarden [33,34] for power-law relations in hierarchical configuration models and epidemic spread in them, respectively, and [23] for its topological properties, such as its connectivity, its clustering, etc. This model can be put into practice by letting a community detection algorithm find the most likely communities, and then rewiring the edges between the communities that were found. This gives rise to the hierarchical configuration model (HCM). Of course, this keeps the edges *inside* communities unchanged, which makes it a slightly unfair comparison. In a second model, we also rewire the edges inside the communities, and call the result HCM*. Figure 5 shows that when we perform percolation on these models, the result is quite similar to that on the original network. In percolation, edges are randomly removed with a certain probability, here denoted by p. Figure 5 shows the size of the largest connected component in the configuration model and both HCM as well as HCM*, and compares this to percolation on the original network. Percolation gives a good impression of the *mesoscopic* properties of a network. One is tempted to conclude that many real-world networks are well described by a hierarchical configuration model. Since the HCM (as well as its close brother HCM*) inherit the analytic tractability from the configuration model structure of the inter-community edges, we can provocatively summarize this statement by saying that *most networks are almost tree-like.*

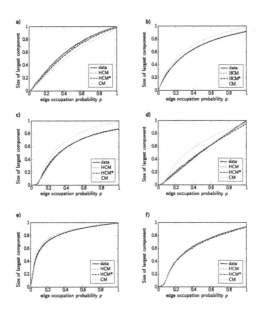

Fig. 5. Percolation on Hierarchical Configuration model, configuration model and real-world networks

Related Models. Many more models have been suggested for real-world networks. We refrain from explaining these, and refer instead to Durrett [8], or the books [19,20] for details.

Acknowledgement. This work is supported by the Netherlands Organisation for Scientific Research (NWO) through VICI grant 639.033.806 and the Gravitation NET-WORKS grant 024.002.003. This work has been for the 43rd International Workshop on Graph-Theoretic Concepts in Computer Science WG2017, held from June 21–23, 2017 in Heeze. RvdH thanks the organisors for giving him the opportunity to speak there.

References

1. Angel, O., van der Hofstad, R., Holmgren, C.: Limit laws for self-loops and multiple edges in the configuration model. arXiv:1603.07172 [math.PR], Preprint (2016)
2. Barabási, A.-L., Albert, R.: Emergence of scaling in random networks. Science **286**(5439), 509–512 (1999)
3. Bender, E., Canfield, E.: The asymptotic number of labelled graphs with given degree sequences. J. Comb. Theory (A) **24**, 296–307 (1978)
4. Bollobás, B.: A probabilistic proof of an asymptotic formula for the number of labelled regular graphs. Eur. J. Comb. **1**(4), 311–316 (1980)
5. Bollobás, B.: Random Graphs. Cambridge Studies in Advanced Mathematics, vol. 73, 2nd edn. Cambridge University Press, Cambridge (2001)
6. Bollobás, B., Riordan, O., Spencer, J., Tusnády, G.: The degree sequence of a scale-free random graph process. Random Struct. Algorithms **18**(3), 279–290 (2001)
7. Britton, T., Deijfen, M., Martin-Löf, A.: Generating simple random graphs with prescribed degree distribution. J. Stat. Phys. **124**(6), 1377–1397 (2006)
8. Durrett, R.: Random Graph Dynamics. Cambridge Series in Statistical and Probabilistic Mathematics. Cambridge University Press, Cambridge (2007)
9. Erdős, P.: Some remarks on the theory of graphs. Bull. Am. Math. Soc. **53**, 292–294 (1947)
10. Erdős, P., Gallai, T.: Graphs with points of prescribed degrees. Mat. Lapok **11**, 264–274 (1960). (Hungarian)
11. Erdős, P., Rényi, A.: On random graphs I. Publ. Math. Debrecen **6**, 290–297 (1959)
12. Erdős, P., Rényi, A.: On the evolution of random graphs. Magyar Tud. Akad. Mat. Kutató Int. Közl. **5**, 17–61 (1960)
13. Erdős, P., Rényi, A.: On the evolution of random graphs. Bull. Inst. Int. Stat. **38**, 343–347 (1961)
14. Erdős, P., Rényi, A.: On the strength of connectedness of a random graph. Acta Math. Acad. Sci. Hungar. **12**, 261–267 (1961)
15. Faloutsos, C., Faloutsos, P., Faloutsos, M.: On power-law relationships of the internet topology. Comput. Commun. Rev. **29**, 251–262 (1999)
16. Federico, L., van der Hofstad, R.: Critical window for connectivity in the configuration model. Combin. Probab. Comput. **26**(5), 660–680 (2017)
17. Gao, P., Wormald, N.: Enumeration of graphs with a heavy-tailed degree sequence. Adv. Math. **287**, 412–450 (2016)
18. Gilbert, E.N.: Random graphs. Ann. Math. Stat. **30**, 1141–1144 (1959)
19. van der Hofstad, R.: Random graphs and complex networks. Cambridge Series in Statistical and Probabilistic Mathematics, vol. 1. Cambridge University Press, Cambridge (2017)
20. van der Hofstad, R.: Random graphs and complex networks, vol. 2 (2018+). In preparation, see http://www.win.tue.nl/~rhofstad/NotesRGCNII.pdf
21. van der Hofstad, R., Hooghiemstra, G., Van Mieghem, P.: Distances in random graphs with finite variance degrees. Random Struct. Algorithms **27**(1), 76–123 (2005)
22. van der Hofstad, R., Hooghiemstra, G., Znamenski, D.: Distances in random graphs with finite mean and infinite variance degrees. Electron. J. Probab. **12**(25), 703–766 (2007). (Electronic)
23. van der Hofstad, R., van Leeuwaarden, J., Stegehuis, C.: Hierarchical configuration model. arXiv:1512.08397 [math.PR], Preprint (2015)

24. Janson, S., Luczak, M.: A new approach to the Giant component problem. Random Struct. Algorithms **34**(2), 197–216 (2009)
25. Kleinberg, J.M.: Navigation in a small world. Nature **406**, 845 (2000)
26. Kleinberg, J.M.: The small-world phenomenon: an algorithm perspective. In: Proceedings of the Twenty-Third Annual ACM Symposium on Principles of Distributed Computing, pp. 163–170, May 2000
27. Krioukov, D., Kitsak, M., Sinkovits, R., Rideout, D., Meyer, D., Boguñá, M.: Network cosmology. Sci. Rep. **2**, Article number 793 (2012)
28. Leskovec, J., Lang, K., Dasgupta, A., Mahoney, M.: Community structure in large networks: natural cluster sizes and the absence of large well-defined clusters. Internet Math. **6**(1), 29–123 (2009)
29. Molloy, M., Reed, B.: A critical point for random graphs with a given degree sequence. Random Struct. Algorithms **6**(2–3), 161–179 (1995)
30. Molloy, M., Reed, B.: The size of the Giant component of a random graph with a given degree sequence. Comb. Probab. Comput. **7**(3), 295–305 (1998)
31. Newman, M.E.J.: Networks: An Introduction. Oxford University Press, Oxford (2010)
32. Newman, M.E.J., Watts, D.: Scaling and percolation in the small-world network model. Phys. Rev. E **60**, 7332–7344 (1999)
33. Stegehuis, C., van der Hofstad, R., van Leeuwaarden, J.: Epidemic spreading on complex networks with community structures. Sci. Rep. **6**, 29748 (2016)
34. Stegehuis, C., van der Hofstad, R., van Leeuwaarden, J.: Power-law relations in random networks with communities. Phys. Rev. E **94**, 012302 (2016)
35. Watts, D.J.: Small Worlds. The Dynamics of Networks Between Order and Randomness. Princeton Studies in Complexity. Princeton University Press, Princeton (1999)
36. Watts, D.J., Strogatz, S.H.: Collective dynamics of 'small-world' networks. Nature **393**, 440–442 (1998)

On Bubble Generators in Directed Graphs

Vicente Acuña[1], Roberto Grossi[2,5], Giuseppe F. Italiano[3],
Leandro Lima[5(✉)], Romeo Rizzi[4], Gustavo Sacomoto[5],
Marie-France Sagot[5], and Blerina Sinaimeri[5]

[1] Center for Mathematical Modeling (UMI 2807 CNRS),
University of Chile, Santiago, Chile
viacuna@dim.uchile.cl
[2] Università di Pisa, Pisa, Italy
grossi@di.unipi.it
[3] Università di Roma "Tor Vergata", Roma, Italy
giuseppe.italiano@uniroma2.it
[4] Università di Verona, Verona, Italy
Romeo.Rizzi@univr.it
[5] Inria Grenoble Rhône-Alpes, CNRS, UMR5558, LBBE,
Université Lyon 1, Villeurbanne, France
{leandro.ishi-soares-de-lima,marie-france.sagot,
blerina.sinaimeri}@inria.fr, sacomoto@gmail.com

Abstract. Bubbles are pairs of internally vertex-disjoint (s,t)-paths with applications in the processing of DNA and RNA data. For example, enumerating alternative splicing events in a reference-free context can be done by enumerating all bubbles in a de Bruijn graph built from RNA-seq reads [16]. However, listing and analysing all bubbles in a given graph is usually unfeasible in practice, due to the exponential number of bubbles present in real data graphs. In this paper, we propose a notion of a bubble generator set, *i.e.* a polynomial-sized subset of bubbles from which all the others can be obtained through the application of a specific symmetric difference operator. This set provides a compact representation of the bubble space of a graph, which can be useful in practice since some pertinent information about all the bubbles can be more conveniently extracted from this compact set. Furthermore, we provide a polynomial-time algorithm to decompose any bubble of a graph into the bubbles of such a generator in a tree-like fashion.

Keywords: Bubbles · Bubble generator set · Bubble space · Decomposition algorithm

1 Introduction

Bubbles are pairs of internally vertex-disjoint (s,t)-paths with applications in the processing of DNA and RNA data. For example, in the genomic context, genome assemblers usually identify and remove bubbles in order to remove sequencing

© Springer International Publishing AG 2017
H.L. Bodlaender and G.J. Woeginger (Eds.): WG 2017, LNCS 10520, pp. 18–31, 2017.
https://doi.org/10.1007/978-3-319-68705-6_2

errors and linearise the graph [10,14,18,22]. However, bubbles can also represent interesting biological events, *e.g.* allelic differences (SNPs and indels) when processing DNA data [7,20,21], and alternative splicing events in RNA data [11,15–17]. Due to their practical relevance, several theoretical studies concerning bubbles were done in the past few years [1,4,13,15,19], usually related to bubble-enumeration algorithms, but the literature regarding this mathematical object remains small when compared to the literature on cycles, *i.e.* undirected eulerian subgraphs, which is a related concept.

In practice, due to the high throughput of modern sequencing machines, the genomic and transcriptomic de Bruijn graphs tend to be huge, usually containing from millions to billions of vertices. As expected, the number of bubbles also tends to be large, exponential in the worst case, and therefore algorithms that deal with them either simplify the graph by removing bubbles, or just analyse a small subset of the bubble space. Such subsets usually correspond to bubbles with some predefined characteristics, and may not be the best representative of the bubble space. More worrying is the fact that all the relevant events described by bubbles that do not satisfy the constraints are lost. On the other hand, any algorithm that tries to be more exhaustive, analysing a big part of the bubble space, will certainly spend a prohibitive amount of time in real data graphs and will not be applicable. This motivates further work for finding efficient ways to represent the information contained in the bubble space. In a graph-theoretical framework, one way to do this is to obtain a compact description of all bubbles.

In this paper, we propose a bubble generator, *i.e.* a "representative set" of the bubbles in a graph that allows to reconstruct all and only the bubbles in a graph. More specifically, we show how to identify, for any given directed graph G, a generator set of bubbles $\mathcal{G}(G)$ which is of polynomial size in the input, and such that any bubble in G can be obtained in a polynomial number of steps by properly combining the bubbles in the generator $\mathcal{G}(G)$ through some suitably defined graph operations. We also propose a polynomial-time decomposition algorithm that, given a bubble B in the graph G, finds a sequence of bubbles from the generator $\mathcal{G}(G)$ whose combination results in B. The latter algorithm can be applied when one needs to know how to decompose a bubble into its elementary parts, which are the bubbles in $\mathcal{G}(G)$, *e.g.* when identifying and decomposing complex alternative splicing events [17] into several elementary alternative splicing events.

This work was inspired by the studies on cycle bases, which represent a compact description of all the cycles in a graph. The study of cycle bases started a long time ago [12] and has attracted much attention in the last fifteen years, leading to many interesting results such as the classification of different types of cycle bases, the generalisation of these notions to weighted and to directed graphs, as well as several complexity results for constructing bases. We refer the interested reader to the books of Deo [5] and Bollobás [2], and to the survey of Kavitha *et al.* [8] for an in-depth coverage of cycle bases. However, it is worth mentioning some characteristics that make the problems related to bubble generators very different (and more difficult) from the ones related to cycle bases.

Indeed, a cycle base in a directed graph contains cycles with orientations that can be arbitrary, so that elements in the base are not even directed cycles in the original graph [9] (if the graph is strongly connected, then it is possible to find a cycle base composed only of directed cycles [6]). On the contrary, bubbles impose a particular orientation of the cycle. Observe that a cycle base composed solely of bubbles cannot be directly translated into a bubble generator, since such set represents the cycle space, which is a superset of the bubble space. In order to obtain a representative set of only the bubble space, it is required to change the symmetric difference operator, *i.e.* the operator used to combine two bubbles. The restriction we impose in this operator is that two bubbles are combinable if the output is also a bubble, *i.e.* the operator is undefined if the output is not a bubble. By imposing such restriction, the bubble space is not closed under the symmetric difference operator, and thus cannot be represented as a vector space over $\mathbb{Z}_2$, as is the case with the cycle space. As such, the algorithms developed for cycle bases in undirected and directed graphs do not apply to our problem with bubbles.

The remainder of the paper is organised as follows. Section 2 present some definitions that will be used throughout the paper. Section 3 introduces the bubble generator. Section 4 presents a polynomial-time algorithm for decomposing any bubble in a graph into elements of the generator set. Finally, we conclude with open problems in Sect. 5.

2   Preliminaries

Throughout the paper, we assume that the reader is familiar with the standard graph terminology, as contained for instance in [3]. A *directed* graph is a pair $G = (V, A)$, where V is the set of vertices, and A is the set of arcs. Given a graph G, we also denote by $V(G)$ the set of vertices of G, and by $A(G)$ the set of arcs of G. For convenience, we set $n = |V(G)|$ and $m = |A(G)|$. In this paper, all graphs considered are directed, unweighted, without parallel arcs and finite. An arc $a = (u, v)$ is said to be incident to vertices u and v. In particular, $a = (u, v)$ is said to be leaving vertex u and entering vertex v. Alternatively, $a = (u, v)$ is an outgoing arc for u and an incoming arc for v. The in-degree of a vertex v is given by the number of arcs entering v, while the out-degree of v is the number of arcs leaving v. The degree of v is the sum of its in-degree and out-degree.

We say that a graph $G' = (V', A')$ is a subgraph of a graph $G = (V, A)$ if $V' \subseteq V$ and $A' \subseteq A$. Given a subset of vertices $V' \subseteq V$, the subgraph of G induced by V', denoted by $G[V']$, has V' as vertex set and contains all arcs of G that have both endpoints in V'. Given a subset of arcs $A' \subseteq A$, the subgraph of G induced by A', denoted by $G[A']$, has A' as arc set and contains all vertices of G that are endpoints of arcs in A'. Given a subset of vertices $V' \subseteq V$ and a subset of arcs $A' \subseteq A$, we denote by $G - V'$ the graph $G[V \setminus V']$ and by $G - A'$ the graph $G[A \setminus A']$. Given two graphs G and H, their union $G \cup H$ is the graph F for which $V(F) = V(G) \cup V(H)$ and $A(F) = A(G) \cup A(H)$. Their intersection $G \cap H$ is the graph F for which $V(F) = V(G) \cap V(H)$ and $A(F) = A(G) \cap A(H)$.

Let s, t be any two vertices in G. A (*directed*) *path* from s to t in G is a sequence of vertices $s = v_1, v_2, \ldots, v_k = t$, such that $(v_i, v_{i+1}) \in A$ for $i = 1, 2, \ldots, k - 1$. We also allow a single vertex to be a path. A path is *simple* if it does not contain repeated vertices. A path from s to t is also referred to as an (s, t)-*path*. The length of a path p is the number of arcs in p and will be denoted by $|p|$. We write $p \subseteq q$ if p is a subpath of q. Given a path p_1 from x to y and a path p_2 from y to z, we denote by $p_1 \cdot p_2$ their concatenation, *i.e.* the path from x to z defined by the path p_1 followed by p_2. For a path $p = v_1, v_2, \ldots, v_k$, we say that the subpath $p_1 = v_1, \ldots, v_i$ ($p_2 = v_j, \ldots, v_k$) is a *prefix* (*suffix*) of p for some $1 \leq i \leq k$ ($1 \leq j \leq k$). Two paths $p = v_1, v_2, \ldots, v_k$ and $q = u_1, u_2, \ldots, u_l$ are vertex disjoint if they share no vertices. Further, if the subpaths $p_1 = v_2, \ldots, v_{k-1}$ of p and $q_1 = u_2, \ldots, u_{l-1}$ of q are vertex disjoint, we say that p and q are internally vertex disjoint. Throughout this paper, all the paths considered will be simple and referred to as paths.

Definition 1. *Given a directed graph G and two vertices $s, t \in V(G)$, not necessarily distinct, an (s, t)-bubble B consists of two (s, t)-paths that are internally vertex disjoint. Vertex s is the source and t is the target of the bubble. If $s = t$ then one of the paths of the bubble has length 0, and therefore B corresponds to a directed cycle. We then say that B is a degenerate bubble.*

In the following, we assume that shortest paths are unique. This is without loss of generality, and indeed there are many standard techniques for achieving this, including perturbing arc weights by infinitesimals. However, for our goal, it suffices to use a "lexicographic ordering". Namely, we define an arbitrary ordering $v_1, \ldots, v_n$ on the vertices of G. A path p is considered lexicographically shorter than a path q if the length of p is strictly smaller than the length of q, or, if p and q have the same length, the sequence of vertices associated to p is lexicographically smaller than the sequence associated to q. We denote this by $p <_{lex} q$.

We denote by $B = (p, q)$ the bubble having p, q as its two internally vertex-disjoint paths, referred to as *legs*. We denote by $\ell(B)$ (resp., by $\mathcal{L}(B)$) the shorter (resp., longer) between the two legs p, q of B. We also denote by $|B|$ the number of arcs of bubble B. Note that $|B| = |\ell(B)| + |\mathcal{L}(B)|$.

Next, we define a total order on the set of bubbles.

Definition 2. *Let B_1 and B_2 be any two bubbles. B_1 is smaller than B_2 (in symbols, $B_1 < B_2$) if one of the following holds: either (i) $\mathcal{L}(B_1) <_{lex} \mathcal{L}(B_2)$; or (ii) $\mathcal{L}(B_1) = \mathcal{L}(B_2)$ and $\ell(B_1) <_{lex} \ell(B_2)$.*

3 The Bubble Generator

As with cycle bases in undirected graphs, we define a symmetric difference operator, but which operands are bubbles. Given two bubbles B_1 and B_2 of a directed graph G, the constrained symmetric difference operator Δ is such that $B_1 \Delta B_2$

is defined if and only if $G[(A(B_1) \cup A(B_2)) \setminus (A(B_1) \cap A(B_2))]$ is a bubble. Otherwise, we say that $B_1 \Delta B_2$ is undefined. If $B_1 \Delta B_2$ is defined, we also say that B_1 and B_2 are *combinable*. Given two combinable bubbles B_1 and B_2, we refer to $B_1 \Delta B_2$ as the *sum of B_1 and B_2*, and denote it also by $B_1 + B_2$. We also say that the bubble $B_1 + B_2$ is *generated* from bubbles B_1 and B_2, and that it can be *decomposed* into the bubbles B_1 and B_2.

Let G be a directed graph and let $\mathcal{B}$ be a set of bubbles in G. The set of all the bubbles that can be generated starting from bubbles in $\mathcal{B}$ is called the *span* of $\mathcal{B}$. A set of bubbles $\mathcal{B}$ is called a *generator* if each bubble in G is spanned by $\mathcal{B}$, *i.e.* it can be recursively decomposed down to bubbles of $\mathcal{B}$. Due to our constrained symmetric difference operator Δ, all subgraphs generated by the elements in $\mathcal{B}$ are necessarily bubbles. Since not all pairs of bubbles of G are combinable, the bubble space is not closed under Δ, and therefore it does not form a vector space over $\mathbb{Z}_2$.

Definition 3. *A bubble B is* composed *if it can be obtained as a sum of two smaller bubbles. Otherwise, the bubble B is called* simple.

For a directed graph G, we denote by $\mathcal{S}(G)$ the set of simple bubbles of G. It is not difficult to see that $\mathcal{S}(G)$ is a generator. For now, we are not able to: (1) prove that $\mathcal{S}(G)$ can be found in polynomial time or if it is $\mathcal{NP}$-Hard to do so; (2) prove that any bubble in G can be obtained in a polynomial number of steps from bubbles in $\mathcal{S}(G)$. Nevertheless, we introduce next another generator $\mathcal{G}(G) \supseteq \mathcal{S}(G)$ which can be found in polynomial time and for which we can prove that any bubble in G can be obtained in a polynomial number of steps from the bubbles in $\mathcal{G}(G)$. Let $p : s = x_0, x_1, \ldots, x_h = t$ be a path from s to t and let $0 \le i \le j \le h$. To ease the notation, we denote by $p_{i,j}$ the subpath of p from x_i to x_j, and refer also to $p_{0,j}$ as $p_{s,j}$ and to $p_{i,h}$ as $p_{i,t}$. The next theorem provides some properties of simple bubbles.

Theorem 1. *Let B be a simple (s,t)-bubble in a directed graph G. The following holds:*

(1) $\ell(B)$ is the shortest path from s to t in G;
(2) Let $\mathcal{L}(B) = s, v_1, \ldots, v_r, t$. Then $s, v_1, \ldots, v_r$ is the shortest path from s to v_r in G.

Proof. Let B be a simple (s,t)-bubble: we show that both conditions (1) and (2) must hold.

We first consider condition (1). If B is degenerate, then it trivially satisfies condition (1). Therefore, assume that B is non-degenerate and, by contradiction, that $\ell(B)$ is not the shortest path from s to t. Let $p^* : s = x_0, x_1, \ldots, x_h = t$ be the shortest path from s to t in G. For $0 \le i \le j \le h$, by subpath optimality, $p^*_{i,j}$ is the shortest path from x_i to x_j. Let k be the smallest index, $0 \le k < h$, for which the arc (x_k, x_{k+1}) does not belong to either one of the legs of B. Such an index k must exist, as otherwise p^* would coincide with a leg of B. Furthermore, let l, $k < l \le h$, be the smallest index greater than k for which $x_l \in V(B)$. Such a vertex x_l must also exist, since $x_h = t \in V(B)$. In other words, x_k is the first vertex of the bubble B where p^* departs from B and x_l, $l > k$, is the first vertex where the shortest path p^* intersects again the bubble B. By definition of x_k and x_l, $p^*_{k,l}$ is internally vertex-disjoint with both legs of B. We now claim that B can be obtained as the sum of two smaller bubbles, thus contradicting our assumption that B is a simple bubble.

To prove the claim, we distinguish two cases, depending on whether x_k and x_l are on the same leg of B or not. Consider first the case when x_k and x_l are on the same leg p of B (see Fig. 1(a)). Let B_1 be the bubble with $\ell(B_1) = p^*_{k,l}$ and $\mathcal{L}(B_1) = p_{k,l}$. First, note that if either $x_k \neq s$ or $x_l \neq t$, then $p_{k,l}$ is a proper subpath of a leg of B. Hence, $|\mathcal{L}(B_1)| = |p_{k,l}| < |\mathcal{L}(B)|$, and $B_1 < B$. Otherwise, suppose $s = x_k$ and $t = x_l$. Then either $\mathcal{L}(B_1) = \ell(B) <_{lex} \mathcal{L}(B)$, or $\mathcal{L}(B_1) = \mathcal{L}(B)$ and $\ell(B_1) = p^*_{k,l} = p^* <_{lex} \ell(B)$. In both cases, $B_1 < B$. Let B_2 be the bubble which is obtained from B by replacing $p_{k,l}$ by $p^*_{k,l}$ (see Fig. 1(a)). Since $p^*_{k,l}$ is the shortest path, by subpath optimality, $p^*_{k,l} <_{lex} p_{k,l}$, thus $B_2 < B$. As a result, B can be obtained as the sum of two smaller bubbles B_1, B_2, thus contradicting the assumption that B is simple.

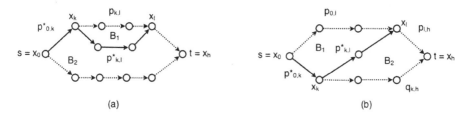

Fig. 1. Case (1) of the proof of Theorem 1. The prefix of the shortest path from s to t is shown as a solid line.

Consider now the case where x_k and x_l are on different legs of B (see Fig. 1(b)). Notice that this means $x_k \neq s$ and $x_l \neq t$. Let p be the leg containing x_l and q the one containing x_k. Note that $p = p_{0,l} \cdot p_{l,h}$ and $q = p^*_{0,k} \cdot q_{k,h}$. Moreover, let B_1 be the bubble such that the two legs of B_1 are $p^*_{0,k} \cdot p^*_{k,l} <_{lex} q$ and $p_{0,l}$, which is a proper subpath of p. Hence, $B_1 < B$. Let B_2 be the bubble such that the two legs of B_2 are $q_{k,h}$, which is a proper subpath of q, and $p^*_{k,l} \cdot p_{l,h} <_{lex} p$. Hence, $B_2 < B$, and $B = B_1 + B_2$, which implies again that B is not simple.

We show now that B satisfies also condition (2). Assume, by contradiction, that B satisfies condition (1) but not (2), and so $p = s, v_1, \ldots, v_r$ (note that p is equal to $\mathcal{L}(B)$ without its last arc) is not the shortest path from s to v_r in G. Let $p^* : s = x_0, \ldots, x_{h-1} = v_r$, $p^* \neq p$, be such a shortest path in G. Similarly to the previous case, let k be the smallest index, $0 \leq k < h - 1$, for which the arc (x_k, x_{k+1}) does not belong to either one of the legs of B, i.e. x_k is the first vertex where the shortest path p^* departs from B. Such an index k must exist, as otherwise p^* would coincide with a leg of B. Let $l, k < l \leq h - 1$, be the smallest index such that $x_l \in V(B)$. Namely, x_l is the first vertex after x_k where the shortest path p^* intersects again bubble B. Such a vertex x_l must always exist, since $x_{h-1} = v_r \in V(B)$. Since $k < l$, we have that $|p_{k,l}^*| \geq 1$. Furthermore, we claim that x_l must be in $\mathcal{L}(B) - \{s, t\}$. If this were not the case, we would have two distinct shortest paths from s to x_l in G ($p_{0,l}^*$ and the subpath of $\ell(B)$ from $s = x_0$ to x_l), which contradicts our assumption that shortest paths are unique.

We again distinguish two cases: when both x_k, x_l belong to $\mathcal{L}(B)$, and when $x_k \in \ell(B)$ and $x_l \in \mathcal{L}(B)$. We set $p = \mathcal{L}(B), q = \ell(B)$.

In the first case (see Fig. 2(a)), let B_1 be the bubble with $\ell(B_1) = \ell(B)$ and $\mathcal{L}(B_1) = p_{0,k}^* \cdot p_{k,l}^* \cdot p_{l,h}$. Since $|p_{k,l}^*| <_{lex} |p_{k,l}|$ then $\mathcal{L}(B_1) <_{lex} \mathcal{L}(B)$, and thus $B_1 < B$. Let B_2 be the bubble with $\ell(B_2) = p_{k,l}^*$, and $\mathcal{L}(B_2) = p_{k,l}$. Since $\mathcal{L}(B_2) \subset \mathcal{L}(B)$ (as $x_k \neq t$), $B_2 < B$. As a result, B can be obtained as the sum of two smaller bubbles B_1, B_2, thus contradicting the assumption that B is simple.

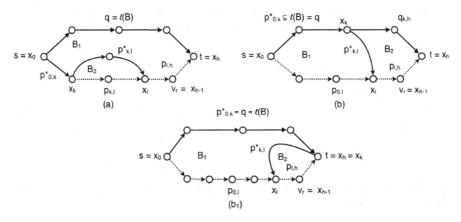

Fig. 2. Case (2) of the proof of Theorem 1. The shortest path from s to t and the prefix of the shortest path from s to v_r are shown as solid lines.

In the second case (see Fig. 2(b)), let B_1 be the bubble with $\ell(B_1) = p_{0,k}^* \cdot p_{k,l}^*$ and $\mathcal{L}(B_1) = p_{0,l}$. Since $\mathcal{L}(B_1) \subset \mathcal{L}(B)$, $B_1 < B$. Let B_2 be the bubble with $\ell(B_2) = q_{k,h}$, and $\mathcal{L}(B_2) = p_{k,l}^* \cdot p_{l,h}$. Since $|\mathcal{L}(B_2)| < |\mathcal{L}(B)|$, $B_2 < B$. Again, B can be obtained as the sum of two smaller bubbles B_1, B_2, thus contradicting the assumption that B is simple. Finally, notice that this includes also the case $x_k = t$ and the argument holds identically with B_2 being a degenerate bubble. For the sake of clarity, we depicted this case separately in Fig. 2(b_1). ∎

Given a directed graph G, we denote by $\mathcal{G}(G)$ the set of bubbles in G satisfying conditions (1) and (2) of Theorem 1.

Remark 1. Conditions (1) and (2) of Theorem 1 are not sufficient to guarantee that a bubble is simple, *e.g.* see Fig. 3. Thus, the generator $\mathcal{G}(G)$ is not necessarily minimal.

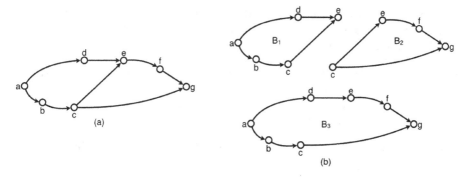

Fig. 3. An example showing that conditions (1) and (2) of Theorem 1 are not sufficient to guarantee that a bubble is simple. (a) A directed graph G. (b) The three bubbles B_1, B_2 and B_3 of G satisfying conditions (1) and (2) of Theorem 1, in which B_1 and B_2 are simple, but B_3 is composed, since $B_1 < B_3$, $B_2 < B_3$ and $B_3 = B_1 + B_2$.

Theorem 2. *Let G be a directed graph. The following holds:*

(1) $\mathcal{G}(G)$ is a generator set for all the bubbles of G;
(2) $|\mathcal{G}(G)| \leq nm$.

Proof. (1) Recall that $\mathcal{S}(G)$ is the set of simple bubbles. By Theorem 1, $\mathcal{S}(G) \subseteq \mathcal{G}(G)$, and thus $\mathcal{G}(G)$ is a generator set for all the bubbles of G.
(2) Since every bubble b in $\mathcal{G}(G)$, with $\ell(b) = s, u_1, \ldots, t$ and $\mathcal{L}(b) = s, v_1, \ldots, v_r, t$, can be uniquely identified by its vertex s and its arc (v_r, t), then the number of bubbles in $\mathcal{G}(G)$ is upper-bounded by nm. ∎

Remark 2. The upper bound given in Theorem 2 is asymptotically tight, as shown by the family of simple directed graphs on vertex set $V_n = \{1, 2, \ldots, n\}$ and all possible $n(n-1)$ arcs in their arc set $A_n = \{(u, v) : u \neq v, u, v \in V\}$.

Remark 3. Given a directed graph G, a naive algorithm to find $\mathcal{G}(G)$ would consist of the following steps. We start with $\mathcal{G}(G)$ as an empty set. We then find all-pairs shortest paths in G (since G is unweighted, this can be done through n BFSs). Finally, denoting, for each vertex $s \in V(G)$ and each arc $(v_r, t) \in A(G)$, by p_1 the shortest path from s to t in G and by p_2 the shortest path from s to v_r in G concatenated with the arc (v_r, t), we add the bubble $b = (p_1, p_2)$ to $\mathcal{G}(G)$ if p_1 and p_2 are internally vertex disjoint. Note that if $s = t$, then b corresponds to a degenerate bubble. A naive implementation of this algorithm takes $\mathcal{O}(n^2 m)$ time.

4   A Polynomial-Time Algorithm for Decomposing a Bubble

The main result of this section is to provide a polynomial-time algorithm for decomposing any bubble of G into bubbles of $\mathcal{G}(G)$. To do so, we make use of a tree-like decomposition. We need to take extra care in this decomposition since a naive approach could generate (several times) all the bubbles that are smaller than B, yielding an exponential number of steps.

Definition 4. *A bubble B is* short *if it satisfies condition (1) of Theorem 1, but not necessarily condition (2). Namely, let $\mathcal{L}(B) = s, v_1, \ldots, v_r, t$ be such that $\ell(B)$ is the shortest path from s to t in G but $s, v_1, \ldots, v_r$ is not necessarily the shortest path from s to v_r in G.*

We next introduce a measure for describing how "close" is a bubble to being short:

Definition 5. *Given an (s,t)-bubble B, let p^* be the shortest path from s to t. We say that B is k-short, for $k \geq 0$, if there is a leg $p \in \{\ell(B), \mathcal{L}(B)\}$ for which p^* and p share a prefix of exactly k arcs.*

Since in our case shortest paths are unique, only one leg of a bubble B can share a prefix with the shortest path p^*. Furthermore, any bubble B is k-short for some k, $0 \leq k \leq |\ell(B)|$. In particular, a bubble is short if and only if it is k-short for $k = |\ell(B)|$.

Definition 6. *Given a k-short bubble, we define the* short residual *of B as follows: $residual_s(B) = |B| - k$.*

Since $0 \leq k \leq |\ell(B)|$, and $|B| = |\ell(B)| + |\mathcal{L}(B)|$, we have that $|\mathcal{L}(B)| \leq residual_s(B) \leq |B|$.

We now present our polynomial time algorithm for decomposing a bubble of the graph G into bubbles of $\mathcal{G}(G)$. In the following, we assume that we have done a preprocessing step to compute all-pairs shortest paths in G in $\mathcal{O}(n(m + n))$ time through n BFSs.

Lemma 1. *Let B be an (s,t)-bubble that is not short. Then, B can be decomposed into two bubbles B_1 and B_2 ($B = B_1 + B_2$), such that: (a) B_1 is short, and (b) $residual_s(B_2) < residual_s(B)$. Moreover, B_1 and B_2 can be found in $\mathcal{O}(n)$ time.*

Proof. Let B be a k-short (s,t)-bubble, $0 \leq k < |\ell(B)|$. Let $p^* : s = x_0, x_1, \ldots, x_h = t$ be the shortest path from s to t in G. To prove (a), we follow a similar approach to Theorem 1. Since B is k-short, there is a leg $p \in \{\ell(B), \mathcal{L}(B)\}$ such that p^* and p share a prefix of exactly k arcs, $0 \leq k < h$. In other terms, leg p starts with arcs (x_0, x_1), ..., (x_{k-1}, x_k), the arc (x_k, x_{k+1}) is not in leg p, i.e., x_k is the first vertex where the shortest path p^* departs from the leg p. Note that as a special case, $k = 0$ and $x_k = x_0 = s$. Let $l, k < l \leq h$, be the smallest index such that $x_l \in V(B)$. Namely, x_l is the first vertex after x_k where the shortest path p^* intersects again the bubble B. Such a vertex x_l must always exist, since $x_h = t \in V(B)$. Since $k < l$, we have that $|p^*_{k,l}| \geq 1$. We have two possible cases: either the vertices x_k and x_l are on the same leg of B (see Fig. 1(a)) or x_k and x_l are on different legs of B (see Fig. 1(b)). In either case, we can decompose B

as $B = B_1 + B_2$, as illustrated in Fig. 1. Note that in both cases, the bubble B_1 is short since one leg of B_1 is a subpath of the shortest path p^*, and hence a shortest path itself by subpath optimality.

Consider now B_2 in Fig. 1. To prove (b), we distinguish among the following three cases: (1) $x_k \neq s$ and vertices x_k and x_l are on the same leg of B; (2) $x_k \neq s$ and vertices x_k and x_l are on different legs of B; (3) $x_k = s$. First, consider case (1) (see Fig. 1(a)) and note that $residual_s(B) = |p_{k,l}| + |p_{l,h}| + |q_{0,h}|$ where q is the other leg of B different from p. Moreover, $residual_s(B_2) = |p_{l,h}| + |q_{0,h}|$. Hence, $residual_s(B) - residual_s(B_2) = |p_{k,l}| \geq |p^*_{k,l}| \geq 1$. Consider now case (2), (see Fig. 1(b)) and note that $residual_s(B) = |p_{0,l}| + |p_{l,h}| + |q_{k,h}|$ and $residual_s(B_2) = |p_{l,h}| + |q_{k,h}|$, and thus $residual_s(B) - residual_s(B_2) = |p_{0,l}| \geq |p^*_{0,k}| + |p^*_{k,l}| \geq 1$. The proof of case (3) is completely analogous to case (1), with $x_k = s$ and $p^*_{0,k} = \emptyset$, and again $residual_s(B) - residual_s(B_2) = |p_{k,l}| \geq |p^*_{k,l}| \geq 1$. In all cases, $residual_s(B) - residual_s(B_2) > 0$, and thus the claim follows. Finally, note that in order to compute B_1 and B_2 from B, it is sufficient to trace the shortest path p^*. Since all shortest paths are pre-computed in a preprocessing step, this can be done in $\mathcal{O}(n)$ time. ∎

Lemma 2. *Any bubble B can be represented as a sum of $\mathcal{O}(n)$ (not necessarily distinct) short bubbles. This decomposition can be found in $\mathcal{O}(n^2)$ time in the worst case.*

Proof. Each time we apply Lemma 1 to a bubble B, we produce in $\mathcal{O}(n)$ time a short bubble B_1 and a bubble B_2 such that $residual_s(B_2) < residual_s(B)$. Since $residual_s(B) \leq |B| \leq n$, the lemma follows. ∎

We next show how to further decompose short bubbles. Before doing that, we define the notion of *residual* for short bubbles, which measures how "close" is a short bubble to being a bubble of our generator set $\mathcal{G}(G)$.

Definition 7. *Let B be a short (s,t)-bubble, let $\ell(B) = p^*_1$ be the shortest path from s to t in G, let $\mathcal{L}(B) = s, v_1, \ldots, v_r, t$ be the other leg of B, let p^*_2 be the shortest path from s to v_r in G, and let p be the longest common prefix between $\mathcal{L}(B) - (v_r, t)$ and p^*_2. Then, the residual of B is defined as $residual(B) = |\mathcal{L}(B)| - 1 - |p|$.*

Since p is a prefix of $\mathcal{L}(B) - (v_r, t)$, we have that $0 \leq |p| \leq |\mathcal{L}(B)| - 1$. Thus, $0 \leq residual(B) \leq |\mathcal{L}(B)| - 1$.

Lemma 3. *Let B be a short (s,t)-bubble such that $residual(B) > 0$. B can be decomposed into two bubbles B_1 and B_2 ($B = B_1 + B_2$) such that B_1 and B_2 are short and $residual(B_1) + residual(B_2) < residual(B)$. Moreover, it is possible to find the bubbles B_1 and B_2 in $\mathcal{O}(n)$ time.*

Proof. Since B is a short (s,t)-bubble, it satisfies condition (1) of Theorem 1. Furthermore, as $residual(B) > 0$, it does not satisfy condition (2). Therefore, there exists two bubbles $B_1 < B$ and $B_2 < B$ such that $B = B_1 + B_2$ (from the proof of Theorem 1). Since $\ell(B)$ is the shortest path from s to t, using arguments similar to the ones in Theorem 1, it can be shown that B can be decomposed into B_1 and B_2 and the only possible cases are the ones depicted in Fig. 2. Note that in all three cases of Fig. 2, each of the bubbles B_1 and B_2 has one leg that is a shortest path. Thus, in all three cases, B_1 and B_2 are short. Moreover, in Fig. 2(a), $residual(B_1) \leq |p_{l,h}| - 1$ and $residual(B_2) \leq |p_{k,l}| - 1$. Therefore, $residual(B_1) + residual(B_2) \leq |p_{l,h}| - 1 + |p_{k,l}| - 1 = residual(B) - 1 < residual(B)$. Similarly, in Fig. 2(b) and (b_1), $residual(B_1) \leq |p_{0,l}| - 1$, $residual(B_2) \leq |p_{l,h}| - 1$, and thus, $residual(B_1) + residual(B_2) \leq |p_{0,l}| - 1 + |p_{l,h}| - 1 = residual(B) - 1 < residual(B)$. In all three cases, B_1 and B_2 are short and $residual(B_1) + residual(B_2) < residual(B)$. The claim thus follows.

Once again, observe that in order to compute B_1 and B_2 from B, it is sufficient to trace the shortest path from s to t. Since all shortest paths are precomputed in a preprocessing step, this can be done in $\mathcal{O}(n)$ time. ∎

Lemma 4. *Any short bubble B has a tree-like decomposition into $\mathcal{O}(n)$ (not necessarily distinct) bubbles from the generator $\mathcal{G}(G)$. This decomposition can be found in $\mathcal{O}(n^2)$ time in the worst case.*

Proof. Each time we apply Lemma 3 to a short bubble B, we produce in $\mathcal{O}(n)$ time two short bubbles B_1 and B_2 such that $residual(B_1) + residual(B_2) < residual(B)$. Since $|\ell(B)| + residual(B) \leq n$, this implies that a short bubble can be decomposed in $\mathcal{O}(n)$ bubbles from the generator set $\mathcal{G}(G)$ in $\mathcal{O}(n^2)$ time. ∎

Theorem 3. *Given a graph G, any bubble B in G can be represented as a sum of $\mathcal{O}(n^2)$ bubbles that belong to $\mathcal{G}(G)$. This decomposition can be found in a total of $\mathcal{O}(n^3)$ time.*

Proof. The theorem follows by Lemmas 2 and 4. ∎

5 Conclusions and Open Problems

Bubbles in de Bruijn graphs represent interesting biological events, like alternative splicing and allelic differences (SNPs and indels). However, the set of all bubbles in a de Bruijn graph built from real data is usually too large to be efficiently enumerated and analysed. Therefore, in this paper we proposed a bubble generator, which is a polynomial-sized subset of the bubble space that can be used to generate all and only the bubbles in a directed graph. The concept of bubble generators is similar to cycle bases, but the algorithms for the latter cannot be applied as black boxes to find the former because the bubble space does not form a vector space. As such, this work describes efficient algorithms to identify, for any given directed graph G, a generator set of bubbles $\mathcal{G}(G)$, and to decompose a given bubble B into bubbles from $\mathcal{G}(G)$.

There remain several theoretical open questions. First, our generator $\mathcal{G}(G)$ is not necessarily minimal, *i.e.* it might happen that there exists three bubbles $B_1, B_2, B_3 \in \mathcal{G}(G)$ such that $B_1 < B_3$, $B_2 < B_3$, and $B_3 = B_1 + B_2$. Is it possible to find in polynomial time a generator $\mathcal{G}'(G)$ that is minimal or even better, to find $\mathcal{S}(G)$? Second, it would be interesting to know if there are polynomial-time algorithms to decompose any bubble of a graph G into bubbles of such generators. Third, it would be interesting to find a generator $\mathcal{G}(G)$ with some additional biologically motivated constraints, such as for example on the maximum length of the legs of a bubble [15]. Given an integer k and a graph G, is it possible to find a generator $\mathcal{G}(G)$ that generates all and only the bubbles of G which have both legs of length at most k? Fourth, are there faster algorithms to find a bubble generator? Fifth, this work is related to the research done in the direction of cycle bases. However, as we already mentioned, our problem displays characteristics that make it very different from the ones related to cycle bases. Thus, it may be of independent interest to further investigate the connections between these problems.

Finally, application of the bubble generator to genomic and transcriptomic graphs must be explored since it is one of the main motivations for this theoretical study. Similarly to the case of cycle bases, the simplest application of the bubble generators is to use it as a preprocessing step in several algorithms to reduce the amount of work to be done. For example, it can remove from the graph all unnecessary arcs (*i.e.* arcs that do not belong to any bubble) in order to lower the running time of an algorithm that is only interested in bubbles. As another example, the polynomial-time decomposition algorithm can be useful in the case where we want to identify and decompose complex alternative splicing events [17] into their elementary parts. However, exploring possible applications of the bubble generator is out of the scope of this paper.

Acknowledgments. V. Acuña was supported by Fondecyt 1140631, CIRIC-INRIA Chile and Basal Project PBF 03. R. Grossi and G.F. Italiano were partially supported by MIUR, the Italian Ministry of Education, University and Research, under the Project AMANDA (Algorithmics for MAssive and Networked DAta). Part of this work was done while G.F. Italiano was visiting Université de Lyon. L. Lima is supported by the Brazilian Ministry of Science, Technology and Innovation (in portuguese, Ministério da Ciência, Tecnologia e Inovação - MCTI) through the National Counsel of Technological and Scientific Development (in portuguese, Conselho Nacional de Desenvolvimento Científico e Tecnológico - CNPq), under the Science Without Borders (in portuguese, Ciências Sem Fronteiras) scholarship grant process number 203362/2014-4. B. Sinaimeri, L. Lima and M.-F. Sagot are partially funded by the French ANR project Aster (2016–2020), and together with V. Acuña, also by the Stic AmSud project MAIA (2016–2017). This work was performed using the computing facilities of the CC LBBE/PRABI.

References

1. Birmelé, E., et al.: Efficient bubble enumeration in directed graphs. In: Calderón-Benavides, L., González-Caro, C., Chávez, E., Ziviani, N. (eds.) SPIRE 2012. LNCS, vol. 7608, pp. 118–129. Springer, Heidelberg (2012). doi:10.1007/978-3-642-34109-0_13
2. Bollobás, B.: Modern Graph Theory. Graduate Texts in Mathematics, vol. 184. Springer-Verlag, Berlin (1998). doi:10.1007/978-1-4612-0619-4

3. Bondy, J.A., Murty, U.S.R.: Graph Theory with Applications. Elsevier, New York (1976)
4. Brankovic, L., Iliopoulos, C.S., Kundu, R., Mohamed, M., Pissis, S.P., Vayani, F.: Linear-time superbubble identification algorithm for genome assembly. Theoret. Comput. Sci. **609**, 374–383 (2016)
5. Deo, N.: Graph Theory with Applications to Engineering and Computer Science. Prentice-Hall series in Automatic Computation. Prentice-Hall, Englewood Cliffs (1974)
6. Gleiss, P.M., Leydold, J., Stadler, P.F.: Circuit bases of strongly connected digraphs. Discuss. Math. Graph Theory **23**(2), 241–260 (2003)
7. Iqbal, Z., Caccamo, M., Turner, I., Flicek, P., McVean, G.: De novo assembly and genotyping of variants using colored de bruijn graphs. Nat. Genet. **44**(2), 226–232 (2012)
8. Kavitha, T., Liebchen, C., Mehlhorn, K., Michail, D., Rizzi, R., Ueckerdt, T., Zweig, K.A.: Cycle bases in graphs characterization, algorithms, complexity, and applications. Comput. Sci. Rev. **3**(4), 199–243 (2009)
9. Kavitha, T., Mehlhorn, K.: Algorithms to compute minimum cycle bases in directed graphs. Theory Comput. Syst. **40**(4), 485–505 (2007)
10. Li, H.: Exploring single-sample SNP and INDEL calling with whole-genome de novo assembly. Bioinformatics **28**(14), 1838–1844 (2012)
11. Lima, L., Sinaimeri, B., Sacomoto, G., Lopez-Maestre, H., Marchet, C., Miele, V., Sagot, M.F., Lacroix, V.: Playing hide and seek with repeats in local and global de novo transcriptome assembly of short RNA-seq reads. Algorithms Mol. Biol. **12**(1), 2:1–2:19 (2017). doi:10.1186/s13015-017-0091-2
12. MacLane, S.: A combinatorial condition for planar graphs. Fundam. Math. **28**, 22–32 (1937)
13. Onodera, T., Sadakane, K., Shibuya, T.: Detecting superbubbles in assembly graphs. In: Darling, A., Stoye, J. (eds.) WABI 2013. LNCS, vol. 8126, pp. 338–348. Springer, Heidelberg (2013). doi:10.1007/978-3-642-40453-5_26
14. Pevzner, P.A., Tang, H., Tesler, G.: De novo repeat classification and fragment assembly. Genome Res. **14**(9), 1786–1796 (2004)
15. Sacomoto, G., Lacroix, V., Sagot, M.-F.: A polynomial delay algorithm for the enumeration of bubbles with length constraints in directed graphs and its application to the detection of alternative splicing in RNA-seq data. In: Darling, A., Stoye, J. (eds.) WABI 2013. LNCS, vol. 8126, pp. 99–111. Springer, Heidelberg (2013). doi:10.1007/978-3-642-40453-5_9
16. Sacomoto, G., Kielbassa, J., Chikhi, R., Uricaru, R., Antoniou, P., Sagot, M.F., Peterlongo, P., Lacroix, V.: KISSPLICE: de-novo calling alternative splicing events from RNA-seq data. BMC Bioinf. **13**(S-6), S5 (2012)
17. Sammeth, M.: Complete alternative splicing events are bubbles in splicing graphs. J. Comput. Biol. **16**(8), 1117–1140 (2009)
18. Simpson, J.T., Wong, K., Jackman, S.D., Schein, J.E., Jones, S.J.M., Birol, I.: ABySS: a parallel assembler for short read sequence data. Genome Res. **19**(6), 1117–1123 (2009)
19. Sung, W.K., Sadakane, K., Shibuya, T., Belorkar, A., Pyrogova, I.: An o(m log m)-time algorithm for detecting superbubbles. IEEE/ACM Trans. Comput. Biol. Bioinf. **12**(4), 770–777 (2015)
20. Uricaru, R., Rizk, G., Lacroix, V., Quillery, E., Plantard, O., Chikhi, R., Lemaitre, C., Peterlongo, P.: Reference-free detection of isolated SNPs. Nucleic Acids Res. **43**(2), e11 (2015)

21. Younsi, R., MacLean, D.: Using 2k+2 bubble searches to find single nucleotide polymorphisms in k-mer graphs. Bioinformatics **31**(5), 642–646 (2015)
22. Zerbino, D., Birney, E.: Velvet: algorithms for de novo short read assembly using de Bruijn graphs. Genome Res. **18**, 821–829 (2008)

Critical Node Cut Parameterized by Treewidth and Solution Size is W[1]-Hard

Akanksha Agrawal[(✉)], Daniel Lokshtanov, and Amer E. Mouawad

Department of Informatics, University of Bergen, Bergen, Norway
{akanksha.agrawal,daniello,a.mouawad}@ii.uib.no

Abstract. In the CRITICAL NODE CUT problem, given an undirected graph G and two non-negative integers k and μ, the goal is to find a set S of exactly k vertices such that after deleting S we are left with at most μ connected pairs of vertices. Back in 2015, Hermelin et al. studied the aforementioned problem under the framework of parameterized complexity. They considered various natural parameters, namely, the size k of the desired solution, the upper bound μ on the number of remaining connected pairs, the lower bound b on the number of connected pairs to be removed, and the treewidth $\mathsf{tw}(G)$ of the input graph G. For all but one combinations of the above parameters, they determined whether CRITICAL NODE CUT is fixed-parameter tractable and whether it admits a polynomial kernel. The only question they left open is whether the problem remains fixed-parameter tractable when parameterized by $k + \mathsf{tw}(G)$. We answer this question in the negative via a new problem of independent interest, which we call SUMCSP. We believe that SUMCSP can be a useful starting point for showing hardness results of the same nature, i.e. when the treewidth of the graph is part of the parameter.

1 Introduction

Consider the following problem, called CRITICAL NODE CUT (or CNC for short). We are given an undirected graph G and two non-negative integers k and μ. The goal is to determine whether there exists a subset of the vertices of G, say S, of size (exactly) k such that, in the graph $G - S$, we are left with at most μ connected pairs of vertices; $G - S$ denotes the graph obtained from G after deleting vertices in S and the edges incident on them. Alternatively, if we let $\mathcal{C}(G-S) = \{C_1, \ldots, C_\ell\}$, for some integer ℓ, denote the set of maximal connected components in $G - S$, the objective is to guarantee that $\sum_{C \in \mathcal{C}(G-S)} \binom{C}{2} \leq \mu$. The CNC problem, having many real-world applications such as controlling the spread of viruses in networks [9], has been investigated from various algorithmic perspectives, e.g. heuristics [12] and approximation algorithms [13]. Since the VERTEX COVER problem is a special case of CNC, i.e. when $\mu = 0$, the problem is clearly NP-complete. On the positive side, it is known that CNC can be solved in polynomial time if we restrict the input graph to trees [4]. More generally, for

---

Due to space limitations most proofs have been omitted.

© Springer International Publishing AG 2017
H.L. Bodlaender and G.J. Woeginger (Eds.): WG 2017, LNCS 10520, pp. 32–44, 2017.
https://doi.org/10.1007/978-3-319-68705-6_3

graphs of bounded treewidth, CNC can be solved in $\mathcal{O}(|V(G)|^{\mathsf{tw}(G)+1})$ time [1], where $\mathsf{tw}(G)$ is the treewidth of G. We refer the reader to [9] for a more extensive survey on CNC and its applications.

Hermelin et al. [9] initiated the study of the parameterized complexity of CNC. In parameterized complexity [6], we are interested in whether the problem can be solved in $f(\kappa) \cdot n^{\mathcal{O}(1)}$ time, for various natural parameters κ and some function f. Alternatively, one can also ask whether or not CNC admits a polynomial kernel for parameter κ, i.e. whether there is an algorithm that reduces any instance of CNC in polynomial time to an equivalent instance of size $\kappa^{\mathcal{O}(1)}$. There are quite a few natural choices for κ in this case and the following choices were considered by Hermelin et al. [9]: The size k of the desired solution, the upper bound μ on the number of remaining connected pairs, the lower bound b on the number of connected pairs to be removed, and the treewidth $\mathsf{tw}(G)$ of the input graph G. For all but one combinations of the parameters, Hermelin et al. determined whether CRITICAL NODE CUT is fixed-parameter tractable (FPT) and whether it admits a polynomial kernel. These results are summarized in Table 1. For more details on parameterized complexity we refer to the books of Downey and Fellows [6], Flum and Grohe [7], Niedermeier [11], and Cygan et al. [3].

Table 1. Summary of results due to Hermelin et al. [9].

Parameter				Result	
k	μ	b	$\mathsf{tw}(G)$	FPT	Polynomial kernel
✓				no	no
	✓			no	no
		✓		yes	no
			✓	no	no
✓	✓			yes	yes
✓		✓		yes	no
✓			✓	**open**	no
	✓	✓		yes	yes
	✓		✓	yes	no
		✓	✓	yes	no
✓	✓	✓		yes	yes
✓	✓		✓	yes	yes
✓		✓	✓	yes	no
	✓	✓	✓	yes	yes
✓	✓	✓	✓	yes	yes

In this work, we complete the table by showing that CNC is W[1]-Hard (or equivalently not likely to be FPT) when parameterized by $k + \mathsf{tw}(G)$. We prove this result via a new problem of independent interest, which we call SumCSP. We believe that SumCSP can be a useful starting point for showing hardness results of the same nature, i.e. when the treewidth of the graph is part of the parameter.

Overview of the Reduction. Our starting point is the PARTITIONED SUB-GRAPH ISOMORPHISM (PSI) problem, which is known to be W[1]-Hard [8,10] even when the pattern graph is 4-regular. The problem is formally defined below.

PARTITIONED SUBGRAPH ISOMORPHISM (PSI)   **Parameter:** $|V(P)|$
Input: A 4-regular pattern graph P with $V(P) = \{p_1, p_2, \cdots, p_\ell\}$, a host graph H, and a coloring function col $: V(H) \to [\ell]$.
Question: Does there exist an injective function $\phi : V(P) \to V(H)$ such that for each $i \in [\ell]$, $\mathrm{col}(\phi(p_i)) = i$ and for each $p_i p_j \in E(P)$, we have $\phi(p_i)\phi(p_j) \in E(H)$?

We reduce PSI to SUMCSP, which is formally defined next.

SUMCSP   **Parameter:** $|A(D)|$
Input: A directed graph D with vertex set $V(D)$ and arc set $A(D)$, vertex weight function $w_V : V(D) \to \mathbb{N}$, arc weight function $w_A : A(D) \to \mathbb{N}$, and a list function $\varphi : A(D) \to 2^{\mathbb{N} \times \mathbb{N}}$ such that for all $a \in A(D)$, and for all $(x, y), (x', y') \in \varphi(a)$ we have $x + y = x' + y' = w_A(a)$.
Question: Does there exists a function $\rho : A(D) \to \mathbb{N} \times \mathbb{N}$ such that for each $a \in A(D)$, $\rho(a) \in \varphi(a)$ and for each $v \in V(D)$, $\sum_{u \in N^+(v)} \mathrm{fir}(\rho(vu)) + \sum_{u \in N^-(v)} \mathrm{sec}(\rho(uv)) = w_V(v)$, where $\mathrm{fir}((x, y)) = x$ and $\mathrm{sec}((x, y)) = y$?

Bodlaender et al. [2] introduced a very closely related problem to show that PLANAR CAPACITATED DOMINATING SET is W[1]-Hard. PLANAR CAPACITATED DOMINATING SET was the first bidimensional problem to be shown W[1]-Hard and the reduction was via an intermediate problem called PLANAR ARC SUP-PLY. The main difference between PLANAR ARC SUPPLY and SUMCSP is the additional constraint we impose using the arc weight function, i.e. the fact that all pairs in $\varphi(a)$, $a \in A(D)$, must sum to $w_A(a)$. This constraint turns out to be crucial for our reduction. Roughly speaking, the reduction from PSI to SUM-CSP constructs a directed graph D whose structure is more of less similar to the pattern graph P (and its size is linear in $|V(P)|$). Edges of H are encoded using the vertex and arc weight functions as well as the function φ. Having established the hardness of SUMCSP, we then reduce SUMCSP to CRITICAL NOTE CUT. Let us first state a formal definition of the latter problem.

CRITICAL NODE CUT (CNC)   **Parameter:** $k + \mathrm{tw}(G)$
Input: An undirected graph G and integers k and μ.
Question: Does there exist a set $S \subseteq V(G)$ of size (exactly) k such that $\sum_{C \in \mathcal{C}(G-S)} \binom{C}{2} \leq \mu$, where $\mathcal{C}(G - S) = \{C_1, \ldots, C_\ell\}$ denotes the set of maximal connected components in $G - S$?

As stated earlier, our reduction from SUMCSP to CNC heavily relies on the arc weight function. Another crucial ingredient is the following proposition (which follows by the convexity of $\frac{x(x-1)}{2}$).

Proposition 1. *Let $x_1, \ldots, x_k$ be non-negative integers and let $x_1 + \ldots + x_k = kn$. Then, $\sum_{i=1}^{i=k} \binom{x_i}{2}$ is minimized if $x_i = n$, for all i. In other words, $\sum_{i=1}^{i=k} \binom{x_i}{2}$ is minimized if $\sum_{i=1}^{i=k} \binom{x_i}{2} = k\binom{n}{2}$.*

At a very high level, starting from an instance of SumCSP, we create a graph G (of bounded treewidth) where an optimal solution for CNC must separate the graph into a fixed number of connected components, all having the same size.

2 Preliminaries

We denote the set of natural numbers by $\mathbb{N}$. For $k \in \mathbb{N}$, by $[k]$ we denote the set $\{1, 2, \cdots, k\}$. For sets X, Y, by $X \times Y$ we denote the set $\{(x, y) \mid x \in X, y \in Y\}$. Furthermore, for $(x, y) \in X \times Y$, we let $\mathsf{fir}((x, y)) = x$ and $\mathsf{sec}((x, y)) = y$, i.e. the first and second coordinate of the (ordered) pair (x, y), respectively.

We use standard terminology from the book of Diestel [5] for graph-related terms that are not explicitly defined here. We consider only finite graphs. For a graph G, by $V(G)$ and $E(G)$ we denote the vertex and edge sets of G, respectively. Similarly, for a directed graph or digraph D, by $V(D)$ and $A(D)$ we denote the vertex and arc sets of D, respectively. For a graph G and $v \in V(G)$, by $N_G(v)$ we denote the set $\{u \in V(G) \mid vu \in E(G)\}$. For a digraph D and $v \in V(D)$, by $N_D^+(v)$ we denote the set $\{u \in V(D) \mid vu \in A(D)\}$, and by $N_D^-(v)$ we denote the set $\{u \in V(D) \mid uv \in A(D)\}$. We drop the subscript G (or D) from $N_G(v)$ (or $N_D^+(v)$, or $N_D^-(v)$) when the context is clear. For a vertex subset $S \subseteq V(G)$, by $G[S]$ we denote the subgraph of G induced by S, i.e. the graph with vertex set S and edge set $\{vu \in E(G) \mid v, u \in S\}$. By $G - S$ we denote the graph $G[V(G) \setminus S]$. A *coloring* of a graph G with $\alpha \in \mathbb{N}$ colors is a function $\varphi : V(G) \to [\alpha]$. A coloring φ of G is said to be a *proper coloring* if for each $uv \in E(G)$, $\varphi(u) \neq \varphi(v)$.

A *path* in a graph is a sequence of vertices $P = v_1, v_2, \cdots, v_\ell$ such that for all $i \in [\ell - 1]$, $v_i v_{i+1} \in E(G)$. We say that such a path is a path between v_1 and v_ℓ or a $v_1 - v_\ell$ path of length $\ell - 1$, and vertices $v_1, v_2, \cdots, v_\ell$ lie on the path P. Two vertices $u, v \in V(G)$ are said to be *connected* if there exists a $u - v$ path in G. A graph is *connected* if there is a path between every pair of vertices. A maximal connected subgraph of G is called a *component* of G. For a pair of vertices $u, v \in V(G)$, by $\mathsf{dist}_G(u, v)$ we denote the length of the shortest path between u and v in G. For a graph G, by G^2 we denote the graph with vertex set $V(G^2) = V(G)$ and edge set $E(G^2) = \{uv \mid \mathsf{dist}_G(u, v) \leq 2\}$.

A *tree decomposition* of a graph is a pair $(\mathcal{X}, \mathcal{T})$, where an element $X \in \mathcal{X}$ is a subset of $V(G)$, called a *bag*, and $\mathcal{T}$ is a rooted tree with vertex set $\mathcal{X}$ satisfying the following properties: (i) $\cup_{X \in \mathcal{X}} X = V(G)$; (ii) For every $uv \in E(G)$, there exists $X \in \mathcal{X}$ such that $u, v \in X$; (iii) For all $X, Y, Z \in \mathcal{X}$, if Y lies on the unique path between X and Z in $\mathcal{T}$, then $X \cap Z \subseteq Y$. For a graph G and its tree decomposition $(\mathcal{X}, \mathcal{T})$, the *width* of the tree decomposition $(\mathcal{X}, \mathcal{T})$ is defined to be $\max_{X \in \mathcal{X}}(|X| - 1)$. The *treewidth* of a graph G, $\mathsf{tw}(G)$, is the minimum of the widths of all its tree decompositions.

3 W[1]-HardNess of SumCSP

Let $(P, H, \mathsf{col} : V(H) \to [\ell])$ be an instance of PSI, where $V(P) = \{p_i \mid i \in [\ell]\}$ and $V(H) = \{h_i \mid i \in [n]\}$. For $i \in [\ell]$, we let $C_i^H = \{h \in V(H) \mid \mathsf{col}(h) = i\}$. We make a few assumption and adopt some conventions that will help simplify the presentation. All numbers that appear in the construction will be represented in binary. We assume that $|V(H)| = n = 2^t$, for some $t \in \mathbb{N}$, i.e. $t = \log n$. Otherwise, if $|V(H)| = 2^{t'} - \delta$, for some $0 < \delta < 2^{t'-1}$, we can construct an equivalent instance $(H', P', \mathsf{col}' : V(H') \to [\ell'])$ of PSI with $|V(H')| = 2^{t'+3}$, where H' is obtained from H by taking the disjoint union of H at most 8 times and adding δ copies of a 4-regular graph on 8 vertices (which exists) to H' and adjusting P to obtain P' and col to obtain col' appropriately. We further assume that for $i, j \in [\ell]$, where $p_i p_j \in E(P)$, there exists $h \in C_i^H$ and $h' \in C_j^H$ such that $hh' \in E(H)$, otherwise $(P, H, \mathsf{col} : V(H) \to [\ell])$ is a no-instance of PSI. Note that P is a 4-regular graph, which implies that it has no isolated vertices. We assume a fixed cyclic ordering $\prec_H$ on the vertices in H and a fixed cyclic ordering $\prec_P$ on the vertices in P. Simply put, we have $h_1 \prec_H \ldots \prec_H h_n \prec_H h_1$ and $p_1 \prec_P \ldots \prec_H p_\ell \prec_H p_1$. With each vertex $h_i \in V(H)$, or equivalently integer $i \in [n]$, we assign two binary strings (or bitstrings for short) B_{h_i} and $\overline{B}_{h_i}$ as follows. We let $\mathbb{B}_i$ denote the binary representation of integer i and $\overline{\mathbb{B}}_i$ denote the (bitwise) complement of $\mathbb{B}_i$. We use $\mathbb{O}_z$ and $\mathbb{1}_z$ to denote the bitstrings of length z consisting of all zeros and all ones, respectively. We let $B_{h_i} = \mathbb{O}_{4t}\mathbb{B}_i\mathbb{O}_{4t}$ and $\overline{B}_{h_i} = \mathbb{O}_{4t}\overline{\mathbb{B}}_i\mathbb{O}_{4t}$. Note that B_{h_i} and $\overline{B}_{h_i}$ are of length $9 \log n = 9t$. The purpose of the additional zero bits is to allow us to "correctly" handle overflows when summing binary numbers. For two bitstrings B and B', we slightly abuse notation and sometimes treat the result of $B + B'$ as another bitstring (obtained after applying the usual binary addition operator) or as an integer (in base 10).

We also assume that, along with instance $(P, H, \mathsf{col} : V(H) \to [\ell])$, we are given a proper coloring $\mathsf{col}_{P^2} : V(P) \to [21]$ of P^2. Observe that such a coloring exists and can be computed in time polynomial in the size of the graph P; the maximum degree of a vertex in P^2 is bounded by 20 and a graph with maximum degree d admits a $d + 1$ proper coloring. For a vertex $p_i \in V(P)$, we let $\mathsf{idx}_i = \mathsf{col}_{P^2}(p_i)$. In what follows, we will always deal with bitstrings of length $21 \cdot 2 \cdot 9 \cdot t$. A *block* consists of $9t$ consecutive bits. We note that two distinct blocks do not intersect in any bit position. Blocks will usually be set to bistrings of the form $\mathbb{O}_{4t}\mathbb{B}_i\mathbb{O}_{4t}$, $\mathbb{O}_{4t}\overline{\mathbb{B}}_i\mathbb{O}_{4t}$, $\mathbb{O}_{4t}\mathbb{1}_t\mathbb{O}_{4t}$, $\mathbb{O}_{9t}$, or $\mathbb{1}_{9t}$, $i \in [n]$. A *group* consists of $2 \cdot 9 \cdot t$ consecutive bits. Two distinct groups do not intersect in any bit position and a group consists of two blocks. Note that we have exactly 21 groups, which is equal to the number of colors in col_{P^2}. The reason why we need col_{P^2} will become clearer later. Intuitively, since we will be encoding the possible edges (from H) between a vertex in P and its four neighbors, we need to make sure that two of its neighbors do not get assigned the same group in a bitstring. Given a bitstring S of length $\gamma \cdot 2 \cdot 9 \cdot t$, for some $\gamma \in \mathbb{N}$, we let $\mathsf{block}[i](S)$ denote the ith block of S, and we let $\mathsf{group}[j](S)$ denote the jth group of S. We also use the notation $\mathsf{group}[i \mid j](S)$ to denote the ith and jth group of S. Finally, we note that, since the length of bitstrings will be bounded by $\mathcal{O}(\log n)$, all numbers in the construction will be bounded by $n^{\mathcal{O}(1)}$. We are now ready to describe the construction of instance $(D, w_V : V(D) \to \mathbb{N}, w_A : A(D) \to \mathbb{N}, \varphi : A(D) \to 2^{\mathbb{N} \times \mathbb{N}})$ of SUMCSP. We start with the description of the edge selection gadget.

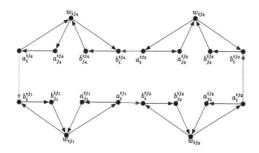

Fig. 1. An illustration of edge selection gadgets and the additional edges between them.

Edge Selection Gagdet. For every (unordered) pair of numbers $i, j \in [\ell]$ such that $p_i p_j \in E(P)$, we add an edge selection gadget E_{ij} (E_{ij} is a graph and not an edge set) to D. Note that both E_{ij} and E_{ji} refer to the same edge selection gadget, which will be responsible for selecting an edge in the host graph H. Moreover, $\mathsf{idx}_i \neq \mathsf{idx}_j$, since col_{P^2} is a proper coloring of P^2. We assume, without loss of generality, that $i < j$. We let $V(E_{ij}) = \{a_i^{ij}, a_j^{ij}, b_i^{ij}, b_j^{ij}, w_{ij}\}$ and $A(E_{ij}) = \{a_i^{ij} a_j^{ij}, a_j^{ij} w_{ij}, w_{ij} a_i^{ij}, b_i^{ij} b_j^{ij}, b_j^{ij} w_{ij}, w_{ij} b_i^{ij}\}$ (see Fig. 1). We now describe the construction of $\varphi : A(D) \to 2^{\mathbb{N} \times \mathbb{N}}$ and $w_A : A(D) \to \mathbb{N}$. We assume that all bitstrings are initialized to $\mathbb{O}_{378t}$. That is, whenever we do not explicitly specify the value of a group (block) in a bitstring, it is set to all zeros.

- Consider $a_i^{ij} a_j^{ij} \in A(E_{ij})$. For each $u \in C_i^H$ and $v \in C_j^H$ such that $uv \in E(H)$, we create a pair of bitstrings $(S_{uv}(a_i^{ij} a_j^{ij}), T_{uv}(a_i^{ij} a_j^{ij}))$ and add it to $\varphi(a_i^{ij} a_j^{ij})$. We set the following groups:
  $\mathsf{group}[\mathsf{idx}_i \mid \mathsf{idx}_j](S_{uv}(a_i^{ij} a_j^{ij})) = B_u \overline{B}_u \mid \overline{B}_v \overline{B}_v$;
  $\mathsf{group}[\mathsf{idx}_i \mid \mathsf{idx}_j](T_{uv}(a_i^{ij} a_j^{ij})) = \overline{B}_u B_u \mid B_v B_v$;
  $\mathsf{group}[\mathsf{idx}_i \mid \mathsf{idx}_j](w_A(a_i^{ij} a_j^{ij})) = \mathbb{O}_{4t} \mathbb{1}_t \mathbb{O}_{4t} \mathbb{O}_{4t} \mathbb{1}_t \mathbb{O}_{4t} \mid \mathbb{O}_{4t} \mathbb{1}_t \mathbb{O}_{4t} \mathbb{O}_{4t} \mathbb{1}_t \mathbb{O}_{4t}$.
- Consider $a_j^{ij} w_{ij} \in A(E_{ij})$. For each $u \in C_i^H$ and $v \in C_j^H$ such that $uv \in E(H)$, we create a pair of bitstrings $(S_{uv}(a_j^{ij} w_{ij}), T_{uv}(a_j^{ij} w_{ij}))$ and add it to $\varphi(a_j^{ij} w_{ij})$. We set the following groups:
  $\mathsf{group}[\mathsf{idx}_i \mid \mathsf{idx}_j](S_{uv}(a_j^{ij} w_{ij})) = B_u \overline{B}_u \mid \mathbb{O}_{9t} \overline{B}_v$;
  $\mathsf{group}[\mathsf{idx}_i \mid \mathsf{idx}_j](T_{uv}(a_j^{ij} w_{ij})) = \overline{B}_u B_u \mid \mathbb{O}_{9t} B_v$;
  $\mathsf{group}[\mathsf{idx}_i \mid \mathsf{idx}_j](w_A(a_j^{ij} w_{ij})) = \mathbb{O}_{4t} \mathbb{1}_t \mathbb{O}_{4t} \mathbb{O}_{4t} \mathbb{1}_t \mathbb{O}_{4t} \mid \mathbb{O}_{4t} \mathbb{O}_t \mathbb{O}_{4t} \mathbb{O}_{4t} \mathbb{1}_t \mathbb{O}_{4t}$.
- Consider $w_{ij} a_i^{ij} \in A(E_{ij})$. For each $u \in C_i^H$ and $v \in C_j^H$ such that $uv \in E(H)$, we create a pair of bitstrings $(S_{uv}(w_{ij} a_i^{ij}), T_{uv}(w_{ij} a_i^{ij}))$ and add it to $\varphi(w_{ij} a_i^{ij})$. We set the following groups:

$\text{group}[\text{idx}_i \mid \text{idx}_j](S_{uv}(w_{ij}a_i^{ij})) = \mathbb{O}_{9t}\overline{B}_u \mid \overline{B}_v B_v;$

$\text{group}[\text{idx}_i \mid \text{idx}_j](T_{uv}(w_{ij}a_i^{ij})) = \mathbb{O}_{9t}B_u \mid B_v B_v;$

$\text{group}[\text{idx}_i \mid \text{idx}_j](w_A(w_{ij}a_i^{ij})) = \mathbb{O}_{4t}\mathbb{O}_t\mathbb{O}_{4t}\mathbb{O}_{4t}\mathbb{1}_t\mathbb{O}_{4t} \mid \mathbb{O}_{4t}\mathbb{1}_t\mathbb{O}_{4t}\mathbb{O}_{4t}\mathbb{1}_t\mathbb{O}_{4t}.$

- Consider $b_i^{ij}b_j^{ij} \in A(E_{ij})$. For each $u \in C_i^H$ and $v \in C_j^H$ such that $uv \in E(H)$, we create a pair of bitstrings $(S_{uv}(b_i^{ij}b_j^{ij}), T_{uv}(b_i^{ij}b_j^{ij}))$ and add it to $\varphi(b_i^{ij}b_j^{ij})$. We set the following groups:

  $\text{group}[\text{idx}_i \mid \text{idx}_j](S_{uv}(b_i^{ij}b_j^{ij})) = \overline{B}_u B_u \mid B_v B_v;$

  $\text{group}[\text{idx}_i \mid \text{idx}_j](T_{uv}(b_i^{ij}b_j^{ij})) = B_u\overline{B}_u \mid \overline{B}_v\overline{B}_v;$

  $\text{group}[\text{idx}_i \mid \text{idx}_j](w_A(b_i^{ij}b_j^{ij})) = \mathbb{O}_{4t}\mathbb{1}_t\mathbb{O}_{4t}\mathbb{O}_{4t}\mathbb{1}_t\mathbb{O}_{4t} \mid \mathbb{O}_{4t}\mathbb{1}_t\mathbb{O}_{4t}\mathbb{O}_{4t}\mathbb{1}_t\mathbb{O}_{4t}.$

- Consider $b_j^{ij}w_{ij} \in A(E_{ij})$. For each $u \in C_i^H$ and $v \in C_j^H$ such that $uv \in E(H)$, we create a pair of bitstrings $(S_{uv}(b_j^{ij}w_{ij}), T_{uv}(b_j^{ij}w_{ij}))$ and add it to $\varphi(b_i^{ij}w_{ij})$. We set the following groups:

  $\text{group}[\text{idx}_i \mid \text{idx}_j](S_{uv}(b_j^{ij}w_{ij})) = \overline{B}_u B_u \mid \mathbb{O}_{9t}B_v;$

  $\text{group}[\text{idx}_i \mid \text{idx}_j](T_{uv}(b_j^{ij}w_{ij})) = B_u\overline{B}_u \mid \mathbb{O}_{9t}\overline{B}_v;$

  $\text{group}[\text{idx}_i \mid \text{idx}_j](w_A(b_j^{ij}w_{ij})) = \mathbb{O}_{4t}\mathbb{1}_t\mathbb{O}_{4t}\mathbb{O}_{4t}\mathbb{1}_t\mathbb{O}_{4t} \mid \mathbb{O}_{4t}\mathbb{O}_t\mathbb{O}_{4t}\mathbb{O}_{4t}\mathbb{1}_t\mathbb{O}_{4t}.$

- Consider $w_{ij}b_i^{ij} \in A(E_{ij})$. For each $u \in C_i^H$ and $v \in C_j^H$ such that $uv \in E(H)$, we create a pair of bitstrings $(S_{uv}(w_{ij}b_i^{ij}), T_{uv}(w_{ij}, b_i^{ij}))$ and add it to $\varphi(w_{ij}b_i^{ij})$. We set the following groups:

  $\text{group}[\text{idx}_i \mid \text{idx}_j](S_{uv}(w_{ij}b_i^{ij})) = \mathbb{O}_{9t}B_u \mid B_v B_v;$

  $\text{group}[\text{idx}_i \mid \text{idx}_j](T_{uv}(w_{ij}, b_i^{ij})) = \mathbb{O}_{9t}\overline{B}_u \mid \overline{B}_v\overline{B}_v;$

  $\text{group}[\text{idx}_i \mid \text{idx}_j](w_A(w_{ij}b_i^{ij})) = \mathbb{O}_{4t}\mathbb{O}_t\mathbb{O}_{4t}\mathbb{O}_{4t}\mathbb{1}_t\mathbb{O}_{4t} \mid \mathbb{O}_{4t}\mathbb{1}_t\mathbb{O}_{4t}\mathbb{O}_{4t}\mathbb{1}_t\mathbb{O}_{4t}.$

Compatibility Between Edge Selection Gadgets. We add edges between various edge selection gadgets to ensure that for each $i \in [\ell]$, the edges selected by the gadgets are incident on the same vertex in C_i^H. The selection of an edge by a gadget will be determined by the pair of number selected from $\varphi(a)$, where $a \in A(E_{ij})$ and $p_i p_j \in E(P)$. For each $p_i \in V(P)$, we have $|N_P(p_i)| = 4$, since P is a 4-regular graph. For $i \in [\ell]$, let $N_P(p_i) = \{p_{j_1}, p_{j_2}, p_{j_3}, p_{j_4}\}$, where we assume a (fixed and cyclic) ordering on the vertices in $N_P(p_i)$ based on the ordering $\prec_P$. That is, we assume $p_{j_1} \prec_P p_{j_2} \prec_P p_{j_3} \prec_P p_{j_4} \prec_P p_{j_1}$. Below we describe the set of arcs added between $E_{ij_1}, E_{ij_2}, E_{ij_3}$ and E_{ij_4}, we call this set A_i. We also describe the values assigned by $w_A(\cdot)$ and $\varphi(\cdot)$ to arcs in A_i (see Fig. 1).

- We add the arc $a_i^{ij_1}b_i^{ij_2}$ to A_i and, for each $u \in C_i^H$, we add a pair of bitstrings $(S_u(a_i^{ij_1}b_i^{ij_2}), T_u(a_i^{ij_1}b_i^{ij_2}))$ to $\varphi(a_i^{ij_1}b_i^{ij_2})$. We set the following groups:

  $\text{group}[\text{idx}_i](S_u(a_i^{ij_1}b_i^{ij_2})) = \overline{B}_u\mathbb{O}_{9t};$

  $\text{group}[\text{idx}_i](T_u(a_i^{ij_1}b_i^{ij_2})) = B_u\mathbb{O}_{9t};$

  $\text{group}[\text{idx}_i](w_A(a_i^{ij_1}b_i^{ij_2})) = \mathbb{O}_{4t}\mathbb{1}_t\mathbb{O}_{4t}\mathbb{O}_{9t}.$

- We add the arc $a_i^{ij_2}b_i^{ij_3}$ to A_i and, for each $u \in C_i^H$, we add a pair of bitstrings $(S_u(a_i^{ij_2}b_i^{ij_3}), T_u(a_i^{ij_2}b_i^{ij_3}))$ to $\varphi(a_i^{ij_2}b_i^{ij_3})$. We set the following groups:

  $\text{group}[\text{idx}_i](S_u(a_i^{ij_2}b_i^{ij_3})) = \overline{B}_u\mathbb{O}_{9t};$

  $\text{group}[\text{idx}_i](T_u(a_i^{ij_2}b_i^{ij_3})) = B_u\mathbb{O}_{9t};$

  $\text{group}[\text{idx}_i](w_A(a_i^{ij_2}b_i^{ij_3})) = \mathbb{O}_{4t}\mathbb{1}_t\mathbb{O}_{4t}\mathbb{O}_{9t}.$

- We add the arc $a_i^{ij_3}b_i^{ij_4}$ to A_i and, for each $u \in C_i^H$, we add a pair of bitstrings $(S_u(a_i^{ij_3}b_i^{ij_4}), T_u(a_i^{ij_3}b_i^{ij_4}))$ to $\varphi(a_i^{ij_3}b_i^{ij_4})$. We set the following groups:
  $\mathsf{group}[\mathsf{idx}_i](S_u(a_i^{ij_3}b_i^{ij_4})) = \overline{B}_u \mathbb{O}_{9t}$;
  $\mathsf{group}[\mathsf{idx}_i](T_u(a_i^{ij_3}b_i^{ij_4})) = B_u \mathbb{O}_{9t}$;
  $\mathsf{group}[\mathsf{idx}_i](w_A(a_i^{ij_3}b_i^{ij_4})) = \mathbb{O}_{4t}\mathbb{1}_t\mathbb{O}_{4t}\mathbb{O}_{9t}$.
- We add the arc $a_i^{ij_4}b_i^{ij_1}$ to A_i and, for each $u \in C_i^H$, we add a pair of bitstrings $(S_u(a_i^{ij_4}b_i^{ij_1}), T_u(a_i^{ij_4}b_i^{ij_1}))$ to $\varphi(a_i^{ij_4}b_i^{ij_1})$. We set the following groups:
  $\mathsf{group}[\mathsf{idx}_i](S_u(a_i^{ij_4}b_i^{ij_1})) = \overline{B}_u \mathbb{O}_{9t}$;
  $\mathsf{group}[\mathsf{idx}_i](T_u(a_i^{ij_4}b_i^{ij_1})) = B_u \mathbb{O}_{9t}$;
  $\mathsf{group}[\mathsf{idx}_i](w_A(a_i^{ij_4}b_i^{ij_1})) = \mathbb{O}_{4t}\mathbb{1}_t\mathbb{O}_{4t}\mathbb{O}_{9t}$.

This completes the description of the vertices and arcs of D, and the functions $w_A : A(D) \to \mathbb{N}$ and $\varphi : A(D) \to 2^{\mathbb{R} \times \mathbb{R}}$. We now move to description of the function $w_V : V(D) \to \mathbb{N}$.

The Vertex Weight Function. Consider $i, j \in [\ell]$, $i < j$, we set $w_V(\cdot)$ as follows.

- For all $u \in \{a_i^{ij}, a_j^{ij}, b_i^{ij}, b_j^{ij}, \}$, we set $w_V(u)$ to be the bitstring X_u of length $378 \log n$, where $\mathsf{group}[\mathsf{idx}_i](X_u) = \mathbb{O}_{4t}\mathbb{1}_t\mathbb{O}_{4t}\mathbb{O}_{4t}\mathbb{1}_t\mathbb{O}_{4t}$ and $\mathsf{group}[\mathsf{idx}_j](X_u) = \mathbb{O}_{4t}\mathbb{1}_t\mathbb{O}_{4t}\mathbb{O}_{4t}\mathbb{1}_t\mathbb{O}_{4t}$.
- For w_{ij}, we set $w_V(w_{ij})$ to be the bitstring $X_{w_{ij}}$ of length $378 \log n$, which we construct as follows. We let Y be the bitstring of length t corresponding to the integer $2^t - 2$, i.e. a bitstring of length t with the last bit set to zero and all other bits set to one. Let Y' to be the bitstring of length $4t$ corresponding to the integer 1, i.e. the bitstring of length $4t$ with the last bit set to one and all other bits set to zero. We set $\mathsf{group}[\mathsf{idx}_i](X_{w_{ij}}) = \mathbb{O}_{4t}\mathbb{1}_t\mathbb{O}_{4t}Y'Y\mathbb{O}_{4t}$ and $\mathsf{group}[\mathsf{idx}_j](X_{w_{ij}}) = \mathbb{O}_{4t}\mathbb{1}_t\mathbb{O}_{4t}Y'Y\mathbb{O}_{4t}$.

This finishes the description of the instance $(D, w_V : V(D) \to \mathbb{N}, w_A : A(D) \to \mathbb{N}, \varphi : A(D) \to 2^{\mathbb{N} \times \mathbb{N}})$ of SumCSP for a given instance $(P, H, \mathsf{col} : V(H) \to [\ell])$ of PSI. Below we state some propositions and lemmata that will be useful in establishing the equivalence of the two instances.

Proposition 2. *Let X, Y be two bitstrings of length $\log q$. Then $X + Y = 2^q - 1$ if and only if $\overline{X} = Y$.*

Proposition 3. *Let X and Y be two bitstrings each of length $42 \cdot 9 \cdot t$ and consisting of 21 groups, where $t = \log n$. Assume that, for each $i \in [21]$, group i in X consists of a bitstring of the form $X_i = \mathbb{O}_{4t}B_x\mathbb{O}_{4t}$ and group i in Y consists of a bitstring of the form $Y_i = \mathbb{O}_{4t}B_y\mathbb{O}_{4t}$, $x, y \in [n]$. Then, $X + Y$ is a bitstring of length $42 \cdot 9 \cdot t$ with the ith group equal to $X_i + Y_i$, $i \in [21]$.*

Lemma 1. *Let $(D, w_V : V(D) \to \mathbb{N}, w_A : A(D) \to \mathbb{N}, \varphi : A(D) \to 2^{\mathbb{N} \times \mathbb{N}})$ be a yes-instance of SumCSP and $\rho : A(D) \to \mathbb{N} \times \mathbb{N}$ be a solution. Consider $p_i, p_{i'}, p_j, p_{j'} \in V(P)$ such that $a_i^{ij}b_i^{ij'}, a_j^{ij}b_j^{i'j} \in A(D)$ and $i < j$. For $u \in C_i^H$ and $v \in C_j^H$, we have $\rho(a_i^{ij}a_j^{ij}) = (S_{uv}(a_i^{ij}a_j^{ij}), T_{uv}(a_i^{ij}a_j^{ij}))$ if and only if $\rho(a_i^{ij}b_i^{ij'}) = (S_u(a_i^{ij}b_i^{ij'}), T_u(a_i^{ij}b_i^{ij'}))$ and $\rho(a_j^{ij}b_j^{i'j}) = (S_v(a_j^{ij}b_j^{i'j}), T_v(a_j^{ij}b_j^{i'j}))$.*

Lemma 2. *Let* $(D, w_V : V(D) \rightarrow \mathbb{N}, w_A : A(D) \rightarrow \mathbb{N}, \varphi : A(D) \rightarrow 2^{\mathbb{N} \times \mathbb{N}})$ *be a yes-instance of* SUMCSP *and* $\rho : A(D) \rightarrow \mathbb{N} \times \mathbb{N}$ *be a solution. Consider* $p_i, p_{i'}, p_j, p_{j'} \in V(P)$ *such that* $a_i^{ij'} b_i^{ij}, a_j^{i'j} b_j^{ij} \in A(D)$ *and* $i < j$. *For* $u \in C_i^H$ *and* $v \in C_j^H$, *we have* $\rho(b_i^{ij} b_j^{ij}) = (S_{uv}(b_i^{ij} b_j^{ij}), T_{uv}(b_i^{ij} b_j^{ij}))$ *if and only if* $\rho(a_i^{ij'} b_i^{ij}) = (S_u(a_i^{ij'} b_i^{ij}), T_u(a_i^{ij'} b_i^{ij}))$ *and* $\rho(a_j^{i'j} b_j^{ij}) = (S_v(a_j^{i'j} b_j^{ij}), T_v(a_j^{i'j} b_j^{ij}))$.

Lemma 3. *Let* $(D, w_V : V(D) \rightarrow \mathbb{N}, w_A : A(D) \rightarrow \mathbb{N}, \varphi : A(D) \rightarrow 2^{\mathbb{N} \times \mathbb{N}})$ *be a yes-instance of* SUMCSP *and* $\rho : A(D) \rightarrow \mathbb{N} \times \mathbb{N}$ *be a solution. Let* $i, j \in [\ell]$, *where* $i < j$ *and* $p_i p_j \in E(P)$, *and let* $u \in C_i^H$ *and* $v \in C_j^H$. *Then, the following three statements are equivalent:*

(1) $\rho(a_i^{ij} a_j^{ij}) = (S_{uv}(a_i^{ij} a_j^{ij}), T_{uv}(a_i^{ij} a_j^{ij}))$;
(2) $\rho(w_{ij} a_i^{ij}) = (S_{uv}(w_{ij} a_i^{ij}), T_{uv}(w_{ij} a_i^{ij}))$;
(3) $\rho(a_j^{ij} w_{ij}) = (S_{uv}(a_j^{ij} w_{ij}), T_{uv}(a_j^{ij} w_{ij}))$.

Lemma 4. *Let* $(D, w_V : V(D) \rightarrow \mathbb{N}, w_A : A(D) \rightarrow \mathbb{N}, \varphi : A(D) \rightarrow 2^{\mathbb{N} \times \mathbb{N}})$ *be a yes-instance of* SUMCSP *and* $\rho : A(D) \rightarrow \mathbb{N} \times \mathbb{N}$ *be a solution. Let* $i, j \in [\ell]$, *where* $i < j$ *and* $p_i p_j \in E(P)$, *and let* $u \in C_i^H$ *and* $v \in C_j^H$. *Then, the following three statements are equivalent:*

(1) $\rho(b_i^{ij} b_j^{ij}) = (S_{uv}(b_i^{ij} b_j^{ij}), T_{uv}(b_i^{ij} b_j^{ij}))$
(2) $\rho(w_{ij} b_i^{ij}) = (S_{uv}^{(w_{ij}, b_i^{ij})}, T_{uv}^{(w_{ij}, b_i^{ij})})$;
(3) $\rho(b_j^{ij} w_{ij}) = (S_{uv}(b_j^{ij} w_{ij}), T_{uv}(b_j^{ij} w_{ij}))$.

Lemma 5. *Let* $(D, w_V : V(D) \rightarrow \mathbb{N}, w_A : A(D) \rightarrow \mathbb{N}, \varphi : A(D) \rightarrow 2^{\mathbb{N} \times \mathbb{N}})$ *be a yes-instance of* SUMCSP *and* $\rho : A(D) \rightarrow \mathbb{N} \times \mathbb{N}$ *be a solution. Let* $i, j \in [\ell]$, $i < j$, $p_i p_j \in E(P)$, *and* $u \in C_i^H$ *and* $v \in C_j^H$. *Then,* $\rho(a_i^{ij} a_j^{ij}) = (S_{uv}(a_i^{ij} a_j^{ij}), T_{uv}(a_i^{ij} a_j^{ij}))$ *if and only if* $\rho(b_i^{ij} b_j^{ij}) = (S_{uv}(b_i^{ij} b_j^{ij}), T_{uv}(b_i^{ij} b_j^{ij}))$.

Lemma 6. $(P, H, col : V(H) \rightarrow [\ell])$ *is a yes-instance of* PSI *if and only if* $(D, w_V : V(D) \rightarrow \mathbb{N}, w_A : A(D) \rightarrow \mathbb{N}, \varphi : A(D) \rightarrow 2^{\mathbb{N} \times \mathbb{N}})$ *is a yes-instance of* SUMCSP.

Theorem 1. SUMCSP *is* W[1]-*Hard when parameterized by the number of vertices in the pattern graph.*

Proof. Let $(D, w_V : V(D) \rightarrow \mathbb{N}, w_A : A(D) \rightarrow \mathbb{N}, \varphi : A(D) \rightarrow 2^{\mathbb{N} \times \mathbb{N}})$ be the contructed instance of SUMCSP given the instance $(P, H, col : V(H) \rightarrow [\ell])$ of PSI. An easy trace of the construction shows that it can be accomplished in time polynomial in $|V(H)|$ and that all the numbers appearing in the construction are bounded by $|V(H)|^{\mathcal{O}(1)}$. Moreover, note that P is 4-regular and therefore $|E(P)| = \mathcal{O}(|V(P)|)$. By construction we have $|A(D)| = \mathcal{O}(|V(P)|)$. These together with Lemma 6 and the W[1]-Hardness of PSI completes the proof. □

4   W[1]-HardNess of CNC

Let $(D, w_V : V(D) \to \mathbb{N}, w_A : A(D) \to \mathbb{N}, \varphi : A(D) \to 2^{\mathbb{N} \times \mathbb{N}})$ be an instance of SumCSP. We let $w_{vmax} = \max_{v \in V(D)}(w_V(v))$, i.e. the maximum weight of a vertex in D, we let $w_{amax} = \max_{a \in A(D)}(w_A(a))$, i.e. the maximum weight of an arc in D, and we let $w_{big} = (w_{amax} \cdot w_{vmax})^{100}$. We assume, without loss of generality, that the number of arcs in D is greater than some constant, say $|A(D)| \geq 50$, $w_{vmax} > 2|A(D)|$, and $w_{amax} > 0$ (otherwise we can increase all numbers in the SumCSP instance appropriately). Moreover, we let $W^\star = (k+3)(w_{big} + w_{vmax} + 2)$. For each vertex $v \in V(D)$, we define a quantity $W_v = W^\star - (k+3)(w_V(v) + 2) = (k+3)(w_{big} + w_{vmax} - w_V(v))$. We shall create an instance (G, k, μ) of CNC, where $k = 2|A(D)|$, $\mu = |V(D)| \cdot \binom{W^\star}{2}$, and $tw(G) = k^{\mathcal{O}(1)}$. We now proceed to the construction of the graph G.

Construction. For each vertex $v \in V(D)$, we create a clique K_v of size $2(k+3)$ and an independent set I_v of size W_v (see Fig. 2). We add all edges between vertices in K_v and vertices in I_v. For each arc $a = uv \in A(D)$, we create a *chain* H_{uv} (which will connect K_u and K_v) as follows. H_{uv} consists of $w_A(a) + 1$ *connecting pairs* of vertices $\mathcal{P}_{uv} = \{p_0, \ldots, p_{w_A(uv)}\}$, i.e. each pair $p_i \in \mathcal{P}_{uv}$ consists of two (independent) vertices $\{p_i^1, p_i^2\}$. Moreover, we have $w_A(a)$ *border walls* $\mathcal{B}_{uv} = \{b_1, \ldots, b_{w_A(a)}\}$, each of size $k+1$, i.e. each wall consists of $k+1$ (independent) vertices. We add all edges between K_u and pair p_0 and we add all edges between K_v and $p_{w_A(uv)}$. Next, we add all edges between p_{i-1} and b_i and all edges between b_i and p_i, for $i \in [w_A(a)]$. We call the pair p_0 the *first pair* of $\mathcal{P}_{uv}$ and denote it by first$(\mathcal{P}_{uv})$. Similarly, we call the pair $p_{w_A(uv)}$ the *last pair* of $\mathcal{P}_{uv}$ and denote it by last$(\mathcal{P}_{uv})$. Then, we sort all entries $(i, j) \in \varphi(a)$ in increasing order based on the first coordinate. Let $\{(i_1, j_1), (i_2, j_2), \ldots, (i_r, j_r)\}$ denote the resulting sorted set. We assume, without loss of generality, that the set contains pairs where no two pairs have the same element in their first (or second) coordinate. This assumption is justified by the fact that for all $(i, j), (i', j') \in \varphi(a)$ we have $i + j = i' + j' = w_A(a)$. We add all edges (if they do not already exist) between K_u and vertices $\{p_{i_1}^1, p_{i_1}^2\}$ and all edges between K_v and vertices $\{p_{i_r}^1, p_{i_r}^2\}$. We call the pair p_{i_1} the *left pair* of $\mathcal{P}_{uv}$ and denote it by left$(\mathcal{P}_{uv})$.

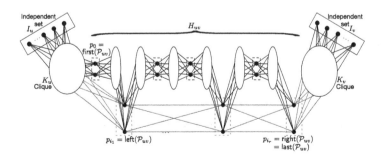

Fig. 2. An illustration of parts of the construction of the graph G.

Similarly, we call the pair p_{i_r} the *right pair* of $\mathcal{P}_{uv}$ and denote it by $\text{right}(\mathcal{P}_{uv})$. Finally, for each two consecutive entries (i, j) and (i', j') in the set $\varphi(a)$, we add all edges between $\{p_i^1, p_i^2\}$ and $\{p_{i'}^1.p_{i'}^2\}$. This completes the construction of the graph G.

Proposition 4. $\text{tw}(G) = k^{\mathcal{O}(1)}$.

Below we prove a series of lemmas that allows us to transform any solution S to an instance (G, k, μ) of CNC into an "equally good" solution S' having some "nice" structural properties. We say that a solution S *splits* a connecting pair $\{p^1, p^2\}$ if $|S \cap \{p^1, p^2\}| = 1$. We let $\mathcal{C}(G - S) = \{C_1, \ldots, C_\ell\}$ denote the maximal connected components in $G - S$. We classify a component $C \in \mathcal{C}(G - S)$ into one of three types. We say C is a *small* component whenever C does not contain any vertices from K_v or I_v, for all $v \in V(D)$. We say C is a *large* component whenever C intersects at least two (distinct) cliques K_u and K_v, $u, v \in V(D)$. Otherwise, C is a *medium* component. Note that, for any $v \in V(D)$, any solution of size k cannot separate $G[V(I_v) \cup V(K_v)]$ into two or more components. Therefore, if $\mathcal{C}(G - S)$ consists of only medium components then $|\mathcal{C}(G - S)|$ is exactly $|V(D)|$ and S includes exactly one connecting pair from each chain H_{uv}, $uv \in A(D)$. We say S is *well structured* whenever $\mathcal{C}(G - S)$ consists of only medium components.

Lemma 7. *Let S be a solution to (G, k, μ) and let $\mathcal{C}(G - S) = \{C_1, \ldots, C_\ell\}$. If $|S \cap \bigcup_{u \in V(D)}(V(I_u) \cup V(K_u))| > 0$ then there exists a solution S' such that $|S'| = |S|$, $\sum_{C' \in \mathcal{C}(G-S')} \binom{C'}{2} \leq \sum_{C \in \mathcal{C}(G-S)} \binom{C}{2}$, and $|S' \cap \bigcup_{u \in V(D)}(V(I_u) \cup V(K_u))| = |S \cap \bigcup_{u \in V(D)}(V(I_u) \cup V(K_u))| - 1$.*

By repeated applications of Lemma 7, we can assume that a solution S does not intersect with $V(I_u) \cup V(K_u)$, for all $u \in V(D)$. Hereafter, we assume that S satisfies this property. We show that S does not intersect with any border walls.

Lemma 8. *Let S be a solution to (G, k, μ) and let $\mathcal{C}(G - S) = \{C_1, \ldots, C_\ell\}$. If $|S \cap \bigcup_{uv \in A(D)} \mathcal{B}_{uv}| > 0$ then there exists a solution S' such that $|S'| = |S|$, $\sum_{C' \in \mathcal{C}(G-S')} \binom{C'}{2} \leq \sum_{C \in \mathcal{C}(G-S)} \binom{C}{2}$, $|S' \cap \bigcup_{u \in V(D)}(V(I_u) \cup V(K_u))| = |S \cap \bigcup_{u \in V(D)}(V(I_u) \cup V(K_u))|$, and $|S' \cap \bigcup_{uv \in A(D)} \mathcal{B}_{uv}|$ is strictly less than $|S \cap \bigcup_{uv \in A(D)} \mathcal{B}_{uv}|$.*

Lemmas 7 and 8 imply that we can always assume that S includes vertices from connecting pairs only. We now proceed to showing that S does not split any connecting pair. We use $\text{split}(G, S)$ to denote the number of connecting pairs split by S in G.

Lemma 9. *Let S be a solution to (G, k, μ) such that $S \subseteq \bigcup_{uv \in A(D)} \mathcal{P}_{uv}$ and let $\mathcal{C}(G - S) = \{C_1, \ldots, C_\ell\}$. If $\text{split}(G, S) > 0$ then there exists a solution S' such that $|S'| = |S|$, $\sum_{C' \in \mathcal{C}(G-S')} \binom{C'}{2} \leq \sum_{C \in \mathcal{C}(G-S)} \binom{C}{2}$, $S' \subseteq \bigcup_{uv \in A(D)} \mathcal{P}_{uv}$, and $\text{split}(G, S') < \text{split}(G, S)$.*

Lemma 10. *Let S be a solution to (G, k, μ) such that $S \subseteq \bigcup_{uv \in A(D)} \mathcal{P}_{uv}$ and $\mathsf{split}(G, S) = 0$. Let $\mathcal{C}(G - S) = \{C_1, \ldots, C_\ell\}$. Assume that $|S \cap \mathcal{P}_{uv}| = 2x \geq 10$, for some $uv \in A(D)$, and hence there exists $u_1 v_1, \ldots, u_{x-1} v_{x-1} \in A(D)$ such that $|S \cap \mathcal{P}_{u_i v_i}| = 0$, for $i \in [x - 1]$. Then, there exists S' such that $|S'| = |S|$, $\sum_{C' \in \mathcal{C}(G-S')} \binom{C'}{2} \leq \sum_{C \in \mathcal{C}(G-S)} \binom{C}{2}$, $S' \subseteq \bigcup_{uv \in A(D)} \mathcal{P}_{uv}$, $\mathsf{split}(G, S') = 0$, $\mathsf{left}(\mathcal{P}_{uv}) \cup \mathsf{right}(\mathcal{P}_{uv}) \cup \mathsf{first}(\mathcal{P}_{uv}) \cup \mathsf{last}(\mathcal{P}_{uv}) \subseteq S'$, $|S' \cap \mathcal{P}_{uv}| = 8$, and $|S' \cap \mathcal{P}_{u_i v_i}| = 2$, for some $i \in [x - 1]$.*

Since k is even, we know (from Lemma 10 and the fact that $\mathsf{split}(G,S) = 0$) that, for all $uv \in A(D)$, $|S \cap \mathcal{P}_{uv}| \in \{0, 2, 4, 6, 8\}$. Moreover, by an argument similar to the one for Lemma 10 we can ensure that if for some $uv \in A(D)$, $|S \cap \mathcal{P}_{uv}| = 8$ then $\mathsf{left}(\mathcal{P}_{uv}) \cup \mathsf{right}(\mathcal{P}_{uv}) \cup \mathsf{first}(\mathcal{P}_{uv}) \cup \mathsf{last}(\mathcal{P}_{uv}) \subseteq S$.

Lemma 11. *Let S be a solution satisfying the following properties: (1) $S \subseteq \bigcup_{uv \in A(D)} \mathcal{P}_{uv}$; (2) $\mathsf{split}(G,S) = 0$; (3) $|S \cap \mathcal{P}_{uv}| \in \{0, 2, 4, 6, 8\}$, for all $uv \in A(D)$; (4) If $|S \cap \mathcal{P}_{uv}| = 8$, for $uv \in A(D)$, then $\mathsf{left}(\mathcal{P}_{uv}) \cup \mathsf{right}(\mathcal{P}_{uv}) \cup \mathsf{first}(\mathcal{P}_{uv}) \cup \mathsf{last}(\mathcal{P}_{uv}) \subseteq S$. Then, there exists a solution S' satisfying the following properties: (i) $S' \subseteq \bigcup_{uv \in A(D)} \mathcal{P}_{uv}$; (ii) $\mathsf{split}(G,S') = 0$; (iii) $|S' \cap \mathcal{P}_{uv}| = 2$, for all $uv \in A(D)$.*

We are now ready to prove the correctness of the reduction, which is implied by Lemmas 12 and 13 below.

Lemma 12. *If $(D, w_V : V(D) \to \mathbb{N}, w_A : A(D) \to \mathbb{N}, \varphi : A(D) \to 2^{\mathbb{N} \times \mathbb{N}})$ is a yes-instance of $\textsc{SumCSP}$ then (G, k, μ) is a yes-instance of $\textsc{CNC}$.*

Proof. Let $\rho : A(D) \to \mathbb{N} \times \mathbb{N}$ be a solution to the $\textsc{SumCSP}$ instance. We construct a solution S to the $\textsc{CNC}$ instance by picking one connecting pair from each chain as follows. Initially, we set $S = \emptyset$. For each $uv \in A(D)$, we let $\mathcal{P}_{uv} = \{p_0, \ldots, p_{w_A(uv)}\}$, we let $\rho(uv) = (x_{uv}, y_{uv})$, and we set $S = S \cup p_{x_{uv}}$. It is not hard to see that $G - S$ consists of exactly $|V(D)|$ components (as we pick one connecting pair from each chain). We associate each component with some vertex $u \in V(D)$. The size of each component is exactly $|V(K_u)| + |V(I_u)| + \sum_{v \in N^+(u)}(k + 3)x_{uv} + \sum_{v \in N^-(u)}(k + 3)y_{vu} = |V(K_u)| + |V(I_u)| + (k + 3)w_V(u) = 2(k + 3) + (k + 3)(w_{big} + w_{vmax} - w_V(v)) + (k + 3)w_V(u) = (k + 3)(w_{big} + w_{vmax} + 2) = W^\star$. $\qquad\square$

Lemma 13. *If (G, k, μ) is a yes-instance of $\textsc{CNC}$ then $(D, w_V : V(D) \to \mathbb{N}, w_A : A(D) \to \mathbb{N}, \varphi : A(D) \to 2^{\mathbb{N} \times \mathbb{N}})$ is a yes-instance of $\textsc{SumCSP}$.*

Proof. Let S be a solution to (G, k, μ). From Lemmas 7 to 11, we know that S must be well structured *i.e.* $S \subseteq \bigcup_{uv \in A(D)} \mathcal{P}_{uv}$, $\mathsf{split}(G,S) = 0$, and $|S \cap \mathcal{P}_{uv}| = 2$, for all $uv \in A(D)$. We assume that the number of components in $G - S$ is exactly $|V(D)|$, otherwise we can find a solution S' with $|S'| = |S|$ and $\sum_{C \in \mathcal{C}(G-S')} \binom{C}{2} \leq \sum_{C \in \mathcal{C}(G-S)} \binom{C}{2}$. Let $\mathcal{C}(G - S) = \{C_1, \ldots, C_{|V(D)|}\}$. Recall that $W^\star = (k + 3)(w_{big} + w_{vmax} + 2)$ and $\mu = |V(D)| \cdot \binom{W^\star}{2}$. Therefore, we have $\sum_{C \in \mathcal{C}(G-S)} \binom{C}{2} \leq |V(D)| \cdot \binom{W^\star}{2}$. Applying Proposition 1, we know that each component in $\mathcal{C}(G-S)$ has $W^\star$ vertices. We associate each component with some vertex $u \in V(D)$. Note that K_u contains $2(k+3)$ vertices and I_u contains $(k+3)(w_{big} + w_{vmax} - w_V(u))$ vertices. Therefore, $W^\star - |V(K_u)| - |V(I_u)| = (k+3)w_V(u)$. Since each chain H_{uv} or H_{vu}, where $uv \in A(D)$ or $vu \in A(D)$, contributes $(k+3)x$ vertices, for some x, to the component associated with u, the sum of those contributions must equal $(k+3)w_V(u)$. This implies that there exists $\rho : A(D) \to \mathbb{N} \times \mathbb{N}$ such that for each $uv \in A(D)$, $\rho(uv) \in \varphi(uv)$ and $\sum_{v \in N^+(u)} \mathsf{fir}(\rho(uv)) + \sum_{v \in N^-(u)} \mathsf{sec}(\rho(vu)) = w_V(u)$. $\square$

Theorem 2. CNC *is* W[1]-Hard *when parameterized by solution size and the treewidth of the input graph.*

References

1. Addis, B., Di Summa, M., Grosso, A.: Removing critical nodes from a graph: complexity results and polynomial algorithms for the case of bounded treewidth. Optimization online (2011). www.optimization-online.org
2. Bodlaender, H.L., Lokshtanov, D., Penninkx, E.: Planar capacitated dominating set is W[1]-hard. In: Chen, J., Fomin, F.V. (eds.) IWPEC 2009. LNCS, vol. 5917, pp. 50–60. Springer, Heidelberg (2009). doi:10.1007/978-3-642-11269-0_4
3. Cygan, M., Fomin, F.V., Kowalik, L., Lokshtanov, D., Marx, D., Pilipczuk, M., Pilipczuk, M., Saurabh, S.: Parameterized Algorithms. Springer, Heidelberg (2015). doi:10.1007/978-3-319-21275-3
4. Di Summa, M., Grosso, A., Locatelli, M.: Complexity of the critical node problem over trees. Comput. Oper. Res. **38**(12), 1766–1774 (2011)
5. Diestel, R.: Graph Theory. Graduate Texts in Mathematics, 4th edn., vol. 173. Springer, Heidelberg (2012)
6. Downey, R.G., Fellows, M.R.: Parameterized Complexity. Springer, Heidelberg (1997). doi:10.1007/978-1-4612-0515-9
7. Flum, J., Grohe, M.: Parameterized Complexity Theory. Texts in Theoretical Computer Science. An EATCS Series. Springer, Heidelberg (2006). doi:10.1007/3-540-29953-X
8. Grohe, M., Marx, D.: On tree width, bramble size, and expansion. J. Comb. Theory Ser. B **99**(1), 218–228 (2009)
9. Hermelin, D., Kaspi, M., Komusiewicz, C., Navon, B.: Parameterized complexity of critical node cuts. Theoret. Comput. Sci. **651**, 62–75 (2016)
10. Marx, D.: Can you beat treewidth? Theory Comput. **6**(1), 85–112 (2010)
11. Niedermeier, R.: Invitation to Fixed-Parameter Algorithms. Oxford Lecture Series in Mathematics and Its Applications. Oxford University Press, Oxford (2006)
12. Ventresca, M.: Global search algorithms using a combinatorial unranking-based problem representation for the critical node detection problem. Comput. Oper. Res. **39**(11), 2763–2775 (2012)
13. Ventresca, M., Aleman, D.: A derandomized approximation algorithm for the critical node detection problem. Comput. Oper. Res. **43**, 261–270 (2014)

Hierarchical Partial Planarity

Patrizio Angelini and Michael A. Bekos$^{(\boxtimes)}$

Institut für Informatik, Universität Tübingen, Tübingen, Germany
{angelini,bekos}@informatik.uni-tuebingen.de

Abstract. In this paper we consider graphs whose edges are associated with a degree of *importance*, which may depend on the type of connections they represent or on how recently they appeared in the scene, in a streaming setting. The goal is to construct layouts in which the readability of an edge is proportional to its importance, that is, more important edges have fewer crossings. We formalize this problem and study the case in which there exist three different degrees of importance. We give a polynomial-time testing algorithm when the graph induced by the two most important sets of edges is biconnected. We also discuss interesting relationships with other constrained-planarity problems.

1 Introduction

Describing a graph in terms of a stream of nodes and edges, arriving and leaving at different time instants, is becoming a necessity for application domains where massive amounts of data, too large to be stored, are produced at a very high rate. The problem of visualizing graphs under this streaming model has been introduced only recently. In particular, the first step in this direction was performed in [8], where the problem of drawing trees whose edges arrive one-by-one and disappear after a certain amount of steps has been studied, from the point of view of the area requirements of straight-line planar drawings. Later on, it was proved [19] that polynomial area could be achieved for trees, tree-maps, and outerplanar graphs if a small number of vertex movements are allowed after each update. The problem has also been studied [14] for general planar graphs, relaxing the requirement that edges have to be straight-line.

In this paper we introduce a problem motivated by this model, and in particular by the fact that the *importance* of vertices and edges in the scene decades with time. In fact, as soon as an edge appears, it is important to let the user clearly visualize it, possibly at the cost of moving "older" edges in the more cluttered part of the layout, which may be unavoidable if the graph is large or dense. The idea is that the user may not need to *see* the connection between two vertices, as she *remembers* it from the previous steps.

Visually, one could associate the decreasing importance of an edge with its fading; theoretically, one could associate it with the fact that it becomes more acceptable to let it participate in some crossings. As a general framework for this kind of problems, we associate a weight $w(e)$ to every edge $e \in E$ and define a function $f : E \times E \to \{\texttt{YES}, \texttt{NO}\}$ that, given a pair of edges e and e', determines

© Springer International Publishing AG 2017
H.L. Bodlaender and G.J. Woeginger (Eds.): WG 2017, LNCS 10520, pp. 45–58, 2017.
https://doi.org/10.1007/978-3-319-68705-6_4

whether it is allowed to have a crossing between e and e' based on their weights. Of course, if no assumption is made on function $f(\cdot)$, this model allows to encode instances of the NP-complete problem WEAK REALIZABILITY [23], in which the pairs of edges that are allowed to cross are explicitly given as part of the input. On the other hand, already the "natural" assumption that, if an edge e is allowed to cross an edge e', then it is also allowed to cross any edge e'' such that $w(e'') \leq w(e')$, could potentially make the problem tractable.

As a first step towards a formalization of this general idea, we introduce problem HIERARCHICAL PARTIAL PLANARITY, which takes as input a graph $G = (V, E = E_p \cup E_s \cup E_t)$ whose edges are partitioned into the *primary* edges in E_p, the *secondary* edges in E_s, and the *tertiary* edges in E_t. The goal is to construct a drawing of G in which the primary edges are crossing-free, the secondary edges can only cross tertiary edges, while these latter edges can also cross one another. We say that any crossing that involves a primary edge or two secondary ones is *forbidden*. We remark that this problem can be easily modeled under the general framework we described above. Namely, we can say that all edges in E_p, E_s, and E_t have weights 4, 2, and 1, respectively, and function $f(\cdot)$ is such that $f(e, e') = \texttt{YES}$ if and only if $w(e) + w(e') \leq 3$.

We observe that our problem is a generalization of the recently introduced PARTIAL PLANARITY problem [2,24], in which the edges of a certain subgraph of a given graph must not be involved in any crossings. An instance of this problem is in fact an instance of our problem only composed of edges in E_p and E_t.

Our main contribution is an $O(|V|^3 \cdot |E_t|)$-time algorithm for HIERARCHICAL PARTIAL PLANARITY when the graph induced by the primary and the secondary edges is biconnected (see Sect. 4). Our result builds upon a formulation of the problem in terms of a *constrained-planarity* problem, which may be interesting in its own. This formulation also allows us to uncover interesting relationships with other graph planarity problems, like PARTIALLY EMBEDDED PLANARITY [5,21] and SIMULTANEOUS EMBEDDING WITH FIXED EDGES [9,12] (see Sect. 3). In Sect. 2 we give definitions, and in Sect. 5 we conclude with open problems.

2 Preliminaries

For the standard definitions on *planar graphs*, on *planar drawings* and their *faces*, on *planar embeddings* and on graph *connectivity*, we point the reader to [22,25]. Let H be a subgraph of a planar graph G, and let $\mathcal{G}$ be a planar embedding of G. The planar embedding of H that is obtained by removing the edges of $G \setminus H$ from $\mathcal{G}$ (and potential isolated vertices) is the *restriction* of $\mathcal{G}$ to H.

The SPQR-tree $\mathcal{T}$ of a biconnected graph G is a labeled tree representing the decomposition of G into its triconnected components [15,16]. Every triconnected component of G is associated with a node μ in $\mathcal{T}$. The triconnected component itself is referred to as the *skeleton* of μ, denoted by G_μ^{skel}. A node $\mu \in \mathcal{T}$ can be of one of four different types. *S-* and *P-nodes* describe series and parallel compositions; *R-nodes* correspond to triconnected structures; *Q-nodes* correspond to the edges of G. The set of leaves of $\mathcal{T}$ coincides with the set of Q-nodes, except for one arbitrary Q-node ρ, which is selected as the root of $\mathcal{T}$. The edges of G_μ^{skel} are called *virtual edges* and each of them represents a subgraph of G; the virtual edge whose subgraph contains the edge correponding to ρ is the *reference edge* of μ and is denoted by $ref(\mu)$. The endvertices of $ref(\mu)$ are the *poles* of μ. The subtree $\mathcal{T}_\mu$ of $\mathcal{T}$ rooted at μ induces a subgraph G_μ^{pert} of G, called *pertinent*. The SPQR-tree of G is unique and can be computed in linear time [20].

3   Relationships to Other Planarity Problems

In this section we define a problem, called FACIAL-CONSTRAINED CORE PLA-NARITY, that will serve as a tool to solve HIERARCHICAL PARTIAL PLANARITY and to uncover interesting relationships with other important graph planarity problems. This problem takes as input a graph $G = (V, E_1 \cup E_2)$ and a set $W \subseteq V \times V$ of pairs of vertices. Let H be the subgraph of G induced by the edges in E_1, which we call *core* of G. The goal is to construct a planar embedding $\mathcal{G}$ of G whose restriction $\mathcal{H}$ to H is such that, for each pair $\langle u, v \rangle \in W$, there exists a face of $\mathcal{H}$ that contains both u and v.

Theorem 1. *Problems* FACIAL-CONSTRAINED CORE PLANARITY *and* HIER-ARCHICAL PARTIAL PLANARITY *are linear-time equivalent.*

Proof (sketch). The correspondence between two instances $\langle G' = (V, E_1 \cup E_2), W \rangle$ and $G = (V, E_p \cup E_s \cup E_t)$ of the two problems is as follows. Graphs G' and G have the same vertex set V; further, $E_1 = E_p$ and $E_2 = E_s$; finally, for each two vertices $u, v \in V$, we have $(u, v) \in E_t$ if and only if $\langle u, v \rangle \in W$. To prove the equivalence note that, in any drawing of G that is a solution for HIER-ARCHICAL PARTIAL PLANARITY, there is no crossing between edges in $E_p \cup E_s$. Thus, graph $\overline{G} = (V, E_p \cup E_s)$, which coincides with G', is planar. Further, since the edges of E_t can cross with each other and with the edges of E_s, the only requirements they impose on the planar embedding of $\overline{G}$ are the same as those imposed by the pairs in W on the possible planar embeddings of G'. Details are given in [1].   □

In PARTIAL PLANARITY [2], given a non-planar graph $G = (V, E)$ and a sub-set $F \subseteq E$ of its edges, the goal is to compute a drawing Γ of G, if any, in which the edges of F are not crossed by any edge of G. Positive and negative results are given in [2] if the graph induced by F is a connected spanning subgraph of G. In [24], the corresponding decision problem is shown to be polynomial-time solvable. By setting $E_p = F$, $E_s = \emptyset$, and $E_t = E \setminus F$, we can model any instance of PARTIAL PLANARITY as an instance of HIERARCHICAL PARTIAL PLANARITY.

Theorem 2. PARTIAL PLANARITY *can be reduced in linear time to* HIERAR-CHICAL PARTIAL PLANARITY.

In PARTIALLY-EMBEDDED PLANARITY [5], given a planar graph G and a planar embedding $\mathcal{H}$ of a subgraph H of G, the goal is to test whether $\mathcal{H}$ can be extended to a planar embedding of G. The problem is linear-time solvable [5] and characterizable in terms of forbidden subgraphs [21]. We prove that HIER-ARCHICAL PARTIAL PLANARITY can be used to encode instances of PARTIALLY-EMBEDDED PLANARITY in which H is biconnected. Note that this special case is a central ingredient in the algorithm in [5] for the general case.

Theorem 3. PARTIALLY-EMBEDDED PLANARITY *with biconnected* H *can be reduced in quadratic time to* HIERARCHICAL PARTIAL PLANARITY.

Proof. Let $\langle G' = (V, E), H, \mathcal{H} \rangle$ be an instance of PARTIALLY-EMBEDDED PLA-NARITY in which H is biconnected. We construct an instance $\langle G = (V, E_1 \cup E_2), W \rangle$ of FACIAL-CONSTRAINED CORE PLANARITY on the same vertex set V as G', as follows. Set E_1 contains all the edges of E that are contained in H; set E_2 contains the other ones, that is, $E_2 = E \setminus E_1$. Finally, for every pair of non-adjacent vertices $\langle u, v \rangle$ that are on the same face of $\mathcal{H}$, we add a pair $\langle u, v \rangle$ to W. This last step requires quadratic time and guarantees that in the solution of FACIAL-CONSTRAINED CORE PLANARITY, for each face f of $\mathcal{H}$, all the vertices of f are incident to the same face f' of the planar embedding of the core of G. These vertices appear in the same order along f and f', since H is biconnected and thus this order is unique. Hence, $\langle G', H, \mathcal{H} \rangle$ is a positive instance if and only if $\langle G, W \rangle$ is. The statement follows by Theorem 1. □

A *simultaneous embedding* of two planar graphs $G_1 = (V, E_1)$ and $G_2 = (V, E_2)$ embeds each graph in a planar way using the same vertex positions; edges are allowed to cross only if they belong to different graphs (see, e.g., [9]). Our problem is related to SIMULTANEOUS EMBEDDING WITH FIXED EDGES (SEFE) [4,6,10–12], in which edges that are *common* to both graphs must be embedded in the same way (and hence, cannot be crossed by other edges). So, these edges correspond to the primary ones. However, to obtain a solution for SEFE, it does not suffice to assume that the *exclusive* edges of G_1 and G_2 are the secondary and tertiary ones, as we could not guarantee that the edges of G_2 do not cross each other. So, in some sense, our problem seems to be more related to *nearly-planar simultaneous embeddings*, where the input graphs are allowed to cross, as long as they avoid some local crossing configurations, e.g., by avoiding triples of mutually crossing edges [17]. Note that the SEFE problem has also been studied in several settings [3,7,13,18]. An interpretation of PARTIAL PLANARITY, which also extends to HIERARCHICAL PARTIAL PLANARITY, in terms of a special version of SEFE, called SUNFLOWER SEFE [9], was already observed in [2].

The algorithm of Sect. 4 is inspired by an algorithm to test whether a pair of graphs admits a SEFE if the common graph is biconnected [6]. The main part of that algorithm is to find an embedding of the common graph in which every pair of vertices that are joined by an exclusive edge are incident to the same face; so, these edges play the role of the pairs in W. In a second step, it checks for crossings between exclusive edges of the same graph. Since the common graph is biconnected, these crossings do not depend on the choice of the embedding.

Thus, for instances of our problem in which the core H of G is biconnected, we can employ the main part of the algorithm in [6] to find a planar embedding of H in which every two vertices that either are joined by an edge of E_2 or form a pair of W are incident to the same face of H; note that it is not even needed to perform the second check for the pairs in W. We extend this result to the case in which H is not biconnected, but it becomes so when adding the edges of E_2. The main difficulty here is to "control" the faces of H by operating on the embeddings of the biconnected graph G composed of H and of the edges of E_2. In Sect. 4 we discuss the problems arising from this and our solution.

4   Biconnected Facial-Constrained Core Planarity

In this section, we give a polynomial-time algorithm for instances $\langle G = (V, E_1 \cup E_2), W \rangle$ of FACIAL-CONSTRAINED CORE PLANARITY in which G is biconnected.

4.1   High-Level Description of the Algorithm

We perform a bottom-up traversal of the SPQR-tree $\mathcal{T}$ of G. At each step of the traversal, we consider a node $\mu \in \mathcal{T}$ and we search for an embedding $\mathcal{G}_\mu^{pert}$ of G_μ^{pert} satisfying the following requirements: **(R.1)** For every pair $\langle x, y \rangle \in W$ such that x and y belong to G_μ^{pert}, vertices x and y lie in the same face of the restriction $\mathcal{H}_\mu^{pert}$ of $\mathcal{G}_\mu^{pert}$ to the part of the core H in G_μ^{pert}. **(R.2)** For every pair $\langle x, y \rangle \in W$ such that exactly one vertex, say x, belongs to G_μ^{pert}, vertex x lies in the outer face of $\mathcal{H}_\mu^{pert}$ (note that y belongs to $G \setminus \mathcal{G}_\mu^{pert}$).

In general, there may exist several "candidate" embeddings of G_μ^{pert} satisfying R.1 and R.2. If there exists none, the instance is negative. Otherwise, we would like to select one of them and proceed with the traversal. However, while it would be sufficient to select *any* embedding of G_μ^{pert} satisfying R.1, it is possible that some of the embeddings satisfying R.2 are "good", in the sense that they can be eventually extended to an embedding of G satisfying both R.1 and R.2, while some others are not. Unfortunately, we cannot determine which ones are good at this stage of the algorithm, as this may depend on the structure of a subgraph that is considered later in the traversal. Thus, we have to maintain succinct information to describe the properties of the embeddings of G_μ^{pert} that satisfy R.1 and R.2, so to group these embeddings into equivalence classes.

We denote by $x_1, \ldots, x_k$ the vertices belonging to pairs $\langle x_i, y_i \rangle \in W$ such that $x_i \in G_\mu^{pert}$ and $y_i \notin G_\mu^{pert}$. By R.2, $x_1, \ldots, x_k$ must lie on the outer face of $\mathcal{H}_\mu^{pert}$. We say that μ is *non-traversable* if there is a cycle C_μ composed of edges of E_1 that contains both poles u and v of μ, at least one edge of G_μ^{pert}, and at least one of $G \setminus G_\mu^{pert}$; see Fig. 1a. Otherwise, μ is *traversable*, i.e., either in G_μ^{pert} or in $G \setminus G_\mu^{pert}$ every path between u and v contains edges of E_2; see Fig. 1b.

Intuitively, when μ is non-traversable, C_μ splits the outer face of $\mathcal{H}_\mu^{pert}$ into two faces f_μ^l and f_μ^r of $\mathcal{H}$ in any planar embedding of G. Hence, R.2 must be refined to take into account the possible partitions of $x_1, \ldots, x_k$ with respect to their incidence to f_μ^l and f_μ^r. In particular, the structure of G_μ^{pert} may enforce dependencies on the relative positions of $x_1, \ldots, x_k$ with respect to f_μ^l and f_μ^r. More precisely, let $\langle x, y \rangle, \langle x', y' \rangle \in W$ be two pairs such that $x, x' \in G_\mu^{pert}$ and

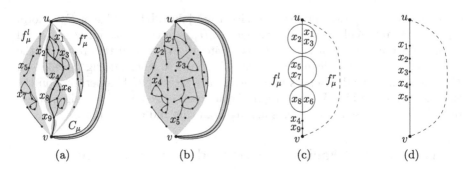

Fig. 1. (a–b) Graph G_{μ}^{pert} when μ is (a) non-traversable and (b) traversable. (c–d) The bags of the nodes in (a) and in (b), respectively. A segment between u and v separates f_{μ}^{l} and f_{μ}^{r}; each bag B_{μ}^{i} is represented by a circle across the segment, with its pockets S_{μ}^{i} and T_{μ}^{i} on the two sides; the vertices in the special bag $\mathcal{B}_{\mu}$ lie along the segment.

$y, y' \notin G_{\mu}^{pert}$. Then, x and x' may be enforced to be incident to the same face, either f_{μ}^{l} or f_{μ}^{r} (see x_1 and x_3 in Fig. 1a), to different faces (see x_2 and x_3 in Fig. 1a), or they may be independent in this respect (see x_1 and x_6 in Fig. 1a).

We encode this information by associating a set of *bags* with μ, which contain vertices $x_1, \ldots, x_k$. Each bag is composed of two *pockets*; all the vertices in a pocket must be incident to the same face of $\mathcal{H}$ in any candidate embedding of G_{μ}^{pert}, while all the vertices in the other pocket must be incident to the other face. Vertices of different bags are independent of each other. For the vertices of $\{x_1, \ldots, x_k\}$ that are incident to both f_{μ}^{l} and f_{μ}^{r} in any embedding (see x_4 in Fig. 1a), we add a *special* bag, composed of a single set containing all such vertices; note that if a vertex of $\{x_1, \ldots, x_k\}$ is a pole of μ, then it belongs to the special bag. See Fig. 1c for the bags of the node in Fig. 1a. When μ is traversable, the outer face of $\mathcal{H}_{\mu}^{pert}$ corresponds to a single face of $\mathcal{H}$ in any planar embedding of G. Thus, we do not need to maintain any information about the relative positions of $x_1, \ldots, x_k$, and we can place all of them in the special bag. An illustration of the bags of the node represented in Fig. 1b is given in Fig. 1d.

If the visit of the root ρ of $\mathcal{T}$ at the end of the bottom-up traversal is completed without declaring $\langle G, W \rangle$ as negative, we have that $G_{\rho}^{pert} = G$ admits a planar embedding satisfying R.1 and thus $\langle G, W \rangle$ is a positive instance.

As anticipated in Sect. 3, we discuss two main problems to extend the algorithm in [6] for SEFE to solve our problem when H is not biconnected.

First, when H is biconnected it is always possible to decide the flip of every child component for every node that is either an R- or a P-node, but not when it is an S-node. However, since no two S-nodes are adjacent in $\mathcal{T}$, this choice is always fixed in the next step of the algorithm (refer to *visible nodes* in [6]). When H is not biconnected, even the flips of the children of R- and P-nodes may be not uniquely determined. So, it may be necessary to defer this choice till the end of the algorithm. Furthermore, in the course of the algorithm, it could be required to make "partial" choices for these flips, in the sense that constraints imposed by the structure of the graph could enforce two or more components to be flipped in the same way (without enforcing, however, a specific flip for them). To encode all the possible flips, we introduced the bags.

Second, the order of the vertices along the faces of H is not unique if H is not biconnected. For FACIAL-CONSTRAINED CORE PLANARITY, this is not an issue, as it is enough that the vertices belonging to the pairs in W share a face. Note that, if we were able to also control these orders, we could provide an algorithm for SEFE when one of the two graphs is biconnected, which would be a significant step ahead for this problem. We recall that an efficient algorithm for this case would imply an efficient algorithm for all the instances in which the common graph is connected (and no restriction on the two input graphs) [4].

4.2   Detailed Description of the Algorithm

Let $\mathcal{T}$ be the SPQR-tree of G, rooted at a Q-node ρ. First, we compute for each node $\mu \in \mathcal{T}$, whether μ is traversable or not. We traverse $\mathcal{T}$ bottom-up to compute for each node μ whether there exists a path composed of edges of H between the poles of μ in G_μ^{pert}, using the same information computed for its children. Then, with a top-down traversal, we search for the path in $G \setminus G_\mu^{pert}$, using the information computed in the first traversal; see also [5].

The main part of our algorithm consists of a bottom-up traversal of $\mathcal{T}$. For a node $\mu \in \mathcal{T}$, let $\langle x_1, y_1 \rangle, \ldots, \langle x_k, y_k \rangle$ be all pairs of W such that $x_i \in G_\mu^{pert}$ and $y_i \notin G_\mu^{pert}$. We denote by $B_\mu^1, \ldots, B_\mu^q$ the bags of μ and by $\mathcal{B}_\mu$ its special bag; these bags determine a partition of the vertices $x_1, \ldots, x_k$ that are required to be on the outer face of $\mathcal{H}_\mu^{pert}$ by R.2. The vertices of each bag $B_\mu^i = \langle S_\mu^i, T_\mu^i \rangle$ are partitioned into its two pockets S_μ^i and T_μ^i; all vertices of S_μ^i must lie in the same face of $\mathcal{H}$, either f_μ^l or f_μ^r, while all those of T_μ^i must lie on the other face.

We first describe an operation, called MERGE-BAGS , to modify the bags of a node μ so to satisfy the constraints that may be imposed by R.1 when there exists a pair $\langle x, y \rangle \in W$ such that $x, y \in G_\mu^{pert}$. Refer to Figs. 2a–2b. In particular, if at least one of x and y belongs to the special bag $\mathcal{B}_\mu$ (see $\langle x_4, x_6 \rangle$ in Fig. 2), or if x and y belong to the same pocket of a bag B_μ^i, then we do not modify any bag. If $x \in S_\mu^i$ and $y \in T_\mu^i$, for some $1 \leq i \leq q$, or vice versa, then we declare the instance negative. Otherwise, we have $x \in B_\mu^i$ and $y \in B_\mu^j$, for some $i \neq j$, and we merge B_μ^i and B_μ^j into a single bag $B_\mu = \langle S_\mu, T_\mu \rangle$, i.e., we merge into S_μ the pockets of B_μ^i and B_μ^j containing x and y, respectively, and we merge into T_μ the other two pockets; see $\langle x_2, x_5 \rangle$ in Fig. 2. We remove $\langle x, y \rangle$ from W and, if there is no other pair in W containing x (resp., y), we remove it from its bag.

At each step, we consider a node μ, with poles u and v, and children $\nu_1, \ldots, \nu_h$. We denote by e_i, for $i = 1, \ldots, h$, the virtual edge of G_μ^{skel} corresponding to ν_i.

Suppose that μ is a Q-node. If any of the two poles of μ belongs to $\{x_1, \ldots, x_k\}$, then we add it to $\mathcal{B}_\mu$, independently of whether μ is traversable or not.

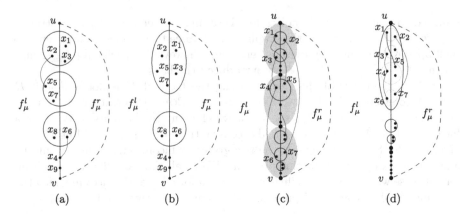

Fig. 2. (a) The bags of a node μ and two pairs $\langle x_2, x_5 \rangle, \langle x_4, x_6 \rangle \in W$ (orange curves). (b) The bags of μ after operation MERGE-BAGS. Pair $\langle x_2, x_5 \rangle$ merged two bags, while $\langle x_4, x_6 \rangle$ did not modify any bag, since $x_4 \in \mathcal{B}_\mu$. (c) Initialization of the bags of an S-node μ. (d) The bags of μ after MERGE-BAGS. The instance is negative, as pair $\langle x_1, x_5 \rangle$ is such that $x_1 \in S_\mu^1$ and $x_5 \in T_\mu^1$. (Color figure online)

Suppose that μ is an S-node. We initialize $\mathcal{B}_\mu = \mathcal{B}_{\nu_1} \cup \cdots \cup \mathcal{B}_{\nu_h}$. Note that if μ is traversable, then all of its children are traversable. So, in this case, we already have that all vertices $x_1, \ldots, x_k$ are in $\mathcal{B}_\mu$. Further, if μ is non-traversable, we add to the set of bags of μ all the non-special bags of its children; see Fig. 2c. Finally, as long as there exists a pair $\langle x, y \rangle \in W$ such that both x and y belong to G_μ^{pert}, we apply operation MERGE-BAGS to $\langle x, y \rangle$. This may result in uncovering a negative instance, but only when μ is non-traversable. See Fig. 2d.

Suppose that μ is an R-node. See Fig. 3a. Let H_μ^{skel} be the graph composed of the vertices of G_μ^{skel} and of the virtual edges corresponding to non-traversable children of μ, plus $ref(\mu)$ if μ is non-traversable; see Fig. 3b. Let $\mathcal{H}_\mu^{skel}$ be the restriction of the unique planar embedding of G_μ^{skel} to H_μ^{skel}. Note that, for each traversable child ν_i of μ, virtual edge e_i is *contained in* one face f_{ν_i} of $\mathcal{H}_\mu^{skel}$; in Fig. 3b, (w_4, w_6) is contained in face $\{w_3, w_4, w_5, w_6\}$. For a non-traversable child ν_i, denote by $f_{\nu_i}^1$ and $f_{\nu_i}^2$ the two faces of $\mathcal{H}_\mu^{skel}$ virtual edge e_i is incident to. For a vertex $x \in V$ that is not in G_μ^{skel}, we denote by $e_\mu(x)$ either the virtual edge e_i, if $x \in G_{\nu_i}^{pert}$, or the reference edge $ref(\mu)$, if $x \in G \setminus G_\mu^{pert}$.

Suppose that μ is non-traversable (i.e., $ref(\mu) \in H_\mu^{skel}$); see Fig. 3c. Let f_μ^l and f_μ^r be the two faces of $\mathcal{H}_\mu^{skel}$ incident to $ref(\mu)$. Any other virtual edge e_i of H_μ^{skel} such that $\{f_{\nu_i}^1, f_{\nu_i}^2\} = \{f_\mu^l, f_\mu^r\}$ is 2-*sided*; see $(w_2, w_3), (v, w_6)$ in Fig. 3c.

We consider each pair $\langle x, y \rangle \in W$ such that $x \in G_\mu^{pert}$ and $y \notin G_\mu^{pert}$. Let $e^x = e_\mu(x)$ and $e^y = e_\mu(y)$. A necessary condition for R.1 and R.2 is that e^y is either contained in or incident to face f_{ν_i} (if ν_i is traversable) or one of $f_{\nu_i}^1$ and $f_{\nu_i}^2$ (if ν_i is non-traversable). Otherwise, we declare the instance negative.

Another constraint imposed by this pair is the following. Suppose that x belongs to a pocket, say S_{ν_i}, of a bag B_{ν_i} of ν_i (this can only happen if ν_i

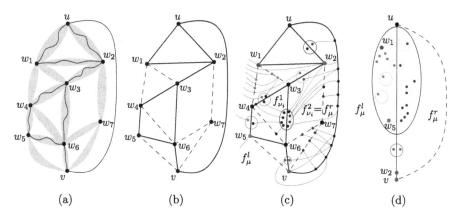

Fig. 3. (a) Graph G_μ^{pert} when μ is a non-traversable R-node. (b) Graph H_μ^{skel} (solid) and the traversable children (dashed) of μ. (c) Blue (green) pockets are associated with f_μ^r (f_μ^l). Red pockets are not associated. Gray pockets belong to 2-sided children, but are associated with f_μ^r and f_μ^l. (d) The bags of μ. (Color figure online)

is non-traversable). If e^x and e^y share exactly one face, say $f_{\nu_i}^1$, then all pairs $\langle x', y' \rangle \in W$ with $x' \in S_{\nu_i}$ must be such that $e_\mu(y')$ is either contained in or incident to f_{ν_1}; also, all the pairs $\langle x'', y'' \rangle \in W$ with $x'' \in T_{\nu_i}$ must be such that $e_\mu(y'')$ is either contained in or incident to f_{ν_2}. This is due to the fact all the vertices in the same pocket must be incident to the same face of $\mathcal{H}_\mu^{skel}$. So, if this is not the case, we declare the instance negative. Otherwise, we *associate* S_{ν_i} with $f_{\nu_i}^1$ and T_{ν_i} with $f_{\nu_i}^2$. If e^x and e^y share both faces $f_{\nu_i}^1$ and $f_{\nu_i}^2$, instead, we have to postpone the association of S_{ν_i} and T_{ν_i}, as at this point we cannot make a unique choice. Note that an association for these pockets may be performed later, due to another pair of W. Suppose now that x belongs to the special bag $\mathcal{B}_{\nu_i}$ of ν_i. Then, we associate $\mathcal{B}_{\nu_i}$ to either f_{ν_i}, if ν_i is traversable, or to both $f_{\nu_i}^1$ and $f_{\nu_i}^2$, if it is non-traversable. This completes the process of pair $\langle x, y \rangle$.

Once all children $\nu_1, \dots, \nu_h$ of μ have been considered, there may still exist pockets that are not associated. Let S_{ν_i} be one of such pockets, and consider each pair $\langle x, y \rangle \in W$ such that $x \in S_{\nu_i}$. Note that $e_\mu(x)$ shares both faces $f_{\nu_i}^1$ and $f_{\nu_i}^2$ with $e_\mu(y)$. If y belongs to a pocket, say T_{ν_j}, that is associated with one of $f_{\nu_i}^1$ and $f_{\nu_i}^2$, say $f_{\nu_i}^1$, then we associate S_{ν_i} with $f_{\nu_i}^1$ and T_{ν_i} with $f_{\nu_i}^2$. In fact, the association of T_{ν_j} with $f_{\nu_i}^1$ implies that y will be incident to $f_{\nu_i}^1$ in any embedding of G that is a solution for $\langle G, W \rangle$. If two pairs determine different associations for S_{ν_i} and T_{ν_i}, we declare the instance negative.

We repeat this process as long as there exist pockets that can be associated. This does not necessarily result in an association for all pockets. Consider any of the remaining pockets S_{ν_i}. If ν_i is not 2-sided, then we associate S_{ν_i} with $f_{\nu_i}^1$ and T_{ν_i} with $f_{\nu_i}^2$, since the effect of this association is limited to G_μ^{pert} and not to $G \setminus G_\mu^{pert}$. Then, we propagate this association to other pockets by performing the procedure described above. We repeat this process until the only pockets that are not associated, if any, belong to bags of 2-sided children of μ.

Based on the association of $\nu_1, \ldots, \nu_h$ with the faces of $\mathcal{H}_\mu^{skel}$, we determine the bags of μ; see Fig. 3d. The special bag $\mathcal{B}_\mu$ of μ contains the poles of μ, if they belong to $\{x_1, \ldots, x_k\}$, and the union of the special bags of the 2-sided children of μ. Next, we create a bag $B_\mu = \langle S_\mu, T_\mu \rangle$, such that S_μ and T_μ contain all the vertices of the pockets associated with f_μ^l and f_μ^r, respectively. Finally, we add to the set of bags of μ the non-special bags of the 2-sided children of μ whose pockets have not been associated with any face of $\mathcal{H}_\mu^{skel}$ (this allows us to postpone their association). Then, we apply operation MERGE-BAGS to all pairs $\langle x, y \rangle \in W$ such that both x and y belong to G_μ^{pert} in order to merge the bags of different 2-sided children of μ (again this may result in uncovering a negative instance). This completes the case in which μ is non-traversable.

In the simpler case in which μ is traversable, reference edge $ref(\mu) \notin H_\mu^{skel}$; hence f_μ^l and f_μ^r do not exist, and none of the children of μ is 2-sided. This implies that performing all the operations described above results in an association of each pocket and of each special bag of the children of μ with some face of $\mathcal{H}_\mu^{skel}$. Recall that, since μ is traversable, μ has only its special bag $\mathcal{B}_\mu$. We add to $\mathcal{B}_\mu$ all the vertices of the pockets and of the special bags that have been associated with the outer face of $\mathcal{H}_\mu^{skel}$. This concludes the R-node case.

Suppose that μ is a P-node. See Fig. 4. We distinguish three cases, based on whether μ has (i) zero, (ii) one, or (iii) more than one non-traversable child.

In Case (i), we have that μ is traversable. So, it has only its special bag $\mathcal{B}_\mu$, in which we add all the vertices of the special bags of its children. Since all virtual edges in G_μ^{skel} are incident to the same face of $\mathcal{H}_\mu^{pert}$, R.1 and R.2 are satisfied.

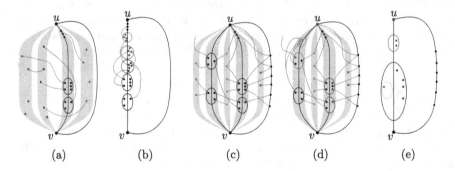

(a)   (b)   (c)   (d)   (e)

Fig. 4. The case in which μ is a P-node with: (a-b) one, and (c-e) more than one non-traversable children. The color-scheme of this figure follows the one of Fig. 3 (Color figure online)

In Case (ii), let ν_1 be the non-traversable child of μ; see Fig. 4a. In this case, μ is non-traversable, since the path of $G \setminus G_\mu^{pert}$ composed of edges of H also belongs to $G \setminus G_\mu^{pert}$. We initialize the set of bags of μ to the one of ν_1. For each traversable child ν_i, with $i = 2, \ldots, h$, we add to μ a new bag B_μ^i, where S_μ^i contains all the vertices in $\mathcal{B}_{\nu_i}$, while T_μ^i is empty; see Fig. 4b. This represents the fact that all the vertices in $G_{\nu_i}^{pert}$ must lie on the same side of the cycle passing through $G_{\nu_1}^{pert}$ and $G \setminus G_\mu^{pert}$ to satisfy R.2. Finally, we apply operation MERGE-BAGS to all pairs $\langle x, y \rangle \in W$ such that both x and y belong to G_μ^{pert}.

Finally, we consider Case (iii), in which μ has more than one non-traversable child; see Figs. 4c–e. We construct an auxiliary graph G_{aux} with a vertex v_i for each child ν_i of μ, which is colored *black* if ν_i is non-traversable and *white* otherwise. Graph G_{aux} also has a vertex v corresponding to $ref(\mu)$, which is colored black if μ is non-traversable and white otherwise. Then, we consider every pair $\langle x, y \rangle \in W$ such that $x \in G_{\nu_i}^{pert}$, for some child ν_i of μ. If $y \in G_{\nu_j}^{pert}$, for some $j \neq i$, then we add edge (v_i, v_j) to G_{aux}, while if $y \in G \setminus G_\mu^{pert}$, then we add edge (v_i, v) to G_{aux}. If G_{aux} has multiple copies of an edge, we keep only one of them. We assume w.l.o.g. that no two white vertices are adjacent in G_{aux}, as otherwise we could contract them to a new white vertex. In fact, the virtual edges representing traversable children of μ corresponding to adjacent white vertices must be contained in the same face of $\mathcal{H}_\mu^{pert}$, due to R.1.

If a white vertex w of G_{aux} has more than two black neighbors, we declare the instance negative, as the virtual edge of the traversable child of μ corresponding to w should share a face in $\mathcal{H}_\mu^{pert}$ with more than two virtual edges representing non-traversable children of μ. If w has at most one black neighbor, we remove w from G_{aux}. Finally, if w has exactly two black neighbors b and b', then we remove w and we add edge (b, b') to G_{aux} (if not present). Once we have considered all white vertices, the resulting graph $\overline{G}_{aux}$ has only black vertices.

We check whether $\overline{G}_{aux}$ is either a cycle through all its vertices or a set of paths (some of which may consist of single vertices). The necessity of this condition can be proved similar to [6]. The only difference is in the edges between black vertices that are introduced due to degree-2 white vertices. Let (b, b') be one of such edges and let w be the white vertex that was adjacent to b and b'. Also, let e_b, $e_{b'}$, and e_w be the virtual edges representing the children of μ (or virtual edge $ref(\mu)$, if μ is non-traversable) corresponding to b, b', and w, respectively. Then, e_b and $e_{b'}$ must share a face in $\mathcal{H}_\mu^{pert}$, and this face must contain e_w, due to R.1 and R.2. If the above condition on $\overline{G}_{aux}$ is not satisfied, then we declare the instance negative; otherwise, we fix an order of the black vertices of $\overline{G}_{aux}$ based either on the cycle or on an arbitrary order of the paths.

We now construct graph H_μ^{skel} in the same way as for the R-node. Since the order of the black vertices of $\overline{G}_{aux}$ induces an order of the virtual edges of H_μ^{skel}, the embedding $\mathcal{H}_\mu^{skel}$ of H_μ^{skel} is fixed. We will again use $\mathcal{H}_\mu^{skel}$ to either determine whether the instance is negative or to construct the bags of μ.

The case in which μ is traversable is identical to the R-node case. When μ is non-traversable, we have $ref(\mu) \in H_\mu^{skel}$, and thus there exist the two faces f_μ^l and f_μ^r incident to $ref(\mu)$. However, since μ has at least two non-traversable children, every two virtual edges share at most one face, and thus no child is 2-sided.

We now consider each traversable child ν_i of μ. Contrary to the R-node case, the face of $\mathcal{H}_\mu^{skel}$ in which ν_i is contained is not necessarily defined, as the embedding of G_μ^{skel} is not unique. Recall that ν_i corresponds to a white vertex v_i of G_{aux}. If v_i has exactly two black neighbors, then they must be connected by an edge in $\overline{G}_{aux}$ after the removal of v_i. So, they are consecutive in the order of the black vertices that we used to construct $\mathcal{H}_\mu^{skel}$. Thus, the two virtual edges of H_μ^{skel} corresponding to them share a face in $\mathcal{H}_\mu^{skel}$, and we say that e_i is *contained* in this particular face. If v_i has exactly one black neighbor in G_{aux}, then e_i may be contained in any of the two faces of $\mathcal{H}_\mu^{skel}$ incident to the virtual edge e corresponding to this black vertex. However, we cannot make a choice at this stage, as this may depend on other pairs whose vertices belong to the subgraph of G represented by e (that is, $G_{\nu_j}^{pert}$, if $e = e_j$, for some $1 \le j \le h$, and $G \setminus G_\mu^{pert}$, if $e = ref(\mu)$). If $e = e_j$, then we add a new bag B_{ν_j} to the child ν_j of μ, so that S_{ν_j} contains all the vertices of the special bag of ν_i, while T_{ν_j} is empty. The association of S_{ν_j} with one of the two faces incident to e_j, to be performed later, will determine the face in which e_i is contained. In the case in which $e = ref(\mu)$, virtual edge e_i should be contained either in f_μ^l or in f_μ^r, but again we cannot determine which of the two. Furthermore, we cannot even delegate this choice to the association of the pockets, since $ref(\mu)$ does not correspond to a child of μ. Thus, we do not associate it to any face, but we will use it to create the bags of μ. Finally, when v_i has no black neighbors, its special bag is empty.

Once all traversable children have been considered, we associate the special bags and the pockets of the non-special bags with the faces of $\mathcal{H}_\mu^{skel}$, as in the R-node case. Then, we construct the bags of μ. We add the poles of μ to its special bag, if they belong to $\{x_1, \ldots, x_k\}$. As in the R-node case, we add to μ a bag B_μ, whose pockets S_μ and T_μ have all the vertices of the special bags and of the pockets associated with f_μ^l and f_μ^r, respectively. Finally, for each traversable child ν_i of μ that has not been associated, we add a new bag B_μ^i so that S_μ^i contains all the vertices of the special bag of ν_i, while T_μ^i is empty. Finally, we apply operation MERGE-BAGS to all pairs $\langle x, y \rangle \in W$ such that both x and y belong to G_μ^{pert}. Hence, R.1 and R.2 are satisfied by any embedding $\mathcal{G}_\mu^{pert}$ of G_μ^{pert} that is described by the bags of μ. This concludes the P-node case.

At the end of the traversal, if root ρ has been visited without declaring the instance negative, the fact that $G_\rho^{pert} = G$ admits a planar embedding satisfying R.1 implies that $\langle G, W \rangle$ is a positive instance. We summarize the above discussion in the following theorem; for details refer to [1].

Theorem 4. *Problem* HIERARCHICAL PARTIAL PLANARITY *can be solved in* $O(|V|^3 \cdot |E_t|)$ *time for instances* $G = (V, E_p \cup E_s \cup E_t)$ *such that the graph induced by the edges in* $E_p \cup E_s$ *is biconnected.*

5   Open Problems

The main open problem raised by our work is to determine the complexity in the general case, where the biconnectivity restriction is relaxed. It is also of interest to broaden the study towards the case in which there exist more than three levels of importance for the edges. As a first step, one could consider the case in which there are four levels and the first two form a biconnected graph. Finally, the relationship with SEFE should be further investigated to understand whether our techniques can be applied to solve some of its open cases.

References

1. Angelini, P., Bekos, M.A.: Hierarchical partial planarity. CoRR, 1707.06844, (2017). http://arxiv.org/abs/1707.06844
2. Angelini, P., Binucci, C., Da Lozzo, G., Didimo, W., Grilli, L., Montecchiani, F., Patrignani, M., Tollis, I.G.: Algorithms and bounds for drawing non-planar graphs with crossing-free subgraphs. Comput. Geom. **50**, 34–48 (2015). doi:10.1016/j.comgeo.2015.07.002
3. Angelini, P., et al.: Simultaneous orthogonal planarity. In: Hu, Y., Nöllenburg, M. (eds.) GD 2016. LNCS, vol. 9801, pp. 532–545. Springer, Cham (2016). doi:10.1007/978-3-319-50106-2_41
4. Angelini, P., Da Lozzo, G., Neuwirth, D.: Advancements on SEFE and partitioned book embedding problems. Theor. Comput. Sci. **575**, 71–89 (2015). doi:10.1016/j.tcs.2014.11.016
5. Angelini, P., Di Battista, G., Frati, F., Jelínek, V., Kratochvíl, J., Patrignani, M., Rutter, I.: Testing planarity of partially embedded graphs. ACM Trans. Algorithms **11**(4), 32 (2015). doi:10.1145/2629341
6. Angelini, P., Di Battista, G., Frati, F., Patrignani, M., Rutter, I.: Testing the simultaneous embeddability of two graphs whose intersection is a biconnected or a connected graph. J. Discrete Algorithms **14**, 150–172 (2012). doi:10.1016/j.jda.2011.12.015.
7. Bekos, M.A., van Dijk, T.C., Kindermann, P., Wolff, A.: Simultaneous drawing of planar graphs with right-angle crossings and few bends. J. Graph Algorithms Appl. **20**(1), 133–158 (2016). doi:10.7155/jgaa.00388
8. Binucci, C., Brandes, U., Di Battista, G., Didimo, W., Gaertler, M., Palladino, P., Patrignani, M., Symvonis, A., Zweig, K.A.: Drawing trees in a streaming model. Inf. Process. Lett. **112**(11), 418–422 (2012). doi:10.1016/j.ipl.2012.02.011
9. Bläsius, T., Kobourov, S.G., Rutter, I.: Simultaneous embedding of planar graphs. In: Tamassia, R. (ed.) Handbook on Graph Drawing and Visualization, pp. 349–381. Chapman and Hall/CRC, London (2013)
10. Bläsius, T., Rutter, I.: Disconnectivity and relative positions in simultaneous embeddings. Comput. Geom. **48**(6), 459–478 (2015). doi:10.1016/j.comgeo.2015.02.002
11. Bläsius, T., Rutter, I.: Simultaneous PQ-ordering with applications to constrained embedding problems. ACM Trans. Algorithms **12**(2), 16:1–16:46 (2016). doi:10.1145/2738054.
12. Braß, P., Cenek, E., Duncan, C.A., Efrat, A., Erten, C., Ismailescu, D., Kobourov, S.G., Lubiw, A., Mitchell, J.S.B.: On simultaneous planar graph embeddings. Comput. Geom. **36**(2), 117–130 (2007). doi:10.1016/j.comgeo.2006.05.006
13. Chan, T.M., Frati, F., Gutwenger, C., Lubiw, A., Mutzel, P., Schaefer, M.: Drawing partially embedded and simultaneously planar graphs. J. Graph Algorithms Appl. **19**(2), 681–706 (2015). doi:10.7155/jgaa.00375
14. Da Lozzo, G., Rutter, I.: Planarity of streamed graphs. In: Paschos, V.T., Widmayer, P. (eds.) CIAC 2015. LNCS, vol. 9079, pp. 153–166. Springer, Cham (2015). doi:10.1007/978-3-319-18173-8_11
15. Di Battista, G., Tamassia, R.: On-line maintenance of triconnected components with SPQR-trees. Algorithmica **15**(4), 302–318 (1996). doi:10.1007/BF01961541
16. Di Battista, G., Tamassia, R.: On-line planarity testing. SIAM J. Comput. **25**(5), 956–997 (1996). doi:10.1137/S0097539794280736

17. Di Giacomo, E., Didimo, W., Liotta, G., Meijer, H., Wismath, S.K.: Planar and quasi-planar simultaneous geometric embedding. Comput. J. **58**(11), 3126–3140 (2015). doi:10.1093/comjnl/bxv048

18. Erten, C., Kobourov, S.G.: Simultaneous embedding of planar graphs with few bends. J. Graph Algorithms Appl. **9**(3), 347–364 (2005). doi:10.7155/jgaa.00113

19. Goodrich, M.T., Pszona, P.: Streamed graph drawing and the file maintenance problem. In: Wismath, S., Wolff, A. (eds.) GD 2013. LNCS, vol. 8242, pp. 256–267. Springer, Cham (2013). doi:10.1007/978-3-319-03841-4_23

20. Gutwenger, C., Mutzel, P.: A linear time implementation of SPQR-trees. In: Marks, J. (ed.) GD 2000. LNCS, vol. 1984, pp. 77–90. Springer, Heidelberg (2001). doi:10.1007/3-540-44541-2_8

21. Jelínek, V., Kratochvíl, J., Rutter, I.: A Kuratowski-type theorem for planarity of partially embedded graphs. Comput. Geom. **46**(4), 466–492 (2013). doi:10.1016/j.comgeo.2012.07.005

22. Kaufmann, M., Wagner, D. (eds.): Drawing Graphs, Methods and Models. LNCS, vol. 2025. Springer, Heidelberg (2001). doi:10.1016/0095-8956(91)90090-7

23. Kratochvíl, J.: String graphs. I. The number of critical nonstring graphs is infinite. J. Comb. Theory, Ser. B **52**(1), 53–66 (1991). doi:10.1016/0095-8956(91)90090-7

24. Schaefer, M.: Picking planar edges; or, drawing a graph with a planar subgraph. In: Duncan, C., Symvonis, A. (eds.) GD 2014. LNCS, vol. 8871, pp. 13–24. Springer, Heidelberg (2014). doi:10.1007/978-3-662-45803-7_2

25. Tamassia, R., Liotta, G.: Graph drawing. In: Goodman, J.E., O'Rourke, J. (eds.) Handbook of Discrete and Computational Geometry, 2nd edn, pp. 1163–1185. Chapman and Hall/CRC, London (2004). doi:10.1201/9781420035315.ch52

On the Relationship Between k-Planar and k-Quasi-Planar Graphs

Patrizio Angelini[1], Michael A. Bekos[1(✉)], Franz J. Brandenburg[2],
Giordano Da Lozzo[3], Giuseppe Di Battista[4], Walter Didimo[5],
Giuseppe Liotta[5], Fabrizio Montecchiani[5], and Ignaz Rutter[6]

[1] Universität Tübingen, Tübingen, Germany
{angelini,bekos}@informatik.uni-tuebingen.de
[2] University of Passau, Passau, Germany
brandenb@fim.uni-passau.de
[3] University of California, Irvine, CA, USA
gdalozzo@uci.edu
[4] Roma Tre University, Rome, Italy
gdb@dia.uniroma3.it
[5] Universitá degli Studi di Perugia, Perugia, Italy
{walter.didimo,giuseppe.liotta,fabrizio.montecchiani}@unipg.it
[6] TU Eindhoven, Eindhoven, The Netherlands
i.rutter@tue.nl

Abstract. A graph is k-planar $(k \geq 1)$ if it can be drawn in the plane such that no edge is crossed $k + 1$ times or more. A graph is k-quasi-planar $(k \geq 2)$ if it can be drawn in the plane with no k pairwise crossing edges. The families of k-planar and k-quasi-planar graphs have been widely studied in the literature, and several bounds have been proven on their edge density. Nonetheless, only trivial results are known about the relationship between these two graph families. In this paper we prove that, for $k \geq 3$, every k-planar graph is $(k + 1)$-quasi-planar.

1 Introduction

Drawings of graphs are used in a variety of application domains, including software engineering, circuit design, computer networks, database design, social sciences, and biology (see e.g. [16,17,27,30,41,43]). The aim of graph visualizations is to clearly convey the structure of the data and their relationships, in order to support users in their analysis tasks. In this respect, and independent of the specific domain, there is a general consensus that graph layouts with many edge crossings are hard to read, as also witnessed by several user studies on the subject (see e.g. [26,38,39,46]). This motivation has generated lots of research on finding bounds on the number of edge crossings in different graph families (see e.g. [37,40,45]) and on the problem of automatically computing graph layouts with as few crossings as possible (see e.g. [5,13]). We recall that, although it is linear-time solvable to decide whether a graph admits a planar drawing (i.e. a drawing without edge crossings) [11,25], minimizing the number of edge crossings is a well-known NP-hard problem [22].

© Springer International Publishing AG 2017
H.L. Bodlaender and G.J. Woeginger (Eds.): WG 2017, LNCS 10520, pp. 59–74, 2017.
https://doi.org/10.1007/978-3-319-68705-6_5

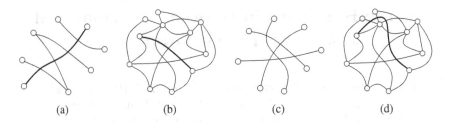

Fig. 1. (a) A crossing configuration that is forbidden in a 3-planar topological graph. (b) A 3-planar topological graph. (c) A crossing configuration that is forbidden in a 4-quasi-planar topological graph. (d) A 4-quasi-planar topological graph obtained from the one of Figure (b) by suitably rerouting the thick edge.

An emerging research area, informally recognized as *beyond planarity* (see e.g. [24,28,31]), concentrates on different models of graph planarity relaxations, which allow edge crossings but forbid specific configurations that would affect the readability of the drawing too much. Forbidden crossing configurations can be, for example, a single edge that is crossed too many times [36], a group of mutually crossing edges [21,42], two edges that cross at a sharp angle [18], a group of adjacent edges crossed by another edge [15], or an edge that crosses two independent edges [6,10,29]. Different models give rise to different families of "beyond planar" graphs. Two of the most popular families introduced in this context are the k-*planar* graphs and the k-*quasi-planar* graphs, which are usually defined in terms of *topological graphs*, i.e., graphs with a geometric representation in the plane with vertices as points and edges as Jordan arcs connecting their endpoints. A topological graph is k-planar ($k \geq 1$) if no edge is crossed $k + 1$ times or more, while it is k-quasi-planar ($k \geq 2$) if it can be drawn in the plane with no k pairwise crossing edges. Figure 1a shows a crossing configuration that is forbidden in a 3-planar topological graph. Figure 1b depicts a 3-planar topological graph that is not 2-planar (e.g., the thick edge is crossed three times). Figure 1c shows a crossing configuration that is forbidden in a 4-quasi-planar topological graph. Figure 1d depicts a 4-quasi-planar topological graph that is not 3-quasi-planar. A graph is k-*planar* (k-*quasi-planar*) if it is isomorphic to some k-planar (k-quasi-planar) topological graph. Clearly, by definition, k-planar graphs are also $(k + 1)$-planar and k-quasi-planar graphs are also $(k + 1)$-quasi-planar. This naturally defines a hierarchy of k-planarity and a hierarchy of k-quasi-planarity. Also, the class of 2-quasi-planar graphs coincides with that of planar graphs. Note that 3-quasi-planar graphs are also called *quasi-planar*.

The k-planarity and k-quasi-planarity hierarchies have been widely explored in graph theory, graph drawing, and computational geometry, mostly in terms of edge density. Pach and Tóth [36] proved that a k-planar *simple* topological graph with n vertices has at most $1.408\sqrt{k}n$ edges. We recall that a topological graph is simple if any two edges cross in at most one point and no two adjacent edges cross. For $k \leq 4$, Pach and Tóth [36] also established a finer bound of $(k + 3)(n - 2)$ on the edge density, and prove its tightness for $k \leq 2$. For $k = 3$, the best known upper bound on the edge density is $5.5n - 11$, which is tight up to an additive constant [7,34].

Concerning k-quasi-planar graphs, a 20-year-old conjecture by Pach et al. [35] asserts that, for every fixed k, the maximum number of edges in a k-quasi-planar graph with n vertices is $O(n)$. However, linear upper bounds have been proven only for $k \leq 4$. Agarwal et al. [3] were the first to prove that 3-quasi-planar simple topological graphs have a linear number of edges. This was generalized by Pach et al. [33], who proved that *all* 3-quasi-planar graphs on n vertices have at most $65n$ edges. This bound was further improved to $8n - O(1)$ by Ackerman and Tardos [2]. For 3-quasi-planar simple topological graphs they also proved a bound of $6.5n - 20$, which is tight up to an additive constant. Ackerman [1] also proved that 4-quasi-planar graphs have at most a linear number of edges. For $k \geq 5$, several authors have shown super-linear upper bounds on the edge density of k-quasi-planar graphs (see, e.g., [14,20,21,35,44]). The most recent results are due to Suk and Walczak [42], who proved that any k-quasi-planar simple topological graph on n vertices has at most $c_k n \log n$ edges, where c_k is a number that depends only on k. For k-quasi-planar topological graphs where two edges can cross in at most t points, they give an upper bound of $2^{\alpha(n)^c} n \log n$, where $\alpha(n)$ is the inverse of the Ackermann function, and c depends only on k and t.

Despite the many papers mentioned above, the relationships between the hierarchies of k-planar and k-quasi-planar graphs have not been studied yet and only trivial results are known. For example, due to the tight bounds on the edge density of 3-planar and 3-quasi-planar simple graphs, it is immediate to conclude that there are infinitely many 3-quasi-planar graphs that are not 3-planar. Also, it can be easily observed that, for $k \geq 1$, every k-planar graph is $(k + 2)$-quasi-planar. Indeed, if a k-planar graph G were not $(k + 2)$-quasi-planar, any topological graph isomorphic to G would contain $k + 2$ pairwise crossing edges; but this would imply that any of these edges is crossed at least $k + 1$ times, thus contradicting the hypothesis that G is k-planar.

Contribution. In this paper we focus on simple topological graphs and prove the first non-trivial inclusion relationship between the k-planarity and the k-quasi-planarity hierarchies. We show that every k-planar graph is $(k + 1)$-quasi-planar, for every $k \geq 3$. In other words, we show that every k-planar simple topological graph can be redrawn to become a $(k + 1)$-quasi-planar simple topological graph ($k \geq 3$). For example, the simple topological graph of Fig. 1b is 3-planar but not 4-quasi-planar. The simple topological graph of Fig. 1d, on the other hand, is a 4-quasi-planar graph obtained from the one of Fig. 1b by rerouting an edge (but it is no longer 3-planar).

The proof of our result is based on the following novel methods: (i) A general purpose technique to "untangle" groups of mutually crossing edges. More precisely, we show how to reroute the edges of a k-planar topological graph in such a way that all vertices of a set of $k + 1$ pairwise crossing edges lie in the same face of their arrangement. (ii) A global edge rerouting technique, based on a matching argument, used to remove all forbidden configurations of $k + 1$ pairwise crossing edges from a k-planar simple topological graph, provided that these edges are "untangled".

The remainder of the paper is structured as follows. In Sect. 2 we give some basic terminology and observations that will be used throughout the paper. Section 3 describes our general proof strategy. Sections 4 and 5 provide details about methods (i) and (ii), respectively. Conclusions and open problems are in Sect. 6. Some proofs and technicalities are omitted in this extended abstract and can be found in [4].

2   Preliminaries

We only consider graphs with no parallel edges no self-loops. Also, we will assume our graphs to be connected, as our results immediately carry over to disconnected graphs. A *topological graph* G is a graph drawn in the plane with vertices represented by points and edges represented by Jordan arcs connecting the corresponding endpoints. In notation and terminology, we do not distinguish between the vertices and edges of a graph, and the points and arcs representing them, respectively. Two edges *cross* if they share one interior point and alternate around this point. A topological graph is *almost simple* if any two edges cross at most once. An almost simple topological graph, such that no two adjacent edges cross each other, is called *simple*. A topological graph divides the plane into topologically connected regions, called *faces*. The unbounded region is the *outer face*. Note that the boundary of a face can contain vertices of the graph and crossing points between edges.

If G and G' are two isomorphic graphs, we write $G \simeq G'$. A graph G' is k-planar (k-quasi-planar) if there exists a k-planar (k-quasi-planar) topological graph $G \simeq G'$.

Given a subgraph X of a graph G, the *arrangement of X*, denoted by A_X, is the arrangement of the curves corresponding to the edges of X. We denote the vertices and edges of X by $V(X)$ and $E(X)$, respectively. A *node* of A_X is either a vertex or a crossing point of X. A *segment* of A_X is a part of an edge of X that connects two nodes, i.e., a maximal uncrossed part of an edge of X. A *fan* is a set of edges that share a common endpoint. A set of k vertex-disjoint mutually crossing edges in a topological graph G is called a *k-crossing*. A k-crossing X is *untangled* if all nodes corresponding to vertices in $V(X)$ are incident to a

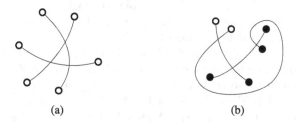

(a)   (b)

Fig. 2. (a) An untangled 3-crossing; all vertices belong to the same face of the arrangement (the outer face). (b) A tangled 3-crossing; the circled vertices and the solid vertices belong to distinct faces of the arrangement.

common face of the arrangement A_X of X. Else, it is *tangled*. For example, the 3-crossing in Fig. 2a is untangled, whereas the 3-crossing in Fig. 2b is tangled. Note that each edge in a $(k+1)$-crossing X crosses each of the remaining k edges in $E(X)$, hence the next observation easily follows.

Observation 1. *Let $G = (V, E)$ be a k-planar simple topological graph and let X be a $(k + 1)$-crossing in G. An edge in $E(X)$ cannot be crossed by any other edge in $E \setminus E(X)$. In particular, for any two $(k + 1)$-crossings $X \neq Y$ in G, $E(X) \cap E(Y) = \emptyset$ holds.*

3   Edge Rerouting Operations and Proof Strategy

We introduce an edge rerouting operation that will be a basic tool for our proof strategy. Let G be a k-planar simple topological graph and consider an untangled $(k + 1)$-crossing X in G. Without loss of generality, the vertices in $V(X)$ lie in the outer face of A_X.

Let $e = \{u, v\} \in E(X)$ and let $w \in V(X) \setminus \{u, v\}$. Let A'_X denote the arrangement obtained from A_X by removing all nodes corresponding to vertices in $V(X) \setminus \{u, v, w\}$, together with their incident segments, and by removing the edge (u, v). The operation of *rerouting* $e = \{u, v\}$ *around* w consists of redrawing e sufficiently close to the boundary of the outer face of A'_X, choosing the routing that passes close to w, in such a way that e crosses the fan incident to w, but not any other edge in $E \setminus E(X)$. See Fig. 3b for an illustration. More precisely, let D be a topological disk that encloses all crossing points of X and such that each edge in $E(X)$ crosses the boundary of D exactly twice. Then, the rerouted edge keeps unchanged the parts of e that go from u to the boundary of D and from v to the boundary of D. We call the unchanged parts of a rerouted edge its *tips* and the remaining part, which routes around w, its *hook*.

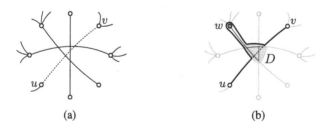

(a)   (b)

Fig. 3. The rerouting operation for dissolving untangled $(k+1)$-crossings. (a) An untangled $(k+1)$-crossing X. (b) The rerouting of the dashed edge (u, v) around the marked vertex w. The arrangement A'_X is thin red, the removed nodes and segments are gray. Note that the dashed curve is part of A'_X. (Color figure online)

Lemma 1. *Let G be a k-planar simple topological graph and let X be an untangled $(k + 1)$-crossing in G. Let $G' \simeq G$ be the topological graph obtained from G*

by rerouting an edge $e = \{u, v\} \in E(X)$ *around a vertex* $w \in V(X) \setminus \{u, v\}$. *Let d be the edge of* $E(X)$ *incident to w. Graph* G' *has the following properties:* (i) *Edges e and d do not cross;* (ii) *The edges that are crossed by e in* G' *but not in G form a fan at w;* (iii) G' *is almost simple.*

Proof. Conditions (i) and (ii) immediately follow from the definition of the rerouting operation and from the fact that edge e can be drawn arbitrarily close to the boundary of the outer face of A'_X. Since G is simple, in order to prove that G' is almost simple, we only need to show that edge e does not cross any other edge more than once. The only part of e that is drawn in G' differently than in G is the one between the intersection points of e and the boundary $B(D)$ of the topological disk D that (a) encloses all crossing points of X and such that (b) each edge in $E(X)$ crosses the boundary of D exactly twice. Since G is simple, by (b) and by the definition of the rerouting operation, the two crossing points between an edge $e' \in E(X)$ and $B(D)$ alternate with the two crossing points between any edge $e' \neq e'' \in E(X)$ and $B(D)$ along $B(D)$. Hence, by redrawing edge e sufficiently close to any of the two parts of $B(D)$ between the two intersection points of edge e and $B(D)$, we encounter each edge in $E(X) \setminus \{e\}$ exactly once. Thus, edge e crosses all the edges in $E(X) \setminus \{e, d\}$ exactly once. This concludes the proof.   □

Lemma 1 does not guarantee that the graph G' is simple. Indeed, if the edge (u, w) or the edge (v, w) existed in G, then the rerouted edge $e = (u, v)$ would cross such an edge. We will show in Sect. 5 how to fix this problem by redrawing (u, w) and (v, w).

We are now ready to describe our general strategy for transforming a k-planar simple topological graph G into a simple topological graph $G' \simeq G$ that is $(k + 1)$-quasi-planar. The idea is to pick from each $(k + 1)$-crossing X in G an edge e_X and a vertex w_X not adjacent to e_X, and to apply the rerouting operation *simultaneously* for all pairs (e_X, w_X), i.e., rerouting e_X around w_X. This operation, which we call *global rerouting*, is well defined since the $(k + 1)$-crossings are pairwise edge-disjoint by Observation 1.

There are however several constraints that have to be satisfied in order for such a global rerouting to have the desired effect. First of all, as mentioned above, the rerouting operation can only be applied to untangled $(k+1)$-crossings. Thus, as a first step, we will show that, in a k-planar simple topological graph, all tangled $(k + 1)$-crossings can be removed, leaving the resulting graph simple and k-planar. More precisely, given a tangled $(k + 1)$-crossing X, it is possible to redraw the whole graph so that either at least two edges of X do not cross or X becomes an untangled $(k + 1)$-crossing, and, furthermore, any two edges cross only if they crossed before the redrawing. The technical details for this operation are described in Sect. 4. Notice that, even assuming that all $(k + 1)$-crossings are untangled, there are further problems that can occur when performing all the rerouting operations independently of each other. Specifically, the resulting topological graph G' may be non-simple and/or the rerouting may create new $(k + 1)$-crossings. We explain how to overcome these issues in Sect. 5.

4 Removing Tangled $(k+1)$-Crossings

The proof of the next lemma describes a technique to "untangle" all $(k+1)$-crossings in a k-planar simple topological graph. This technique is of general interest, as it gives more insights into the structure of k-planar simple topological graphs.

Lemma 2. *Let G be a k-planar simple topological graph. There exists a k-planar simple topological graph $G' \simeq G$ without tangled $(k+1)$-crossings.*

Proof. We first show how to untangle a $(k+1)$-crossing X in a k-planar simple topological graph G by neither creating new $(k+1)$-crossings nor introducing new crossings.

Let X be a tangled $(k+1)$-crossing and let A_X be its arrangement. For each face f of A_X, let V_f denote the subset of vertices of $V(X)$ incident to f. Since in X all vertices have degree 1, the sets V_f form a partition of $V(X)$.

For each inner face f of A_X, let G_f denote the subgraph of G consisting of the vertices of V_f, and of the vertices and edges of G that lie in the interior of f. Refer to Fig. 4a for an illustration. Since G is k-planar and X is a $(k+1)$-crossing, there exists no crossing between a segment in G_f and a segment not in G_f. Therefore, the boundary of f corresponds to the boundary of a topological disk D_f so that G_f is k-planarly embedded inside D_f: only the vertices of V_f lie on the boundary of D_f. For the external face h, graph G_h consists of the vertices of V_h, and of the vertices and edges of G that lie outside A_X. In this case, the topological disk D_h is obtained after a suitable inversion of G_h, if needed. We can rearrange and deform each D_f such that: (i) the vertices in $V(X)$ lie on a common circle C; (ii) for any two distinct faces f and g of A_X, the topological disks D_f and D_g do not intersect; (iii) the interior of the circle C is empty. Then, the $k+1$ edges of X are redrawn as straight-line segments inside C. This construction implies that X is untangled (and some of its edges may not cross anymore). Also, each subgraph G_f remains topologically equivalent to its initial drawing. Thus, two

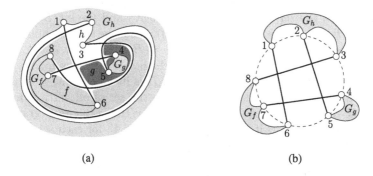

(a)   (b)

Fig. 4. Illustration of the untangling procedure in the proof of Lemma 2: (a) A 3-planar simple topological graph with a 4-crossing X (thicker edges). (b) The topological graph resulting from the procedure that untangles X.

edges cross after the transformation only if they crossed before, which ensures that the resulting graph is simple and no new $(k + 1)$-crossing is created.

We apply the above transformation sequentially to each subgraph G_f of the new topological graph until all $(k + 1)$-crossings are untangled. This concludes the proof.   □

An illustration of the untangling procedure described in the proof of Lemma 2 is given in Fig. 4. Figure 4a shows an example of a 3-planar simple topological graph G with a tangled 4-crossing X; the edges of X are thicker. Faces f, g, and h are the three faces of A_X whose union contains the vertices of $V(X)$. Subgraphs G_f, G_g, and G_h are schematically depicted. Figure 4b shows G after the transformation that untangles X.

5   Removing Untangled $(k + 1)$-Crossings

Let G be a k-planar simple topological graph with $k \geq 3$. By Lemma 2, we may assume that G has no tangled $(k + 1)$-crossings. In Sect. 5.1, we show how to transform G into a (possibly not almost simple) $(k + 1)$-quasi-planar topological graph $G' \simeq G$. Then, in Sect. 5.2, we describe how to make G' simple without introducing $(k + 1)$-crossings.

5.1   Obtaining $(k + 1)$-Quasi-Planarity

We first show the existence of a global rerouting such that no two edges of G are rerouted around the same vertex (Lemma 5). Then, we show that any global rerouting of G with this property yields a topological graph G' with no $(k + 1)$-crossings (Lemma 9).

The existence of this global rerouting is proved by defining a bipartite graph composed of the vertices of G and of its $(k + 1)$-crossings, and by showing that a matching covering all the $(k + 1)$-crossings always exists. Let G be a k-planar simple topological graph and let S be the set of $(k + 1)$-crossings of G. We define a bipartite graph $H = (A \cup B, E)$, where $E \subseteq A \times B$, as follows. For each $(k + 1)$-crossing $X \in S$, the set A contains a vertex $v(X)$ and the set B contains the endpoints of $E(X)$ (that is, $B = \bigcup_{X \in S} V(X)$). Also, the set E contains an edge between a vertex $v(X) \in A$ and a vertex $v \in B$ if and only if $v \in V(X)$. A *matching from A into B* is a set $M \subseteq E$ such that each vertex in A is incident to exactly one edge in M and each vertex in B is incident to at most one edge in M. For a subset $A' \subseteq A$, let $N(A')$ denote the set of all vertices in B that are adjacent to a vertex in A'. We recall that, by Hall's theorem, graph H has a matching from A into B if and only if $|N(A')| \geq |A'|$ for each set $A' \subseteq A$. We have the following (the proof is omitted).

Lemma 3. *The graph $H = (A \cup B, E)$ is a simple bipartite planar graph. Also, each vertex in A has degree $2k + 2$.*

Lemma 4. *The graph $H = (A \cup B, E)$ has a matching from A into B.*

Proof. Let $A' \subseteq A$ and let H' be the subgraph of H induced by $A' \cup N(A')$. Since the vertices in A have degree $2k+2$, by Lemma 3, we have $|E(H')| = (2k+2)|A'|$. Also, since H (and thus H') is bipartite planar, by Lemma 3, we have $|E(H')| \leq 2(|A'| + N(A')) - 4$ [32, Corollary 1.2]. Thus, $|N(A')| \geq k|A'| + 2 > |A'|$, and Hall's theorem applies.   □

Lemma 5. *Let G be a k-planar simple topological graph. It is possible to perform a global rerouting on G such that no two edges are rerouted around the same vertex.*

Proof. Let $S = \{X_1, X_2, \ldots, X_h\}$, with $h > 0$, be the set of $(k+1)$-crossings of G. By Lemma 4, it is possible to assign a vertex $v_i \in V(X_i)$ to each $(k+1)$-crossing X_i in such a way that no two distinct $(k+1)$-crossings are assigned the same vertex. The statement follows by considering a global rerouting such that, for each $(k+1)$-crossing X_i, any edge in X_i not incident to v_i is rerouted around v_i.   □

Let G' be a topological graph obtained from G by performing a global rerouting as in Lemma 5. We prove that G' has no $(k+1)$-crossings. To this aim, we first give some important structural insights for G' (Lemmas 6–8).

Lemma 6. *Let e and d be two edges that cross in G' but not in G. Then, one of e and d has been rerouted around an endpoint of the other.*

Proof. Since e and d do not cross in G, we may assume that one of them, say e, has been rerouted. Suppose first that the hook of e crosses d. We claim that e has been rerouted around an endpoint of d. In fact, if d has not been rerouted, then the claim is trivially true; see Fig. 5b. On the other hand, if d has been rerouted, then the crossing with e must be on a tip of d, and not on its hook, since no two edges have been rerouted around the same vertex in the global rerouting; see Fig. 5c. Thus, the claim follows. Suppose now that a tip of e crosses d. Then, this crossing must be with the hook of d, and the same argument applies to prove that d has been rerouted around an endpoint of e.   □

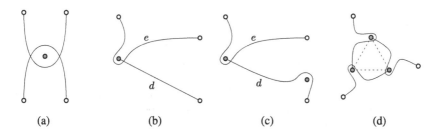

(a)   (b)   (c)   (d)

Fig. 5. (a) Two edges rerouted around the same vertex. (b)–(c) Different cases of edges that do not cross before a global rerouting operation but cross afterwards. The vertices used for rerouting are filled green. (d) A 3-crossing arising from redrawing three edges. (Color figure online)

The next two lemmas can be proved by using Lemma 6.

Lemma 7. *Every non-rerouted edge e is crossed by at most three rerouted edges in G'. Further, if e is crossed by exactly three rerouted edges, then two of them have been rerouted around distinct endpoints of e.*

Lemma 8. *If G' contains a $(k+1)$-crossing X', then X' contains at most one edge that has not been rerouted.*

Altogether the above lemmas can be used to prove the following.

Lemma 9. *Graph G' does not contain any $(k+1)$-crossing.*

Proof. Assume for a contradiction that G' contains a $(k+1)$-crossing X'. By Lemma 8, X' contains at most one non-rerouted edge.

Suppose that X' contains such an edge e. By Lemma 7, there are at most three rerouted edges crossing e in G'. Since $k \geq 3$, there are exactly three such rerouted edges, say d, h, and l. By Lemma 7, two of them, say d and h, have been rerouted around (distinct) endpoints of e. Thus, d and h do not cross in G, by Observation 1, as they belong to different $(k+1)$-crossings. Hence, they can cross in G' only if one of them has been rerouted around an endpoint of the other, by Lemma 6. This is not possible as d and h, belonging to the same $(k+1)$-crossing in G, do not share an endpoint.

Suppose that X' contains only rerouted edges. Let e be any edge of X' and let w be the vertex used for rerouting e. Since at most one edge in X' can be incident to w and since $k \geq 3$, there are two edges d, h in X' that have been rerouted around distinct endpoints of e. As in the previous case, we can prove that d and h do not cross.   □

For $k = 2$, Lemma 9 does not hold, as some 3-crossings may still appear after the global rerouting; see Fig. 5d for an illustration and refer to Sect. 6 for a discussion.

5.2   Obtaining Simplicity

Lemmas 2, 5, and 9 imply that, for $k \geq 3$, any k-planar simple topological graph G can be redrawn so that the resulting topological graph $G' \simeq G$ contains no $(k+1)$-crossings and no two edges are rerouted around the same vertex. However, the graph G' may not be simple or even almost simple. We first show how to remove pairs of edges crossing more than once from G', without introducing $(k+1)$-crossings, thus leading to a $(k+1)$-quasi-planar almost-simple graph (Lemma 11). Then we show how to remove crossings between edges incident to a common vertex, still without introducing $(k+1)$-crossings (Lemma 12). We will use the following auxiliary lemma.

Lemma 10. *The graph G' is almost-simple if and only if there is no pair of edges such that either of them is rerouted around an endpoint of the other.*

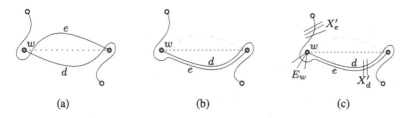

(a)   (b)   (c)

Fig. 6. (a) A double crossing between two edges e, d, due to a global rerouting. (b) Solving the configuration in (a) by redrawing e. (c) Edges crossing e after the transformation.

Lemma 11. *There is a $(k + 1)$-quasi-planar almost simple topological graph $G^* \simeq G'$.*

Proof. We assume that G' is not almost simple, as otherwise the statement would follow with $G^* = G'$. By Lemma 10, there exist pairs of edges in which each of the two edges has been rerouted around an endpoint of the other; see Fig. 6a. For each such pair e, d, we remove the two crossings by redrawing one of the two edges, say e, by following d between the two crossings. More precisely, we redraw the tip of e crossed by the hook of d by following the tip of d crossed by the hook of e, without crossing it; see Fig. 6b. In the following we prove that the graph G^* obtained by applying this operation for all the pairs does not contain new $(k + 1)$-crossings and is almost simple.

Observe first that each edge tip is involved in at most one pair, since no two edges are rerouted around the same vertex. Thus, no tip of an edge is transformed twice in G^* and no two tips of transformed edges cross each other. Also, a transformed edge can participate to a $(k+1)$-crossing in G^* only through its tips. Hence, any $(k+1)$-crossing in G^* would contain exactly one transformed edge. We prove that this is not possible.

Let e be an edge that has been redrawn due to a double crossing with an edge d, and let X_e and X_d be the $(k + 1)$-crossings of G containing e and d, respectively. The edges crossing e in G^* are (see Fig. 6c): (i) a set X'_d of edges in X_d crossing the tip of d that has been used to redraw e; (ii) a set X'_e of edges in X_e crossing the tip of e not crossed by d; (iii) a set E_w of edges incident to the vertex w around which e has been rerouted (and thus they cross the hook of e). Note that X'_d contains all the edges that cross e in G^* and not in G'. These edges do not cross those in X'_e, since they are non-rerouted edges belonging to distinct $(k + 1)$-crossings of G. Also, they do not cross edges in E_w, since X_d does not contain any edge incident to w other than d. Finally, there are at most $k - 1$ edges in X'_d, since X_d contains $k+1$ edges and at least two of them do not cross e, namely d and the edge incident to the endpoint of e around which d has been rerouted. Thus, the edges in X'_d are not involved in any $(k + 1)$-crossing with e. To see that the same holds for the edges in X'_e and in E_w, note that any $(k + 1)$-crossing in G^* involving these edges and e would also appear in G' (because the tip of e crossed by X'_e has not been transformed), which is however $(k + 1)$-quasi-planar.

To prove that G^* is almost simple, we show that any edge d' in X'_d is crossed only once by e. Recall that d' does not cross e in G'. Also, since each tip is involved in at most one transformation, d' crosses the tip of d (and hence the tip of e that has been redrawn) only once. On the other hand, it could be that also the other tip of e has been transformed by following the tip of an edge h and that this transformation introduced a new crossing between e and d'. But then d' would cross both d and h in G', and hence by Lemma 6 also in G. This is however not possible, since both d and h have been rerouted, and hence they belong to different $(k+1)$-crossings in G. □

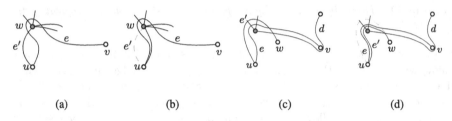

Fig. 7. (a), (c) Topological graphs that are not simple. (b), (d) Avoiding the non-simplicity in (a) and (c) by redrawing one of the two edges.

Lemma 12. *There is a $(k+1)$-quasi-planar simple topological graph $\overline{G} \simeq G^*$.*

Proof. We assume that G^* is not simple, as otherwise the statement would follow with $\overline{G} = G^*$. Let $e = (u, v)$ and $e' = (u, w)$ be two crossing edges that share an endpoint u. Since G is simple, at least one of them has been redrawn, say e.

We distinguish two cases, based on whether (i) e has been rerouted but not transformed afterwards, or (ii) e has also been transformed, due to a double crossing.

In case (i), edge e crosses e' with its hook, see Fig. 7a. We redraw e' by following e till reaching u, as in Fig. 7b. This guarantees that e and e' no longer cross and that e' does not cross any edge twice, since e' crosses only edges that cross the tip of e incident to u. Also, no $(k+1)$-crossing is introduced. Indeed, any new $(k+1)$-crossing should contain e', but then also e would be part of this $(k+1)$-crossing, which is impossible since e and e' do not cross and G^* does not contain $(k+1)$-crossings.

In case (ii), let d be the edge that crosses e twice in G'; note that d has not been transformed, see Fig. 7c. Recall that, by Lemma 10, e and d have been rerouted one around an endpoint of the other. Suppose that the endpoint of e around which d has been rerouted is v, the case in which it is u can be treated analogously. This implies that e' crosses a tip of d, and therefore e' and d are part of a $(k+1)$-crossing in G, namely the one that caused the rerouting of d. We redraw the part of e' from u to its crossing point with e by following e, without crossing it, and leave the rest of e' unchanged, as in Fig. 7d. This guarantees that e and e' no longer cross, and that any new crossing of e' is with an edge that also crosses e. As in case (i), this implies that e' does not cross any edge twice and that no new $(k+1)$-crossing has been generated.   □

The next theorem summarizes the main result of the paper.

Theorem 1. *Let G be a k-planar simple topological graph, with $k \geq 3$. Then, there exists a $(k+1)$-quasi-planar simple topological graph $\overline{G} \simeq G$.*

Proof. First recall that, by Lemma 2, we can assume that G does not contain any tangled $(k+1)$-crossing. We apply Lemma 5 to compute a global rerouting for G in which no two edges are rerouted around the same vertex. By Lemma 9, the resulting topological graph $G' \simeq G$ is $(k+1)$-quasi-planar. Also, by Lemma 11, if G' is not almost simple, then it is possible to redraw some of its edges in such a way that the resulting topological graph $G^* \simeq G'$ is almost simple and remains $(k+1)$-quasi-planar. Finally, by Lemma 12, if G^* is not simple, then it can be made so, again by redrawing some of its edges, while maintaining $(k+1)$-quasi-planarity. This concludes the proof that there exists a $(k+1)$-quasi-planar simple topological graph $\overline{G} \simeq G$.   □

6   Conclusions and Open Problems

We proved that, for any $k \geq 3$, the family of k-planar graphs is included in the family of $(k+1)$-quasi-planar graphs. This result represents the first non-trivial relationship between the k-planar and the k-quasi-planar graph hierarchies, and contributes to the literature that studies the connection between different families of beyond planar graphs (see, e.g. [9,10,12,19]). Our proof strategy does not immediately apply to the case of $k = 2$, due to the possible existence of three rerouted edges that are pairwise crossing after a global rerouting (as in Fig. 5d). This problem has been recently solved by Hoffmann and Tóth [23], who proved that 2-planar graphs are quasi-planar.
  Interesting problems remain open. Among them:

(i) For $k \geq 3$, one can also ask whether the family of k-planar graphs is included in the family of k-quasi-planar graphs. For $k = 2$ the answer is trivially negative, as 2-quasi-planar graphs coincide with the planar graphs. On the other hand, optimal 3-planar graphs are known to be (3-)quasi-planar [8]. We recall that an n-vertex 3-planar graph is optimal if it has $5.5n - 11$ edges. For sufficiently large values of k, one can even investigate whether every k-planar simple topological (sparse) graph G is $f(k)$-quasi-planar, for some function $f(k) = o(k)$.
(ii) One can study *non-inclusion* relationships between the k-planar and the k-quasi-planar graph hierarchies, other than those that are easily derivable from the known edge density results. For example, for any given $k > 3$, can we establish an integer function $h(k)$ such that some $h(k)$-planar graph is not k-quasi-planar?

Acknowledgements. The research in this paper started at the Dagstuhl Seminar 16452 "Beyond-Planar Graphs: Algorithmics and Combinatorics". We thank all participants, and in particular Pavel Valtr and Raimund Seidel, for useful discussions on the topic.

References

1. Ackerman, E.: On the maximum number of edges in topological graphs with no four pairwise crossing edges. Discrete Comput. Geom. **41**(3), 365–375 (2009)
2. Ackerman, E., Tardos, G.: On the maximum number of edges in quasi-planar graphs. J. Comb. Theory Ser. A **114**(3), 563–571 (2007)
3. Agarwal, P.K., Aronov, B., Pach, J., Pollack, R., Sharir, M.: Quasi-planar graphs have a linear number of edges. Combinatorica **17**(1), 1–9 (1997)
4. Angelini, P., Bekos, M.A., Brandenburg, F.J., Da Lozzo, G., Di Battista, G., Didimo, W., Liotta, G., Montecchiani, F., Rutter, I.: On the relationship between k-planar and k-quasi planar graphs. CoRR, abs/1702.08716 (2017)
5. Angelini, P., Da Lozzo, G., Di Battista, G., Frati, F., Patrignani, M., Roselli, V.: Relaxing the constraints of clustered planarity. Comput. Geom. **48**(2), 42–75 (2015)
6. Bekos, M.A., Cornelsen, S., Grilli, L., Hong, S.-H., Kaufmann, M.: On the recognition of fan-planar and maximal outer-fan-planar graphs. In: Duncan, C., Symvonis, A. (eds.) GD 2014. LNCS, vol. 8871, pp. 198–209. Springer, Heidelberg (2014). doi:10.1007/978-3-662-45803-7_17
7. Bekos, M.A., Kaufmann, M., Raftopoulou, C.N.: On the density of non-simple 3-planar graphs. In: Hu, Y., Nöllenburg, M. (eds.) GD 2016. LNCS, vol. 9801, pp. 344–356. Springer, Cham (2016). doi:10.1007/978-3-319-50106-2_27
8. Bekos, M.A., Kaufmann, M., Raftopoulou, C.N.: On optimal 2- and 3-planar graphs. In: Aronov, B., Katz, M.J. (eds.) SoCG 2017. LIPIcs, vol. 77, p. 16:1. Schloss Dagstuhl - Leibniz-Zentrum fuer Informatik (2017)
9. Binucci, C., Chimani, M., Didimo, W., Gronemann, M., Klein, K., Kratochvíl, J., Montecchiani, F., Tollis, I.G.: Algorithms and characterizations for 2-layer fan-planarity: from caterpillar to stegosaurus. J. Graph Algorithms Appl. **21**(1), 81–102 (2017)
10. Binucci, C., Di Giacomo, E., Didimo, W., Montecchiani, F., Patrignani, M., Symvonis, A., Tollis, I.G.: Fan-planarity: properties and complexity. Theor. Comput. Sci. **589**, 76–86 (2015)
11. Booth, K.S., Lueker, G.S.: Testing for the consecutive ones property, interval graphs, and graph planarity using PQ-tree algorithms. J. Comput. Syst. Sci. **13**(3), 335–379 (1976)
12. Brandenburg, F.J., Didimo, W., Evans, W.S., Kindermann, P., Liotta, G., Montecchiani, F.: Recognizing and drawing IC-planar graphs. Theor. Comput. Sci. **636**, 1–16 (2016)
13. Buchheim, C., Chimani, M., Gutwenger, C., Jünger, M., Mutzel, P.: Crossings and planarization. In: Tamassia, R. (ed.) Handbook on Graph Drawing and Visualization, pp. 43–85. Chapman and Hall/CRC, Boca Raton (2013)
14. Capoyleas, V., Pach, J.: A Turán-type theorem on chords of a convex polygon. J. Comb. Theory Ser. B **56**(1), 9–15 (1992)
15. Cheong, O., Har-Peled, S., Kim, H., Kim, H.: On the number of edges of fan-crossing free graphs. Algorithmica **73**(4), 673–695 (2015)

16. Di Battista, G., Eades, P., Tamassia, R., Tollis, I.G.: Graph Drawing. Prentice Hall, Upper Saddle River (1999)
17. Didimo, W., Liotta, G.: Mining graph data. In: Graph Visualization and Data Mining, pp. 35–64. Wiley, Hoboken (2007)
18. Didimo, W., Liotta, G.: The crossing-angle resolution in graph drawing. In: Pach, J. (ed.) Thirty Essays on Geometric Graph Theory, pp. 167–184. Springer, New York (2013). doi:10.1007/978-1-4614-0110-0_10
19. Eades, P., Liotta, G.: Right angle crossing graphs and 1-planarity. Discrete Appl. Math. **161**(7–8), 961–969 (2013)
20. Fox, J., Pach, J.: Coloring K_k-free-free intersection graphs of geometric objects in the plane. Eur. J. Comb. **33**(5), 853–866 (2012)
21. Fox, J., Pach, J., Suk, A.: The number of edges in k-quasi-planar graphs. SIAM J. Discrete Math. **27**(1), 550–561 (2013)
22. Garey, M.R., Johnson, D.S.: Crossing number is NP-complete. SIAM J. Algebraic Discrete Methods **4**(3), 312–316 (1983)
23. Hoffmann, M., Tóth, C.D.: Two-planar graphs are quasiplanar. CoRR, abs/1705.05569. Accepted at MFCS 2017 (2017)
24. Hong, S., Tokuyama, T.: Algorithmics for beyond planar graphs. NII Shonan Meeting Seminar 089, 27 November–1 December 2016
25. Hopcroft, J.E., Tarjan, R.E.: Efficient planarity testing. J. ACM **21**(4), 549–568 (1974)
26. Huang, W., Hong, S., Eades, P.: Effects of sociogram drawing conventions and edge crossings in social network visualization. J. Graph Algorithms Appl. **11**(2), 397–429 (2007)
27. Jünger, M., Mutzel, P. (eds.): Graph Drawing Software. Springer, Heidelberg (2003). doi:10.1007/978-3-642-18638-7
28. Kaufmann, M., Kobourov, S., Pach, J., Hong, S.: Beyond planar graphs: algorithmic and combinatorics. Dagstuhl Seminar 16452, 6–11 November 2016
29. Kaufmann, M., Ueckerdt, T.: The density of fan-planar graphs. CoRR, abs/1403.6184 (2014)
30. Kaufmann, M., Wagner, D. (eds.): Drawing Graphs. LNCS, vol. 2025. Springer, Heidelberg (2001). doi:10.1007/3-540-44969-8
31. Liotta, G.: Graph drawing beyond planarity: some results and open problems. In: ICTCS 2014. CEUR Workshop Proceedings, vol. 1231, pp. 3–8. CEUR-WS.org (2014)
32. Nishizeki, T., Chiba, N.: Planar Graphs: Theory and Algorithms. North-Holland Mathematics Studies. Elsevier Science (1988)
33. Pach, J., Radoičić, R., Tóth, G.: Relaxing planarity for topological graphs. In: Akiyama, J., Kano, M. (eds.) JCDCG 2002. LNCS, vol. 2866, pp. 221–232. Springer, Heidelberg (2003). doi:10.1007/978-3-540-44400-8_24
34. Pach, J., Radoičić, R., Tardos, G., Tóth, G.: Improving the crossing lemma by finding more crossings in sparse graphs. Discrete Comput. Geom. **36**(4), 527–552 (2006)
35. Pach, J., Shahrokhi, F., Szegedy, M.: Applications of the crossing number. Algorithmica **16**(1), 111–117 (1996)
36. Pach, J., Tóth, G.: Graphs drawn with few crossings per edge. Combinatorica **17**(3), 427–439 (1997)
37. Pach, J., Tóth, G.: Thirteen problems on crossing numbers. Geombinatorics **9**, 194–207 (2000)
38. Purchase, H.C.: Effective information visualisation: a study of graph drawing aesthetics and algorithms. Interact. Comput. **13**(2), 147–162 (2000)

39. Purchase, H.C., Carrington, D.A., Allder, J.: Empirical evaluation of aesthetics-based graph layout. Empirical Softw. Eng. **7**(3), 233–255 (2002)
40. Schaefer, M.: The graph crossing number and its variants: a survey. Electron. J. Combin. **DS21**, 1–100 (2014)
41. Sugiyama, K.: Graph Drawing and Applications for Software and Knowledge Engineers. World Scientific, Singapore (2002)
42. Suk, A., Walczak, B.: New bounds on the maximum number of edges in k-quasi-planar graphs. Comput. Geom. **50**, 24–33 (2015)
43. Tamassia, R. (ed.): Handbook of Graph Drawing and Visualization. Chapman and Hall/CRC, Boca Raton (2013)
44. Valtr, P.: Graph drawing with no k pairwise crossing edges. In: DiBattista, G. (ed.) GD 1997. LNCS, vol. 1353, pp. 205–218. Springer, Heidelberg (1997). doi:10.1007/3-540-63938-1_63
45. Vrťo, I.: Crossing numbers bibliography. www.ifi.savba.sk/~imrich
46. Ware, C., Purchase, H.C., Colpoys, L., McGill, M.: Cognitive measurements of graph aesthetics. Inf. Vis. **1**(2), 103–110 (2002)

Extension Complexity of Stable Set Polytopes of Bipartite Graphs

Manuel Aprile[1], Yuri Faenza[2], Samuel Fiorini[3], Tony Huynh[3(✉)], and Marco Macchia[3]

[1] École Polytechnique Fédérale de Lausanne (EPFL), Lausanne, Switzerland
manuel.aprile@epfl.ch
[2] IEOR Department, Columbia University, New York, USA
yf2414@columbia.edu
[3] Université Libre de Bruxelles, Brussels, Belgium
{sfiorini,mmacchia}@ulb.ac.be, tony.bourbaki@gmail.com

Abstract. The *extension complexity* $\mathsf{xc}(P)$ of a polytope P is the minimum number of facets of a polytope that affinely projects to P. Let G be a bipartite graph with n vertices, m edges, and no isolated vertices. Let $\mathsf{STAB}(G)$ be the convex hull of the stable sets of G. It is easy to see that $n \leqslant \mathsf{xc}(\mathsf{STAB}(G)) \leqslant n + m$. We improve both of these bounds. For the upper bound, we show that $\mathsf{xc}(\mathsf{STAB}(G))$ is $O(\frac{n^2}{\log n})$, which is an improvement when G has quadratically many edges. For the lower bound, we prove that $\mathsf{xc}(\mathsf{STAB}(G))$ is $\Omega(n \log n)$ when G is the incidence graph of a finite projective plane. We also provide examples of 3-regular bipartite graphs G such that the edge vs stable set matrix of G has a fooling set of size $|E(G)|$.

1 Introduction

A polytope $Q \subseteq \mathbb{R}^p$ is an *extension* of a polytope $P \subseteq \mathbb{R}^d$ if there exists an affine map $\pi : \mathbb{R}^p \to \mathbb{R}^d$ with $\pi(Q) = P$. The *extension complexity* $\mathsf{xc}(P)$ of P is the minimum number of facets of any extension of P. If Q is an extension of P such that Q has significantly fewer facets than P, then it is advantageous to run linear programming algorithms over Q instead of P.

One example of a polytope that admits a much more compact representation in a higher dimensional space is the spanning tree polytope, $\mathsf{P_{sp.trees}}(G)$. Edmonds' [5] classic description of $\mathsf{P_{sp.trees}}(G)$ has $2^{\Omega(|V|)}$ facets. However, Wong [16] and Martin [11] proved that for every connected graph $G = (V, E)$,

$$|E| \leqslant \mathsf{xc}(\mathsf{P_{sp.trees}}(G)) \leqslant O(|V| \cdot |E|).$$

Fiorini et al. [7] were the first to show that many polytopes arising from NP-hard problems (such as the stable set polytope) do indeed have high extension complexity. Their results answer an old question of Yannakakis [17] and do not rely on any complexity assumptions such as $\mathsf{P} \neq \mathsf{NP}$.

© Springer International Publishing AG 2017
H.L. Bodlaender and G.J. Woeginger (Eds.): WG 2017, LNCS 10520, pp. 75–87, 2017.
https://doi.org/10.1007/978-3-319-68705-6_6

On the other hand, Rothvoß [12] proved that the perfect matching polytope of the complete graph K_n has extension complexity at least $2^{\Omega(n)}$. This is somewhat surprising since the maximum weight matching problem can be solved in polynomial-time via Edmond's blossom algorithm [4]. By now many accessible introductions to extension complexity are available (see [1,2,9,13]).

Let $G = (V, E)$ be a (finite, simple) graph with $n := |V|$ and $m := |E|$. The *stable set polytope* of G, denoted $\mathsf{STAB}(G)$, is the convex hull of the characteristic vectors of stable sets of G. As previously mentioned, $\mathsf{STAB}(G)$ can have very high extension complexity. In [7], it is proved that if G is obtained from a complete graph by subdividing each edge twice, then $\mathsf{xc}(\mathsf{STAB}(G))$ is at least $2^{\Omega(\sqrt{n})}$. Very recently, Göös et al. [8] improved this to $2^{\Omega(n/\log n)}$, via a different class of graphs. For perfect graphs, Yannakakis [17] proved an upper bound of $n^{O(\log n)}$, and it is an open problem whether Yannakakis' upper bound can be improved to a polynomial bound.

In this paper we restrict our attention to bipartite graphs. Let $G = (V, E)$ be a bipartite graph with n vertices, m edges and no isolated vertices. By total unimodularity,

$$\mathsf{STAB}(G) = \{x \in \mathbb{R}^V \mid x_u \geqslant 0 \text{ for all } u \in V,\ x_u + x_v \leqslant 1 \text{ for all } uv \in E\},$$

and so $n \leqslant \mathsf{xc}(\mathsf{STAB}(G)) \leqslant n+m$. In this case $\mathsf{xc}(\mathsf{STAB}(G))$ lies in a very narrow range, and it is a good test of current methods to see if we can improve these bounds.

The situation is analogous to what happens with the spanning tree polytope of (arbitrary) graphs, where as previously mentioned, we also know that $\mathsf{xc}(P_{\mathsf{sp.trees}}(G))$ lies in a very narrow range. Indeed, a notorious problem of Goemans (see [10]) is to improve the known bounds for $\mathsf{xc}(P_{\mathsf{sp.trees}}(G))$, but this is still wide open.

However, for the stable set polytopes of bipartite graphs, we are able to give an improvement. Our main results are the following.

Theorem 1. *For all bipartite graphs G with n vertices, the extension complexity of $\mathsf{STAB}(G)$ is $O(n^2/\log n)$.*

Note that Theorem 1 is an improvement over the obvious upper bound when G has quadratically many edges.

Theorem 2. *There exists an infinite class $\mathcal{C}$ of bipartite graphs such that every n-vertex graph in $\mathcal{C}$ has extension complexity $\Omega(n \log n)$.*

These are the first known examples of stable set polytopes of bipartite graphs where the extension complexity is more than linear in the number of vertices. For instance, $\mathsf{xc}(\mathsf{STAB}(K_{n/2,n/2})) = \Theta(n)$. To the best of our knowledge, even for general perfect graphs G, the previous best lower bound for $\mathsf{xc}(\mathsf{STAB}(G))$ was the trivial bound $|V(G)|$.

Paper Organization. In Sect. 2 we define rectangle covers and fooling sets and we give examples of 3-regular graphs with tight fooling sets. We prove Theorem 1 in Sect. 3 and Theorem 2 in Sect. 4. In Sect. 5 we show that it is impossible to prove a better lower bound with the approach in Sect. 4. Thus, to further improve the lower bound, different methods (or different graphs) are required.

2   Rectangle Covers and Fooling Sets

Consider a polytope $P := \text{conv}(X) = \{x \in \mathbb{R}^d \mid Ax \geq b\}$, where $X := \{x^{(1)}, \ldots, x^{(n)}\} \subseteq \mathbb{R}^d$, $A \in \mathbb{R}^{m \times d}$ and $b \in \mathbb{R}^m$. The *slack matrix* of P (with respect to the chosen inner and outer descriptions of the polytope) is the matrix $S \in \mathbb{R}_{\geq 0}^{m \times n}$ having rows indexed by the inequalities $A_1 x \geq b_1, \ldots, A_m x \geq b_m$ and columns indexed by the points $x^{(1)}, \ldots, x^{(n)}$, defined as $S_{ij} := A_i x^{(j)} - b_i \geq 0$.

Yannakakis [17] proved that the extension complexity of P equals the non-negative rank of S. In this work, we only rely on a lower bound that follows directly from this fact. For a matrix M, we define the *support* of M as $\text{supp}(M) := \{(i,j) \mid M_{ij} \neq 0\}$. A *rectangle* is any set of the form $R = I \times J$, with $R \subseteq \text{supp}(M)$. A *size-k rectangle cover* of M is a collection $R_1, \ldots, R_k$ of rectangles such that $\text{supp}(M) = R_1 \cup \cdots \cup R_k$. The *rectangle covering bound* of M is the minimum size of a rectangle cover of M, and is denoted $\text{rc}(M)$.

Theorem 3. (Yannakakis, [17]). *Let P be a polytope with $\dim(P) \geq 1$ and let S be any slack matrix of P. Then, $\text{xc}(P) \geq \text{rc}(S)$.*

A *fooling set* for M is a set of entries $F \subseteq \text{supp}(M)$ such that $M_{i\ell} \cdot M_{kj} = 0$ for all distinct $(i,j), (k,\ell) \in F$. The largest size of a fooling set of M is denoted by $\text{fool}(M)$. Clearly, $\text{rc}(M) \geq \text{fool}(M)$.

Let G be a bipartite graph. The *edge vs stable set matrix* of G, denoted $M(G)$, is the 0/1 matrix with a row for each edge of G, a column for each stable set of G, and a 1 in position (e, S) if and only if $e \cap S = \varnothing$ (as usual, we regard edges as pairs of vertices). We say that G has a *tight fooling set* if $M(G)$ has a fooling set of size $|E(G)|$. Note that if G has a tight fooling set, then the non-negative rank of $M(G)$ is exactly $|E(G)|$. Also observe that the property of having a tight fooling set is closed under taking subgraphs.

It is easy to check that even cycles have tight fooling sets. We now give an infinite family of 3-regular graphs that have tight fooling sets. A graph is C_4-*free* if it does not contain a cycle of length four.

Theorem 4. *Let $G = (V, E)$ be a 3-regular, C_4-free bipartite graph. Then G has a tight fooling set.*

Proof. For $X \subseteq V$, we let $N(X)$ denote the set of neighbours of X. Let $V = A \cup B$ be a bipartition of the vertex set, and let $\phi : E \to \{1, 2, 3\}$ be a proper edge coloring of G, which exists by 3-regularity and König's edge-coloring theorem (see e.g. [14, Theorem 20.1]). For each vertex $a \in A$, we name its neighbors $a_1, a_2, a_3 \in B$ so that $\phi(aa_i) = i$. For each $a \in A$, consider the following stable sets:

$$S_{aa_1} := A \setminus \{a\}$$
$$S_{aa_2} := \{a_1\} \cup \{a' \in A \mid a' \notin N(a_1)\}$$
$$S_{aa_3} := B \setminus \{a_3\}.$$

This defines a stable set S_e disjoint from e, for every edge $e \in E$. Since ϕ is proper, no two of these stable sets are equal. We claim that $\{(e, S_e) \mid e \in E\}$ is a fooling set in the edge vs stable set matrix of G.

Let e and f be distinct edges. We want to show that S_e intersects f or S_f intersects e. Consider the following three cases. Let $e = aa_i$, where $i = \phi(e)$.

Case 1. If $\phi(e) = 1$, then $S_e = S_{aa_1}$ intersects f unless $f = aa_i$ for some $i \in \{2, 3\}$. In both cases we have $a_1 \in S_f \cap e$.

Case 2. If $\phi(e) = 3$, then $S_e = S_{aa_3}$ intersects f unless $f = a'a_3$ for some $a' \in A$. Either $\phi(f) = 1$ and S_f intersects e (as in Case 1), or $\phi(f) = 2$. In the last case, since G is C_4-free, we have $a \notin N(a_1')$. It follows that $S_f = S_{a'a_3} = S_{a'a_2'}$ intersects e.

Case 3. If $\phi(e) = 2$, then we may also assume $\phi(f) = 2$ since otherwise by exchanging the roles of e and f we are back to one of the previous cases. Let a' denote the endpoint of f in A, so that $f = a'a_2'$. Because ϕ is proper, $a' \neq a$ and $a_1' \neq a_1$. Since G is C_4-free, we have $a \notin N(a_1')$ or $a' \notin N(a_1)$. Hence, $a \in S_f \cap e$ or $a' \in S_e \cap f$. □

Note that there are infinitely many 3-regular, C_4-free bipartite graphs. For example, we can take a hexagonal grid on a torus.

3   An Improved Upper Bound

In this section we prove Theorem 1. We use the following result of Martin [11].

Lemma 5. *If Q is a nonempty polyhedron, $\gamma \in \mathbb{R}$, and*

$$P = \{x \mid \langle x, y \rangle \leq \gamma \text{ for every } y \in Q\},$$

then $\mathsf{xc}(P) \leq \mathsf{xc}(Q) + 1$.

The *edge polytope* $\mathsf{P}_{\mathsf{edge}}(G)$ of a graph G is the convex hull of the incidence vectors in $\mathbb{R}^{V(G)}$ of all edges of G. The second ingredient we need is the following bound on the extension complexity of the edge polytope of all n-vertex graphs due to Fiorini et al. [6, Lemma 3.4]. This bound follows from a nice result of Tuza [15], which states that every n-vertex graph can be covered with a set of bicliques of total weight $O(n^2/\log n)$, where the weight of a biclique is its number of vertices.

Lemma 6. *For every graph G with n vertices, $\mathsf{xc}(\mathsf{P}_{\mathsf{edge}}(G)) = O(n^2/\log n)$.*

Proof of Theorem 1. Let $G = (V, E)$. Since

$$\mathsf{STAB}(G) = \mathbb{R}_{\geq 0}^V \cap \{x \in \mathbb{R}^V \mid \langle x, y \rangle \leq 1 \text{ for every } y \in \mathsf{P}_{\mathsf{edge}}(G)\},$$

By Lemmas 5 and 6, the extension complexity of $\mathsf{STAB}(G)$ is $O(n^2/\log n)$. □

4   An Improved Lower Bound

In this section we prove Theorem 2. The examples we use to prove our lower bound are incidence graphs of finite projective planes. We will not use any theorems from projective geometry, but the interested reader can refer to [3].

Let q be a prime power, $\mathsf{GF}(q)$ be the field with q elements, and $\mathsf{PG}(2,q)$ be the projective plane over $\mathsf{GF}(q)$. The *incidence graph* of $\mathsf{PG}(2,q)$, denoted $\mathcal{I}(q)$, is the bipartite graph with bipartition $(\mathcal{P}, \mathcal{L})$, where $\mathcal{P}$ is the set of points of $\mathsf{PG}(2,q)$, $\mathcal{L}$ is the set of lines of $\mathsf{PG}(2,q)$, and $\mathsf{p} \in \mathcal{P}$ is adjacent to $\ell \in \mathcal{L}$ if and only if the point p lies on the line ℓ. For example, $\mathsf{PG}(2,2)$ and its incidence graph $\mathcal{I}(2)$ are depicted in Fig. 1.

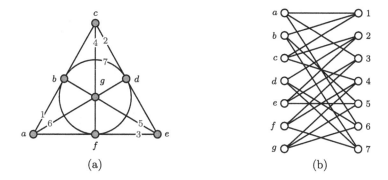

(a)   (b)

Fig. 1. $\mathsf{PG}(2,2)$ and its incidence graph $\mathcal{I}(2)$.

Before proving Theorem 2 we gather a few lemmas on binomial coefficients. The first two are well-known, so we omit the easy proofs.

Lemma 7. *For all integers h and c with $h \geqslant c \geqslant 0$*

$$\sum_{j=c}^{h} \binom{j}{c} = \binom{h+1}{c+1}.$$

Lemma 8. *For all positive integers $x, y,$ and h,*

$$\sum_{j=0}^{h} \binom{x+j}{j} \binom{h+y-j}{h-j} = \binom{x+y+h+1}{h}.$$

Lemma 9. *Let q, c, t be positive integers with $c + t \leqslant q + 1$. Then*

$$t \sum_{k=c}^{q+1-t} \frac{1}{k} \binom{q+1-t-c}{k-c} \binom{q}{k}^{-1} = \binom{t+c-1}{t}^{-1} \leqslant \frac{1}{c}.$$

Proof. We have that

$$
t \sum_{k=c}^{q+1-t} \frac{1}{k} \binom{q+1-t-c}{k-c} \binom{q}{k}^{-1}
$$

$$
= \frac{t(q+1-t-c)!}{q!} \sum_{k=c}^{q+1-t} \frac{(k-1)!(q-k)!}{(k-c)!(q+1-t-k)!}
$$

$$
= \frac{t(q+1-t-c)!}{q!}(c-1)!(t-1)! \sum_{k=c}^{q+1-t} \binom{k-1}{c-1}\binom{q-k}{t-1}.
$$

Moreover,

$$
\sum_{k=c}^{q+1-t} \binom{k-1}{c-1}\binom{q-k}{t-1} = \sum_{j=0}^{q+1-t-c} \binom{c-1+j}{c-1}\binom{q-c-j}{t-1}
$$

$$
[h = q+1-t-c, \; x = c-1, \; y = t-1] = \sum_{j=0}^{h} \binom{x+j}{j}\binom{h+y-j}{h-j}
$$

$$
[\text{by Lemma 8}] = \binom{x+y+h+1}{h}
$$

$$
= \binom{q}{q+1-t-c}.
$$

We conclude that

$$
t \sum_{k=c}^{q+1-t} \frac{1}{k} \binom{q+1-t-c}{k-c} \binom{q}{k}^{-1} = \frac{t(q+1-t-c)!}{q!} \frac{q!(c-1)!(t-1)!}{(q+1-t-c)!(t+c-1)!}
$$

$$
= \binom{t+c-1}{t}^{-1}.
$$

The number of t-subsets of a set of size $t + c - 1$ is at least c, since it includes all t-subsets containing a fixed set of size $t - 1$. Hence, $\binom{t+c-1}{t}^{-1} \leqslant \frac{1}{c}$.   □

From the definition of $\mathsf{PG}(2,q)$ it follows that that $\mathcal{I}(q)$ is $(q+1)$-regular, $|V(\mathcal{I}(q))| = 2(q^2 + q + 1)$, and $|E(\mathcal{I}(q))| = (q+1)(q^2 + q + 1)$. Let $n = q^2 + q + 1$ and note that $\mathcal{I}(q)$ has $2n$ vertices. We let $\mathcal{P}$ and $\mathcal{L}$ denote the set of points and lines of $\mathsf{PG}(2,q)$. We also use the fact that $\mathcal{I}(q)$ is C_4-free.

We denote the edge vs stable set incidence matrix of $\mathcal{I}(q)$ by S_q. Each 1-entry of S_q is of the form (e, S) where $e \in E$, $S \subseteq V$ is a stable set, and $e \cap S = \varnothing$. To prove Theorem 2 we will assign weights to the 1-entries of S_q in such a way that the total weight is at least $\Omega(n \log n)$, while the weight of every rectangle is at most 1. The only entries that will receive non-zero weight are what we call *special entries*, which we now define.

Definition 10. A 1-entry of S_q is *special* if it has the form $(e, S(X))$ where

- $e = \mathsf{p}\ell$ with $\mathsf{p} \in \mathcal{P}, \ell \in \mathcal{L}$,
- $X \subseteq N(\ell) \setminus \{\mathsf{p}\}$, X non-empty,
- $S(X) = X \cup (\mathcal{L} \setminus N(X))$.

We also need the following compact representation of maximal rectangles.

Definition 11. Let R be a maximal rectangle. Then R is determined by a pair $(\mathcal{P}_R, \mathcal{L}_R)$ with $\mathcal{P}_R \subseteq \mathcal{P}$, $\mathcal{L}_R \subseteq \mathcal{L}$, where the rows of R are all the edges between $\mathcal{P}_R$ and $\mathcal{L}_R$ and the columns of R are all the stable sets $S \subseteq V \setminus (\mathcal{P}_R \cup \mathcal{L}_R)$.

We are now ready to prove Theorem 2 in the following form.

Theorem 2. *Let q be a prime power and $n = q^2 + q + 1$. Then there exists a constant $c > 0$ such that*

$$\mathsf{xc}(\mathrm{STAB}(\mathcal{I}(q))) \geqslant cn \log n.$$

Proof. Let $n = q^2 + q + 1$. Let $V = \mathcal{P} \cup \mathcal{L}$ be the vertices of $\mathcal{I}(q)$, and E be the edges of $\mathcal{I}(q)$. To each special entry $(e, S(X))$ we assign the weight

$$w(e, S(X)) = \frac{1}{|X| \binom{q}{|X|}(q+1)}.$$

All other entries of S_q receive weight zero.

Claim 12. $w(S_q) := \sum_{(e,S)} w(e, S) \geqslant cn \log n$ *for some constant c.*

Subproof. We have that

$$\sum_{(e,S)} w(e, S) = \sum_{(e, S(X)) \text{ special}} w(e, S(X)) = \sum_{e \in E} \sum_{k=1}^{q} \binom{q}{k} \frac{1}{k \binom{q}{k}(q+1)}$$

$$= \frac{|E|}{q+1} \sum_{k=1}^{q} \frac{1}{k} = n \sum_{k=1}^{q} \frac{1}{k} > cn \log n.$$

The claim follows. ∎

Let $R = (\mathcal{P}_R, \mathcal{L}_R)$ be an arbitrary maximal rectangle. We finish the proof by showing that $w(R) := \sum_{(e,S) \in R} w(e, S) \leqslant 1$. Together with Claim 12 this clearly implies Theorem 2. We will need the following obvious but useful Claim.

Claim 13. *A special entry $(\mathsf{p}\ell, S(X))$ is covered by $R = (\mathcal{P}_R, \mathcal{L}_R)$ if and only if $X \cap \mathcal{P}_R = \varnothing$, $\mathcal{L}_R \subseteq N(X)$, $\mathsf{p} \in \mathcal{P}_R$, and $\ell \in \mathcal{L}_R$.*

We consider two cases. First suppose that $\mathcal{L}_R = \{\ell\}$ for some ℓ. Then the only special entries covered by R are of the form $(\mathsf{p}\ell, S(X))$, with $X \subseteq N(\ell) \setminus \mathcal{P}_R$. Let $N(\ell) \cap \mathcal{P}_R = \{\mathsf{p}_1, \ldots, \mathsf{p}_t\}$, where $1 \leqslant t \leqslant q + 1$. To compute $w(R)$ we have to sum over all edges $p_i\ell$ and over all subsets $X \subseteq N(\ell) \setminus \{p_1, \ldots, p_t\}$. It follows that

$$
\begin{aligned}
w(R) &= \sum_{i=1}^{t} \sum_{k=1}^{q+1-t} \binom{q+1-t}{k} \frac{1}{k\binom{q}{k}(q+1)} \\
&= t \sum_{k=1}^{q+1-t} \frac{(q+1-t)!}{k!(q+1-t-k)!} \frac{k!(q-k)!}{kq!(q+1)} \\
&= \frac{t(q+1-t)!(t-1)!}{(q+1)!} \sum_{k=1}^{q+1-t} \binom{q-k}{q+1-t-k} \frac{1}{k} \\
&= \frac{1}{\binom{q+1}{t}} \sum_{k=1}^{q+1-t} \binom{q-k}{t-1} \frac{1}{k} \leqslant \frac{1}{\binom{q+1}{t}} \sum_{j=t-1}^{q-1} \binom{j}{t-1} = \frac{1}{\binom{q+1}{t}} \binom{q}{t} \leqslant 1,
\end{aligned}
$$

where the last equality follows from Lemma 7.

The remaining case is if $|\mathcal{L}_R| \geqslant 2$. For $\ell \in \mathcal{L}_R$ such that $(p\ell, S(X))$ is covered by R for some p, X, define

$$k_\ell = \min\{|X| \mid \text{there exist } p, X : (p\ell, S(X)) \text{ is a special entry covered by } R\}.$$

Claim 14. *Let $(p\ell, S(X))$ be a special entry covered by R such that $|X| = k_\ell$. Then for each p', Y such that R covers $(p'\ell, S(Y))$, we have $X \subseteq Y$.*

Subproof. For each $\ell' \in \mathcal{L}_R \setminus \{\ell\}$ (there is at least one since $|\mathcal{L}_R| > 1$), we have $\ell' \in N(X)$ by Claim 13. That is, there is $x = x(\ell') \in X$ adjacent to ℓ'. Similarly, since $\ell' \in N(Y)$, there is $y = y(\ell') \in Y$ adjacent to ℓ'. Now, if $x(\ell') \neq y(\ell')$, then $\mathcal{I}(q)$ contains a 4-cycle, which is a contradiction. Hence we must have $x(\ell') = y(\ell')$ for all $\ell' \in \mathcal{L}_R \setminus \{\ell\}$. Now if there is an $x \in X$ such that $x \neq x(\ell')$ for every $\ell' \in \mathcal{L}_R \setminus \{\ell\}$, then $(p\ell, S(X \setminus \{x\}))$ is still covered by R, contradicting the minimality of X. We conclude $X \subseteq Y$, as required. ∎

Now fix $\ell \in \mathcal{L}_R$, and let

$$w(\ell) = \sum \{w(p\ell, S(X)) \mid (p\ell, S(X)) \text{ special}\}.$$

Claim 15. *For every $\ell \in \mathcal{L}_R$,*

$$w(\ell) \leqslant \frac{1}{(q+1)k_\ell}.$$

Subproof. Let $N(\ell) \cap \mathcal{P}_R = \{p_1, \ldots, p_t\}$, where $1 \leqslant t \leqslant q+1$. Let X be such that $(p\ell, S(X))$ is a special entry covered by R and $|X| = k_\ell$. By Claim 14, the only special entries appearing in the above sum are of the form $(p_i\ell, S(Y))$ where $i \in [t]$ and $X \subseteq Y \subseteq (\mathcal{P} \setminus \mathcal{P}_R) \cap N(\ell)$. Therefore

$$w(\ell) \leqslant t \sum_{k=k_\ell}^{q+1-t} \binom{q+1-t-k_\ell}{k-k_\ell} \frac{1}{k\binom{q}{k}(q+1)} \leqslant \frac{1}{(q+1)k_\ell},$$

where the last inequality follows from Lemma 9 with $c = k_\ell$. ∎

Claim 16. *For every $\ell \in \mathcal{L}_R$, $|\mathcal{L}_R| \leqslant (q+1)k_\ell$.*

Subproof. Again, let X be such that $(p\ell, S(X))$ is covered by R and assume that $|X| = k_\ell$. By Claim 13, we have $\mathcal{L}_R \subseteq N(X)$.

Hence $|\mathcal{L}_R| \leqslant |N(X)| \leqslant (q+1)|X| = (q+1)k_\ell$. ∎

By Claims 15 and 16, for every $\ell \in \mathcal{L}_R$, $w(\ell) \leqslant \frac{1}{|\mathcal{L}_R|}$. But clearly $w(R) = \sum_{\ell \in \mathcal{L}_R} w(\ell)$, and so $w(R) \leqslant 1$, as required. This completes the entire proof. □

5   A Small Rectangle Cover of the Special Entries

In this section we show that the submatrix of special entries considered in the previous section has a rectangle cover of size $O(n \log n)$. Combined with Theorem 2, this implies that a minimal set of rectangles that cover all the special entries always has size $\Theta(n \log n)$. Thus, to improve our bound, we must consider a different set of entries of the slack matrix, or use a different set of graphs.

This cover will be built from certain labeled trees which we now define. Note that the *length* of a path is its number of edges.

Definition 17. For every integer $k \geqslant 1$, we build a tree $T(k)$ recursively:

- The tree $T(1)$ consists of a root r and a single leaf attached to it.
- For $k > 1$, we construct $T(k)$ by first identifying one end of a path P_1 of length $k_1 := \lceil \frac{k}{2} \rceil$ to another end of a path P_2 of length $k_2 := \lfloor \frac{k}{2} \rfloor$ along a root vertex r. Let λ_i be the end of P_i that is not r. We then attach a copy of $T(k_i)$ to λ_{3-i}, identifying λ_{3-i} with the root of $T(k_i)$. We call P_1 and P_2 the *main* paths of $T(k)$.

The next Lemma follows easily by induction on k.

Lemma 18. *For all $k \geqslant 1$,*

(i) *$T(k)$ has $O(k \log k)$ vertices;*
(ii) *$T(k)$ has k leaves;*
(iii) *every path from the root r to a leaf has length k.*

Definition 19. We recursively define a labeling $\varphi_k : V(T(k)) \setminus \{r\} \to [k]$ as follows:

- Let v be the non-root vertex of $V(T(1))$ and set $\varphi_1(v) := 1$.
- For $k > 1$, let P_1 and P_2 be the main paths of $T(k)$. We name the vertices of P_1 as $r, v_1, \ldots, v_{\lceil \frac{k}{2} \rceil}$ and P_2 as $r, v_{\lceil \frac{k}{2} \rceil + 1}, \ldots, v_k$, where these vertices are listed according to their order along P_1 and P_2. Set $k_1 := \lceil \frac{k}{2} \rceil$ and $k_2 := \lfloor \frac{k}{2} \rfloor$. Note that $V(T(k)) = \bigcup_{i=1,2}(V(P_i) \cup V(B_i))$, where B_i is a copy of the tree $T(k_{3-i})$. We define (Fig. 2)

$$\varphi_k(v) = \begin{cases} i, & \text{if } v = v_i \\ \varphi_{k_2}(v) + k_1, & \text{if } v \in V(B_1) \setminus V(P_1) \\ \varphi_{k_1}(v), & \text{if } v \in V(B_2) \setminus V(P_2) \end{cases}$$

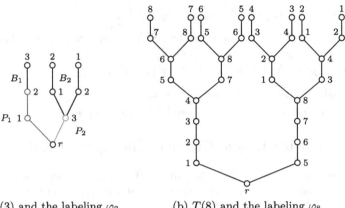

(a) $T(3)$ and the labeling φ_3 (b) $T(8)$ and the labeling φ_8

Fig. 2. Examples of the labeling function.

For each vertex $v \in T(k)$ we let $P(v)$ be the path in $T(k)$ from r to v.

Lemma 20. *Let φ_k, B_1, and B_2 be as in Definition 19.*

(i) *If L is the set of leaves of $T(k)$, then $\varphi_k(L \cap V(B_1)) = \{\lceil \frac{k}{2} \rceil + 1, \ldots, k\}$ and $\varphi_k(L \cap V(B_2)) = \{1, \ldots, \lceil \frac{k}{2} \rceil\}$.*

(ii) *For every leaf λ of $T(k)$, $\varphi_k(V(P(\lambda)) \setminus \{r\}) = [k]$.*

(iii) *Each label $i \in [k]$ occurs at most $\lceil \log k \rceil + 1$ times in the labeling of $T(k)$.*

Proof. We proceed by induction on k. Property (i) follows directly from the recursive definition of the labeling φ_k.

For (ii), let λ be a leaf and let the (ordered) vertices of $P(\lambda)$ be $r, p_1, \ldots, p_k = \lambda$. Suppose that $\lambda \in V(B_i)$. Then $P(\lambda) := P_i \cup P'$, where P_i is a main path of $T(k)$ and P' is the path in B_i going from the root of B_i to λ. Property (ii) now follows by induction and the definition of φ_k.

For (iii), first suppose that the label i is in $[k_1]$. Then i appears exactly once in the labeling of the main path P_1 of $T(k)$, it does not in the labeling of the nodes $V(P_2) \cup (V(B_1) \setminus V(P_1))$, and, by the inductive step, it occurs $\lceil \log \lceil \frac{k}{2} \rceil \rceil + 1 = \lceil \log k \rceil$ times in $\varphi_k(B_2)$. The thesis follows. A similar argument settles the remaining case $i \in [k] \setminus [k_1]$.   □

Henceforth, we simplify notation and denote the labeling φ_k of $T(k)$ as φ. We now recall some notation from the previous section. Let q be a prime power and S_q be the edge vs stable set incidence matrix of $\mathcal{I}(q)$.

A maximal rectangle $R = (\mathcal{P}_R, \mathcal{L}_R)$ is *centered* if $|\mathcal{L}_R| \geqslant 2$ and there is a point $\mathsf{c} \in \mathcal{P} \setminus \mathcal{P}_R$ such that c is incident to all lines in $\mathcal{L}_R$. We call c the *center* of R. Note that the center is unique and its existence implies that $|\mathcal{L}_R| \leqslant q + 1$.

One way to create centered rectangles is as follows. Let ℓ be a line, c be a point on ℓ, and $Y \subseteq N(\ell)$ with $\mathsf{c} \in Y$. We let $\boxed{\mathsf{c}, \ell, Y}$ be the centered rectangle $R = (\mathcal{P}_R, \mathcal{L}_R)$ where $\mathcal{P}_R = N(\ell) \setminus Y$ and $\mathcal{L}_R = N(\mathsf{c})$. Note that a special entry of the form $(\mathsf{p}\ell, S(X))$ is covered by the centered rectangle $\boxed{\mathsf{c}, \ell, Y}$ if and only if $\mathsf{p} \notin Y$ and $\mathsf{c} \in X \subseteq Y$.

We now fix a line $\ell \in \mathsf{PG}(2, q)$ and let $N(\ell) = \{\mathsf{p}_1, \ldots, \mathsf{p}_{q+1}\}$. We will use the labeling φ of $T(q+1)$ to provide a collection of centered rectangles that cover all special entries of the form $(\mathsf{p}\ell, S(X))$. Recall that for a vertex v of $T(q+1)$, $P(v)$ denotes the path in $T(q+1)$ from r to v. If v is neither the root nor a leaf of $T(q+1)$, we define

$$Y(v) := \{\mathsf{p}_{\varphi(u)} \mid u \text{ is a non-root vertex of } P(v)\}.$$

Lemma 21. *Fix a line $\ell \in \mathsf{PG}(2, q)$ and let $N(\ell) = \{\mathsf{p}_1, \ldots, \mathsf{p}_{q+1}\}$. Let $\mathcal{R}_\ell$ be the collection of all centered rectangles $\boxed{\mathsf{p}_{\varphi(v)}, \ell, Y(v)}$ where v ranges over all non-root, non-leaf vertices of $T(q+1)$. Then every special entry (e, S) with ℓ incident to e is covered by some rectangle $R \in \mathcal{R}_\ell$.*

Proof. Let $(\mathsf{p}_i\ell, S(X))$ be such a special entry and let λ be the (unique) leaf of $T(q+1)$ such that $\varphi(\lambda) = i$. Name the vertices of $P(\lambda)$ as $r, u_1, \ldots, u_{q+1} = \lambda$ (ordered away from the root).

Define $j = \max\{i \mid \mathsf{p}_{\varphi(u_i)} \in X\}$. Since $\mathsf{p}_{\varphi(\lambda)} \notin X$, note $j < q + 1$. By Lemma 20, $X \subseteq Y(u_j)$. Also, by construction, $\mathsf{p}_{\varphi(u_j)} \in X$ and $\mathsf{p} \notin Y(u_j)$. We conclude that the centered rectangle $\boxed{\mathsf{p}_{\varphi(u_j)}, \ell, Y(u_j)}$ covers the special entry $(\mathsf{p}_i\ell, S(X))$, as required. $\qquad\square$

By Lemma 21, for each line ℓ, there is a set $\mathcal{R}_\ell$ of $O(q \log q)$ centered rectangles that cover all special entries of the form $(\mathsf{p}\ell, S(X))$. By taking the union of all $\mathcal{R}_\ell$, we get a cover $\mathcal{R}$ of size $O(nq \log q)$ for all the special entries. To prove the main theorem of this section, we now reduce the size of $\mathcal{R}$ by a factor of q.

Theorem 22. *There is a set of $O(n \log n)$ centered rectangles that cover all the special entries.*

Proof. If $R_1 := \boxed{\mathsf{c}, \ell_1, Y_1}, \ldots, R_k := \boxed{\mathsf{c}, \ell_k, Y_k}$ are centered rectangles with the same center c, we let $\sum_{i=1}^{k} R_i = R$ be the maximal rectangle with $\mathcal{P}_R = \bigcup_{i=1}^{k} N(\ell_i) \setminus \bigcup_{i=1}^{k} Y_i$ and $\mathcal{L}_R = N(\mathsf{c})$. Note that $\sum_{i=1}^{k} R_i$ is also a centered rectangle with center c.

Claim 23. *If $R_1 := \boxed{\mathsf{c}, \ell_1, Y_1}, \ldots, R_k := \boxed{\mathsf{c}, \ell_k, Y_k}$ are centered rectangles such that $\ell_1, \ldots, \ell_k$ are all distinct, then $\sum_{i=1}^{k} R_i$ covers all special entries covered by $\bigcup_{i=1}^{k} R_i$.*

Subproof. Let $(\mathsf{p}\ell, S(X))$ be a special entry covered by some $\boxed{\mathsf{c}, \ell_j, Y_j}$. Clearly $\mathsf{c} \in X \subseteq Y_j \subseteq \bigcup_{i=1}^{k} Y_i$. By contradiction, suppose $\mathsf{p} \in \bigcup_{i=1}^{k} Y_i$. Since $\mathsf{p} \notin Y_j$, $\mathsf{p} \in Y_{j'} \subseteq N(\ell_{j'})$ for some $j' \neq j$. But then $\mathsf{c}\ell_j\mathsf{p}\ell_{j'}$ is a 4-cycle in $\mathcal{I}(q)$, which is a contradiction. Hence the entry $(\mathsf{p}\ell, S(X))$ is also covered by $\sum_{i=1}^{k} R_i$. $\qquad\blacksquare$

We iteratively use Claim 23 to reduce the number of rectangles in our covering $\mathcal{R}$. For each point c, name the $q+1$ lines through c as $\ell, \ell_1, \ldots, \ell_q$, so that among $\mathcal{R}_\ell, \mathcal{R}_{\ell_1}, \ldots, \mathcal{R}_{\ell_q}$, the collection $\mathcal{R}_\ell$ has the most rectangles with center c. Note that, by Lemma 20, $\mathcal{R}_\ell$ contains $O(\log q)$ rectangles with center c.

Fix $i \in [q]$ and for each rectangle $R \in \mathcal{R}_{\ell_i}$ with center c choose a rectangle $f_i(R)$ with center c in $\mathcal{R}_\ell$ such that $f_i(R) \neq f_i(R')$ if $R \neq R'$. For each $R \in \mathcal{R}_\ell$ we let

$$f^{-1}(R) = \{R\} \cup \bigcup_{i=1}^q \{R' \in \mathcal{R}_{\ell_i} \mid f_i(R') = R\}.$$

We then remove all rectangles with center c that appear in $\mathcal{R}_\ell, \mathcal{R}_{\ell_1}, \ldots, \mathcal{R}_{\ell_q}$ and replace them with all rectangles of the form $\sum_{R' \in f^{-1}(R)} R'$, where R ranges over all rectangles in $\mathcal{R}_\ell$ with center c. In doing so, we obtain at most $O(\log q) = O(\log n)$ rectangles with center c. Repeating for every $\mathsf{c} \in \mathcal{P}$ gives us $O(n \log n)$ rectangles in total. □

Acknowledgement. We thank Monique Laurent and Ronald de Wolf for bringing the topic of this paper to our attention. We also acknowledge support from ERC grant *FOREFRONT* (grant agreement no. 615640) funded by the European Research Council under the EU's 7th Framework Programme (FP7/2007-2013) and *Ambizione* grant PZ00P2 154779 *Tight formulations of 0–1 problems* funded by the Swiss National Science Foundation. Finally, we also thank the five anonymous referees for their constructive comments.

References

1. Conforti, M., Cornuéjols, G., Zambelli, G.: Extended formulations in combinatorial optimization. Ann. Oper. Res. **204**, 97–143 (2013)
2. Conforti, M., Cornuéjols, G., Zambelli, G.: Integer programming. Graduate Texts in Mathematics, vol. 271. Springer, Cham (2014)
3. Coxeter, H.S.M.: Projective Geometry, Revised reprint of the 2 edn. Springer, New York (1994)
4. Edmonds, J.: Paths, trees, and flowers. Canad. J. Math. **17**, 449–467 (1965)
5. Edmonds, J.: Matroids and the greedy algorithm. Math. Program. **1**, 127–136 (1971)
6. Fiorini, S., Kaibel, V., Pashkovich, K., Theis, D.O.: Combinatorial bounds on nonnegative rank and extended formulations. Discrete Math. **313**(1), 67–83 (2013)
7. Fiorini, S., Massar, S., Pokutta, S., Tiwary, H.R., Wolf, R.D.: Exponential lower bounds for polytopes in combinatorial optimization. J. ACM **62**(2), 17–23 (2015)
8. Göös, M., Jain, R., Watson, T.: Extension complexity of independent set polytopes. In: 2016 IEEE 57th Annual Symposium on Foundations of Computer Science (FOCS), pp. 565–572. IEEE (2016)
9. Kaibel, V.: Extended formulations in combinatorial optimization. arXiv preprint arXiv:1104.1023 (2011)
10. Khoshkhah, K., Theis, D.O.: Fooling sets and the spanning tree polytope. arXiv preprint arXiv:1701.00350 (2017)
11. Martin, R.K.: Using separation algorithms to generate mixed integer model reformulations. Oper. Res. Lett. **10**(3), 119–128 (1991)

12. Rothvoß, T.: The matching polytope has exponential extension complexity. In: STOC 2014–Proceedings of 2014 ACM Symposium on Theory of Computing, pp. 263–272. ACM, New York (2014)
13. Roughgarden, T.: Communication complexity (for algorithm designers). arXiv preprint arXiv:1509.06257 (2015)
14. Schrijver, A.: Combinatorial Optimization. Polyhedra and Efficiency. Springer, Heidelberg (2003)
15. Tuza, Z.: Covering of graphs by complete bipartite subgraphs: complexity of 0-1 matrices. Combinatorica **4**(1), 111–116 (1984)
16. Wong, R.T.: Integer programming formulations of the traveling salesman problem. In: Proceedings of 1980 IEEE International Conference on Circuits and Computers, pp. 149–152 (1980)
17. Yannakakis, M.: Expressing combinatorial optimization problems by linear programs. J. Comput. System Sci. **43**(3), 441–466 (1991)

On the Number of Labeled Graphs
of Bounded Treewidth

Julien Baste[1], Marc Noy[2], and Ignasi Sau[1,3(✉)]

[1] CNRS, LIRMM, Université de Montpellier, Montpellier, France
{baste,sau}@lirmm.fr
[2] Department of Mathematics, Barcelona Graduate School of Mathematics,
Universitat Politècnica de Catalunya, Barcelona, Catalonia
marc.noy@upc.edu
[3] Departamento de Matemática, Universidade Federal do Ceará, Fortaleza, Brazil

Abstract. Let $T_{n,k}$ be the number of labeled graphs on n vertices and treewidth at most k (equivalently, the number of labeled partial k-trees). We show that

$$\left(c \frac{k\, 2^k n}{\log k} \right)^n 2^{-\frac{k(k+3)}{2}} k^{-2k-2} \leqslant T_{n,k} \leqslant \left(k\, 2^k n \right)^n 2^{-\frac{k(k+1)}{2}} k^{-k},$$

for $k > 1$ and some explicit absolute constant $c > 0$. Disregarding lower-order terms, the gap between the lower and upper bound is of order $(\log k)^n$. The upper bound is a direct consequence of the well-known formula for the number of labeled k-trees, while the lower bound is obtained from an explicit construction. It follows from this construction that both bounds also apply to graphs of pathwidth and proper-pathwidth at most k.

Keywords: Treewidth · Partial k-trees · Enumeration · Pathwidth · Proper-pathwidth

1 Introduction

Given an integer $k > 0$, a *k-tree* is a graph that can be constructed starting from a $(k+1)$-clique and iteratively adding a vertex connected to k vertices that form a clique. They are natural extensions of trees, which correspond to 1-trees. A formula for the number of labeled k-trees on n vertices was first found by Beineke and Pippert [1], and alternative proofs were given by Moon [18] and Foata [8].

Theorem 1. *The number of n-vertex labeled k-trees is equal to*

$$\binom{n}{k}(kn - k^2 + 1)^{n-k-2}. \tag{1}$$

The second author was partially supported by MTM2014-54745-P, and the other two by the DE-MO-GRAPH grant ANR-16-CE40-0028.

H.L. Bodlaender and G.J. Woeginger (Eds.): WG 2017, LNCS 10520, pp. 88–99, 2017.
https://doi.org/10.1007/978-3-319-68705-6_7

A *partial k-tree* is a subgraph of a k-tree. For integers n, k with $0 < k+1 \leqslant n$, let $T_{n,k}$ denote the number of n-vertex labeled partial k-trees. While the number of n-vertex labeled k-trees is given by Theorem 1, it appears that very little is known about $T_{n,k}$. Indeed, to the best of our knowledge, only the cases $k = 1$ (forests) and $k = 2$ (series-parallel graphs) have been studied. The number of n-vertex labeled forests is asymptotically $T_{n,1} \sim \sqrt{e}n^{n-2}$ [21], and the number of n-vertex labeled series-parallel graphs is asymptotically $T_{n,2} \sim g \cdot n^{-5/2}\gamma^n n!$ for some explicit constants g and $\gamma \approx 9.07$ [2].

Partial k-trees are exactly the graphs of *treewidth* at most k. Let us recall the definition of treewidth. A *tree-decomposition* of width k of a graph $G = (V, E)$ is a pair $(T, \mathcal{B})$, where T is a tree and $\mathcal{B} = \{B_t \mid B_t \subseteq V, t \in V(T)\}$ such that:

1. $\bigcup_{t \in V(T)} B_t = V$.
2. For every edge $\{u, v\} \in E$ there is a $t \in V(T)$ such that $\{u, v\} \subseteq B_t$.
3. $B_i \cap B_\ell \subseteq B_j$ for all $\{i, j, \ell\} \subseteq V(T)$ such that j lies on the unique path from i to ℓ in T.
4. $\max_{t \in V(T)} |B_t| = k + 1$.

The sets of $\mathcal{B}$ are called *bags*. The *treewidth* of G, denoted by $\mathbf{tw}(G)$, is the smallest integer k such that there exists a tree-decomposition of G of width k. If T is a path, then $(T, \mathcal{B})$ is also called a *path-decomposition*. The *pathwidth* of G, denoted by $\mathbf{pw}(G)$, is the smallest integer k such that there exists a path-decomposition of G of width k.

The following lemma is well-known and a proof can be found, for instance, in [16].

Lemma 1. *A graph has treewidth at most k if and only if it is a partial k-tree.*

In this article we are interested in counting n-vertex labeled graphs that have treewidth at most k. By Lemma 1, this number is equal to $T_{n,k}$, and actually our approach relies heavily on the definition of partial k-trees.

An easy upper bound on $T_{n,k}$ is obtained as follows. Since every partial k-tree is a subgraph of a k-tree, and a k-tree has exactly $kn - k(k + 1)/2$ edges, Theorem 1 gives

$$T_{n,k} \leqslant 2^{kn - \frac{k(k+1)}{2}} \binom{n}{k} (kn - k^2 + 1)^{n-k-2}. \tag{2}$$

Simple calculations yield, disregarding lower-order terms, that

$$T_{n,k} \leqslant (k2^k n)^n 2^{-\frac{k(k+1)}{2}} k^{-k} \leqslant (k2^k n)^n. \tag{3}$$

We can provide a lower bound with the following construction. Starting from an $(n - k + 1)$-vertex forest, we add $k - 1$ *apices*, that is, $k - 1$ vertices with an arbitrary neighborhood in the forest. Every graph created in this way has exactly n vertices and is of treewidth at most k, since adding an apex increases treewidth by at most one. The number of labeled forests on $n - k + 1$ vertices is at least the number of trees on $n - k + 1$ vertices, which is well-known to be $(n - k + 1)^{n-k-1}$. Since each apex can be connected to the ground forest in 2^{n-k+1} different ways, we obtain

$$T_{n,k} \geqslant (n - k + 1)^{n-k-1} 2^{(k-1)(n-k+1)}. \tag{4}$$

If we assume that n/k tends to infinity then asymptotically

$$T_{n,k} \geqslant \left(2^{k-1} n\right)^{n-o(1)}. \tag{5}$$

We conclude that $T_{n,k}$ is essentially between $(2^k n)^n$ and $(k 2^k n)^n$. These bounds differ by a factor k^n. For constant k this does not matter much since (except when $k = 1, 2$) we do not have a precise estimate on $T_{n,k}$. However, when k goes to infinity, the gap k^n is quite significant. Our main result considerably reduces the gap.

Theorem 2. *For integers n, k with $k \geq 2$ and $k + 1 \leqslant n$, the number $T_{n,k}$ of n-vertex labeled graphs with treewidth at most k satisfies*

$$T_{n,k} \geqslant \left(\frac{1}{128e} \cdot \frac{k 2^k n}{\log k}\right)^n 2^{-\frac{k(k+3)}{2}} k^{-2k-2}. \tag{6}$$

It follows that $T_{n,k}$ is asymptotically between $\left(\frac{k}{\log k} 2^k n\right)^n$ and $(k 2^k n)^n$ when n and k grow. Thus the gap is now of order $(\log k)^n$ instead of k^n.

In order to prove Theorem 2, we present in Sect. 2 an algorithmic construction of a family of n-vertex labeled partial k-trees, which is inspired by the definition of k-trees. When exhibiting such a construction toward a lower bound, one has to play with the trade-off of, on the one hand, constructing as many graphs as possible and, on the other hand, being able to bound the number of duplicates; we perform this analysis in Sect. 3. Namely, we first count the number of elements created by the construction, and then we bound the number of times that the same element may have been created. We conclude in Sect. 4 with some remarks and a discussion of further research.

2   The Construction

Let n and k be fixed positive integers with $0 < k \leqslant n - 1$. In this section we construct a set $\mathcal{R}_{n,k}$ of n-vertex labeled partial k-trees. We let $R_{n,k} = |\mathcal{R}_{n,k}|$. In Sect. 2.1 we introduce some notation and definitions used in the construction, in Sect. 2.2 we describe the construction, and in Sect. 2.3 we prove that the treewidth of the graphs generated this way is indeed at most k. In fact, we prove a stronger property, namely that the graphs we construct have *proper-pathwidth* at most k, where the proper-pathwidth, defined later, is a graph invariant that is at least the pathwidth, which is at least the treewidth.

2.1 Notation and Definitions

For the construction, we use a *labeling function* σ defined by a permutation of $\{1, \ldots, n\}$ with the constraint that $\sigma(1) = 1$. Inspired by the definition of k-trees, we will introduce vertices $\{v_1, v_2, \ldots, v_n\}$ one by one following the order $\sigma(1), \sigma(2), \ldots, \sigma(n)$ given by σ. If $i, j \in \{1, \ldots, n\}$, then i is called the *index* of $v_{\sigma(i)}$, the vertex $v_{\sigma(i)}$ is the i-th introduced vertex and, if $j < i$, the vertex $v_{\sigma(j)}$ is said to be *to the left* of $v_{\sigma(i)}$.

In order to build explicitly a class of partial k-trees, for every $i \geqslant k + 1$ we define:

1. A set $A_i \subseteq \{j \mid j < i\}$ of *active* vertices, corresponding to the clique to which a new vertex can be connected in the definition of k-trees, such that $|A_i| = k$.
2. A vertex $a_i \in A_i$, called the *anchor*, whose role will be described in the next paragraph.
3. An element $f(i) \in A_i$, called the *frozen* vertex, which corresponds to a vertex that will not be active anymore.
4. A set $N(i) \subseteq A_i$, which corresponds to the indices of the neighbors of $v_{\sigma(i)}$ to the left.

The construction works with *blocks* of size s, for some integer s depending of n and k, to be specified later. Namely, we insert the vertices by consecutive blocks of size s, with the property that all vertices of the same block share the same anchor and are adjacent to it.

In the description of the construction, we use the term *choose* for the elements for which there are several choices, which will allow us to provide a lower bound on the number of elements in $\mathcal{R}_{n,k}$. This will be the case for the functions σ, f, and N. As will become clear later (see Sect. 3), once σ, f, and N are fixed, all the other elements of the construction are uniquely defined.

For every index $i \geqslant k + 2$, we impose that

$$|N(i)| \geqslant \frac{k+1}{2},$$

in order to have simultaneously enough choices for $N(i)$ and enough choices for the frozen vertex $f(i)$, which will be chosen among the vertices in $N(i-1)$. On the other hand, as will become clear later, the role of the anchor vertices is to determine uniquely the vertices belonging to "its" block. To this end, when a new block starts, its anchor is defined as the smallest currently active vertex.

2.2 Description of the Construction

We say that a triple (σ, f, N), with σ a permutation of $\{1, \ldots, n\}$, $f : \{k+2, \ldots, n\} \rightarrow \{1, \ldots, n\}$, and $N : \{2, \ldots, n\} \rightarrow 2^{\{1, \ldots, n\}}$, is *constructible* if it is one of the possible outputs of the following algorithm:

Choose σ, a permutation of $\{1,\dots,n\}$ such that $\sigma(1) = 1$.
for $i=2$ **to** k **do**
   ⌊ Choose $N(i) \subseteq \{j \mid j < i\}$, such that $1 \in N(i)$.
for $i=k+1$ **do**
   Define $A_{k+1} = \{j \mid j < k+1\}$.
   Define $a_{k+1} = 1$.
   ⌊ Choose $N(k+1) \subseteq \{j \mid j < i\}$, such that $1 \in N(k+1)$.
for $i=k+2$ **to** n **do**
   if $i \equiv k+2 \pmod{s}$ **then**
      Define $f(i) = a_{i-1}$.
      Define $A_i = (A_{i-1} \setminus \{f(i)\}) \cup \{i-1\}$.
      Define $a_i = \min A_i$.
      Choose $N(i) \subseteq A_i$ such that $a_i \in N(i)$ and $|N(i)| \geqslant \frac{k+1}{2}$; cf. Fig. 1.
   else
      Choose $f(i) \in (A_{i-1} \setminus \{a_{i-1}\}) \cap N(i-1)$.
      Define $A_i = (A_{i-1} \setminus \{f(i)\}) \cup \{i-1\}$.
      Define $a_i = a_{i-1}$.
      ⌊ Choose $N(i) \subseteq A_i$ such that $a_i \in N(i)$ and $|N(i)| \geqslant \frac{k+1}{2}$; cf. Fig. 2.

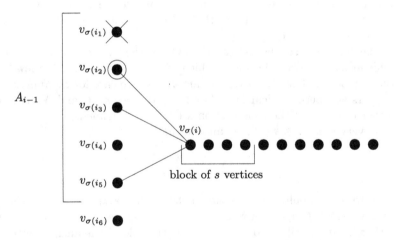

Fig. 1. Introduction of $v_{\sigma(i)}$ with $k+2 \leqslant i \leqslant n$ and $i \equiv k+2 \pmod{s}$, $s = 4$, and $k = 5$. We assume that $i_1 < i_2 < i_3 < i_4 < i_5 < i_6 < i$, and note that $i_5 = i-2$ and $i_6 = i-1$. We have defined $f(i) = v_{\sigma(i_1)}$ and $a_i = v_{\sigma(i_2)}$. The frozen vertex $f(i)$ is marked with a cross, and the anchor a_i is marked with a circle. We choose $N(i) = \{i_2, i_3, i_5\}$.

Let (σ, f, N) be a constructible triple. We define the graph $G(\sigma, f, N) = (V, E)$ such that $V = \{v_i \mid i \in \{1,\dots,n\}\}$, and $E = \{\{v_{\sigma(i)}, v_{\sigma(j)}\} \mid j \in N(i)\}$. Note that, given (σ, f, N), the graph $G(\sigma, f, N)$ is well-defined. We denote by $\mathcal{R}_{n,k}$ the set of all graphs $G(\sigma, f, N)$ such that (σ, f, N) is constructible.

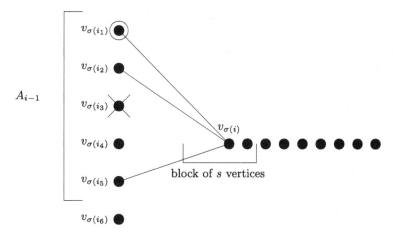

Fig. 2. Introduction of $v_{\sigma(i)}$ with $k+2 \leqslant i \leqslant n$ and $i \not\equiv k+2 \pmod{s}$, $s = 4$, and $k = 5$. We assume that $i_1 < i_2 < i_3 < i_4 < i_5 < i_6 < i$, and note that $i_5 = i - 2$ and $i_6 = i - 1$. We have defined $a_i = a_{i-1} = v_{\sigma(i_1)}$. The frozen vertex $f(i)$ is marked with a cross, and the anchor a_i is marked with a circle. We choose $f(i) = v_{\sigma(i_3)}$, assuming $v_{\sigma(i_3)}$ is a neighbor of $v_{\sigma(i_5)}$, and $N(i) = \{i_1, i_2, i_5\}$.

2.3 Bounding the Treewidth

We start by defining the notion of proper-pathwidth of a graph. This parameter was introduced by Takahashi *et al.* [22] and its relation with search games has been studied in [23].

Let G be a graph and let $\mathcal{X} = \{X_1, X_2, \ldots, X_r\}$ be a sequence of subsets of $V(G)$. The *width* of $\mathcal{X}$ is $\max_{1 \leqslant i \leqslant r} |X_i| - 1$. $\mathcal{X}$ is called a *proper-path decomposition* of G if the following conditions are satisfied:

1. For any distinct i and j, $X_i \nsubseteq X_j$.
2. $\bigcup_{i=1}^{r} X_i = V(G)$.
3. For every edge $\{u, v\} \in E(G)$, there exists an i such that $u, v \in X_i$.
4. For all a, b, and c with $1 \leqslant a \leqslant b \leqslant c \leqslant r$, $X_a \cap X_c \subseteq X_b$.
5. For all a, b, and c with $1 \leqslant a < b < c \leqslant r$, $|X_a \cap X_c| \leqslant |X_b| - 2$.

The *proper-pathwidth* of G, denoted by $\mathbf{ppw}(G)$, is the minimum width over all proper-path decompositions of G. Note that if $\mathcal{X}$ satisfies only conditions 1-4 above, then $\mathcal{X}$ is a path-decomposition as defined in Sect. 1.

From the definitions, for any graph G it clearly holds that

$$\mathbf{ppw}(G) \geqslant \mathbf{pw}(G) \geqslant \mathbf{tw}(G). \tag{7}$$

Let us show that any element of $\mathcal{R}_{n,k}$ has proper-pathwidth at most k. Let (σ, f, N) be constructible such that $G(\sigma, f, N) \in \mathcal{R}_{n,k}$ and let A_i be defined as in Subsect. 2.2. We define for every $i \in \{k+1, \ldots, n\}$ the bag $X_i = \{v_{\sigma(j)} \mid j \in A_i \cup \{i\}\}$. The sequence $\mathcal{X} = \{X_{k+1}, X_{k+2}, \ldots, X_n\}$ is a path-decomposition satisfying the five conditions of the above definition, and for every $i \in \{k + 1, \ldots, n\}$, $|X_i| = k + 1$. It follows that $G(\sigma, f, N)$ has proper-pathwidth at most k, so it also has treewidth at most k, and therefore $G(\sigma, f, N)$ is a partial k-tree by Lemma 1.

3   Proof of the Main Result

In this section we analyze our construction and give a lower bound on $R_{n,k}$. We first start by counting the number of constructible triples (σ, f, N) generated by the algorithm, and then we provide an upper bound on the number of duplicates. Finally, we determine the best choice for the parameter s defined in the construction.

3.1   Number of Constructible Triples (σ, F, N)

We proceed to count the number of constructible triples (σ, f, N) created by the algorithm given in Subsect. 2.2. As σ is a permutation of $\{1, \ldots, n\}$ with the constraint that $\sigma(1) = 1$, there are $(n - 1)!$ distinct possibilities for the choice of σ. The function f can take more than one value only for $k + 2 \leqslant i \leqslant n$ and $i \not\equiv k + 2 \pmod{s}$. This represents $n - (k + 1) - \lceil \frac{n - (k+1)}{s} \rceil$ cases. In each of these cases, there are at least $\frac{k-1}{2}$ distinct possible values for $f(i)$. Thus, we have at least $(\frac{k-1}{2})^{(n-(k+1)-\lceil \frac{n-(k+1)}{s} \rceil)}$ distinct possibilities for the choice of f. For every $i \in \{2, \ldots, k + 1\}$, $N(i)$ can be chosen as any subset of $i - 1$ vertices containing the fixed vertex $v_{\sigma(1)}$. This yields $\prod_{i=2}^{k+1} 2^{i-2} = 2^{\frac{k(k-1)}{2}}$ ways to define N over $\{2, \ldots, k + 1\}$. For $i \geqslant k + 2$, $N(i)$ can be chosen as any subset of size at least $\frac{k+1}{2}$ of a set of k elements with one element that is imposed. This results in $\sum_{i=\lceil \frac{k+1}{2} \rceil}^{k} \binom{k-1}{i-1} \geqslant 2^{k-2}$ possible choices for $N(i)$. Thus, we have $2^{\frac{k(k+1)}{2}} 2^{(n-(k+1))(k-2)}$ distinct possibilities to construct N.

By combining everything, we obtain at least

$$(n - 1)! \left(\frac{k-1}{2} \right)^{n - (k+1) - \lceil \frac{n-(k+1)}{s} \rceil} 2^{\frac{k(k-1)}{2}} 2^{(n-(k+1))(k-2)} \tag{8}$$

distinct possible constructible triples (σ, f, N).

3.2   Bounding the Number of Duplicates

Let H be an element of $\mathcal{R}_{n,k}$. Our objective is to obtain an upper bound on the number of constructible triples (σ, f, N) such that $H = G(\sigma, f, N)$.

Given H, we start by reconstructing σ. Firstly, we know by construction that $\sigma(1) = 1$. Secondly, we know that $f(k + 2) = 1$ and so, for every $i > k + 1$, $1 \notin A_i$, implying that $1 \notin N(i)$. It follows that the only neighbors of $v_{\sigma(1)}$ are the vertices $\{v_{\sigma(i)} \mid 1 < i \leqslant k + 1\}$. So the set of images under σ of $\{2, \ldots, k + 1\}$

is uniquely determined. Then we guess the function σ over this set $\{2, \ldots, k+1\}$. Overall, this results in $k!$ possible guesses for σ.

Thirdly, assume that we have correctly guessed σ on $\{1, \ldots, k+1+ps\}$ for some non-negative integer p with $k+1+ps < n$. Then $a_{k+1+ps+1}$ is the smallest active vertex that is adjacent to at least one element that is still not introduced after step $k+1+ps$. Then the neighbors of $a_{k+1+ps+1}$ over the elements that are not introduced yet after step $k+1+ps$ are the elements whose indices are between $k+1+ps+1$ and $k+1+(p+1)s$, and these vertices constitute the next block of the construction; see Fig. 3 for an illustration. As before, the set of images by σ of $\{k+1+ps+1, \ldots, k+1+(p+1)s\}$ is uniquely determined, and we guess σ over this set. We have at most $s!$ possible such guesses. Fourthly, if $k+1+(p+1)s > n$ (that is, for the last block, which may have size smaller than s), we have $t!$ possible guesses with $t = n - (k+1) - s\lfloor \frac{n-(k+1)}{s} \rfloor$.

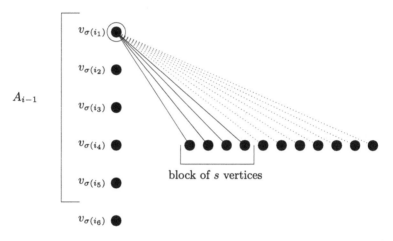

Fig. 3. The current anchor $v_{\sigma(i_1)}$ is connected to all the s vertices of the current block but will not be connected to any of the remaining non-introduced vertices.

We know that the first, the second, and the fourth cases can occur only once in the construction, and the third case can occur at most $\lfloor \frac{n-(k+1)}{s} \rfloor$ times. Therefore, an upper bound on the number of distinct possible guesses for σ is $k!(s!)^{\lfloor \frac{n-(k+1)}{s} \rfloor} t!$, where $t = n - (k+1) - s\lfloor \frac{n-(k+1)}{s} \rfloor$.

Let us now fix σ. Then the function N is uniquely determined. Indeed, for every $i \in \{1, \ldots, n\}$, $N(i)$ corresponds to the neighbors of $v_{\sigma(i)}$ to the left. It remains to bound the number of possible functions f. In order to do this, we define for every $i > 1$, $D_i = \{j \in N(i) \mid \forall j' > i, \{v_{\sigma(j)}, v_{\sigma(j')}\} \notin E(H)\}$. Then, for every $i \geqslant k+2$, by definition of $f(i)$, $f(i) \in D_{i-1}$. Moreover, for $i, j > k+1$ with $i \neq j$, it holds that, by definition of D_i and D_j, $D_i \cap D_j = \emptyset$. Indeed, assume w.l.o.g. that $i < j$, and suppose for contradiction that there exists $a \in D_i \cap D_j$. As $a \in D_j$, it holds that $a \in N(j)$, but as $a \in D_i$, for every $j' > i$, $a \notin N(j')$, hence $a \notin N(j)$, a contradiction.

We obtain that the number of distinct functions f is bounded by $\prod_{i=k+1}^{n} |D_i|$. As $D_i \cap D_j = \emptyset$ for every $i, j \geq k+1$ with $i \neq j$ and $D_i \subseteq \{1, \ldots, n\}$ for every $i \geq k+1$, we have that $\sum_{i=k+1}^{n} |D_i| \leq n$. Let $I = \{i \in \{k+1, \ldots, n\} \mid |D_i| \geq 2\}$, and note that $|I| \leq k$. By the previous discussion, it holds that $\sum_{i \in I} |D_i| \leq 2k$. So it follows that, by using Cauchy-Schwarz inequality,

$$\prod_{i=k+1}^{n} |D_i| = \prod_{i \in I} |D_i| \leq \left(\frac{\sum_{i \in I} |D_i|}{k}\right)^k \leq \left(\frac{2k}{k}\right)^k = 2^k. \tag{9}$$

To conclude, the number of constructible triples that can give rise to H is at most $2^k (s!)^{\lfloor \frac{n-(k+1)}{s} \rfloor} t!$ where $t = n - (k+1) - s\lfloor \frac{n-(k+1)}{s} \rfloor$. Thus, we obtain that

$$R_{n,k} \geq \frac{(n-1)! \left(\frac{k-1}{2}\right)^{n-(k+1)-\lceil \frac{n-(k+1)}{s} \rceil} 2^{\frac{k(k-1)}{2}} 2^{(n-(k+1))(k-2)}}{2^k k! (s!)^{\lfloor \frac{n-(k+1)}{s} \rfloor} (n - (k+1) - s\lfloor \frac{n-(k+1)}{s} \rfloor)!}. \tag{10}$$

For better readability, we bound separately each of the terms on the right-hand side:

- $(n-1)! \geq \left(\frac{n}{e}\right)^n 2^{-n}$, $2^{\frac{k(k-1)}{2}} 2^{(n-(k+1))(k-2)} \geq 2^{kn - \frac{k(k+3)}{2}} 2^{-2n}$.
- $(k-1)^{(n-(k+1)-\lceil \frac{n-(k+1)}{s} \rceil)} \geq 2^{-n} k^{(n-\frac{n}{s}-k-2)}$, since $k \geq 2$.
- $2^k k! \leq 2^n k^k$, $(s!)^{\lfloor \frac{n-(k+1)}{s} \rfloor} (n - (k+1) - s\lfloor \frac{n-(k+1)}{s} \rfloor)! \leq s^n$.

Applying these relations to (10) gives

$$R_{n,k} \geq \left(\frac{1}{64e} \cdot \frac{k2^k n}{k^{1/s} s}\right)^n 2^{-\frac{k(k+3)}{2}} k^{-2k-2}. \tag{11}$$

3.3 Choosing the Parameter s

We now discuss how to choose the size s of the blocks in the construction. In order to obtain the largest possible lower bound for $R_{n,k}$, we would like to choose s minimizing the factor $k^{1/s} s$ in the denominator of (11). To be as general as possible, assume that s is a function $s(n,k)$ that may depend on n and k, and we define $t(n,k) = \frac{s(n,k)}{\log k}$. With this definition, it follows that

$$\log\left(k^{\frac{1}{s(n,k)}} s(n,k)\right) = \frac{\log k}{s(n,k)} + \log s(n,k) = \frac{1}{t(n,k)} + \log t(n,k) + \log\log k. \tag{12}$$

It is elementary that the minimum of $\frac{1}{t(n,k)} + \log t(n,k)$ is achieved for $t(n,k) = 1$. Thus, we obtain that $s(n,k) = \log k$ is the function that maximizes the lower bound given by Eq. (11). Therefore, we obtain that

$$R_{n,k} \geq \left(\frac{1}{128e} \cdot \frac{k2^k n}{\log k}\right)^n 2^{-\frac{k(k+3)}{2}} k^{-2k-2}, \tag{13}$$

concluding the proof of Theorem 2, where we assume that $k \geq 2$.

4   Concluding Remarks and Further Research

Comparing Eqs. (3) and (6), there is still a gap of $(128e \cdot \log k)^n$ in the dominant term of $T_{n,k}$, and closing this gap remains an open problem. The factor $(\log k)^n$ appears because, in our construction, when a new block starts, we force the frozen vertex to be the previous anchor. Therefore, this factor is somehow artificial, and we believe that it could be avoided.

One way to improve the upper bound would be to show that *every* partial k-tree with n vertices and m edges can be extended to at least a large number $\alpha(n, m)$ of k-trees, and then use *double counting*. This is the approach taken in [19] for bounding the number of planar graphs, but so far we have not been able to obtain a significant improvement using this technique.

As mentioned before, our results also apply to other relevant graph parameters such as pathwidth and proper-pathwidth. For both parameters, besides improving the lower bound given by our construction, it may be also possible to improve the upper bound given by Eq. (3). For proper-pathwidth, a modest improvement can be obtained as follows. It follows easily from the definition of proper-pathwidth that the edge-maximal graphs of proper-pathwidth k, which we call *proper linear k-trees*, can be constructed starting from a $(k + 1)$-clique and iteratively adding a vertex v_i connected to a clique K_{v_i} of size k, with the constraints that $v_{i-1} \in K_{v_i}$ and $K_{v_i} \setminus \{v_{i-1}\} \subseteq K_{v_{i-1}}$. From this observation, and taking into account that the order of the first k vertices is not relevant and that there are $2k$ initial cliques giving rise to the same graph, it follows that the number of n-vertex labeled proper linear k-trees is equal to

$$n! k^{n-k-1} \frac{1}{(2k)k!}. \tag{14}$$

From this and the fact that a k-tree has $kn - \frac{k(k+1)}{2}$ edges, an easy calculation yields that the number of n-vertex labeled graphs of proper-pathwidth at most k is at most $\left(\frac{k 2^k n}{c} \right)^n$, for some absolute constant $c \geqslant 1.88$.

It would be interesting to count graphs of bounded "width" in other cases. For instance, branchwidth seems to be a good candidate, as it is known that, if we denote by $\mathbf{bw}(G)$ the branchwidth of a graph G and $|E(G)| \geqslant 3$, then $\mathbf{bw}(G) \leqslant \mathbf{tw}(G) + 1 \leqslant \frac{3}{2}\mathbf{bw}(G)$ [20]. Other relevant graph parameters are cliquewidth, rankwidth, tree-cutwidth, or booleanwidth. For any of these parameters, a first natural step would be to find a "canonical" way to build such graphs, as in the case of partial k-trees.

Our results find algorithmic applications, specially in the area of Parameterized Complexity [6]. When designing a parameterized algorithm, usually a crucial step is to solve the problem at hand restricted to graphs decomposable along small separators by performing dynamic programming (see [14] for a recent example). For instance, precise bounds on $T_{n,k}$ are useful when dealing with the TREEWIDTH-k VERTEX DELETION problem, which has recently attracted significant attention in the area [9,12,15]. In this problem, given a graph G and a fixed integer $k > 0$, the objective is to remove as few vertices from G as possible in order to obtain a graph of treewidth at most k. When solving TREEWIDTH-k VERTEX DELETION by dynamic programming, the natural approach is to enumerate, for any partial solution at a given separator of the decomposition, all possible graphs of treewidth at most k that are "rooted" at the separator. In this setting, the value of $T_{n,k}$, as well as an explicit construction to generate such graphs, may be crucial in order to speed-up the running time of the algorithm. Other recent algorithmic applications of knowing the precise number of graphs of bounded treewidth are finding path- or tree-decompositions with minimum number of bags [4] and subgraph embedding problems on sparse graphs [5].

Finally, a challenging open problem is to count the number of *unlabeled* partial k-trees, for which nothing is known except for some results concerning random models [3,13,17]. Note that the number of unlabeled k-trees was an open problem for long time, until it was recently solved by Gainer-Dewar [10] (see also [7,11]).

Acknowledgement. We would like to thank Dimitrios M. Thilikos for pointing us to the notion of proper-pathwidth, and the anonymous referees for helpful remarks that improved the presentation of the paper and for suggesting several relevant references.

References

1. Beineke, L.W., Pippert, R.E.: The number of labeled k-dimensional trees. J. Comb. Theory **6**(2), 200–205 (1969)
2. Bodirsky, M., Giménez, O., Kang, M., Noy, M.: Enumeration and limit laws for series-parallel graphs. Eur. J. Comb. **28**(8), 2091–2105 (2007)
3. Bodlaender, H., Kloks, T.: Only few graphs have bounded treewidth. Technical report RUU-CS-92-35, Utrecht University. Department of Computer Science (1992)
4. Bodlaender, H.L., Nederlof, J.: Subexponential time algorithms for finding small tree and path decompositions. CoRR, abs/1601.02415 (2016)
5. Bodlaender, H.L., Nederlof, J., van der Zanden, T.C.: Subexponential time algorithms for embedding H-minor free graphs. In: Proceedings of the 43rd International Colloquium on Automata, Languages, Programming (ICALP), volume 55 of LIPIcs, pp. 9:1–9:14 (2016)
6. Cygan, M., Fomin, F.V., Kowalik, Ł., Lokshtanov, D., Marx, D., Pilipczuk, M., Pilipczuk, M., Saurabh, S.: Parameterized Algorithms. Springer, Cham (2015). doi:10.1007/978-3-319-21275-3
7. Drmota, M., Jin, E.Y.: An asymptotic analysis of labeled and unlabeled k-trees. Algorithmica **75**(4), 579–605 (2016)
8. Foata, D.: Enumerating k-trees. Discret. Math. **1**(2), 181–186 (1971)
9. Fomin, F.V., Lokshtanov, D., Misra, N., Saurabh, S.: Planar $\mathcal{F}$-deletion: approximation, kernelization and optimal FPT algorithms. In: Proceedings of the 53rd Annual IEEE Symposium on Foundations of Computer Science (FOCS), pp. 470–479 (2012)
10. Gainer-Dewar, A.: Γ-species and the enumeration of k-trees. Electron. J. Comb. **19**(4), P45 (2012)
11. Gainer-Dewar, A., Gessel, I.M.: Counting unlabeled k-trees. J. Comb. Theory, Ser. A **126**, 177–193 (2014)

12. Gajarský, J., Hlinený, P., Obdrzálek, J., Ordyniak, S., Reidl, F., Rossmanith, P., Villaamil, F.S., Sikdar, S.: Kernelization using structural parameters on sparse graph classes. J. Comput. Syst. Sci. **84**, 219–242 (2017)
13. Gao, Y.: Treewidth of Erdős-Rényi random graphs, random intersection graphs, and scale-free random graphs. Discret. Appl. Math. **160**(4–5), 566–578 (2012)
14. Jansen, B.M.P., Lokshtanov, D., Saurabh, S.: A near-optimal planarization algorithm. In: Proceedings of the 25th Annual ACM-SIAM Symposium on Discrete Algorithms (SODA), pp. 1802–1811 (2014)
15. Kim, E.J., Langer, A., Paul, C., Reidl, F., Rossmanith, P., Sau, I., Sikdar, S.: Linear kernels and single-exponential algorithms via protrusion decompositions. ACM Trans. Algorithms **12**(2), 21 (2016)
16. Kloks, T.: Treewidth, Computations and Approximations. LNCS, vol. 842. Springer, Heidelberg (1994). doi:10.1007/BFb0045375
17. Mitsche, D., Perarnau, G.: On the treewidth and related parameters of random geometric graphs. In: Proceedings of the 29th International Symposium on Theoretical Aspects of Computer Science (STACS). LIPIcs, vol. 14, pp. 408–419 (2012)
18. Moon, J.W.: The number of labeled k-trees. J. Comb. Theory **6**(2), 196–199 (1969)
19. Osthus, D., Prömel, H.J., Taraz, A.: On random planar graphs, the number of planar graphs and their triangulations. J. Comb. Theory, Ser. B **88**(1), 119–134 (2003)
20. Robertson, N., Seymour, P.D.: Graph minors. X. Obstructions to tree-decomposition. J. Comb. Theory, Seri. B **52**(2), 153–190 (1991)
21. Takács, L.: On the number of distinct forests. SIAM J. Discret. Math. **3**(4), 574–581 (1990)
22. Takahashi, A., Ueno, S., Kajitani, Y.: Minimal acyclic forbidden minors for the family of graphs with bounded path-width. Discret. Math. **127**(1–3), 293–304 (1994)
23. Takahashi, A., Ueno, S., Kajitani, Y.: Mixed searching and proper-path-width. Theoret. Comput. Sci. **137**(2), 253–268 (1995)

Uniquely Restricted Matchings
and Edge Colorings

Julien Baste[1], Dieter Rautenbach[2], and Ignasi Sau[1,3(✉)]

[1] CNRS, LIRMM, Université de Montpellier, Montpellier, France
{baste,sau}@lirmm.fr
[2] Institute of Optimization and Operations Research, Ulm University, Ulm, Germany
dieter.rautenbach@uni-ulm.de
[3] Departamento de Matemática, Universidade Federal do Ceará, Fortaleza, Brazil

Abstract. A matching in a graph is *uniquely restricted* if no other matching covers exactly the same set of vertices. This notion was defined by Golumbic, Hirst, and Lewenstein and studied in a number of articles. Our contribution is twofold. We provide approximation algorithms for computing a uniquely restricted matching of maximum size in some bipartite graphs. In particular, we achieve a ratio of 5/9 for subcubic bipartite graphs, improving over a 1/2-approximation algorithm proposed by Mishra. Furthermore, we study the *uniquely restricted chromatic index* of a graph, defined as the minimum number of uniquely restricted matchings into which its edge set can be partitioned. We provide tight upper bounds in terms of the maximum degree and characterize all extremal graphs. Our constructive proofs yield efficient algorithms to determine the corresponding edge colorings.

Keywords: Uniquely restricted matching · Bipartite graph · Approximation algorithm · Edge coloring · Subcubic graph

1 Introduction

Matchings in graphs are among the most fundamental and well-studied objects in combinatorial optimization [25,32]. While classical matchings lead to many efficiently solvable problems, more restricted types of matchings [30] are often intractable; induced matchings [3,6,7,10,11,15,18–20,26] being a prominent example. Here we study the so-called uniquely restricted matchings, which were introduced by Golumbic et al. [14] and studied in a number of papers [13,21–23,27,29]. We also consider the corresponding edge coloring notion.

Before we explain our contribution and discuss related research, we collect some terminology and notation (cf. e.g. [8] for undefined terms). We consider finite, simple, and undirected graphs. A *matching* in a graph G [25] is a set of pairwise non-adjacent edges of G. For a matching M, let $V(M)$ be the set of vertices incident with an edge in M. A matching M in a graph G is *induced* [10]

This work has been supported by the DE-MO-GRAPH grant ANR-16-CE40-0028.

H.L. Bodlaender and G.J. Woeginger (Eds.): WG 2017, LNCS 10520, pp. 100–112, 2017.
https://doi.org/10.1007/978-3-319-68705-6_8

if the subgraph $G[V(M)]$ of G induced by $V(M)$ is 1-regular. Golumbic et al. [14] define a matching M in a graph G to be *uniquely restricted* if there is no matching M' in G with $M' \neq M$ and $V(M') = V(M)$, that is, no other matching covers exactly the same set of vertices. It is easy to see that a matching M in G is uniquely restricted if and only if there is no M-*alternating cycle* in G, which is a cycle in G that alternates between edges in M and edges not in M. Let the *matching number* $\nu(G)$, the *strong matching number* $\nu_s(G)$, and the *uniquely restricted matching number* $\nu_{ur}(G)$ of G be the maximum size of a matching, an induced matching, and a uniquely restricted matching in G, respectively. Since every induced matching is uniquely restricted, we obtain

$$\nu_s(G) \leq \nu_{ur}(G) \leq \nu(G)$$

for every graph G.

It is worth mentioning that, as discussed by Golumbic et al. [14], maximum uniquely restricted matchings in bipartite graphs correspond to largest possible upper triangular submatrices that can be obtained by permuting rows and columns of a given matrix. Upper triangular submatrices are interesting objects, since they allow the associated systems of linear equations to be solved quickly; see [14] for more details.

Each type of matching naturally leads to an edge coloring notion. For a graph G, let $\chi'(G)$ be the *chromatic index* of G, which is the minimum number of matchings into which the edge set $E(G)$ of G can be partitioned. Similarly, let the *strong chromatic index* $\chi'_s(G)$ [11] and the *uniquely restricted chromatic index* $\chi'_{ur}(G)$ of G be the minimum number of induced matchings and uniquely restricted matchings into which the edge set of G can be partitioned, respectively. A partition of the edges of a graph G into uniquely restricted matchings is a *uniquely restricted edge coloring* of G. Another related notion are *acyclic edge colorings*, which are partitions of the edge set into matchings such that the union of every two of the matchings is a forest. The minimum number of matchings in an acyclic edge coloring of a graph G is its *acyclic chromatic index* $a'(G)$ [1,12]. Exploiting the obvious relations between the different edge coloring notions, we obtain

$$\chi'(G) \leq a'(G) \leq \chi'_{ur}(G) \leq \chi'_s(G) \tag{1}$$

for every graph G.

Stockmeyer and Vazirani [30] showed that computing the strong matching number is NP-hard. Their result was strengthened in many ways, and also restricted graph classes where the strong matching number can be determined efficiently were studied [3,6,7,26]. Golumbic et al. [14] showed that it is NP-hard to determine $\nu_{ur}(G)$ for a given bipartite or split graph G. Mishra [27] strengthened this by showing that it is not possible to approximate $\nu_{ur}(G)$ within a factor of $O(n^{\frac{1}{3}-\epsilon})$ for any $\epsilon > 0$, unless NP = ZPP, even when restricted to bipartite, split, chordal or comparability graphs of order n. Furthermore, he showed that $\nu_{ur}(G)$ is APX-complete for subcubic bipartite graphs.

On the positive side, Golumbic et al. [14] described efficient algorithms that determine $\nu_{ur}(G)$ for cacti, threshold graphs, and proper interval graphs. Solving a problem from [14], Francis et al. [13] described an efficient algorithm for $\nu_{ur}(G)$ in interval graphs. Solving yet another problem from [14], Penso et al. [29] showed that the graphs G with $\nu(G) = \nu_{ur}(G)$ can be recognized in polynomial time. Complementing his hardness results, Mishra [27] proposed a 2-approximation algorithm for cubic bipartite graphs.

While $\chi'(G)$ of a graph G of maximum degree Δ is either Δ or $\Delta + 1$ [31], Erdős and Nešetřil (see, e.g., [11]) conjectured $\chi'_s(G) \leq \frac{5}{4}\Delta^2$, and much of the research on the strong chromatic index is motivated by this conjecture. Building on earlier work of Molloy and Reed [28], Bruhn and Joos [4] showed $\chi'_s(G) \leq 1.93\Delta^2$ provided that Δ is sufficiently large. For further results on the strong chromatic index we refer to [2,11,16,17].

Fiamčik [12] and Alon et al. [1] conjectured that every graph of maximum degree Δ has an acyclic edge coloring using no more than $\Delta + 2$ colors. See [5,9] for further references and the currently best known results concerning general graphs and graphs of large girth.

In view of the famous open conjectures on $\chi'_s(G)$ and $a'(G)$, the inequality chain (1) motivates to study upper bounds on $\chi_{ur}(G)$ in terms of the maximum degree Δ of a graph G.

Our contribution is twofold. We present approximation algorithms for $\nu_{ur}(G)$ in some bipartite graphs in Sect. 2 and tight bounds on $\chi'_{ur}(G)$ in Sect. 3.

Improving on Mishra's approximation algorithm [27], we describe a 5/9-approximation algorithm for computing $\nu_{ur}(G)$ of a given bipartite subcubic graph G. Our algorithm requires some complicated preprocessing based on detailed local analysis, and due to space limitations its proof can be found only in the full version of this article [arXiv:1611.06815]. In order to illustrate our general approach in a cleaner setting, we describe here in full detail an approximation algorithm for C_4-free bipartite graphs of arbitrary maximum degree.

Concerning the uniquely restricted chromatic index, we achieve best-possible upper bounds in terms of the maximum degree, and even characterize all extremal graphs. Since our proofs are constructive, it is easy to extract efficient algorithms finding the corresponding edge colorings.

We conclude with some open problems in Sect. 4.

2   Approximation Algorithms for Bipartite Graphs

Before we proceed to our main result in this section, Theorem 2, we describe an approximation algorithm for the C_4-free case. The proof of the next lemma contains the main algorithmic ingredients. Note that the size of the smaller partite set in a bipartite graph is always an upper bound on the uniquely restricted matching number.

For an integer k, let $[k]$ denote the set of positive integers between 1 and k. For a graph G, let $n(G)$ denote its number of vertices.

Lemma 1. *Let $\Delta \geq 3$ be an integer. If G is a connected C_4-free bipartite graph of maximum degree at most Δ with partite sets A and B such that every vertex in A has degree at least 2, and some vertex in B has degree less than Δ, then G has a uniquely restricted matching M of size at least $\frac{(\Delta-1)^2+(\Delta-2)}{(\Delta-1)^3+(\Delta-2)}|A|$. Furthermore, such a matching can be found in polynomial time.*

Proof: We give an algorithmic proof of the lower bound such that the running time of the corresponding algorithm is polynomial in $n(G)$, which immediately implies the second part of the statement. Therefore, let G be as in the statement. Throughout the execution of our algorithm, as illustrated in Fig. 1, we maintain a pair (U, M) such that

(a) U is a subset of $V(G)$,
(b) M is a uniquely restricted matching with $V(M) \subseteq U$,
(c) every vertex in $B \cap U$ has all its neighbors in $A \cap U$,
(d) every vertex in $B \setminus U$ has a neighbor in $A \setminus U$,
(e) if
  s vertices in $A \cap U$ are incident with an edge in M,
  d vertices in $A \cap U$ are not incident with an edge in M but have a neighbor in $B \setminus U$, and
  f vertices in $A \cap U$ are neither incident with an edge in M nor have a neighbor in $B \setminus U$, then

$$(\Delta - 1)^2\Big((\Delta - 2)s - (d + f)\Big) \geq (\Delta - 2)f. \tag{2}$$

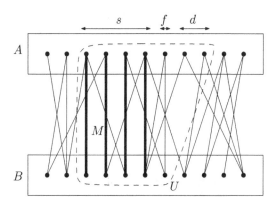

Fig. 1. Example for $\Delta = 3$ of the parameters defined in the proof of Lemma 1. The set U is dashed, and the uniquely restricted matching M corresponds to the thicker edges.

Initially, let U and M be empty sets. Note that properties (a) to (e) hold.
  As long as U is a proper subset of $V(G)$, we iteratively replace the pair (U, M) with a pair (U', M') such that U is a proper subset of U', M is a proper subset

of M', and properties (a) to (e) hold for (U', M'). Let s', d', and f' denote the updated values considered in (e). Once $U = V(G)$, we have $s = |M|$, $d = 0$, and $f = |A| - |M|$, and (2) implies the stated lower bound on $|M|$.

We proceed to the description of the extension operations. Therefore, suppose that U is a proper subset of $V(G)$. Since G is connected, and some vertex in B has degree less than Δ, some vertex u in $B \backslash U$ has less than Δ neighbors in $A \backslash U$, that is, if $d_{\bar{U}}(u) = |N_G(u) \backslash U|$, then $1 \le d_{\bar{U}}(u) \le \Delta - 1$, where the existence of u and the first inequality follow from property (d). We choose $u \in B \backslash U$ such that $d_{\bar{U}}(u)$ is as small as possible.

Case 1: $d_{\bar{U}}(u) = 1$.

Let v be the unique neighbor of u in $A \backslash U$. Let $\{u_1, \dots, u_k\}$ be the set of all vertices u in $B \backslash U$ with $N_G(u) \backslash U = \{v\}$, and note that $1 \le k \le \Delta$. Let $U' = U \cup \{u_1, \dots, u_k, v\}$. For some integer $\ell \le k$, we may assume that $\{u_1, \dots, u_\ell\}$ is the set of those u_i with $i \in [k]$ such that u_i has a neighbor in $A \cap U$, and no neighbor of u_i in $A \cap U$ is incident with M. Note that every vertex u_i with $i \in [k] \backslash [\ell]$ either has no neighbor in $A \cap U$ or has some neighbor in $A \cap U$ that is incident with M.

First, suppose that $\ell \ge 2$. Let M' arise from M by adding, for every $i \in [\ell]$, an edge between u_i and a neighbor w_i of u_i in $A \cap U$. Note that all these neighbors w_i in $A \cap U$ are distinct. Indeed, if two vertices u_i and u_j have a common neighbor w in $A \cap U$, then the set of vertices $\{v, u_i, u_j, w\}$ would induce a C_4 in G. Note also that M' is indeed a uniquely restricted matching, as if there exists an edge $u_i w_j$ with $i, j \in [\ell]$ and $i \ne j$ that could potentially create an M'-alternating cycle, then the set of vertices $\{v, u_i, u_j, w_j\}$ would again induce a C_4 in G. Clearly, replacing (U, M) with (U', M'), we maintain properties (a) to (d), and $s' = s + \ell$. Let n_d be the number of vertices in $A \cap U$ that are not incident with an edge in M', have a neighbor in $B \backslash U$, but do not have a neighbor in $B \backslash U'$; note that each such vertex has a neighbor in the set $\{u_1, \dots, u_k\}$. As every vertex in $\{u_1, \dots, u_k\}$ is neighbor of v and of a vertex incident with an edge in M', it holds that $n_d \le k(\Delta - 2) \le \Delta(\Delta - 2)$. If v has a neighbor in $B \backslash U'$, then $d' = d - n_d + 1$ and $f' = f + n_d$, and, if v has no neighbor in $B \backslash U'$, then $d' = d - n_d$ and $f' = f + n_d + 1$. In both cases $d' + f' = d + f + 1$ and $f' \le f + n_d + 1$. Since $\frac{(\Delta-1)^2}{\Delta-2}\left((\Delta - 2)\ell - 1\right) \ge \Delta(\Delta - 2) + 1 \ge n_d + 1$, property (e) is maintained.

Next, suppose that $\ell \le 1$. Let M' arise from M by adding the edge $u_1 v$. Clearly, replacing (U, M) with (U', M'), we maintain properties (a) to (d), and $s' = s + 1$. Defining n_d exactly as above, we obtain $n_d \le k(\Delta - 2) + \ell \le \Delta(\Delta - 2) + 1$, $d' = d - n_d$, and $f' = f + n_d$. Since $\frac{(\Delta-1)^2}{\Delta-2}(\Delta - 2) \ge \Delta(\Delta - 2) + 1 \ge n_d$, property (e) is maintained.

Case 2: $2 \le d_{\bar{U}}(u) \le \Delta - 1$.

Let $\{v_1, \dots, v_k\} = N_G(u) \backslash U$ and let $U' = U \cup \{u, v_1, \dots, v_k\}$. Note that $2 \le k \le \Delta - 1$.

First, suppose that u has a neighbor v in $A \cap U$, and that no neighbor of u in $A \cap U$ is incident with M. Let M' arise from M by adding the edge uv.

Clearly, replacing (U, M) with (U', M'), we maintain properties (a) to (c), and $s' = s + 1$. Let us prove that property (d) is also maintained. Since G has no C_4 and $k \geq 2$, no vertex in $B \setminus U$ that is distinct from u can have more than one neighbor among $v_1, \ldots, v_k$. Since we are in Case 2, every vertex in $B \setminus U$ has more than one neighbor in $A \setminus U$, hence property (d) remains true. Similarly as above, let n_d be the number of vertices in $A \cap U$ that are not incident with an edge in M', have a neighbor in $B \setminus U$, but do not have a neighbor in $B \setminus U'$. Note that $n_d \leq \Delta - k - 1$, $d' = d + k - n_d - 1$, and $f' = f + n_d$. Since $\frac{(\Delta-1)^2}{\Delta-2}\left((\Delta - 2) - (k - 1)\right) \geq \Delta - k - 1 \geq n_d$, property (e) is maintained.

Next, suppose that u has no neighbor in $A \cap U$ or some neighbor of u in $A \cap U$ is incident with M. Let M' arise from M by adding the edge uv_1. Clearly, replacing (U, M) with (U', M'), we again maintain properties (a) to (d), and $s' = s + 1$. Note that, in the case where u has a neighbor in $A \cap U$, v_1 does not have neighbors in $V(M)$ because of property (c), which guarantees that M' is indeed a uniquely restricted matching. Defining n_d exactly as above, we obtain $n_d \leq \Delta - k - 1$. Indeed, if u has no neighbor in $A \cap U$, then $n_d = 0$. On the other hand, if u has a neighbor in $A \cap U$ that is incident with M, then $n_d \leq \Delta - k - 1$. As $k \leq \Delta - 1$, in both cases it holds that $n_d \leq \Delta - k - 1$. Also, we get that $d' = d + k - n_d + 1$ and $f' = f + n_d$, and the same calculation as above implies that property (e) is maintained.

Since the considered cases exhaust all possibilities, and in each case we described an extension that maintains the relevant properties, the proof is complete up to the running time of the algorithm, which we proceed to analyze. One can easily check that each extension operation takes time $O(\Delta n)$, where $n = n(G)$. As in each extension operation, the size of U is incremented by at least one, it follows that the overall running time of the algorithm is $O(\Delta n^2)$. □

With Lemma 1 at hand, we proceed to our first approximation algorithm.

Theorem 1. *Let $\Delta \geq 3$ be an integer. For a given connected C_4-free bipartite graph G of maximum degree at most Δ, one can find in polynomial time a uniquely restricted matching M of G of size at least $\frac{(\Delta-1)^2+(\Delta-2)}{(\Delta-1)^3+(\Delta-2)}\nu_{ur}(G)$.*

Proof: Let $\alpha = \frac{(\Delta-1)^2+(\Delta-2)}{(\Delta-1)^3+(\Delta-2)}$ and let $\mathcal{G}$ be the set of all C_4-free bipartite graphs G of maximum degree at most Δ such that every component of G has a vertex of degree less than Δ. First, we prove that, for every given graph G in $\mathcal{G}$, one can find in polynomial time a uniquely restricted matching M of size at least $\alpha \nu_{ur}(G)$. Therefore, let G be in $\mathcal{G}$.

If G has a vertex u of degree 1, and v is the unique neighbor of u, then let $G' = G - \{u, v\}$. Clearly, $\nu_{ur}(G') \geq \nu_{ur}(G) - 1$, and if M' is a uniquely restricted matching of G', then $M' \cup \{uv\}$ is a uniquely restricted matching of G. Note that G' belongs to $\mathcal{G}$. Let G'' be the graph obtained from G' by removing every isolated vertex. Clearly, $\nu_{ur}(G'') \geq \nu_{ur}(G')$, if M'' is a uniquely restricted matching of G'', then M'' is a uniquely restricted matching of G', and G'' belongs to $\mathcal{G}$.

Iteratively repeating these reductions, we efficiently obtain a set M_1 of edges of G as well as a subgraph G_2 of G such that $G_2 \in \mathcal{G}$, $\nu_{ur}(G_2) \geq \nu_{ur}(G) - |M_1|$,

$M_1 \cup M_2$ is a uniquely restricted matching of G for every uniquely restricted matching M_2 of G_2, and either $n(G_2) = 0$ or $\delta(G_2) \geq 2$. Note that if G has minimum degree at least 2, then we may choose M_1 empty and G_2 equal to G. Now, by suitably choosing the bipartition of each component K of G_2, and applying Lemma 1 to K, one can determine in polynomial time a uniquely restricted matching M_2 of G_2 with $|M_2| \geq \alpha \nu_{ur}(G_2)$. Since the set $M_1 \cup M_2$ is a uniquely restricted matching of G of size at least $|M_1| + \alpha \nu_{ur}(G_2) \geq |M_1| + \alpha(\nu_{ur}(G) - |M_1|) \geq \alpha \nu_{ur}(G)$, the proof of our claim about $\mathcal{G}$ is complete.

Now, let G be a given connected C_4-free bipartite graph of maximum degree at most Δ. If G is not Δ-regular, then $G \in \mathcal{G}$, and the desired statement already follows. Hence, we may assume that G is Δ-regular, which implies that its two partite sets A and B are of the same order. By [29], we can efficiently decide whether $\nu_{ur}(G) = \nu(G)$. Furthermore, if $\nu_{ur}(G) = \nu(G)$, then, again by [29], we can efficiently determine a maximum matching that is uniquely restricted. Hence, we may assume that $\nu_{ur}(G) < \nu(G)$. This implies that $\nu_{ur}(G) < |A|$, and, hence, there is some vertex $u \in V(G)$ with $\nu_{ur}(G - u) = \nu_{ur}(G)$. Since $G - u \in \mathcal{G}$ for every vertex u of G, considering the $n(G)$ induced subgraphs $G - u$ for $u \in V(G)$, one can determine in polynomial time a uniquely restricted matching M of G with $|M| \geq \max\{\alpha \nu_{ur}(G - u) : u \in V(G)\} = \alpha \nu_{ur}(G)$. The desired statement follows.

Our next result shows that – at least for $\Delta = 3$ – C_4-freeness is not an essential assumption. The proof can be found in [arXiv:1611.06815].

Theorem 2. *For a given connected subcubic bipartite graph G, one can find in polynomial time a uniquely restricted matching of G of size at least $\frac{5}{9}\nu_{ur}(G)$.*

We believe that Theorem 2 extends to larger maximum degrees, that is, the conclusion of Theorem 1 should hold without the assumption of C_4-freeness.

3   Upper Bounds on $\chi'_{ur}(G)$

Our first result in this section applies to general graphs, and its proof relies on a natural greedy strategy. Faudree et al. [10] conjectured $\chi'_s(G) \leq \Delta^2$ for a bipartite graph G of maximum degree Δ, and our Theorem 3 can be considered a weak version of this conjecture. Theorem 4 below shows that excluding the unique extremal graph from Theorem 3, the uniquely restricted chromatic index of bipartite graphs drops considerably.

Theorem 3. *If G is a connected graph of maximum degree at most Δ, then $\chi'_{ur}(G) \leq \Delta^2$ with equality if and only if G is $K_{\Delta,\Delta}$.*

Proof: Since no two edges of $K_{\Delta,\Delta}$ form a uniquely restricted matching in this graph, we obtain $\chi'_{ur}(K_{\Delta,\Delta}) = |E(K_{\Delta,\Delta})| = \Delta^2$. Now, let G be a connected graph of maximum degree at most Δ. We first show that $\chi'_{ur}(G) \leq \Delta^2$. In a second step, we show that $\chi'_{ur}(G) < \Delta^2$ provided that G is not $K_{\Delta,\Delta}$.

We consider the vertices of G in some linear order, say $u_1, \ldots, u_n$. For i from 1 up to n, we assume that the edges of G incident with vertices in $\{u_1, \ldots, u_{i-1}\}$ have already been colored, and we color all edges between u_i and $\{u_{i+1}, \ldots, u_n\}$ using distinct colors, and avoiding any color that has already been used on a previously colored edge incident with some neighbor of u_i. Since u_i has at most Δ neighbors, each of which is incident with at most Δ edges, this procedure requires at most Δ^2 many distinct colors.

Suppose, for a contradiction, that some color class M is not a uniquely restricted matching in G. Since M is a matching by construction, there is an M-alternating cycle C. Let $C : u_{r_1} u_{s_1} u_{r_2} u_{s_2} \ldots u_{r_k} u_{s_k} u_{r_1}$ be such that r_1 is the minimum index of any vertex on C, and $u_{r_1} u_{s_k} \in M$. These choices trivially imply $r_1 < s_1$ and $r_1 < r_2$. If $r_2 > s_1$, then $u_{r_1} u_{s_k} \in M$ implies that, when coloring the edge $u_{s_1} u_{r_2}$, some edge incident with the neighbor u_{r_1} of u_{s_1} would already have been assigned the color of the edges in M, and the above procedure would have avoided this color on $u_{s_1} u_{r_2}$. Therefore, since $u_{r_1} u_{s_k} \in M$ and $u_{r_2} u_{s_1} \in M$, the coloring rules imply $r_2 < s_1$, that is, $r_1 < r_2 < s_1$. Now, suppose that $r_i < r_{i+1} < s_i$ for some $i \in [k-1]$. Since $u_{r_{i+1}} u_{s_i} \in M$ and $u_{r_{i+2}} u_{s_{i+1}} \in M$, the coloring rules imply in turn

- $r_{i+2} < s_{i+1}$, since otherwise we would have colored $u_{r_{i+2}} u_{s_{i+1}}$ differently, and
- $r_{i+1} < r_{i+2}$, since otherwise we would have colored $u_{r_{i+1}} u_{s_i}$ differently.

It follows that $r_{i+1} < r_{i+2} < s_{i+1}$, where we identify r_{k+1} with r_1. Now, by an inductive argument, we obtain $r_1 < r_2 < \cdots < r_k < r_1$, which is a contradiction.

Altogether, we obtain $\chi'_{ur}(G) \leq \Delta^2$.

Now, let G be distinct from $K_{\Delta, \Delta}$, and we want to prove that $\chi'_{ur}(G) < \Delta^2$. Among all uniquely restricted edge colorings of G using colors in $[\Delta^2]$, we choose a coloring for which the number of edges with color 1 is as small as possible. Clearly, we may assume that some edge uv has color 1, as otherwise we already have that $\chi'_{ur}(G) < \Delta^2$.

If there is a color α in $[\Delta^2] \setminus \{1\}$ such that no edge incident with a neighbor of u has color α, then changing the color of uv to α yields a uniquely restricted edge coloring of G with less edges of color 1, which is a contradiction. In view of the maximum degree, this implies that every vertex in $N_G[u]$ has degree Δ, the set $N_G(u)$ is independent, and, for every color α in $[\Delta^2]$, there is exactly one edge incident with a neighbor of u that has color α.

Since G is not $K_{\Delta, \Delta}$, some neighbor x of u has a neighbor y that does not lie in $N_G(v)$. Without loss of generality, let ux have color 2, and let xy have color 3. Let M be the set of edges with color 3.

If G does not contain an M-alternating path of odd length (number of edges) at least 3 between x and a vertex in $N_G(v) \setminus \{u\}$ that contains the edge xy, then changing the color of uv to 3 yields a uniquely restricted edge coloring of G with less edges of color 1, which is a contradiction. Hence, G contains such a path, which implies that two edges incident with neighbors of y have color 3.

If there is a color α in $[\Delta^2] \setminus \{1\}$ such that no edge incident with a neighbor of y has color α, then changing the color of xy to α and the color of uv to 3 yields a uniquely restricted edge coloring of G with less edges of color 1, which is a contradiction. Similarly as above, this implies that, for every color α in $[\Delta^2] \setminus \{1,3\}$, there is exactly one edge incident with a neighbor of y that has color α. Now, changing the color of uv to 2, the color of ux to 3, and the color of xy to 2 yields a uniquely restricted edge coloring of G with less edges of color 1, which is a contradiction. This completes the proof. □

As observed above, the proof of Theorem 3 is algorithmic; the simple greedy strategy considered in its first half efficiently constructs uniquely restricted edge colorings using at most Δ^2 colors. Furthermore, also its second half can be turned into an efficient algorithm that finds uniquely restricted edge colorings using at most $\Delta^2 - 1$ colors for connected graphs of maximum degree Δ that are distinct from $K_{\Delta,\Delta}$; the different cases considered in the proof correspond to simple manipulations of a given uniquely restricted edge coloring that iteratively reduce the number of edges of color 1 down to 0. Golumbic et al. [14] showed that deciding whether a given matching is uniquely restricted can be done in polynomial time, and their algorithm can be used to decide which of the simple manipulations can be executed.

Our next goal is to improve Theorem 3 for bipartite graphs. The following proof was inspired by Lovász's [24] elegant proof of Brooks' Theorem.

Lemma 2. *If G is a connected bipartite graph of maximum degree at most $\Delta \geq 4$ that is distinct from $K_{\Delta,\Delta}$, and M is a matching in G, then M can be partitioned into at most $\Delta - 1$ uniquely restricted matchings in G.*

Proof: Let A and B be the partite sets of G, and let $R = V(G) \setminus V(M)$. Note that M is perfect if and only if R is empty. Whenever we consider a coloring of the edges in M, and α is one of the colors, let M_α be the set of edges in M colored with α.

First, we assume that R is empty, and that G is not Δ-regular. By symmetry, we may assume that some vertex a in A has degree less than Δ. Let $ab \in M$. Let T be a spanning tree of G that contains the edges in M. Contracting within T the edges from M, rooting the resulting tree at the vertex corresponding to the edge ab, and considering a breadth-first search order, we obtain the existence of a linear order $a_1 b_1, \ldots, a_n b_n$ of the edges in M such that $ab = a_n b_n$, and, for every $i \in [n-1]$, there is an edge between $\{a_i, b_i\}$ and $\{a_{i+1}, b_{i+1}, \ldots, a_n, b_n\}$. Since a_n has degree less than Δ, this implies that, for every $i \in [n]$, some vertex u_i in $\{a_i, b_i\}$ has at most $\Delta - 2$ neighbors in $\{a_1, b_1, \ldots, a_{i-1}, b_{i-1}\}$. Now, we color the edges in M greedily in the above linear order. Specifically, for every i from 1 up to n, we color the edge $a_i b_i$ with some color α in $[\Delta - 1]$ such that, for every $j \in [i-1]$, for which $u_i \in \{a_i, b_i\}$ has a neighbor in $\{a_j, b_j\}$, the edge $a_j b_j$ is not colored with α. By the degree condition on u_i, such a coloring exists. Suppose, for a contradiction, that M_α is not uniquely restricted for some color α in $[\Delta-1]$. Let the edge $a_i b_i$ in M_α be such that it belongs to some M_α-alternating cycle C, and, subject to this condition, the index i is maximum. If the neighbor of u_i on C outside of $\{a_i, b_i\}$ is in $\{a_j, b_j\}$, then the choice of the edge $a_i b_i$ implies $j < i$, and the coloring rule implies that the edge $a_j b_j$ is not colored with α, which is a contradiction. Altogether, the statement follows.

Next, we assume that R is non-empty. Let K be a component of $G - R$. Let M_K be the set of edges in M that lie in K. Since G is connected, the graph K is not Δ-regular. Therefore, proceeding exactly as above, we obtain a coloring of the edges in M_K using the colors in $[\Delta - 1]$ such that each color class is a uniquely restricted matching in K. If $K_1, \ldots, K_k$ are the components of $G - R$, and M_i is a uniquely restricted matching in K_i for every $i \in [k]$, then $M_1 \cup \cdots \cup M_k$ is a uniquely restricted matching in G. Therefore, combining the colorings within the different components, we obtain that also in this case the statement follows.

At this point, we may assume that G is Δ-regular, and that M is perfect.

Next, we assume that there are two distinct edges e and e' in M such that $V(\{e, e'\})$ is a vertex cut of G. This implies that we can partition the set $M \setminus \{e, e'\}$ into two non-empty sets M_1 and M_2 such that there is no edge between $V(M_1)$ and $V(M_2)$. For $i \in [2]$, let G_i be the subgraph of G induced by $V(\{e, e'\} \cup M_i)$. Since G is connected, the graph G_i is not Δ-regular. In view of the above, this implies that there is a coloring c_i of the edges of the perfect matching $\{e, e'\} \cup M_i$ of G_i using the colors in $[\Delta - 1]$ such that each color class of c_i is a uniquely restricted matching in G_i. If $c_i(e) \neq c_i(e')$ for both i in $[2]$, then we may assume that c_1 and c_2 assign the same colors to e and e', and it is easy to verify that the common extension c of c_1 and c_2 to M has the property that every color class of c is a uniquely restricted matching in G. Hence, we may assume that necessarily $c_1(e) = c_1(e')$. Note that this implies in particular that at least one of the two possible edges between $V(\{e\})$ and $V(\{e'\})$ is missing.

Let $c_1(e) = \alpha$. Let $e = ab$, $e' = a'b'$, and $U = \{a, b, a', b'\}$. For every vertex $u \in U$, let $C_1(u)$ be the set of colors β for which M_1 contains an edge vw with $c_1(vw) = \beta$ such that u is adjacent to v or w. If there is some $u \in U$ and some color $\beta \in ([\Delta - 1] \setminus \{\alpha\}) \setminus C_1(u)$, then changing the color of the unique edge in $\{e, e'\}$ incident with u from α to β yields a coloring c_1' of the edges in $\{e, e'\} \cup M_1$ using the colors in $[\Delta - 1]$ such that each color class of c_1' is a uniquely restricted matching in G_1. Furthermore, $c_1'(e) \neq c_1'(e')$, which is a contradiction. This implies that $[\Delta - 1] \setminus \{\alpha\} \subseteq C_1(u)$ for every $u \in U$. In particular, each vertex u in U has at least $\Delta - 2$ neighbors in $V(M_1)$, and, hence, at most one neighbor in $V(M_2)$. Let $C_2(u)$ for $u \in U$ be defined analogously as above. Clearly, the set $C_2(a) \cup C_2(a')$ contains at most two distinct colors. Since $\Delta - 1 \geq 3$, we may assume that c_2 is such that the set $C_2(a) \cup C_2(a')$ does not contain the color α. Now, let c_2' be a coloring of the edges in $\{e, e'\} \cup M_2$ that coincides with c_2 on M_2 and colors e and e' with color α. It is easy to see that each color class of c_2' is a uniquely restricted matching in G_2. Let c be the common extension of c_1 and c_2' to M. Suppose, for a contradiction, that the color class M_β of c is not uniquely restricted for some color β in $[\Delta - 1]$. Clearly, we have $\beta = \alpha$. Let C be an M_α-alternating cycle in G. It is easy to see that C contains both edges e and e' Furthermore, since at least one of the two possible edges between $\{a, b\}$ and $\{a', b'\}$ is missing, it follows that C contains an edge between $\{a, a'\}$ and $V(M_2)$. Since c coincides with c_2 on M_2, and $C_2(a) \cup C_2(a')$ does not contain α, we obtain a contradiction.

Altogether, we may assume that there are no two distinct edges e and e' in M such that $V(\{e, e'\})$ is a vertex cut of G.

Now, we show the existence of three edges ab, $a'b'$, and $a''b''$ in M such that some of the two possible edges between $\{a', b'\}$ and $\{a'', b''\}$ is missing, and either a is adjacent to b' as well as b'' or b is adjacent to a' as well as a''. Therefore, let $a_1 b_1$ be an edge in M. Let $a_2 b_2, \ldots, a_\Delta b_\Delta$ be the edges in M such that $N_G(a_1) = \{b_1, \ldots, b_\Delta\}$. We may assume that $\{a_2, b_2, \ldots, a_\Delta, b_\Delta\}$ induces a complete bipartite graph $K_{\Delta-1, \Delta-1}$; otherwise, we find the three edges with the desired properties. Since G is not $K_{\Delta, \Delta}$, the vertex b_1 is non-adjacent to some vertex a_i in $\{a_2, \ldots, a_\Delta\}$. Now, if $a_j \in \{a_2, \ldots, a_\Delta\} \setminus \{a_i\}$, then one of the two possible edges between $\{a_1, b_1\}$ and $\{a_i, b_i\}$ is missing, and b_j is adjacent to a_1 as well as a_i. Altogether, we obtain three edges ab, $a'b'$, and $a''b''$ in M with the desired properties.

By symmetry, we may assume that a is adjacent to b' and b'', and a' is non-adjacent to b''. In view of the above, the graph $G' = G - V(\{a'b', a''b''\})$ is connected, and $M' = M \setminus \{a'b', a''b''\}$ is a perfect matching of G'. Let T' be a spanning tree of G' that contains the edges in M'. Contracting within T' the edges from M', rooting the resulting tree in the vertex corresponding to the edge ab, and considering a breadth-first search order, we obtain the existence of a linear order $a_3 b_3, \ldots, a_n b_n$ of the edges in M' such that $ab = a_n b_n$, and, for every $i \in [n-1] \setminus [2]$, there is an edge between $\{a_i, b_i\}$ and $\{a_{i+1}, b_{i+1}, \ldots, a_n, b_n\}$. Now, we color the edges in M greedily in the linear order $a_1 b_1, a_2 b_2, a_3 b_3, \ldots, a_n b_n$, where $a_1 b_1 = a''b''$ and $a_2 b_2 = a'b'$. Note that, for every $i \in [n-1] \setminus [2]$, some vertex u_i in $\{a_i, b_i\}$ has at most $\Delta - 2$ neighbors in $\{a_1, b_1, \ldots, a_{i-1}, b_{i-1}\}$. We color $a_1 b_1$ and $a_2 b_2$ with the same color. For every i from 3 up to $n-1$, we color the edge $a_i b_i$ with a color α in $[\Delta - 1]$ such that, for every $j \in [i-1]$, for which u_i has a neighbor in $\{a_j, b_j\}$, the edge $a_j b_j$ is not colored with α. By the degree condition on u_i, such a coloring exists. Finally, since a_n has neighbors in the two edges $a_1 b_1$ and $a_2 b_2$ that are colored with the same color, there is some color α in $[\Delta - 1]$ for which no edge $a_i b_i$ with $i \in [n-1]$ such that a_n is adjacent to b_i, is colored with α, and we color the edge $a_n b_n$ with that color α. Suppose, for a contradiction, that M_β is not uniquely restricted for some color β in $[\Delta - 1]$. Let the edge $a_i b_i$ in M_β be such that it belongs to some M_β-alternating cycle C, and, subject to this condition, the index i is maximum. Since a' is non-adjacent to b'', we have $i \geq 3$. Let $u_n = a_n$. If the neighbor of u_i on C outside of $\{a_i, b_i\}$ is in $\{a_j, b_j\}$, then the choice of the edge $a_i b_i$ implies $j < i$, and the coloring rule implies that the edge $a_j b_j$ is not colored with β, which is a contradiction. This completes the proof.   □

Lemma 2 fails for $\Delta = 3$; the matching $\{a_1 b_1, a_2 b_2, a_3 b_3, a_4 b_4, a_5 b_5\}$ of the graph G in Fig. 2 cannot be partitioned into two uniquely restricted matchings. Note that the matching $\{a_1 b_3, a_2 b_1, a_3 b_5, a_4 b_2, a_5 b_4\}$ though is the union of the two uniquely restricted matchings $\{a_1 b_3, a_3 b_5\}$ and $\{a_2 b_1, a_4 b_2, a_5 b_4\}$.

Lemma 2 also fails for non-bipartite graphs; in fact, if G arises from the disjoint union of two copies of K_Δ by adding a perfect matching M, then every partition of M into uniquely restricted matchings requires Δ sets.

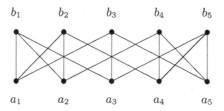

Fig. 2. A bipartite graph G.

With Lemma 2 at hand, the proof of our final result is easy.

Theorem 4. *If G is a connected bipartite graph of maximum degree at most $\Delta \geq 4$ that is distinct from $K_{\Delta,\Delta}$, then $\chi'_{ur}(G) \leq \Delta^2 - \Delta$.*

Proof: Since G is bipartite, its edge set can be partitioned into Δ matchings [25]. By Lemma 2, each of these matchings can be partitioned into $\Delta - 1$ uniquely restricted matchings. This completes the proof. $\square$

Note that the graph G in Fig. 2 also satisfies $\chi'_{ur}(G) \leq \Delta^2 - \Delta = 9 - 3 = 6$. In fact, the uniquely restricted matchings $\{a_1b_1, a_4b_2, a_5b_4\}$, $\{a_1b_2, a_2b_4, a_5b_5\}$, $\{a_2b_1, a_3b_3, a_4b_5\}$, $\{a_1b_3, a_4b_4\}$, $\{a_2b_2, a_3b_5\}$, and $\{a_3b_1, a_5b_3\}$ partition $E(G)$.

4   Concluding Remarks

Our results motivate several open problems. As stated above, we believe that the conclusion of Theorem 1 holds without the assumption of C_4-freeness. We also believe that better approximation factors are possible, and that approximation lower bounds in terms of the maximum degree could be proved. One could study the approximability of the uniquely restricted matching number in other classes of graphs. Finally, complexity results concerning the uniquely restricted chromatic index should be provided.

References

1. Alon, N., Sudakov, B., Zaks, A.: Acyclic edge colorings of graphs. J. Graph Theory **37**, 157–167 (2001)
2. Andersen, L.: The strong chromatic index of a cubic graph is at most 10. Discret. Math. **108**, 231–252 (1992)
3. Brandstädt, A., Mosca, R.: On distance-3 matchings and induced matchings. Discret. Appl. Math. **159**, 509–520 (2011)
4. Bruhn, H., Joos, F.: A stronger bound for the strong chromatic index. arXiv:1504.02583 (2015)
5. Cai, X., Perarnau, G., Reed, B., Watts, A.: Acyclic edge colourings of graphs with large girth. arXiv:1411.3047 (2014)
6. Cameron, K.: Induced matchings. Discret. Appl. Math. **24**, 97–102 (1989)

7. Cameron, K.: Induced matchings in intersection graphs. Discret. Math. **278**, 1–9 (2004)
8. Diestel, R.: Graph Theory. Graduate Texts in Mathematics. Springer, Heidelberg (2005)
9. Esperet, L., Parreau, A.: Acyclic edge-coloring using entropy compression. Eur. J. Comb. **34**, 1019–1027 (2013)
10. Faudree, R., Gyárfás, A., Schelp, R., Tuza, Z.: Induced matchings in bipartite graphs. Discret. Math. **78**, 83–87 (1989)
11. Faudree, R., Schelp, R., Gyárfás, A., Tuza, Z.: The strong chromatic index of graphs. Ars Combinatoria **29B**, 205–211 (1990)
12. Fiamčik, J.: The acyclic chromatic class of a graph. Mathematica Slovaca **28**, 139–145 (1978)
13. Francis, M.C., Jacob, D., Jana, S.: Uniquely restricted matchings in interval graphs. arXiv:1604.07016 (2016)
14. Golumbic, M., Hirst, T., Lewenstein, M.: Uniquely restricted matchings. Algorithmica **31**, 139–154 (2001)
15. Henning, M., Rautenbach, D.: Induced matchings in subcubic graphs without short cycles. Discret. Math. **35–316**, 165–172 (2014)
16. Hocquard, H., Montassier, M., Raspaud, A., Valicov, P.: On strong edge-colouring of subcubic graphs. Discret. Appl. Math. **161**, 2467–2479 (2013)
17. Hocquard, H., Ochem, P., Valicov, P.: Strong edge-colouring and induced matchings. Inf. Process. Lett. **113**, 836–843 (2013)
18. Horák, P., Qing, H., Trotter, W.: Induced matchings in cubic graphs. J. Graph Theory **17**, 151–160 (1993)
19. Joos, F., Rautenbach, D., Sasse, T.: Induced matchings in subcubic graphs. SIAM J. Discret. Math. **28**, 468–473 (2014)
20. Kang, R., Mnich, M., Müller, T.: Induced matchings in subcubic planar graphs. SIAM J. Discret. Math. **26**, 1383–1411 (2012)
21. Levit, V., Mandrescu, E.: Very well-covered graphs of girth at least four and local maximum stable set greedoids. Discret. Math. Algorithms Appl. **3**, 245–252 (2011)
22. Levit, V.E., Mandrescu, E.: Local maximum stable sets in bipartite graphs with uniquely restricted maximum matchings. Discret. Appl. Math. **132**, 163–174 (2003)
23. Levit, V.E., Mandrescu, E.: Unicycle graphs and uniquely restricted maximum matchings. Electron. Notes Discret. Math. **22**, 261–265 (2005)
24. Lovász, L.: Three short proofs in graph theory. J. Comb. Theory Ser. B **19**, 269–271 (1975)
25. Lovász, L., Plummer, M.: Matching Theory. North-Holland, Amsterdam (1986)
26. Lozin, V.: On maximum induced matchings in bipartite graphs. Inf. Process. Lett. **81**, 7–11 (2002)
27. Mishra, S.: On the maximum uniquely restricted matching for bipartite graphs. Electron. Notes Discret. Math. **37**, 345–350 (2011)
28. Molloy, M., Reed, B.: A bound on the strong chromatic index of a graph. J. Comb. Theory Ser. B **69**, 103–109 (1997)
29. Penso, L., Rautenbach, D., Souza, U.: Graphs in which some and every maximum matching is uniquely restricted. arXiv:1504.02250 (2015)
30. Stockmeyer, L., Vazirani, V.: NP-completeness of some generalizations of the maximum matching problem. Inf. Process. Lett. **15**, 14–19 (1982)
31. Vizing, V.: On an estimate of the chromatic class of a p-graph. Diskretnyj Analiz **3**, 25–30 (1964)
32. Yu, Q.R., Liu, G.: Graph Factors and Matching Extensions. Springer; Higher Education Press, Berlin; Beijing (2009)

Defective Coloring on Classes of Perfect Graphs

Rémy Belmonte[1(✉)], Michael Lampis[2], and Valia Mitsou[3]

[1] University of Electro-Communications, Chofu, Tokyo 182-8585, Japan
remy.belmonte@uec.ac.jp
[2] LAMSADE, CNRS, UMR 7243, Université Paris-Dauphine,
PSL Research University, 75016 Paris, France
michail.lampis@dauphine.fr
[3] LIRIS, CNRS, UMR 5205, Université de Lyon,
Université Lyon 1, 69622 Villeurbanne, Lyon, France
vmitsou@liris.cnrs.fr

Abstract. In DEFECTIVE COLORING we are given a graph G and two integers χ_d, Δ^* and are asked if we can χ_d-color G so that the maximum degree induced by any color class is at most Δ^*. We show that this natural generalization of COLORING is much harder on several basic graph classes. In particular, we show that it is NP-hard on split graphs, even when one of the two parameters χ_d, Δ^* is set to the smallest possible fixed value that does not trivialize the problem ($\chi_d = 2$ or $\Delta^* = 1$). Together with a simple treewidth-based DP algorithm this completely determines the complexity of the problem also on chordal graphs.

We then consider the case of cographs and show that, somewhat surprisingly, DEFECTIVE COLORING turns out to be one of the few natural problems which are NP-hard on this class. We complement this negative result by showing that DEFECTIVE COLORING is in P for cographs if either χ_d or Δ^* is fixed; that it is in P for trivially perfect graphs; and that it admits a sub-exponential time algorithm for cographs when both χ_d and Δ^* are unbounded.

1 Introduction

In this paper we study the computational complexity of DEFECTIVE COLORING, which is also known in the literature as IMPROPER COLORING: given a graph and two parameters χ_d, Δ^* we want to color the graph with χ_d colors so that every color class induces a graph with maximum degree at most Δ^*. DEFECTIVE COLORING is a very natural generalization of GRAPH COLORING, which corresponds to the case $\Delta^* = 0$. As a result, since the introduction of this problem more than thirty years ago [2,13] a great deal of research effort has been devoted to its study. Among the topics that have been investigated are its extremal properties [1,10,18,20,30,31], especially on planar graphs and graphs on surfaces [4,12,14,25], as well as its asymptotic behavior on random graphs

V. Mitsou—This work has been supported by the ANR-14-CE25-0006 project of the French National Research Agency.

© Springer International Publishing AG 2017
H.L. Bodlaender and G.J. Woeginger (Eds.): WG 2017, LNCS 10520, pp. 113–126, 2017.
https://doi.org/10.1007/978-3-319-68705-6_9

[28,29]. Lately, the problem has attracted renewed interest due to its applicability to communication networks, with the coloring of the graph modeling the assignment of frequencies to nodes and Δ^* representing some amount of tolerable interference. This has led to the study of the problem on Unit Disk Graphs [24] as well as various classes of grids [3,5,7]. Weighted generalizations have also been considered [6,23]. More background can be found in the survey by Frick [17] or Kang's Ph.D. thesis [27].

Our main interest in this paper is to study the computational complexity of DEFECTIVE COLORING through the lens of structural graph theory, that is, to investigate the classes of graphs for which the problem becomes tractable. Since DEFECTIVE COLORING generalizes GRAPH COLORING we immediately know that it is NP-hard already in a number of restricted graph classes and for small values of χ_d, Δ^*. Nevertheless, the fundamental question we would like to pose is what is the *additional* complexity brought to this problem by the freedom to produce a slightly improper coloring. In other words, we ask what are the graph classes where even though GRAPH COLORING is easy, DEFECTIVE COLORING is still hard (and conversely, what are the classes where both are tractable). Though some results of this type are already known (for example Cowen et al. [14] prove that the problem is hard even on planar graphs for $\chi_d = 2$), this question is not well-understood. Our focus on this paper is to study DEFECTIVE COLORING on subclasses of perfect graphs, which are perhaps the most widely studied class of graphs where GRAPH COLORING is in P. The status of the problem appears to be unknown here, and in fact its complexity on interval and even proper interval graphs is explicitly posed as an open question in [24].

Our results revolve around two widely studied classes of perfect graphs: split graphs and cographs. For split graphs we show not only that DEFECTIVE COLORING is NP-hard, but that it remains NP-hard even if either χ_d or Δ^* is a constant with the smallest possible non-trivial value ($\chi_d \geq 2$ or $\Delta^* \geq 1$). To complement these negative results we provide a treewidth-based DP algorithm which runs in polynomial time if both χ_d and Δ^* are constant, not only for split graphs, but also for chordal graphs. This generalizes a previous algorithm of Havet et al. [24] on interval graphs.

We then go on to show that DEFECTIVE COLORING is also NP-hard when restricted to cographs. We note that this result is somewhat surprising since relatively few natural problems are known to be hard for cographs. We complement this negative result in several ways. First, we show that DEFECTIVE COLORING becomes polynomially solvable on trivially perfect graphs, which form a large natural subclass of cographs. Second, we show that, unlike the case of split graphs, DEFECTIVE COLORING is in P on cographs if either χ_d or Δ^* is fixed. Both of these results are based on dynamic programming algorithms. Finally, by combining the previous two algorithms with known facts about the relation between χ_d and Δ^* we obtain a sub-exponential time algorithm for DEFECTIVE COLORING on cographs. We note that the existence of such an algorithm for split graphs is ruled out by our reductions, under the Exponential Time Hypothesis.

Table 1. Summary of results

Chordal graphs	Cographs
NP-hard on Split if $\chi_d \geq 2$	NP-hard
Theorem 10	Theorem 2
NP-hard on Split if $\Delta^* \geq 1$	In P if χ_d or Δ^* is fixed
Theorem 9	Theorems 5,6
In P if χ_d, Δ^* fixed	Solvable in $n^{O(n^{4/5})}$
Theorem 13	Theorem 7
In P on Trivially perfect for any χ_d, Δ^*	
Theorem 4	

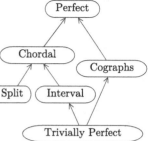

Table 1 summarizes our results. For the reader's convenience, it also depicts an inclusion diagram for the graph classes that we mention.

2 Preliminaries and Definitions

We use standard graph theory terminology, see e.g. [16]. In particular, for a graph $G = (V, E)$ and $u \in V$ we use $N(u)$ to denote the set of neighbors of u, $N[u]$ denotes $N(u) \cup \{u\}$, and for $S \subseteq V$ we use $G[S]$ to denote the subgraph induced by the set S. A proper coloring of G with χ colors is a function $c : V \to \{1, \ldots, \chi\}$ such that for all $i \in \{1, \ldots, \chi\}$ the graph $G[c^{-1}(i)]$ is an independent set. We will focus on the following generalization of coloring:

Definition 1. *If χ_d, Δ^* are positive integers then a (χ_d, Δ^*)-coloring of a graph $G = (V, E)$ is a function $c : V \to \{1, \ldots, \chi_d\}$ such that for all $i \in \{1, \ldots, \chi_d\}$ the maximum degree of $G[c^{-1}(i)]$ is at most Δ^*.*

We call the problem of deciding if a graph admits a (χ_d, Δ^*)-coloring, for given parameters χ_d, Δ^*, DEFECTIVE COLORING. For a graph G and a coloring function $c : V \to \mathbb{N}$ we say that the *deficiency* of a vertex u is $|N(u) \cap c^{-1}(c(u))|$, that is, the number of its neighbors with the same color. The deficiency of a color class i is defined as the maximum deficiency of any vertex colored with i.

We recall the following basic facts about DEFECTIVE COLORING:

Lemma 1 [27]. *For any χ_d, Δ^* and any graph $G = (V, E)$ with $\chi_d \cdot \Delta^* \geq |V|$ we have that G admits a (χ_d, Δ^*)-coloring.*

Lemma 2 [27]. *If G admits a (χ_d, Δ^*)-coloring then $\omega(G) \leq \chi_d \cdot (\Delta^* + 1)$.*

Let us now also give some quick reminders regarding the definitions of the graph classes we consider in this paper.

A graph $G = (V, E)$ is a *split* graph if $V = K \cup S$ where K induces a clique and S induces an independent set. A graph G is *chordal* if it does not contain any induced cycles of length four or more. It is well known that all split graphs are chordal; furthermore it is known that the class of chordal graphs contains the class of interval graphs, and that chordal graphs are perfect. For more information on these standard containments see [11].

A graph is a *cograph* if it is either a single vertex, or the disjoint union of two cographs, or the complete join of two cographs [33]. A graph is *trivially perfect* if in all induced subgraphs the maximum independent set is equal to the number of maximal cliques [21]. Trivially perfect graphs are exactly the cographs which are chordal [34], and hence are a subclass of cographs, which are a subclass of perfect graphs. We recall that GRAPH COLORING is polynomial-time solvable in all the mentioned graph classes, since it is polynomial-time solvable on perfect graphs [22], though of course for all these classes simpler and more efficient combinatorial algorithms are known.

We will also use the notion of treewidth for the definition of which we refer the reader to [9,15].

3   NP-Hardness on Cographs

In this section we establish that DEFECTIVE COLORING is already NP-hard on the very restricted class of cographs. To this end, we show a reduction from 4-PARTITION.

Definition 2. *In* 4-PARTITION *we are given a set A of $4n$ elements, a size function $s : A \to \mathbb{N}^+$ which assigns a value to each element, and a target integer B. We are asked if there exists a partition of A into n sets of four elements (quadruples), such that for each set the sum of its elements is exactly B.*

4-PARTITION has long been known to be strongly NP-hard, that is, NP-hard even if all values are polynomially bounded in n. In fact, the reduction given in [19] establishes the following, slightly stronger statement.

Theorem 1. 4-PARTITION *is strongly NP-complete if A is given to us partitioned into four sets of equal size A_1, A_2, A_3, A_4 and any valid solution is required to place exactly one element from each $A_i, i \in \{1, \ldots, 4\}$ in each quadruple.*

Theorem 2. DEFECTIVE COLORING *is NP-complete even when restricted to complete k-partite graphs. Therefore,* DEFECTIVE COLORING *is NP-complete on cographs.*

Proof. We start our reduction from an instance of 4-PARTITION where the set of elements A is partitioned into four equal-size sets as in Theorem 1. We first transform the instance by altering the sizes of all elements as follows: for each element $x \in A_i$ we set $s'(x) := s(x) + 5^i B + 5^5 n^2 B$ and we also set $B' := B + B \cdot \sum_{i=1}^{4} 5^i + 4 \cdot 5^5 n^2 B$. After this transformation our instance is "ordered", in the sense that all elements of A_{i+1} have strictly larger size than all elements of A_i. Furthermore, it is not hard to see that the answer to the problem did not change, as any quadruple that used one element from each A_i and summed up to B now sums up to B'. In addition, we observe that in the new instance the condition that exactly one element must be used from each set is imposed by the element sizes themselves: a quadruple that contains two or more elements of A_4 will have sum strictly more than B', while one containing no elements of A_4 will have sum strictly less than B'. Similar reasoning can then be applied to A_3, A_2. We note that the element sizes now have the extra property that $s'(x) \in (B'/4 - 5B'/n^2, B'/4 + 5B'/n^2)$.

We now construct an instance of DEFECTIVE COLORING as follows. We set $\Delta^* = B'$ and $\chi_d = n$. To construct the graph G, for each element $x \in A_2 \cup A_3 \cup A_4$ we create an independent set of $s'(x)$ new vertices which we will call V_x. For each element $x \in A_1$ we construct two independent sets of $s'(x)$ new vertices each, which we will call V_x^1 and V_x^2. Finally, we turn the graph into a complete $5n$-partite graph, that is, we add all possible edges while retaining the property that all sets V_x and V_x^i remain independent.

Let us now argue for the correctness of the reduction. First, suppose that there exists a solution to our (modified) 4-PARTITION instance where each quadruple sums to B'. Number the quadruples arbitrarily from 1 to n and consider the i-th quadruple $(x_i^1, x_i^2, x_i^3, x_i^4)$ where we assume that for each $j \in \{1, \ldots, 4\}$ we have $x_i^j \in A_j$. Hence, $s'(x_i^1)$ is minimum among the sizes of the elements of the quadruple. We now use color i for all the vertices of the sets $V_{x_i^j}$ for $j \in \{2, 3, 4\}$ as well as the sets $V_{x_i^1}^1, V_{x_i^1}^2$. We continue in this way using a different color for each quadruple and thus color the whole graph (since the quadruples use all the elements of A). We observe that for any color i the vertices with maximum deficiency are those from $V_{x_i^1}^1$ and $V_{x_i^1}^2$, and all these vertices have deficiency exactly $x_i^1 + x_i^2 + x_i^3 + x_i^4 = B'$. Hence, this is a valid solution.

For the converse direction of the reduction, suppose we are given a (χ_d, Δ^*)-coloring of the graph we constructed. We first need to argue that such a coloring must have a very special structure. In particular, we will claim that in such a coloring each independent set V_x or V_x^i must be monochromatic. Towards this end we formulate a number of claims.[1]

Claim 1 ($\star$). *Every color i is used on at most $5B'/4 + 25B'/n^2$ vertices.*

Because of the previous claim, which states that no color appears too many times, we can also conclude that no color appears too few times.

Claim 2 ($\star$). *Every color i is used on at least $5B'/4 - 50B'/n$ vertices.*

Given the above bounds on the size of each color class we can now conclude that each color appears in exactly five independent sets V_x.

Claim 3 ($\star$). *For each color i the graph induced by $c^{-1}(i)$ is complete 5-partite.*

Claim 4 ($\star$). *In any valid solution every maximal independent set of G is monochromatic.*

---

[1] Due to space restrictions, the proofs of statements marked with ($\star$) are omitted.

We are now ready to complete the converse direction of the reduction. Consider the vertices of $c^{-1}(i)$, for some color i. By the previous sequence of claims we know that they appear in and fully cover 5 independent sets V_x or V_x^i. We claim that for each $j \in \{2, 3, 4\}$ any color i is used in exactly one V_x with $x \in A_j$. This can be seen by considering the deficiency of the vertices of the smallest independent set where i appears. The deficiency of these vertices is equal to $x_i^1 + x_i^2 + x_i^3 + x_i^4$, which are the sizes of the four larger independent sets. By the construction of the modified 4-PARTITION instance, any quadruple that contains two elements of A_4 will have sum strictly greater than B'. Hence, these elements must be evenly partitioned among the color classes, and with similar reasoning the same follows for the elements of A_3, A_2 and A_1.

We thus arrive at a situation where each color i appears in the independent sets $V_{x_i^4}, V_{x_i^3}, V_{x_i^2}$ as well as two of the "small" independent sets. Recall that all "small" independent sets were constructed in two copies of the same size V_x^1, V_x^2. We would now like to ensure that all color classes contain one small independent set of the form $V_{x_i^1}^1$. If we achieve this the argument will be complete: we construct the quadruple $(x_i^4, x_i^3, x_i^2, x_i^1)$ from the color class i, and the deficiency of the vertices of the remaining small independent set ensures that the sum of the elements of the quadruple is at most B'. By constructing n such quadruples we conclude that they all have sum exactly B', since the sum of all elements of the 4-PARTITION instance is (without loss of generality) exactly nB'.

To ensure that each color class contains an independent set V_x^1 we first observe that we can always exchange the colors of independent sets V_x^1 and V_x^2, since they are both of equal size (and monochromatic). Construct now an auxiliary graph with χ_d vertices, one for each color class and a directed edge for each $x \in A_1$. Specifically, if for $x \in A_1$ the independent set V_x^1 is colored i and the set V_x^2 is colored j we place a directed edge from i to j (note that this does not rule out the possibility of self-loops). In the auxiliary directed graph, each vertex that does not have a self-loop is incident on two directed edges. We would like all such vertices to end up having out-degree 1, because then each color class would contain an independent set of the form V_x^1. The main observation now is that in each connected component in the underlying undirected graph that contains a vertex u with out-degree 0 there must also exist a vertex v of out-degree 2. Exchanging the colors of V_x^1 and V_x^2 corresponds to flipping the direction of an edge in the auxiliary graph. Hence, we can take a maximal directed path starting at v and flip all its edges, while maintaining a valid coloring of the original graph. This decreases the number of vertices with out-degree 0 and therefore repeating this process completes the proof.   □

4   Polynomial Time Algorithm on Trivially Perfect Graphs

In this section, we complement the NP-completeness proof from Sect. 3 by giving a polynomial time algorithm for DEFECTIVE COLORING on the class of trivially perfect graphs, a well-studied subclass of cographs and interval graphs. We will heavily rely on the following equivalent characterization of trivially perfect graphs given by Golumbic [21]:

Theorem 3. *A graph is trivially perfect if and only if it is the comparability graph of a rooted tree.*

In other words, for every trivially perfect graph G, there exists a rooted tree T such that making every vertex in the tree adjacent to all of its descendants yields a graph isomorphic to G. We refer to T as the *underlying rooted tree* of G. We recall that it is known how to obtain T from G in polynomial (in fact linear) time [34].

We are now ready to describe our algorithm. The following observation is one of its basic building blocks.

Lemma 3. *Let $G = (V, E)$ be a trivially perfect graph, T its underlying rooted tree, and $u \in V$ be an ancestor of $v \in V$ in T. Then $N[v] \subseteq N[u]$.*

Theorem 4. DEFECTIVE COLORING *can be solved in polynomial time on trivially perfect graphs.*

Proof. Given a trivially perfect graph $G = (V, E)$ with underlying rooted tree $T = (V, E')$ and two non-negative integers χ_d and Δ^*, we compute a coloring of G using at most χ_d colors and with deficiency at most Δ^* as follows. First, we partition the vertices of T (and therefore of G) into sets $V_1, \ldots, V_\ell$, where ℓ denotes the height of T, such that V_1 contains the leaves of T and, for every $2 \leq i \leq \ell$, V_i contains the leaves of $T \setminus (\bigcup_{j=1}^{i-1} V_j)$. Observe that each set V_i is an independent set in G. We now start our coloring by giving all vertices of V_1 color 1. Then, for every $2 \leq i \leq \ell$, we color the vertices of V_i by giving each of them the lowest color available, i.e., we color each vertex u with the lowest j such that u has at most Δ^* descendants colored j. If for some vertex no color is available, that is, its subtree contains at least $\Delta^* + 1$ vertices colored with each of the colors $\{1, \ldots, \chi_d\}$, then we return that G does not admit a (χ_d, Δ^*)-coloring.

This procedure can clearly be performed in polynomial time and, if it returns a solution, it uses at most χ_d colors. Furthermore, whenever the procedure uses color i on a vertex u it is guaranteed that u has deficiency at most Δ^* among currently colored vertices. Since any neighbor of u that is currently colored with i must be a descendant of u, by Lemma 3 this guarantees that the deficiency of all vertices remains at most Δ^* at all times.

It now only remains to prove that the algorithm concludes that G cannot be colored with χ_d colors and deficiency Δ^* only when no such coloring exists. For this we will rely on the following claim which states that any valid coloring can be "sorted".

Claim. If G admits a (χ_d, Δ^*)-coloring, then there exists a (χ_d, Δ^*)-coloring c of G such that, for every two vertices $u, v \in V(G)$, if v is a descendant of u, then $c(v) \leq c(u)$.

It follows from the previous claim that if a (χ_d, Δ^*)-coloring exists, then a sorted (χ_d, Δ^*)-coloring exists where ancestors always have colors at least as high as their descendants. We can now argue that our algorithm also produces a sorted coloring, with the extra property that whenever it sets $c(u) = i$ we know that *any* sorted (χ_d, Δ^*)-coloring of G must give color at least i to u. This can be shown by induction on i: it is clear for the vertices of V_1 to which the algorithm gives color 1; and if the algorithm assigns color i to u, then u has $\Delta^* + 1$ descendants which (by inductive hypothesis) must have color at least $i - 1$ in any valid sorted coloring of G. □

5   Algorithms on Cographs

In this section we present algorithms that can solve DEFECTIVE COLORING on cographs in polynomial time if either Δ^* or χ_d is bounded; both algorithms rely on dynamic programming. After presenting them we show how their combination can be used to obtain a sub-exponential time algorithm for DEFECTIVE COLORING on cographs.

5.1   Algorithm for Small Deficiency

We now present an algorithm that solves DEFECTIVE COLORING in polynomial time on cographs if Δ^* is bounded. Before we proceed, let us sketch the main ideas behind the algorithm. Given a (χ_d, Δ^*)-coloring c of a graph G, we define the *type* of a color class i, as the pair of two integers (s_i, d_i) where $s_i := \min\{|c^{-1}(i)|, \Delta^* + 1\}$ and d_i is the maximum degree of $G[c^{-1}(i)]$. In other words, the type of a color class is characterized by its size (up to value $\Delta^* + 1$) and the maximum deficiency of any of its vertices. We observe that there are at most $(\Delta^* + 1)^2$ possible types in a valid (χ_d, Δ^*)-coloring, because s_i only takes values in $\{1, \ldots, \Delta^* + 1\}$ and d_i in $\{0, \ldots, \Delta^*\}$.

We can now define the *signature* of a coloring c as a tuple which contains one element for every possible color type (s, d). This element is the number of color classes in c that have type (s, d), and hence is a number in $\{0, \ldots, \chi_d\}$. We can conclude that there are at most $(\chi_d + 1)^{(\Delta^* + 1)^2}$ possible signatures that a valid (χ_d, Δ^*)-coloring can have. Our algorithm will maintain a binary table which states for each possible signature if the current graph admits a (χ_d, Δ^*)-coloring with this signature. The obstacle now is to describe a procedure which, given two such tables for graphs G_1, G_2 is able to generate the table of admissible signatures for their union and their join.

Theorem 5 ($\star$). *There is an algorithm which decides if a cograph admits a* (χ_d, Δ^*)-coloring in time $O^* \left(\chi_d^{O((\Delta^*)^4)} \right)$.

5.2   Algorithm for Few Colors

In this section we provide an algorithm that solves DEFECTIVE COLORING in polynomial time on cographs if χ_d is bounded. The type of a color class i is defined in a similar manner as in the first paragraph of Sect. 5.1, with the only difference that the first coordinate of the output pair takes values in $\{0, \ldots, \Delta^* + 1\}$. The signature S of a coloring c is now a function $S : \{1, \ldots, \chi_d\} \rightarrow \{0, \ldots, \Delta^* + 1\} \times \{0, \ldots, \Delta^*\}$, which takes as input a color class and returns its type. Once again, we should maintain a table T of size less than $(\Delta^* + 2)^{2\chi_d}$ for which $T(S) = 1$ if and only if there is a (χ_d, Δ^*)-coloring of signature S for the current graph G. As in the previous section, we shall describe how to compute table T of a graph G which is the union or the join of two graphs G_1 and G_2 whose tables T_1 and T_2 are known.

Theorem 6 ($\star$). *There is an algorithm which decides if a cograph admits a (χ_d, Δ^*)-coloring in time $O^*\left((\Delta^*)^{O(\chi_d)}\right)$.*

5.3   Sub-Exponential Time Algorithm

We now combine the algorithms of Sects. 5.1 and 5.2 in order to obtain a sub-exponential time algorithm for cographs.

Theorem 7 ($\star$). DEFECTIVE COLORING *can be solved in time* $n^{O\left(n^{4/5}\right)}$ *on cographs.*

6   Split and Chordal Graphs

In this section we present the following results: first, we show that DEFECTIVE COLORING is hard on split graphs even when Δ^* is a fixed constant, as long as $\Delta^* \geq 1$; the problem is of course in P if $\Delta^* = 0$. Then, we show that DEFECTIVE COLORING is hard on split graphs even when χ_d is a fixed constant, as long as $\chi_d \geq 2$; the problem is of course trivial if $\chi_d = 1$. These results completely describe the complexity of the problem when one of the two relevant parameters is fixed. We then give a treewidth-based procedure through which we obtain a polynomial-time algorithm even on chordal graphs when χ_d, Δ^* are bounded (in fact, the algorithm is FPT parameterized by $\chi_d + \Delta^*$). Hence these results give a complete picture of the complexity of the problem on chordal graphs: the problem is still hard when one of χ_d, Δ^* is bounded, but becomes easy if both are bounded.

Let us also remark that both of the reductions we present are linear. Hence, under the Exponential Time Hypothesis [26], they establish not only NP-hardness, but also unsolvability in time $2^{o(n)}$ for DEFECTIVE COLORING on split graphs, for constant values of χ_d or Δ^*. This is in contrast with the results of Sect. 5.3 on cographs.

6.1   Hardness for Bounded Deficiency

In this section we show that DEFECTIVE COLORING is NP-hard for any fixed value $\Delta^* \geq 1$. We first show hardness for $\Delta^* = 1$, then we tweak our reduction in order to make it work for larger Δ^*.

We will reduce from 3CNFSAT. Suppose we are given a CNF formula f where $X = \{x_1, \ldots, x_n\}$ are the variables and $C = \{c_1, \ldots, c_m\}$ are the clauses and each clause contains exactly 3 literals. We construct a split graph $G = (V, E)$, where $\{U, Z\}$ is a partition of V with U inducing a clique of $4n$ vertices and Z inducing an independent set of $m + 4n$ vertices, such that having a satisfying assignment $s : X \rightarrow \{T, F\}$ for f implies a $(2n, 1)$-coloring $c : V \rightarrow \{1, \ldots, 2n\}$ for G and vice versa.

The construction is as follows. For every variable $x_i, i \in \{1, \ldots, n\}$ we construct a set of four vertices $U_i = \{u_i^A, u_i^B, u_i^C, u_i^D\}$ which are part of the clique vertices U (that is, for all $i \in \{1, \ldots, n\}$ and $k \in \{A, B, C, D\}$, vertices u_i^k are fully connected). For each $i \in \{1, 2, \ldots, n\}$, we also construct four vertices $Z_i = \{z_i^A, z_i^B, z_i^C, z_i^D\}$ in the independent set Z. Furthermore, for each clause $c_j, j \in \{1, \ldots, m\}$ we construct a vertex v_j in the independent set. Lastly we add every edge between U and Z save for the following non-edges: for every $i \in \{1, \ldots, n\}, k \in \{A, B, C, D\}$, z_i^k does not connect to vertices $u_i^{k'}, k' \neq k$ and for every $i \in \{1, \ldots, n\}, j \in \{1, \ldots, m\}$, if clause c_j contains variable x_i then: if x_i appears positive then v_j does not connect to u_i^A, u_i^B, whereas if it appears negative then v_j does not connect to u_i^A, u_i^C. This completes the construction.

Lemma 4. *Given a satisfying assignment $s : X \rightarrow \{T, F\}$ for f we can always construct a $(2n, 1)$-coloring $c : V \rightarrow \{1, \ldots, 2n\}$.*

Proof. Let us first assign colors to the clique vertices. We are going to use two distinct colors for every quadruple U_i. The way we choose to color vertices in U_i should depend on the assignment $s(x_i)$: if $s(x_i) = T$ then $c(u_i^A) = c(u_i^B) = 2i - 1$ and $c(u_i^C) = c(u_i^D) = 2i$; if $s(x_i) = F$ then $c(u_i^A) = c(u_i^C) = 2i - i$ and $c(u_i^B) = c(u_i^D) = 2i$. Observe that we have consumed the entire supply of the $2n$ available colors on coloring U and for every color $l \in \{1, \ldots, 2n\}$ we have that $|c^{-1}(l) \cap U| = 2$.

In order to finish coloring the independent set Z, we can only reuse colors that have already appeared in U. If for some $z \in Z$ there exists a color l such that $c^{-1}(l) \cap N(z) = \emptyset$, that is if both vertices of color l in U are non-neighbors of z, then we can assign $c(z) = l$. Remember that a vertex in Z_i is a non-neighbor of exactly three vertices in U_i, thus two of them should be using the same color. Additionally, if s is a satisfying assignment for f, then for every c_j there is at least one satisfied literal, say $(\neg)x_i$ and by the construction of G and the assignment of colors on U we should be able once again to find two vertices in U_i having the same color that v_j does not connect to, these should be u_i^A and, depending on s, either u_i^B or u_i^C.   □

Lemma 5. *Given a $(2n, 1)$-coloring $c : V \rightarrow \{1, \ldots, 2n\}$ of G, we can produce a satisfying assignment $s : X \rightarrow \{T, F\}$ for f.*

Proof. First, observe that, since $\Delta^* = 1$, for any color $l \in \{1, \ldots, 2n\}$ we have that $|c^{-1}(l) \cap U| \leq 2$. Since there are at most $2n$ colors in use and $|U| = 4n$, that means that the color classes of c should induce a matching of size $2n$ in the clique. The above imply that for any $z \in Z$ with $c(z) = l$ there exist $u, u' \in U$ with $c(u) = c(u') = l$ which are non-neighbors of z.

We can now make the following claim:

Claim. For any $u, u' \in U$, if $c(u) = c(u')$ then there exists i such that $u, u' \in U_i$.

Proof. This is a consequence of vertices in Z_i having exactly three non-neighbors in U all of them belonging to U_i. More precisely, for any $k \in \{A, B, C, D\}$, $c(z_i^k) = l$ for some color l implies that $\exists k_1, k_2 \neq k$ such that $c(u_i^{k_1}) = c(u_i^{k_2})(= l)$. Similarly, the fact that $c(z_i^{k_1}) = l'$ for some color l' together with the fact that $|c^{-1}(l) \cap U| = 2$ gives us that $c(u_i^k) = c(u_i^{k'})(= l')$, where of course $u_i^k, u_i^{k'}$ are the only vertices of U colored l'. □

The above claim directly provides the assignment: if $c(u_i^A) = c(u_i^B)$ then set $s(x_i) = T$, else $s(x_i) = F$.

Claim. The assignment s as described above satisfies f.

Proof. By construction, for all $j \in \{1, \ldots, m\}$, vertex v_j should be a non-neighbor to six vertices of U. At least two of them, call them u, u' should have the same color as v_j. From the previous claim, u, u' should belong to the same group U_i. By construction $u = u_i^A$ and $u' \in \{u_i^B, u_i^C\}$. Consider that $u' = u_i^B$ (similar arguments hold when $u' = u_i^C$). Since $c(u_i^A) = c(u_i^B)$, the assignment should set $s(x_i) = T$. Observe now that c_j, which by construction contains literal x_i, should be satisfied.
This concludes the proof. □

Lemmata 4 and 5 prove the following Theorem:

Theorem 8. DEFECTIVE COLORING *is NP-hard on split graphs for* $\Delta^* = 1$.

To show hardness for $\Delta^* \geq 2$, all we need to do is slightly change the above construction so that we are now forced to create bigger color classes. Namely, we add $2(\Delta^* - 1)n$ more vertices to U which we divide into $2n$ sets U_i^D and U_i^A and which we fully connect to each other and to previous vertices of U. We also remove vertices z_i^B, z_i^C from Z_i. Last, for $k \in \{A, D\}$, we connect vertices of U_i^k to all vertices in Z save for the following: U_i^D does not connect to z_i^A and U_i^A does not connect to z_i^D and to v_j if variable x_i appears in clause c_j.

Lemma 6 ($\star$). *Given a satisfying assignment s for f we can always construct a $(2n, \Delta^*)$-coloring c.*

Lemma 7 ($\star$). *Given a $(2n, \Delta^*)$-coloring c, we can produce a satisfying assignment s for f.*

The main theorem of this section follows from Lemmata 6, 7 and Theorem 8.

Theorem 9. DEFECTIVE COLORING *is NP-hard on split graphs for any fixed* $\Delta^* \geq 1$.

6.2   Hardness for Bounded Number of Colors

Theorem 10 $(\star)$. DEFECTIVE COLORING *is NP-complete on split graphs for every fixed value of* $\chi_d \geq 2$.

6.3   A Dynamic Programming Algorithm

In this section we present an algorithm which solves the problem efficiently on chordal graphs when χ_d and Δ^* are small. Our main tool is a treewidth-based procedure, as well as known connections between the maximum clique size and treewidth of chordal graphs.

Theorem 11 $(\star)$. DEFECTIVE COLORING *can be solved in time* $(\chi_d \Delta^*)^{O(tw)}$ $n^{O(1)}$ *on any graph* G *with* n *vertices if a tree decomposition of width* tw *of* G *is supplied with the input.*

We now recall the following theorem connecting $\omega(G)$ and $tw(G)$ for chordal graphs.

Theorem 12 [8,32]. *In chordal graphs* $\omega(G) = tw(G) + 1$. *Furthermore, an optimal tree decomposition of a chordal graph can be computed in polynomial time.*

Together with Lemma 2 this gives the following algorithm for chordal graphs.

Theorem 13. DEFECTIVE COLORING *can be solved in time* $(\chi_d \Delta^*)^{O(\chi_d \Delta^*)}$ $n^{O(1)}$ *in chordal graphs.*

Proof. We use Theorem 12 to compute an optimal tree decomposition of the input graph and its maximum clique size. If $\omega(G) > \chi_d(\Delta^* + 1)$ then we can immediately reject by Lemma 2. Otherwise, we know that $tw(G) \leq \chi_d(\Delta^* + 1)$ from Theorem 12, so we apply the algorithm of Theorem 11. $\square$

7   Conclusions

Our results indicate that DEFECTIVE COLORING is significantly harder than GRAPH COLORING, even on classes where the latter is easily in P. Though we have completely characterized the complexity of the problem on split and chordal graphs, its tractability on interval and proper interval graphs remains an interesting open problem as already posed in [24].

Beyond this, the results of this paper point to several potential future directions. First, the algorithms we have given for cographs are both XP parameterized by χ_d or Δ^*. Is it possible to obtain FPT algorithms? On a related question, is it possible to obtain a faster sub-exponential time algorithm for DEFECTIVE COLORING on cographs? Second, is it possible to find other natural classes of graphs, beyond trivially perfect graphs, which are structured enough to make DEFECTIVE COLORING tractable? Finally, in this paper we have not considered the question of approximation algorithms. Though in general DEFECTIVE COLORING is likely to be quite hard to approximate (as a consequence of the hardness of GRAPH COLORING), it seems promising to also investigate this question in classes where GRAPH COLORING is in P.

References

1. Achuthan, N., Achuthan, N.R., Simanihuruk, M.: On minimal triangle-free graphs with prescribed k-defective chromatic number. Discret. Math. **311**(13), 1119–1127 (2011)
2. Andrews, J.A., Jacobson, M.S.: On a generalization of chromatic number. Congr. Numer. **47**, 33–48 (1985)
3. Araújo, J., Bermond, J.-C., Giroire, F., Havet, F., Mazauric, D., Modrzejewski, R.: Weighted improper colouring. J. Discret. Algorithms **16**, 53–66 (2012)
4. Archdeacon, D.: A note on defective colorings of graphs in surfaces. J. Graph Theory **11**(4), 517–519 (1987)
5. Archetti, C., Bianchessi, N., Hertz, A., Colombet, A., Gagnon, F.: Directed weighted improper coloring forcellular channel allocation. Discret. Appl. Math. **182**, 46–60 (2015)
6. Bang-Jensen, J., Halldórsson, M.M.: Vertex coloring edge-weighted digraphs. Inf. Process. Lett. **115**(10), 791–796 (2015)
7. Bermond, J.-C., Havet, F., Huc, F., Sales, C.L.: Improper coloring of weighted grid and hexagonal graphs. Discret. Math. Algorithms Appl. **2**(3), 395–412 (2010)
8. Bodlaender, H.L.: A partial k-arboretum of graphs with bounded treewidth. Theor. Comput. Sci. **209**(1–2), 1–45 (1998)
9. Bodlaender, H.L., Koster, A.M.C.A.: Combinatorial optimization on graphs of bounded treewidth. Comput. J. **51**(3), 255–269 (2008)
10. Borodin, O.V., Kostochka, A.V., Yancey, M.: On 1-improper 2-coloring of sparse graphs. Discret. Math. **313**(22), 2638–2649 (2013)
11. Brandstädt, A., Le, V.B., Spinrad, J.P.: Graph classes: a survey. SIAM (1999)
12. Choi, I., Esperet, L.: Improper coloring of graphs on surfaces. arXiv e-prints, March 2016
13. Cowen, L.J., Cowen, R.H., Woodall, D.R.: Defective colorings of graphs in surfaces: partitions into subgraphs of bounded valency. J. Graph Theory **10**(2), 187–195 (1986)
14. Cowen, L., Goddard, W., Jesurum, C.E.: Defective coloring revisited. J. Graph Theory **24**(3), 205–219 (1997)
15. Cygan, M., Fomin, F.V., Kowalik, L., Lokshtanov, D., Marx, D., Pilipczuk, M., Pilipczuk, M., Saurabh, S.: Parameterized Algorithms. Springer, Cham (2015)
16. Diestel, R.: Graph Theory. Graduate Texts in Mathematics, vol. 173, 4th edn. Springer, Heidelberg (2012)
17. Frick, M.: A survey of (m, k)-colorings. Ann. Discret. Math. **55**, 45–57 (1993)
18. Frick, M., Henning, M.A.: Extremal results on defective colorings of graphs. Discret. Math. **126**(1–3), 151–158 (1994)
19. Garey, M.R., Johnson, D.S.: Computers and Intractability: A Guide to the Theory of NP-Completeness. W. H. Freeman & Co., New York (1979)
20. Goddard, W., Honghai, X.: Fractional, circular, and defective coloring of series-parallel graphs. J. Graph Theory **81**(2), 146–153 (2016)
21. Golumbic, M.C.: Trivially perfect graphs. Discret. Math. **24**(1), 105–107 (1978)
22. Grötschel, M., Lovász, L., Schrijver, A.: Geometric Algorithms and Combinatorial Optimization, vol. 4060 XII, p. 362 S. Springer, Berlin (1988)
23. Gudmundsson, B.A., Magnússon, T.K., Sæmundsson, B.O.: Bounds and fixed-parameter algorithms for weighted improper coloring. Electron. Notes Theor. Comput. Sci. **322**, 181–195 (2016)

24. Havet, F., Kang, R.J., Sereni, J.-S.: Improper coloring of unit disk graphs. Networks **54**(3), 150–164 (2009)
25. Havet, F., Sereni, J.-S.: Improper choosability of graphs and maximum average degree. J. Graph Theory **52**(3), 181–199 (2006)
26. Impagliazzo, R., Paturi, R., Zane, F.: Which problems have strongly exponential complexity? J. Comput. Syst. Sci. **63**(4), 512–530 (2001)
27. Kang, R.J.: Improper colourings of graphs. Ph.D. thesis, University of Oxford, UK (2008)
28. Kang, R.J., McDiarmid, C.: The t-improper chromatic number of random graphs. Comb. Probab. Comput. **19**(1), 87–98 (2010)
29. Kang, R.J., Müller, T., Sereni, J.-S.: Improper colouring of (random) unit disk graphs. Discret. Math. **308**(8), 1438–1454 (2008)
30. Kim, J., Kostochka, A.V., Zhu, X.: Improper coloring of sparse graphs with a given girth, I:(0, 1)-colorings of triangle-free graphs. Eur. J. Comb. **42**, 26–48 (2014)
31. Kim, J., Kostochka, A.V., Zhu, X.: Improper coloring of sparse graphs with a given girth, II: constructions. J. Graph Theory **81**(4), 403–413 (2016)
32. Robertson, N., Seymour, P.D.: Graph minors. II. Algorithmic aspects of tree-width. J. Algorithms **7**(3), 309–322 (1986)
33. Seinsche, D.: On a property of the class of n-colorable graphs. J. Comb. Theory Ser. B **16**(2), 191–193 (1974)
34. Jing-Ho, Y., Jer-Jeong, C., Chang, G.J.: Quasi-threshold graphs. Discret. Appl. Math. **69**(3), 247–255 (1996)

Token Sliding on Chordal Graphs

Marthe Bonamy[1] and Nicolas Bousquet[2(✉)]

[1] CNRS, LaBRI, Université de Bordeaux, Bordeaux, France
marthe.bonamy@labri.fr
[2] CNRS, G-SCOP, Univ. Grenoble Alpes, Grenoble, France
nicolas.bousquet@grenoble-inp.fr

Abstract. Let I be an independent set of a graph G. Imagine that a token is located on every vertex of I. We can now move the tokens of I along the edges of the graph as long as the set of tokens still defines an independent set of G. Given two independent sets I and J, the TOKEN SLIDING problem consists in deciding whether there exists a sequence of independent sets which transforms I into J so that every pair of consecutive independent sets of the sequence can be obtained via a single token move. This problem is known to be PSPACE-complete even on planar graphs.

In [9], Demaine et al. asked whether the TOKEN SLIDING problem can be decided in polynomial time on interval graphs and more generally on chordal graphs. Yamada and Uehara [20] showed that a polynomial time transformation can be found in proper interval graphs.

In this paper, we answer the first question of Demaine et al. and generalize the result of Yamada and Uehara by showing that we can decide in polynomial time whether an independent set I of an interval graph can be transformed into another independent set J. Moreover, we answer similar questions by showing that: (i) determining if there exists a token sliding transformation between every pair of k-independent sets in an interval graph can be decided in polynomial time; (ii) deciding this latter problem becomes co-NP-hard and even co-W[2]-hard (parameterized by the size of the independent set) on split graphs, a sub-class of chordal graphs.

1 Introduction

Reconfiguration problems consist in finding step-by-step transformations between two feasible solutions such that all intermediate results are also feasible. Reconfiguration problems model dynamic situations where a given solution is in place and has to be modified, but no property disruption can be afforded. Two types of questions naturally arise when we deal with reconfiguration problems: (i) when can we ensure that there exist such a transformation? (ii) What is the complexity of finding such a reconfiguration? In the last few years reconfiguration problems received a lot of attention for various different problems such as

N. Bousquet—Supported by ANR Projects STINT (ANR-13-BS02-0007) and LabEx PERSYVAL-Lab (ANR-11-LABX-0025-01).

© Springer International Publishing AG 2017
H.L. Bodlaender and G.J. Woeginger (Eds.): WG 2017, LNCS 10520, pp. 127–139, 2017.
https://doi.org/10.1007/978-3-319-68705-6_10

proper colorings [1,11], Kempe chains [4,12], satisfiability [14] or shortest paths [5]. For a complete survey on reconfiguration problems, the reader is referred to [19]. In this paper our reference problem is independent set.

In the whole paper, $G = (V, E)$ is a graph where n denotes the size of V and k is an integer. For standard definitions and notations on graphs, we refer the reader to [10]. A k-*independent set* of G is a subset $S \subseteq V$ of size k of pairwise non-incident vertices. The k-independent set reconfiguration graph is a graph where vertices are k-independent sets and two independent sets are incident if they are "close" to each other. In the last few years, three possible definitions of adjacency between independent sets have been introduced. In the *Token Addition Removal* (TAR) model [2,17,18], two independent sets I, J are adjacent if they differ on exactly one vertex (*i.e.* if there exists a vertex u such that $I = J \cup \{u\}$ or the other way round). In the *Token Jumping* (TJ) model [7,16,17], two independent sets are adjacent if one can be obtained from the other by replacing a vertex with another one (in particular it means that we only look at independent sets of a given size). In the *Token Sliding* (TS) model, first introduced in [15], tokens can be moved along edges of the graph, *i.e* vertices can only be replaced with vertices which are adjacent to them (see [6] for a general overview of the results for all these models).

In this paper we concentrate on the Token Sliding (TS) model. Given a graph G, the *k-TS reconfiguration graph of G*, denoted $TS_k(G)$, is the graph whose vertices are k-independent sets of G and where two independent sets are incident if we can transform one into the other by sliding a token along an edge. More formally, I and J are adjacent in $TS_k(G)$ if $J \setminus I = \{u\}$, $I \setminus J = \{v\}$ and (u, v) is an edge of G. We then say that the token on u is slid onto v (or, informally, that u is slid onto v or that we slide from u to v). Hearn and Demaine proved in [15] that deciding if two independent sets are in the same connected component of $TS_k(G)$ is PSPACE-complete, even for planar graphs. On the positive side, Kaminski et al. gave a linear-time algorithm to decide this problem for cographs (which are characterized as P_4-free graphs) [17]. Bonsma et al. [7] showed that we can decide in polynomial time if two independent sets are in the same connected component for claw-free graphs. Demaine et al. [9] described a quadratic algorithm deciding if two independent sets lie in the same connected component for trees. Yamada and Uehara showed in [20] that a polynomial transformation exists in proper interval graphs.

Our Contribution. In their paper, Demaine et al. [9] asked if determining whether two independent sets are in the same connected component of $TS_k(G)$ can be decided in polynomial time for interval graphs and then more generally for chordal graphs. An interval graph is a graph which can be represented as an intersection graph of intervals in the real line. Chordal graphs, that are graphs without induced cycle of length at least 4, strictly contain interval graphs. In this paper, we prove the first conjecture. Moreover we answer related questions by proving that (i) the connectivity of $TS_k(G)$ can be decided in polynomial time if G is an interval graph (ii) deciding if $TS_k(G)$ is connected for chordal graphs is co-NP-hard and co-$W[2]$-hard parameterized by the size k of the independent set even for split graphs, a subclass of chordal graphs. More formally, the paper is devoted to proving the three following results:

First we answer a question raised by Demaine et al. [9] and Yamada and Uehara [20]:

Theorem 1. *Given an interval graph G and two independent sets I and J of size k, one can decide in polynomial time if I and J are in the same connected component of $TS_k(G)$.*

We then show that the method can be adapted in order to prove the following:

Theorem 2. *Given an interval graph G and an integer k, the connectivity of $TS_k(G)$ can be decided in polynomial time.*

In light of the two first results, we ask the following question. The *clique-tree degree* of a chordal graph G is the smallest maximum degree of a clique-tree of G.

Question 1. Let D be a fixed constant. For any integer k and any chordal graph G of clique-tree degree at most D, can the connectivity of $TS_k(G)$ be decided in polynomial time?

We finally prove the following hardness result.

Theorem 3. *The following problem is co-NP-hard and co-W[2]-hard parameterized by the size of the independent set.*

TOKEN SLIDING IN SPLIT GRAPHS
Input: *A split graph G, an integer k.*
Output: *YES if and only if $TS_k(G)$ is connected.*

A problem is *FPT* parameterized by k if it can be decided in time $f(k) \cdot n^c$ where c is a constant and n is the size of the instance. An FPT algorithm is deterministic, thus we have FPT = co-FPT. Moreover, the class $W[2]$ is conjectured to strictly contain the class. In particular it means that this problem is unlikely to be solved in FPT-time parameterized by the size of the solution. For more information on parameterized complexity the reader is referred to [13].

2   Hardness Results

This section is devoted to prove Theorem 3. A graph $G = (V, E)$ is a *split graph* if the vertices of G can be partitioned into two sets V_1, V_2 such that the graph induced by V_1 is a clique and the graph induced by V_2 is an independent set. There is no restriction on the edges between V_1 and V_2. One can easily notice that split graphs do not contain any induced cycle of length at least 4 since two non-adjacent vertices of such a cycle would belong to V_1, a contradiction. Thus split graphs are chordal graphs.

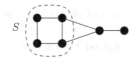

Fig. 1. The set S is blocking: no vertex in $N(S)$ is in the neighborhood of precisely one vertex of S. Note however that S is not a dominating set.

Let $G = (V, E)$ be a graph. The *neighborhood* of a vertex v, denoted by $N(v)$ is the set of vertices which are adjacent to v. The *closed neighborhood of v*, denoted by $N[v]$ is the set $N(v) \cup \{v\}$. A subset S of vertices is a *dominating set* of G if $V = \bigcup_{s \in S} N[s]$. Deciding the existence of a dominating set of size k (where k is part of the input) is NP-complete and $W[2]$-complete when parameterized by k.

Let S be a subset of vertices. A vertex x is a *private neighbor* of $s \in S$ if $x \in N(s)$ and $x \notin N[s']$ for every $s' \in S \setminus \{s\}$. We then say that s has a private neighbor *with respect to S*. A set S is *blocking* if no vertex of S has a private neighbor with respect to S. A graph G is *k-blocking* if it contains a blocking set of size at most k. In Fig. 1, the set S is blocking, thus the graph is 4-blocking. Let us consider the following problem which is a variation of the domination problem:

DOMINATING SET IN NON-BLOCKING GRAPHS
Input: An integer k, a graph G such that G is not $(2k - 1)$-blocking.
Output: YES if and only if G has a dominating set of size k.

Lemma 1. DOMINATING SET IN NON-BLOCKING GRAPHS *is NP-complete and* $W[2]$*-complete.*

Proof. Checking whether a set is dominating can be performed in polynomial time. Thus DOMINATING SET IN NON-BLOCKING GRAPHS is in NP. In the remainder part of this proof, we argue that DOMINATING SET IN NON-BLOCKING GRAPHS is NP-complete. Let us show that there exists a polynomial-time reduction from DOMINATING SET to DOMINATING SET IN NON-BLOCKING GRAPHS.

Let $G = (V, E)$ be a graph with vertex set $\{v_1, \ldots, v_n\}$ and k be an integer. If $k \geq n$ or $k \leq 3$, the problem can be decided in polynomial time. Thus we can assume that $n > k$ and that $k \geq 4$.

We will construct a new graph G' such that:

1. G' has no blocking set of size at most $2k - 1$.
2. G' has a dominating set of size at most k if and only if G has.

The graph G' is constructed as follows.

- Let H be the disjoint union of k copies $G_1, \ldots, G_k$ of G. We denote by v_ℓ^i the copy of v_ℓ in G_i. Let H' be the graph obtained from H by adding an edge between any two vertices that are copies of the same vertex or of two adjacent vertices in G (i.e. we add the edges (v_ℓ^i, v_ℓ^j) and (v_ℓ^i, v_m^j) for any $i, j \leq k$ and $\ell, m \leq n$ such that $(v_\ell, v_m) \in E(G)$).

- Let U be the disjoint union of n induced paths on $4k$ vertices, and for every $i \in \{1, \ldots, n\}$, let w_i be an arbitrary endpoint of the ith path. Let U' be the disjoint union of $k(k-1)$ copies $U_{i,j}$ $(i, j \in \{1, \ldots, k\}$ with $i \neq j)$ of U. We denote by $w_\ell^{i,j}$ the copy of $w_\ell \in U$ in $U_{i,j}$.

- Let G' be the graph obtained from the disjoint union of H' and U' by adding edges as follows. For every i, j, p, we add an edge between every vertex of G_i and every vertex of $U_{i,j}$ for every $j \neq i$, and between v_ℓ^i and $w_\ell^{j,i}$ for every $j \neq i$. In other words, $w_\ell^{i,j}$ is adjacent to all the vertices of G_i and exactly one vertex of G_j for $j \neq i$, namely v_ℓ^j. The other vertices in $U_{i,j}$ are adjacent only to all the vertices of G_i.

Let us briefly explain why (1) and (2) hold. For (1), we essentially show that the lengths of the paths ensure that blocking sets must contain vertices in G'. The vertices of G' are k copies of the vertices of G. One can easily prove that any blocking set contains at least two vertices in each G_i. Thus a blocking set of size at most $2k - 1$ contain at most one vertex in some G_i, which leads to a contradiction. For (2), we show that (1) and the sparsity between the sets H' and U' will ensure a k-dominating set of G' is contained in H' and then is a dominating set of G.

Note that the graph G' can be constructed in polynomial time and that the parameter does not change. Due to space restriction, the proofs of the two following claims are not included in this extended abstract. A complete version can be found on [3].

Claim. The graph G' does not contain any blocking set of size at most $2k - 1$.

Claim. The graph G has a dominating set of size k if and only if G' has a dominating set of size k.

The combination of the two claims ensure that Lemma 1 holds.

2.1 Split Graphs

A k-independent set I is *frozen* in $TS_k(G)$ if I is an isolated vertex of $TS_k(G)$. In other words, no token of I can be slid. Note that I is frozen if and only if I is blocking. Indeed, if a vertex u of I has a private neighbor v then the independent set $I \cup v \setminus u$ is incident to I in $TS_k(G)$: the token of u can be slid onto v. Conversely, if I and J are incident in $TS_k(G)$ then a vertex of I has a private neighbor. Indeed, assume that J is incident to I and let $u = I \setminus J$ and $v = J \setminus I$. Then (u, v) is an edge. Since J is an independent set, no vertex of $I \setminus u$ is incident to v, and then v is a private neighbor of u.

Theorem 4. Token Sliding in Split Graphs *is co-NP-hard.*

Proof. Let us prove that there is a reduction from DOMINATING SET IN NON-BLOCKING GRAPHS to TOKEN SLIDING IN SPLIT GRAPHS such that the first instance if positive if and only if the second one is negative. Since DOMINATING SET IN NON-BLOCKING GRAPHS is NP-complete by Lemma 1, it implies that TOKEN SLIDING IN SPLIT GRAPHS is co-NP-hard.

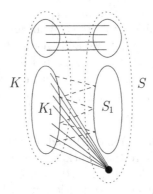

Fig. 2. K is the clique and S is the independent set. The dashed part corresponds to the edges of the graph G. Then we add $k + 1$ vertices on each side (left one inducing a clique) with a matching between the two sides. Finally the bottommost vertex of S is the vertex w_{n+k+2} which is connected to all the vertices of K_1.

Let G be a graph on vertex set $\{v_1, \ldots, v_n\}$ that is not $(2k - 1)$-blocking. We construct a split graph G' on vertex set $K \cup S$ where K induces a clique of size $n + k + 1$ and S induces an independent set of size $n + k + 2$ (Fig. 2). The vertices of K are denoted by $u_1, \ldots, u_{n+k+1}$ and the vertices of S are denoted by $w_1, \ldots, w_{n+k+2}$. Moreover we will denote by K_1 (resp. S_1) the subset of K (resp. S) $u_1, \ldots, u_n$ (resp. $w_1, \ldots, w_n$). The edges of the graph G' are the following:

- For every $i, j \leq n$, u_i is incident to w_j if $i = j$ or (v_i, v_j) is an edge of G. In other words, the graph induced by (K_1, S_1) simulates the incidences in the graph G.
- For every $n + k + 1 \geq j \geq n + 1$, w_j is the unique neighbor of u_j in the independent set.
- The vertex w_{n+k+2} is connected to the whole set $u_1, \ldots, u_n$.

The proofs that the graph $TS_{k+1}(G')$ is connected if and only if G has no dominating set of size k can be found in [3].

In the proof of Theorem 4, the difference between the size of the dominating set of G in the original graph and the size of the independent sets we want to slide in the split graph is one. As a by-product, we immediately obtain the following corollary:

Corollary 1. TOKEN SLIDING IN SPLIT GRAPHS *is co-W[2]-hard.*

2.2 Bipartite Graphs

Theorem 5. TOKEN SLIDING IN BIPARTITE GRAPHS *is co-NP-hard and co-W[2]-hard.*

The construction and the proofs are largely inspired from the proof of Theorem 4 with slight modifications. The complete proof of this theorem can be found in [3].

3 Interval Graphs

This section is devoted to the proof of Theorem 2. A graph G is an *interval graph* if G can be represented as an intersection of segments on the line. More formally, each vertex can be represented with a pair (a, b) (where $a \leq b$) and vertices $u = (a, b)$ and $v = (c, d)$ are adjacent if the intervals (a, b) and (c, d) intersect. Let $u = (a, b)$ be a vertex; a is the *left extremity* of u and b the *right extremity* of u. The left and right extremities of u are denoted by respectively $l(u)$ and $r(u)$. Given an interval graph, a representation of this graph as the intersection of intervals in the plane can be found in $\mathcal{O}(|V| + |E|)$ time by ordering the maximal cliques of G (see for instance [8]). Actually, interval graphs admit clique paths and are thus a special case of chordal graphs. Using small perturbations, we can moreover assume that all the intervals start and end at distinct points of the line. In the remainder of this section we assume that we are given a representation of the interval graph on the real line.

3.1 Basic Facts on Independent Sets in Interval Graphs

Leftmost independent set. Let $G = (V, E)$ be an interval graph with its representation on the line. There are two natural orders on the vertices of an interval graph: the *left order* denoted by $\prec_l$ and the *right order* denoted by $\prec_r$. We have $u \prec_l v$ if and only if $l(u) \leq l(v)$ Note that by our small perturbations assumption, we never have $l(u) = l(v)$, thus $\prec_l$ defines a total order on V. Similarly, $u \prec_r v$ if $r(u) \leq r(v)$. Note that these two orders do not necessarily coincide. We denote by $\alpha(G)$ the maximum size of an independent set of G. The value $\alpha(G)$ is called the *independence number* of G.

Let u and w be two vertices of G. The graph G_u is the graph induced by all the vertices v such that $l(v) > r(u)$. In other words, G_u is the graph induced by the intervals located at the right of u that do not intersect the interval u (see Fig. 3 for an illustration). Similarly, G^w is the graph induced by all the vertices v such that $r(v) < l(w)$ (the graph induced by the intervals located at the left of w that do not intersect the interval w). The graph G_u^w is the graph induced by all the vertices v where $u \prec_l v \prec_l w$ and both (u, v) and (v, w) are not edges. In other words, G_u^w is the graph induced by the intervals located between u and w. We can alternatively define G_u^w as the graph induced by the vertices both in G_u and G^w.

We define the *first vertex* of G, denoted fv(G), as the vertex u that maximizes the number of vertices in G_u. Note that for any two vertices u and v such that

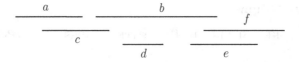

Fig. 3. The leftmost independent set is $\{a, d, e\}$ while the rightmost one is $\{c, d, f\}$. The graph G_c is restricted to the vertices d, e, f and G_c^e is restricted to the single vertex d.

$r(u) < r(v)$, we have that G_v is an induced subgraph of G_u. In other words, the first vertex of G can be equivalently defined as the vertex u with minimum $r(u)$. The *leftmost independent set* of G, denoted $LM(G)$, is $\emptyset$ if G is empty and $\{u\} \cup LM(G_u)$ otherwise, where $u = \mathrm{fv}(G)$.

Given an independent set $u_1, \ldots, u_k$, if $l(u_i) < l(u_j)$ then $r(u_i) < l(u_j)$. So there is a natural order on the vertices of an independent set which corresponds to both $\prec_l$ and $\prec_r$. In the following, we say that the independent set is *ordered* if $r(u_i) < l(u_{i+1})$ for every $i \leq k - 1$. Moreover, by abuse of notation, when we say that $l_1, \ldots, l_p$ is the leftmost independent set, we assume that the vertices are ordered.

Remark 1. Let $l_1, \ldots, l_p$ be the leftmost independent set of G. If $\{u_1, \ldots, u_k\}$ is an ordered independent set of size k, then for every $j < k$, $r(u_j) \geq r(l_j)$.

From the definition of $\mathrm{fv}(G)$, it follows that $\alpha(G) \geq 1 + \alpha(G_{\mathrm{fv}(G)})$. We first point out that the way $G_{\mathrm{fv}(G)}$ is constructed from G is quite specific.

Remark 2. The graph induced in G by $\{\mathrm{fv}(G)\} \cup N(\mathrm{fv}(G))$ is a clique.

Indeed, every interval in $N(\mathrm{fv}(G))$ starts before $r(\mathrm{fv}(G))$, and ends after $r(\mathrm{fv}(G))$, thus they all overlap at that point. Note that $\{\mathrm{fv}(G)\} \cup N(\mathrm{fv}(G))$ is even the unique maximal clique containing $\mathrm{fv}(G)$. We obtain that $\alpha(G) \leq 1 + \alpha(G_{\mathrm{fv}(G)})$, hence the following remark.

Remark 3. The leftmost independent set of G has size $\alpha(G)$.

Remark 3 ensures that (i) $TS_k(G)$ is empty if k is larger than the size of the leftmost independent set (ii) $TS_k(G)$ is connected iff one can transform any independent set of size k into the k first vertices of the leftmost independent set. Our proof technique will precisely be based on point (ii).

Token moves. By Remark 1, we can define the *first vertex* of an independent set I as the minimum vertex of I for both $\prec_l$ and $\prec_r$. More generally, for any integer ℓ, the *ℓth vertex* of I is the ℓth vertex u of I for both $\prec_l$ and $\prec_r$. The integer ℓ is said to be the *position* of u in I.

Before stating the next lemma, let us first make an observation on the representation of an independent set reconfiguration. We can represent a reconfiguration sequence as a sequence of independent sets $I_1, \ldots, I_m$ where $|I_i \setminus I_{i+1}| =$

$|I_{i+1} \setminus I_i| = 1$ and the unique vertex of $I_i \setminus I_{i+1}$ is incident to the unique vertex of $I_{i+1} \setminus I_i$. Let us denote by u the unique vertex in $I_i \setminus I_{i+1}$ and by v the unique vertex in $I_{i+1} \setminus I_i$. Thus the adjacency between these two independent sets can also be represented as the edge (u, v). We say that a token is *slid from u to v* and that (u, v) is *the move* from I_i to I_{i+1}. Thus the reconfiguration of an independent set can be seen either as a sequence of adjacent independent sets in $TS_k(G)$ or as a sequence of moves.

Let I_0 be an independent set and $u \in I_0$. Let $I_0, \ldots, I_s$ be a reconfiguration sequence of independent sets. We define the *token of origin u* in I_k as follows: the token of origin u in I_0 is u. For t the token of origin u in I_{k-1}, the token of origin u in I_k is either t (if $t \in I_k$) or the vertex v such that (t, v) is the move from I_{k-1} to I_k (if $t \notin I_k$).

On an interval graph, the *first token* of I is the token on the first vertex of I. More generally, there is a natural order on the tokens on vertices of I, as there is a natural order on the vertices of an independent set. The ℓth *token* is the token on the ℓth vertex of I. The next remark ensures that the order is preserved by token sliding, i.e. we cannot "permute" tokens during a reconfiguration of an interval graph. A similar remark was observed in [20].

Remark 4. Let G be an interval graph and I, J be two k-independent sets of G. Let u be the ℓth vertex of I. For any reconfiguration from I to J, the ℓth token of J is the token of origin u.

Consider two k-independent sets I and J in G and a reconfiguration sequence $\mathcal{P}$ from I to J. The *leftmost location* of the ℓth token in the sequence $\mathcal{P}$ is the vertex u with minimal $r(u)$ among all vertices that host the ℓth token in an independent set of $\mathcal{P}$ (the *rightmost location* is the symmetric, with maximal $l(u)$). If we delete the first vertex of every independent set in the sequence and omit the moves where the first token is slid, we obtain a reconfiguration sequence from $I \setminus u$ to $J \setminus v$ in G, where u and v respectively denotes the first vertices of I and J. The following lemma ensures that such a sequence also exists in a well-chosen subgraph of G.

Lemma 2. *Let I and J be two k-independent sets such that there exists a transformation from I to J and let $\ell < k$. Let w be the leftmost location of the ℓth token in the sequence between I and J. In G_w, there exists a sequence between I minus its ℓ first vertices and J minus its ℓ first vertices.*

As an immediate corollary, we obtain the following lemmas, which we state formally as we use them several times.

Lemma 3. *Let I and J be two independent sets such that there exists a transformation from I to J. Let u, v, w be the first token of I, the first token of J and the leftmost location of the first token in the sequence, respectively. There exists a transformation from $I \setminus u$ into $J \setminus v$ in G_w.*

Lemma 3 is in particular true if the first vertex is never slid. In other words, if there is a sequence from I to J where the first vertex u of I is never slid, then there is a transformation from $I \setminus u$ to $J \setminus u$ in G_u.

The following lemma also is a corollary of Lemma 2 if we consider the interval graph from the right to the left rather than from the left to the right (note that the $\prec_r$ order then becomes $\prec_l$ order).

Lemma 4. *Let I and J be two k-independent sets such that there exists a sequence $\mathcal{P}$ from I to J. Let w be the rightmost location of the second token all along the sequence between I and J. The first token of I and the first token of J are in the same connected component of G^w.*

We finally make an easy observation that will be widely used all along the proof.

Remark 5. Let u and v be two vertices and I and J be two k-independent sets of G_u that are in the same connected component of $TS_k(G_u)$. If $r(v) < l(u)$, then $I \cup \{v\}$ and $J \cup \{v\}$ are in the same connected component of $TS_{k+1}(G)$.

3.2 Reachability

Let H be an interval graph and I an independent set of H of size k. We denote by $\mathcal{C}_{H,k}(I)$ the connected component of I in $TS_k(H)$. Given an independent set J, we denote by $\mathrm{fv}(J)$ the first vertex of J. Let

$$RM(I, H) = \max_{\prec_l} \Big\{ \mathrm{fv}(J) \ : \ J \in \mathcal{C}_{H,k}(I) \Big\}.$$

The value of $RM(I, H)$ is the rightmost possible (for $\prec_l$) first vertex of J amongst all the independent sets J in $\mathcal{C}_{H,k}(I)$. Therefore, if we try to push the first vertex to the right, we cannot push it further than $RM(I, H)$.

We define the symmetric notion consisting in pushing the first vertex to the left (for $\prec_r$).

$$LM(I, H) = \min_{\prec_r} \Big\{ \mathrm{fv}(J) \ : \ J \in \mathcal{C}_{H,k}(I) \Big\}.$$

Note that when we want to push an independent to the left, we want to minimize $\prec_r$ while when we want to push an independent to the right we want to maximize $\prec_l$. The proof of the following lemma can be found in [3].

Lemma 5. *Let I_1 and I_2 be two k-independent sets of an interval graph G, with $k \geq 2$. The independent sets I_1 and I_2 are in the same connected component of $TS_k(G)$ if and only if:*

1. *$LM(I_1, G)$ and $LM(I_2, G)$ are the same, and*
2. *The independent sets $I_1 \setminus \{\mathrm{fv}(I_1)\}$ and $I_2 \setminus \{\mathrm{fv}(I_2)\}$ are in the same connected component of $TS_{k-1}(G_{LM(I_1,G)})$.*

Lemma 5 guarantees us that there is a polynomial-time algorithm to decide whether two k-independent sets I_1 and I_2 of some interval graph G are in the same connected component of $TS_k(G)$, if and only if there is a polynomial-time algorithm to compute $LM(I_1, G)$. The rest of this subsection is precisely devoted to this.

Let I be a k-independent set of an interval graph H. A subset J of I is a *right subset* of I if J corresponds to the restriction of I to some H_b: in other words, when J contains a vertex $x \in I$, then J also contains all the vertices of I larger than x (for both $\prec_r$ and $\prec_l$). Let us prove that we can compute in polynomial time $RM(J, H_b)$ and $LM(J, H_b)$ for any right subset J of I included in H_b. Note that there are only k right subsets of I (the empty one is not interesting), and n choices of b. We proceed by dynamic programming, and simply argue how to obtain the values for the right subset of I of size p assuming constant-time access to the values for the right subsets of size at most $p-1$. For an independent set J of size 1, we can compute $RM(J, H_b)$ and $LM(J, H_b)$ in quadratic time (linear in the number of edges) by Lemma 6, as follows. Due to space restriction, all the proofs of this section can be found in the long version.

Lemma 6. *Let H be a graph, and u and v be two vertices of H. The independent sets $\{u\}$ and $\{v\}$ are in the same connected component of $TS_1(H)$ if and only if u and v are in the same connected component of H.*

In other words, to compute $RM(J, H_b)$ and $LM(J, H_b)$ for an independent set J of size 1, it is enough to compute the leftmost and rightmost vertices in the connected component of H_b that contains the only element of J. These values can indeed be computed in polynomial time.

Now, to compute $LM(I, H)$ when I contains at least two vertices, it may not suffice to push the first element of I to the left. We may need to push the rest of I to the right, then try again to push the first element to the left, and keep going as long as progress is made. Pushing the rest of I to the right means reconfiguring $I \setminus \{fv(I)\}$ into an independent set whose first vertex is $RM(I \setminus \{fv(I)\}, H_{fv(I)})$. Trying again to push $fv(I)$ to the left means, according to Lemma 6, looking at the leftmost element of its connected component in $H^{RM(I \setminus \{fv(I)\}, H_{fv(I)})}$. Making progress means being able to slide the token from $fv(I)$ to a vertex with smaller right index – possibly in more than one step – without reconfiguring the rest of I. Let us consider what happens when no progress is made.

Lemma 7. *Let H be an interval graph and I be a k-independent set of H, with $k \geq 2$. If the vertex $fv(I)$ is the leftmost element of its connected component in $H^{RM(I \setminus \{fv(I)\}, H_{fv(I)})}$, then $LM(I, H) = fv(I)$.*

Lemma 7 guarantees us that the procedure informally described will output $LM(I, H)$ after a linear number of steps. Let us now argue how to obtain the value of $RM(I, H)$.

Lemma 8. *Let H be an interval graph and I be a k-independent set of H, with $k \geq 2$. The rightmost vertex (for $\prec_l$) of the connected component of $H^{RM(I \setminus \{LM(I,H)\}, H_{LM(I,H)})}$ that contains $LM(I, H)$ is $RM(I, H)$.*

We are ready to formalize the algorithm, and introduce Procedure 1.

Let us first point out that by definition, r can only decrease, so r takes at most n different values, and each iteration is done in $O(m)$ operations. Therefore,

Procedure 1. Computing $(LM(J, H_b), RM(J, H_b))$ assuming constant-time access to values for proper right subsets of J.

$c := \text{fv}(J)$.
$r := +\infty$.
while $r \neq c$ **do**
  $r = c$.
  $j = RM(J \setminus \{\text{fv}(J)\}, H_c)$
  $c =$ the minimal vertex (for $\prec_r$) in the connected component of c in H_b^j.
  $d =$ the maximal vertex (for $\prec_l$) in the connected component c in H_b^j.
end while
return (c, d)

assuming constant-time access to values for proper right subsets of the independent set being considered, Procedure 1 runs in $O(n \cdot m)$. However, as pointed out earlier, for a given input interval graph and k-independent set, there are at most $k \cdot n$ possible combinations of right subsets and subgraphs that we might run Procedure 1 on. Lemmas 7 and 8 guarantee that once Procedure 1 ends, the variables c and d correspond to $LM(J, H_b)$ and $RM(J, H_b)$, respectively. Therefore, for any interval graph H on n vertices and any k-independent set J, we can compute $(LM(J, H), RM(J, H))$ in $O(k \cdot n^2 \cdot m)$ operations. And then by Lemma 5, we can compute whether any two k-independent sets I and J of some interval graph G are in the same connected component of $TS_k(G)$ in $O(k \cdot n^2 \cdot m)$ operations.

3.3 Connectivity

We established in Sect. 3.2 that we can decide in polynomial time whether two given independent sets of some interval graph can be reconfigured one into the other through a series of token slidings. Roughly speaking, the technique consists in pointing out that there is a natural total order on the k-independent sets, computing the leftmost independent set that can be reached from each, and checking whether the two match. Now, we have a broader goal: decide whether it holds that any k-independent set can be reconfigured into any other. To that purpose, we push a bit further the notions of $LM(I, H)$ and $RM(I, H)$ introduced in Sect. 3.2. Informally, instead of handling a specific independent set, we will think about the worst possible behaviour of an independent set that we could be handling. With these generalizations of $LM(I, H)$ and $RM(I, H)$ we can show using similar arguments that it is possible to decide in polynomial time the TS-connectivity problem on interval graphs. All the details can be found in [3].

References

1. Bonamy, M., Bousquet, N.: Recoloring bounded treewidth graphs. Electron. Notes Discret. Math. **44**, 257–262 (2013). (LAGOS 2013)

An *interval graph* is an intersection graph of intervals in the line. More formally, vertices are intervals and two vertices are incident if their respective intervals intersect. Let H be an interval graph with its representation. Let $u \in V(H)$. The *left extremity of* u is the leftmost point p of the line such that u contains p. The *right extremity of* u is the rightmost point p of the line such that u contains p.

Let G be an EPG graph with its representation on the grid. In what follows, we will always denote by roman letters $a, b, \ldots$ the rows of the grid and by greek letters $\alpha, \beta, \ldots$ the columns of the grids. Given a row a (resp. column α) of the grid, the row $a - 1$ (resp. $\alpha - 1$) denotes the row under a (resp. at the left of α). Given a row a and a column α, we will denote by (a, α) the grid point at the intersection of a and α. By abuse of notation, we will also denote by α (for a given row a) the point at the intersection of row a and column α. Let $u \in V(G)$. We denote by P_u the path representing u on the grid. The vertex u of G *intersects* a row (resp. column) if P_u contains at least one edge of it.

3 Typed Intervals and Projection Graphs

Let G be a B_2-EPG graph with its representation on the grid. Free to slightly modify the representation of G, we can assume that the path associated to every vertex has exactly 2 bends. Indeed, if there is a vertex u such that P_u has less than two bends, let (a, α) be one of the two extremities of P_u. Up to a rotation of the grid, we can assume that the unique edge of P_u incident to (a, α) is the horizontal edge c between (a, α) and $(a, \alpha + 1)$. Then create a new column β between α and $\alpha + 1$, and replace the edge e by two edges, one between α and β on row a, and another one going up at β. This transformation does not modify the graph G. So we will assume in the following that for every vertex u, the path P_u has exactly two bends.

A *Z-vertex* of G is a vertex that intersects two rows and one column. A *U-vertex* is a vertex that intersects one row and two columns. The *index* of a vertex u is the set of rows intersected by u. The vertex u *contains a in its index* if a is in the index of u.

Let u be a vertex containing a in its index. The *extremities* of u on a are the points of the row a on which P_u stops or bends. Since P_u has at most two bends, P_u has exactly two extremities on a and the subpath of P_u on row a is the interval of a between these two extremities. The *a-interval* of u, denoted by P_u^a, is the interval between the two extremities of u on a. Note that since the index of u contains a, P_u^a contains at least one edge. Let $\alpha \leq \beta$ be two points of a. The path P_u^a *intersects non-trivially* $[\alpha, \beta]$ if $P_u^a \cap [\alpha, \beta]$ is not empty or reduced to a single point. The path P_u^a *weakly contains* $[\alpha, \beta]$ if $[\alpha, \beta]$ is contained in P_u^a.

3.1 Typed Intervals

Knowing that $P_u^a = [\alpha, \beta]$ is not enough to understand the structure of P_u. Indeed whether P_u stops at α, or bends (upwards or downwards) at α, affects the neighborhood of u in the graph. To catch this difference we introduce typed intervals that contain information on the "possible bends" on the extremities of the interval.

We define three *types* namely the *empty type* •, the *d-type* ↕ and the *u-type* ↕. A *typed point* (on row a) is a pair x, α where x is a type and α is the point at the intersection of row a and column α. A *typed interval* (for row a) is a pair of typed points (x, α) and (y, β) (on row a) with $\alpha \leq \beta$ denoted by $[x\alpha, y\beta]$. Informally, a typed interval is an interval $[\alpha, \beta]$ and indications on the structure of the bends on the extremities. A typed interval t is *proper* if $\alpha \neq \beta$ or if $\alpha = \beta$, $x = y$, and $x \in \{↕, ↕\}$.

Let $t = [x\alpha, y\beta]$ and $t' = [x'\alpha', y'\beta']$ be two typed intervals on a row a. Denote by $z\gamma$ one of the endpoints of t'. We say that t is *coherent* with the endpoint $z\gamma$ of t' if one of the following holds: (i) γ is included in the open interval (α, β), or (ii) $z = •$, and $[\alpha, \beta]$ contains the edge of $[\alpha', \beta']$ adjacent to γ, or (iii) $z \neq •$, and $z\gamma \in \{x\alpha, y\beta\}$. We can remark in particular that if t is coherent with an endpoint $z\gamma$, then γ is in the closed interval $[\alpha, \beta]$. The typed interval t *contains* t' if $[\alpha', \beta'] \subset [\alpha, \beta]$, and t is coherent with both endpoints of t'. The typed interval t *intersects* t' if $[\alpha, \beta]$ intersects non trivially $[\alpha', \beta']$ (i.e. the intersection contains at least one grid-edge), or t is coherent with an endpoint of t', or t' is coherent with an endpoint of t. Note that, if t' contains t then in particular it intersects t.

Let u be a vertex containing a in its index. The *t-projection of u on a* is the typed interval $[x\alpha, y\beta]$ where α, β are the extremities of P_u^a and the type of an extremity $\gamma \in \{\alpha, \beta\}$ is • if P_u stops at γ and ↕ (resp. ↕) if P_u^a bends downwards (resp. upwards) at γ. Note that this typed interval is proper since it contains at least one edge. The path P_u *contains* (resp. *intersects*) a typed interval t (on a) if the t-projection of u on a contains t (resp. intersects t). By abuse of notation we say that u contains or intersects t. Note that by definition, the path P_u contains the t-projection of u on a. Moreover if a vertex u contains a typed interval $t = [x\alpha, y\beta]$, then the path P_u weakly contains $[\alpha, \beta]$. If u intersects t, then the path P_u intersects the segment $[\alpha, \beta]$ (possibly on a single point) (Fig. 2).

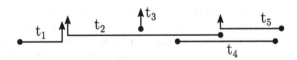

Fig. 2. Examples of typed intervals on the same row. In this example, the interval t_3 is reduced to a single point. The interval t_2 is coherent with the right extremity of t_1, the extremities of t_3 and the left extremity of t_4. It is not coherent with the extremities of t_5. Moreover, t_2 intersects t_1, t_3 and t_4 but not t_5. And t_2 contains t_3.

The following simple lemma motivates the introduction of typed intervals. Due to space restrictions, the following proof cannot be included in this extended abstract. A proof can be found in the full version [5].

Lemma 1. *Let G be a B_2-EPG graph, let u, v be two vertices whose index contain a, and t be a proper typed interval of a. If u contains t and v intersects t, then u and v are adjacent.*

Lemma 2. *Let G be a B_2-EPG graph. If the t-projections of u and v on a intersect, then uv is an edge of G. Moreover if u and v are two vertices containing a in their index and have no other row in common, then uv is an edge of G if and only if the t-projections of u and v on a intersect.*

Proof. Let u and v be two vertices containing a in their index. The first part of the statement is just a corollary of Lemma 1 since v intersects the t-projection of u on a.

Assume that there is no other row b contained in the index of both u and v. And suppose moreover that the t-projections of u and v on a do not intersect. Suppose by contradiction that u and v are adjacent in G. The two vertices cannot share a common edge on row a, otherwise their projections would intersect. By assumption they cannot share an edge on another row. Consequently, they must have a common edge e on a column, and let α this column. By symmetry, we can assume that e is below the row a. Since the paths P_u and P_v have at most two bends, both path must bend downwards at the intersection of row a and column α, to intersect the edge e. However, this implies that the t-projections of u and v on a intersect, a contradiction. $\qquad\square$

3.2 Projection Graphs

Let Y be the subset of vertices of G such that all the vertices of Y contain a in their index. *The projection graph of Y on a is the graph on vertex set Y such that there is an edge between u and v if and only if the t-projections of u and v on a intersect.* Note that the projection of Y is not necessarily an interval graph since it can contain an induced C_4 (see Fig. 3). We say that a set of vertices Y is a *clique on a* if the t-projection of Y on a is a clique. Lemma 2, ensures that a clique on a is indeed a clique in the graph G.

In the very simple case where the representation uses only two rows a and b, we have the following lemma:

Lemma 3. *Let G be a B_2-EPG graph and G_{ab} be the subset of vertices with index $\{a, b\}$. Then G_{ab} induces a 2-track interval graph.*

A proof of the following statement can be found in the full version [5]. Let us end this section with two lemmas that will be widely used all along the paper.

Lemma 4. *Let G be a B_2-EPG graph and Y be a subset of vertices whose index contain a. Suppose that the projection of Y on a is a clique. Then there is a proper typed interval t such that:*

Fig. 3. In this example, the projection graph on row a of the four vertices is an induced cycle of length four.

- *all vertices of Y contain t,*
- *if u is a vertex with index $\{a\}$ or $\{a, c\}$ where c is not in the index of any vertex of Y, then u is complete to Y if and only if u intersects t.*

Proof. Let α be the rightmost left extremity of an a-interval of Y, and β be the leftmost right extremity of an a-interval of Y. Since the projection of Y on a is a clique, $\alpha \leq \beta$. Let Y_α be the set of vertices of Y whose a-segment have left extremity α, and Y_β be the set of vertices of Y whose a-segment have right extremity β. We define the typed interval $t = [x\alpha, y\beta]$ where x is equal to ↕ (resp. ↨) if all the vertices of Y_α bend upwards (resp. downwards) at α, and is equal to • otherwise. Similarly, y is equal to ↕ (resp. ↨) if all the vertices of Y_β bend upwards (resp. downwards) at α, and is equal to • otherwise. Let us prove that t satisfies the conclusion of the lemma.

One can easily check that, by construction, all the vertices of Y contain the typed interval t. Indeed $[\alpha, \beta]$ is contained in all the intervals P_v^a for $v \in Y$ by definition of α and β. Let us prove now by contradiction that the typed interval t is proper. If $\alpha \neq \beta$ then t is proper, so we can suppose that $\alpha = \beta$. Now assume that the types x and y are distinct or equal to • . Up to symmetry, we can assume that $x \neq$ ↕ and $y \neq$ ↨ . Consequently, there exists a vertex $v_1 \in Y$ such that $P_{v_1}^a$ has left extremity α and either starts on α or bends downwards on α. Similarly, there exists a vertex $v_2 \in Y$ such that $P_{v_2}^a$ has right extremity $\beta = \alpha$ and either ends on α or bends upwards on α. But then the t-projection of v_1 and v_2 do not intersect, a contradiction since the t-projection of Y on a is a clique.

Let us finally prove the second point. Suppose that u intersects t, then for all $y \in Y$, y contains t, and by Lemma 2 u and y are adjacent in G. Assume now that u does not intersect t. Let us prove that there exists $y \in Y$ that is not incident to u. Either $P_u^a = [\alpha', \beta']$ does not intersect $[\alpha, \beta]$ or it intersects it on exactly one vertex. We moreover know that if $\alpha = \beta$ then P_u^a does not contain the edge at the left and at the right of α (otherwise u would intersect t). So, up to symmetry, we can assume that $\beta' \leq \alpha$. If $\beta' < \alpha$ then let v be a vertex such that P_v^a has leftmost extremity α (such a vertex exists by definition of α). The projections of u and v on a do not intersect and Lemma 2 ensures that u is not adjacent to v.

Assume now that P_u^a intersects $[\alpha, \beta]$ on a single point. Since P_u^a contains at least one edge, this point is either α or β, α say. Let x_u be the type of β'. Since u does not intersect t, this means that either $x_u \neq x$ or $x_u = x$ and $x_u = \bullet$. Up to symmetry, we can assume $x_u \neq \updownarrow$ and $x \neq \updownarrow$. So there exists $v \in Y$ such that P_v^a has a left extremity at α and P_v either bends upwards at α or has no bend at α. Since $x_u \neq \updownarrow$, P_u^a has right extremity α and either bends downwards at α or has no bend at α. So the projections of u and v on a does not intersect. By Lemma 1, u and v are not adjacent in G.   □

The flavour of the following proof is similar to the proof of Lemma 4. The proof of the lemma can be found in the full version [5].

Lemma 5. *Let G be a B_2-EPG graph and Y be a subset of vertices with index $\{a, b\}$. Suppose that the projection of Y on a is not clique. Then there is a proper typed interval t such that:*

- *All vertices of Y intersect t,*
- *Let u be a vertex with index $\{a\}$ or $\{a, c\}$ where $c \neq a, b$. Then u is complete to Y if and only if u contains t.*

4   Maximum Clique in B_2-EPG Graphs

4.1   Graphs with Z-Vertices

We start with the case where the graph only contains Z-vertices. We will show in Sect. 4.2 that it is possible to treat independently Z-vertices and U-vertices. The remaining part of Sect. 4.1 is devoted to prove the following theorem.

Theorem 1. *Let G be a B_2-EPG graph with a representation containing only Z-vertices. The size of a maximum clique can be computed in polynomial time.*

Note that, up to rotation of the representation of G, this theorem also holds for U-vertices. In other words, the size of a maximum clique can be computed in polynomial time if the graph only contains U-vertices.

The proof of Theorem 1 is divided in three steps. We will first define a notion of good subgraphs of G, and prove that:

- there is a polynomial number of good subgraphs of G and,
- a maximum clique of a good graph can be computed in polynomial time,
- and any maximal clique of G is contained in a good subgraph.

Recall that a clique is *maximum* if its size is maximum. And it is *maximal* if it is maximal by inclusion. The first point is an immediate corollary of the definition of good graphs. The proof of the second point consists in decomposing good graphs into sets on which a maximum clique can be computed efficiently. The proof of the third point, the most complicated, will be divided into several lemmas depending on the structure of the maximal clique we are considering.

An induced subgraph H of G is a *good graph* if one of the following holds:

(I) there are two rows a and b, and two proper typed intervals t_a and t_b on a and b respectively such that H is the subgraph induced by vertices v such that, v contains t_a, or v contains t_b, or v intersects both t_a and t_b,

(II) or there are three rows a, b, and c, and three proper typed intervals t_a, t_b, and t_c on a, b, c respectively such that H is the subgraph induced by vertices v such that, either v contains t_a, or v contains t_b, or v intersects t_b and contains t_c.

Lemma 6. *Let G be a B_2-EPG graph. There are $\mathcal{O}(n^6)$ good graphs, and a maximum clique of a good graph can be computed in polynomial time.*

The proof is in the full version [5] of the article. The remaining part of Sect. 4.1 is devoted to prove that any maximal clique of G is contained in a good graph.

Lemma 7. *Let G be a B_2-EPG graph containing only Z-vertices, and X be a clique of G. Assume that there are two rows a and b such that every horizontal segment of X is included in either a or b, then X is contained in a good graph.*

Proof. By taking two typed intervals t_a and t_b consisting of the whole rows a and b, the clique X is clearly contained in a good graph of type (I).   □

We say that a set of vertices X *intersects* a column α (resp. a row a) if at least one vertex of X intersects the column α (resp. the row a). If X is a clique of G, and a, b two rows of the grid, we denote by X_{ab} the subset of vertices of X with index $\{a, b\}$.

The combination of the three following lemmas directly implies Theorem 1. The three proofs are based on the same technique. The main idea consists in using the tools of Sect. 3 to find typed intervals containing the vertices in a clique X. Only the proof of the second lemma is given, the proof of the two others can be found in the full version [5].

Lemma 8. *Let G be a B_2-EPG graph containing only Z-vertices, and X be a clique of G. If there are two rows a and b such that the projection graphs of X_{ab} on a and b are not cliques, then X is included in a good graph.*

Lemma 9. *Let G be a B_2-EPG graph containing only Z-vertices, and X be a clique of G. If there are two rows a and b such that the projection graph of X_{ab} on a is not a clique, then there is a good graph containing X.*

Proof. Let X be a clique satisfying this property for rows a and b. We can assume that the projection graph of X_{ab} on b is a clique since otherwise we can apply Lemma 8. Let X_a be the set of vertices of $X \setminus X_{ab}$ intersecting row a and not row b, and X_b be the set of vertices of $X \setminus X_{ab}$ intersecting row b and not row a.

First note that $X = X_a \cup X_b \cup X_{ab}$. Otherwise there would exist a vertex w of index $\{c, d\}$ such that $\{c, d\} \cap \{a, b\} = \emptyset$ in X. Since w is complete to X_{ab}, w would intersect all the vertices of X_{ab} on their vertical part. But the projection graph of X_{ab} on a is not a clique, consequently X_{ab} intersects at least two columns. Thus a vertex of X_{ab} does not intersect the unique column of w, a contradiction.

Suppose first that the projection graph of $X_b \cup X_{ab}$ on b is a clique. By Lemma 4 applied to X_b on b, there exists a proper typed interval t_b on row b such that every vertex of X_b contains t_b. By Lemma 5 applied to X_{ab} on a, there exists a proper typed interval t_a on row a such that every vertex of X_a contains t_a. So all the vertices of X are contained in the good graph of type (I) defined by t_a and t_b.

Suppose now that the projection graph of $X_b \cup X_{ab}$ on b is not a clique. Then there are two vertices u, v of X_b such that the t-projections of u and v on b do not intersect. By Lemma 1, P_u and P_v intersect another row c, and since the projection graph of X_{ab} on b is a clique, $c \neq a$. Since the projection graph of X_{bc} on b is not a clique, and $c \neq a$, we can assume that the projection graph of X_{bc} on c is a clique since otherwise Lemma 8 can be applied on X_{bc}. So by Lemma 4 applied to X_{bc} on c, there exists a typed interval t_c contained in every vertex of X_{bc}. By Lemma 5 applied to X_{bc} on b (resp. X_{ab} on a), we know that there exists a typed interval t_b (resp. t_a) satisfying the two conditions of Lemma 5. Now we divide $X_b \cup X_{ab}$ into two subclasses: X_{bc} and $Y_b = (X_b \cup X_{ab}) \setminus X_{bc}$.

We have $X = Y_b \cup X_{bc} \cup X_a$. Vertices of X_a contain t_a. Vertices of X_{bc} intersect t_b and contain t_c. Vertices of Y_b contain t_b. This proves that X is included in the good graph of type (II) defined by t_a, t_b and t_c on respectively rows a, b, c. □

Lemma 10. *Let G be a B_2-EPG graph containing only Z-vertices, and X be a clique of G. If for every pair of rows a, b, the projection graphs of X_{ab} on both a and b are cliques, then there is a good graph containing X.*

4.2   General B_2-EPG Graphs

In Sect. 4.1, we have seen how to compute a maximum clique in a graph containing only Z-vertices. This section is devoted to prove that we can actually separate the graph in order to assume the graph only contains Z-vertices or U-vertices. We start by proving two lemmas showing that the existence of U-vertices puts some constraints on the Z-vertices that can be appear in a clique. We will then use these two lemmas to prove our main theorem.

Lemma 11. *Let G be a B_2-EPG graph with a representation, and X be a clique of G, then:*

- *either there are 3 rows intersecting all the U-vertices of X*
- *or there are three columns intersecting all the Z-vertices of X*

Proof. Let u_1, u_2, u_3, and u_4 be four U-vertices of X intersecting pairwise different rows. Let us prove that there are three columns containing every Z-vertex of X.

First assume that there are three columns α, β, γ such that, the set of columns intersected by u_i is in $\{\alpha, \beta, \gamma\}$ for every $i \leq 4$. Let us prove that these three columns intersect every Z-vertex of X. Assume by contradiction that there exists v in X that does not intersect α, β and γ. Then for every i, P_v shares an edge with P_{u_i} on a horizontal segment. Since all the u_i have disjoint index, this would imply that v intersects four different rows, a contradiction.

So we can assume that u_1, u_2, u_3, u_4 intersect at least four columns. Let α and β be the columns of u_1. We can assume w.l.o.g. that u_2 intersects the columns α and γ, with $\gamma \neq \alpha, \beta$. And that u_3 intersects a fourth column $\delta \neq \alpha, \beta, \gamma$. So both u_3 and u_4 must intersect α since they must intersect both u_1 and u_2. Let τ be the second column intersected by u_4. Then any Z-vertex of X intersects one of α, δ, τ. Indeed, suppose by contradiction that a Z-vertex v of X does not intersect one of these columns. Since P_v does not intersect P_{u_3} and P_{u_4} on their vertical intervals, it shares an edge with P_{u_3} and P_{u_4} on their two horizontal parts. Since u_1, u_2, u_3, u_4 have pairwise different index, P_v that intersect the row of u_3 and the row of u_4, share an edge with P_{u_1} on the column β and P_{u_2} on the column γ since v does not intersect column α. However, a Z-vertex intersects a single column, a contradiction.   □

In Sect. 3, we have introduced typed intervals. These typed interval defines intervals on a given row. In the following claim, we need two typed of typed interval: horizontal and vertical typed interval. An *horizontal typed interval* is a typed interval as defined in Sect. 3. A *vertical typed interval* is a typed interval of the graph after a rotation, i.e. the graph where rows become columns and columns become rows.

Lemma 12. *Let G be a B_2-EPG graph with its representation, and X be a clique of G containing only U-vertices with the same index $\{a\}$. There exists a set S_t of at most three typed intervals such that:*

- *S_t contains exactly one horizontal typed interval, and at most two vertical typed intervals,*
- *every vertex of X contains all the typed intervals of S_t,*
- *a Z-vertex u is complete to X if and only if u intersects one of the typed intervals of S_t.*

Proof. Since X is a clique of G and X only contains U-vertices of index $\{a\}$, Lemma 2 ensures that the projection graph of X on a is a clique. By Lemma 4 applied to X on a, there is a typed interval t such that every vertex of X contains t, and, if u is a vertex containing a in its index, and u is complete to X, then u must intersect t. The typed interval t is the unique horizontal typed interval of S_t.

Suppose that there is a column α, such that every vertex of X intersects α. Since all the vertices of X intersect the same row a and X is a clique, the projection graph of X on the column α is a clique. Indeed, since all the vertices of X intersect column α and row a, all of them must bend on the point (a, α). Either they all bend on the same direction on column α, say upwards, and then they all contain the edge of the column α between a and $a+1$, and the projection graph is a clique. Or, some vertices of X bend upwards and other downwards on (a, α). Since X is a clique, they all come from the same direction on row a and then their t-projections on α pairwise intersect.

By Lemma 4 applied to X on column α, there exists a vertical typed interval t_α satisfying both properties of Lemma 4. Since every U-vertex intersects two columns, there are at most two columns α, β for which every vertex of X intersects these columns. Let S_t be the set composed of t and the typed intervals t_α and t_β if they exist.

Let us prove that S_t satisfies the conclusion of the lemma. By construction S_t contains exactly one horizontal typed interval and at most two vertical typed intervals. By definitions of t and t_α, t_β, every vertex v of X contains the typed intervals in S_t. Let us finally show the last point. Let u be a Z-vertex. If u intersects a typed interval in S_t, then by Lemma 1, u is complete to X. Conversely, suppose that u is complete to X. If u contains a in its index, then Lemma 4 ensures that u intersects t since vertices of X all have index $\{a\}$. Assume now that the index of u does not contain a. So u intersects al the vertices of X on its unique column. Let γ be the unique column intersected by u. All the vertices of X must intersect γ since otherwise u cannot be complete to X. Then $\gamma \in \{\alpha, \beta\}$, and w.l.o.g., we can assume $\gamma = \alpha$. Then Lemma 4 ensures that u intersects t_α since the unique column of u is α.   □

The two previous lemmas are the main ingredients to prove that a maximum clique in B_2-EPG graphs can be computed in polynomial time. The idea of the algorithm is, using Lemma 12, to guess some typed intervals contained in the U-vertices of the clique. Lemma 11 ensures that we do not have to guess too many intervals. Once we have guessed these intervals, we are left with a subgraph which is actually the join of two subgraphs, one with only Z-vertices, and another with only U-vertices. Then the maximum clique is obtained by applying Theorem 1 to each of the components. The details of the proof can be found in the full version of the article [5]

Theorem 2. *Given a B_2-EPG graph G with its representation, there is a polynomial time algorithm computing the maximum clique of G.*

5   Colorings and χ-boundedness

In what follows we denote by $\chi(G)$ the *chromatic number* of G, i.e. the minimum number of colors needed to properly color the graph G. And we denote by $\omega(G)$ the maximum size of a clique of G.

Lemma 13. *Let G be a B_k-EPG graph on n vertices. There are at most $(k + 1)(\omega(G) - 1)n$ edges in G.*

The idea behind the proof is essentially to remark that on each row and on each column, the graph induced by the segments of paths forms an interval graph. Again, the proof of this lemma and its corollary are in the full version [5].

A graph is *k-degenerate* if there is an ordering $v_1, \ldots, v_n$ of the vertices such that for every i, $|N(v_i) \cap \{v_{i+1}, \ldots, v_n\}| \le k$. A k-degenerate graph is obviously $(k + 1)$-colorable. Lemma 13 immediately implies the following:

Corollary 1. *Let G be a B_k-EPG graph:*

- *The graph G is $\left(2(k+1)\omega - 1\right)$-degenerate.*
- $\chi(G) \leq 2(k+1)\omega(G)$.
- *There is a polynomial time $2(k+1)$-approximation algorithm for the coloring problem without knowing the representation of G.*
- *Every graph of B_k-EPG contains a clique or a stable set of size at least*
$$\sqrt{\frac{n}{2(k+1)}}.$$

References

1. Alcón, L., Bonomo, F., Durán, G., Gutierrez, M., Mazzoleni, M.P., Ries, B., Valencia-Pabon, M.: On the bend number of circular-arc graphs as edge intersection graphs of paths on a grid. Discrete Appl. Math. (2016)
2. Asinowski, A., Ries, B.: Some properties of edge intersection graphs of single-bend paths on a grid. Discrete Math. **312**(2), 427–440 (2012)
3. Flavia, B., Mazzoleni, M.P., Maya, S.: Clique coloring -EPG graphs. Discrete Math. **340**(5), 1008–1011 (2017)
4. Bougeret, M., Bessy, S., Gonçalves, D., Paul, C.: On independent set on B1-EPG graphs. In: Sanità, L., Skutella, M. (eds.) WAOA 2015. LNCS, vol. 9499, pp. 158–169. Springer, Cham (2015). doi:10.1007/978-3-319-28684-6_14
5. Bousquet, N., Heinrich, M.: Computing maximum cliques in b_2-epg graphs. arXiv preprint arXiv:1706.06685 (2017)
6. Cameron, K., Chaplick, S., Hoáng, C.T.: Edge intersection graphs of -shaped paths in grids. Discrete Appl. Math. **210**, 185–194 (2016). LAGOS 2013: Seventh Latin-American Algorithms, Graphs, and Optimization Symposium, Playa del Carmen, México (2013)
7. Cohen, E., Golumbic, M.C., Ries, B.: Characterizations of cographs as intersection graphs of paths on a grid. Discrete Appl. Math. **178**, 46–57 (2014)
8. Epstein, D., Golumbic, M.C., Morgenstern, G.: Approximation algorithms for B_1-EPG graphs. In: Dehne, F., Solis-Oba, R., Sack, J.-R. (eds.) WADS 2013. LNCS, vol. 8037, pp. 328–340. Springer, Heidelberg (2013). doi:10.1007/978-3-642-40104-6_29
9. Francis, M.C., Gonçalves, D., Ochem, P.: The maximum clique problem in multiple interval graphs. Algorithmica **71**(4), 812–836 (2015)
10. Francis, M.C., Lahiri, A.: VPG and EPG bend-numbers of Halin graphs. Discrete Appl. Math. **215**, 95–105 (2016)
11. Golumbic, M.C., Lipshteyn, M., Stern, M.: Edge intersection graphs of single bend paths on a grid. Networks **54**(3), 130–138 (2009)
12. Gyárfás, A., Lehel, J.: Covering and coloring problems for relatives of intervals. Discrete Math. **55**(2), 167–180 (1985)
13. Heldt, D., Knauer, K., Ueckerdt, T.: Edge-intersection graphs of grid paths: the bend-number. Discrete Appl. Math. **167**, 144–162 (2014)
14. Heldt, D., Knauer, K., Ueckerdt, T.: On the bend-number of planar and outerplanar graphs. Discrete Appl. Math. **179**, 109–119 (2014)
15. Pergel, M., Rzążewski, P.: On edge intersection graphs of paths with 2 bends. In: Heggernes, P. (ed.) WG 2016. LNCS, vol. 9941, pp. 207–219. Springer, Heidelberg (2016). doi:10.1007/978-3-662-53536-3_18
16. Trotter, W.T., Harary, F.: On double and multiple interval graphs. J. Graph Theory **3**(3), 205–211 (1979)

Intersection Graphs of Rays and Grounded Segments

Jean Cardinal[1], Stefan Felsner[2], Tillmann Miltzow[3(✉)], Casey Tompkins[4], and Birgit Vogtenhuber[5]

[1] Université libre de Bruxelles (ULB), Brussels, Belgium
jcardin@ulb.ac.be
[2] Institut für Mathematik, Technische Universität Berlin (TU), Berlin, Germany
felsner@math.tu-berlin.de
[3] Institute for Computer Science and Control,
Hungarian Academy of Sciences (MTA SZTAKI), Budapest, Hungary
t.miltzow@gmail.com
[4] Alfréd Rényi Institute of Mathematics, Hungarian Academy of Sciences,
Budapest, Hungary
ctompkins496@gmail.com
[5] Institute of Software Technology, Graz University of Technology, Graz, Austria
bvogt@ist.tugraz.at

Abstract. We consider several classes of intersection graphs of line segments in the plane and prove new equality and separation results between those classes. In particular, we show that:

- intersection graphs of grounded segments and intersection graphs of downward rays form the same graph class,
- not every intersection graph of rays is an intersection graph of downward rays, and
- not every outer segment graph is an intersection graph of rays.

The first result answers an open problem posed by Cabello and Jejčič. The third result confirms a conjecture by Cabello. We thereby completely elucidate the remaining open questions on the containment relations between these classes of segment graphs. We further characterize the complexity of the recognition problems for the classes of outer segment, grounded segment, and ray intersection graphs. We prove that these recognition problems are complete for the existential theory of the reals. This holds even if a 1-string realization is given as additional input.

1 Introduction

Intersection graphs encode the intersection relation between objects in a collection. More precisely, given a collection $\mathcal{A}$ of sets, the induced intersection graph has the collection $\mathcal{A}$ as the set of vertices, and two vertices $A, B \in \mathcal{A}$ are adjacent whenever $A \cap B \neq \emptyset$. Intersection graphs have drawn considerable attention

T. Miltzow—Supported by the ERC grant PARAMTIGHT: "Parameterized complexity and the search for tight complexity results", no. 280152.

© Springer International Publishing AG 2017
H.L. Bodlaender and G.J. Woeginger (Eds.): WG 2017, LNCS 10520, pp. 153–166, 2017.
https://doi.org/10.1007/978-3-319-68705-6_12

in the past thirty years, to the point of constituting a whole subfield of graph theory (see, for instance, the book from McKee and McMorris [20]). The roots of this subfield can be traced back to the properties of interval graphs – intersection graphs of intervals on a line – and their role in the discovery of the linear structure of bacterial genes by Benzer in 1959 [1].

We consider *geometric* intersection graphs, that is, intersection graphs of simple geometric objects in the plane, such as curves, disks, or segments. While early investigations of such graphs are a half-century old [28], the modern theory of geometric intersection graphs was established in the nineties by Kratochvíl [14,15], and Kratochvíl and Matoušek [16,17]. They introduced several classes of intersection graphs that are the topic of this paper. Geometric intersection graphs are now ubiquitous in discrete and computational geometry, and deep connections to other fields such as complexity theory [19,24,25] and order dimension theory [7,8,10] have been established.

We will focus on the following classes of intersection graphs, most of which are subclasses of intersection graphs of line segments in the plane, or *segment (intersection) graphs*.

Grounded Segment Graphs. Given a *grounding line* ℓ, we call a segment s a *grounded segment* if one of its endpoints, called the base point, is on ℓ and the interior of s is above ℓ. A graph G is a *grounded segment graph* if it is the intersection graph of a collection of grounded segments (w.r.t. the same grounding line ℓ).

Outer Segment Graphs. Given a *grounding circle* $\mathcal{C}$, a segment s is called an *outer segment* if exactly one of its endpoints, called the base point, is on $\mathcal{C}$ and the interior of s is inside $\mathcal{C}$. A graph G is an *outer segment graph* if it is the intersection graph of a collection of outer segments (w.r.t. the same grounding circle $\mathcal{C}$).

Ray Graphs and Downward Ray Graphs. A graph G is a *ray graph* if it is the intersection graph of rays (halflines) in the plane. A ray r is called a *downward ray* if its apex is above all other points of r. A graph G is a *downward ray graph* if it is the intersection graph of a collection of downward rays. It is not difficult to see that every ray graph is also an outer segment graph: consider a grounding circle at infinity. Similarly, one can check that every downward ray graph is a grounded segment graph.

String Graphs. String graphs are defined as intersection graphs of collections of simple curves in the plane with no three intersecting in the same point. We consider here *1-string graphs*, defined as intersection graphs of strings that pairwise intersect at most once. In particular, we define *outer 1-string graphs* and *grounded 1-string graphs* in the same way as for segments.

In a recent paper, Cabello and Jejčič initiated a comprehensive study aiming at refining our understanding of the containment relations between classes of geometric intersection graphs involving segments, disks, and strings [3]. They introduced and solved many questions about the containment relations between various classes. In particular, they proved proper containment between intersection graphs of segments with k or $k + 1$ distinct lengths, intersection graphs of disks with k or $k + 1$ distinct radii, and intersection graphs of outer strings and outer segments. In their conclusion [3], they left two natural questions open:

- Is the class of ray graphs a proper subclass of outer segment graphs?
- Is the class of downward ray graphs a proper subclass of grounded segment graphs?

In this contribution, we answer the first question in the positive, thereby proving a conjecture of Cabello. We also give a negative answer to the second question by showing that downward rays and grounded segments yield the same class of intersection graphs. Moreover, we show that downward ray graphs are a proper subclass of ray graphs, and thereby completely settle the remaining open questions on the containment relations between these classes of segment graphs. We also complete the picture by giving computational hardness results on stretchability questions for the classes of grounded 1-string and outer 1-string graphs, where we ask for a representation of the same intersection graph by grounded segments and outer segments, respectively. This strengthens the result of Cabello and Jejčič on the separation between outer string and outer segment graphs. A schematic description of the established inclusion relations between the graph classes we consider is given in Fig. 1.

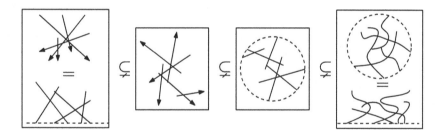

Fig. 1. Inclusion relations between the considered graph classes.

Previous Work and Motivation. The understanding of the inclusion properties and the complexity of the recognition problem for classes of geometric intersection graphs have been the topic of numerous previous works.

Early investigations of string graphs date back to Sinden [28], and Ehrlich et al. [9]. Kratochvíl [14] initiated a systematic study of string graphs, including the complexity-theoretic aspects [15]. It is only relatively recently, however, that the recognition problem for string graphs has been identified as NP-complete [25]. NP membership is far from obvious, given that there exist string graphs requiring exponential-size representations [16].

Intersection graphs of line segments were extensively studied by Kratochvíl and Matoušek [17]. In particular, they proved that the recognition of such graphs was complete for the existential theory of the reals. A key construction used in their proof is the *Order-forcing Lemma*, which permits the embedding of pseudo-line arrangements as segment representations of graphs. Some of our constructions can be seen as extensions of the Order-forcing Lemma to grounded and outer segment representations.

Outer segment graphs form a natural subclass of outer string graphs as defined by Kratochvíl [14]. They also naturally generalize the class of *circle graphs*, which are intersection graphs of chords of a circle [22].

A recent milestone in the field of segment intersection graphs is the proof of Scheinerman's conjecture by Chalopin and Gonçalves [6], stating that planar graphs form a subclass of segment graphs. It is also known that outerplanar graphs form a proper subclass of circle graphs [30], hence of outer segment graphs. Cabello and Jejčič [3] proved that a graph is outerplanar if and only if its 1-subdivision is an outer segment graph.

Intersection graphs of rays in two directions have been studied by Soto and Telha [29]. They show connections with the jump number of some posets and hitting sets of rectangles. The class has been further studied by Shrestha et al. [27], and Mustaţă et al. [21]. The results include polynomial-time recognition and isomorphism algorithms. This is in contrast with our hardness result for arbitrary ray graphs.

Properties of the chromatic number of geometric intersection graphs have been studied as well. For instance, Rok and Walczak proved that outer string graphs are χ-bounded [23], and Kostochka and Nešetřil [12,13] studied the chromatic number of ray graphs in terms of the girth and the clique number.

The complexity of the maximum clique and independent set problems on classes of segment intersection graphs is also a central topic of study. It has been shown recently, for instance, that the maximum clique problem is NP-hard on ray graphs [2], and that the maximum independent set problem is polynomial-time tractable on outer segment graphs [11].

Organization of the Paper. In the next section, we give some basic definitions and observations. We also provide a short proof of the equality between the classes of downward ray and of grounded segment graphs.

In Sect. 3, we introduce the *Cycle Lemma*, a construction that will allow us to control the order of the slopes of the rays in a representation of a ray graph, and the order in which the segments are attached to the grounding line or circle in representations of grounded segment and outer segment graphs.

In Sect. 4, we show how to use the Cycle Lemma to encode the pseudoline stretchability problem in the recognition problem for outer segment, grounded segment, and ray graphs. We thereby prove that those problems are complete for the existential theory of the reals.

Finally, in Sect. 5, we establish two new separation results. First, we prove that ray graphs form a proper subclass of outer segment graphs, proving

Cabello's conjecture. Then we prove that downward ray graphs form a proper subclass of ray graphs.

2 Preliminaries

We first give a short proof of the equality between the classes of downward ray and grounded segment graphs, thereby answering Cabello and Jejčič's second question. The proof is illustrated in Fig. 2a.

Lemma 1 (Downward Rays = Grounded Segments). *A graph G can be represented as a grounded segment graph if and only if it can be represented by downward rays.*

Proof. Consider a coordinate system where the grounding line is the x-axis, and take the projective transformation defined in homogeneous coordinates by

$$\begin{pmatrix} x \\ y \\ 1 \end{pmatrix} \mapsto \begin{pmatrix} x \\ -1 \\ y \end{pmatrix}.$$

This projective transformation is a bijective mapping from the projective plane to itself, which maps grounded segments to downward rays. In the plane, it can be seen as mapping the points (x, y) with $y > 0$ to $(x/y, -1/y)$. Since projective transformations preserve the incidence structure, the equivalence of the graph classes follows. □

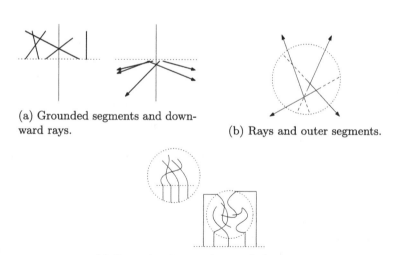

(a) Grounded segments and downward rays.

(b) Rays and outer segments.

(c) Outer 1-strings and grounded 1-strings.

Fig. 2. Simple transformations between different graph representations.

The proofs of the following two lemmas are not difficult, and omitted. They are illustrated in Fig. 2b and c.

Lemma 2 (Ray Characterization). *A graph G can be represented as an outer segment graph with all intersections of the supporting lines inside the grounding circle C if and only if it can be represented by rays.*

Note that it is tempting to try to find a projective transformation that maps the unit circle S^1 to infinity in a way that outer segments become rays. As we will show later, outer segments and rays represent different graph classes. Thus such a mapping is impossible. With the help of Möbius transformations it is possible to find a mapping that maps the unit circle S^1 to infinity. However, outer segments then become connected parts of hyperbolas instead of straightline rays.

Recall that we define *grounded 1-string graphs* and *outer 1-string graphs* in an analogous way to the corresponding segment graphs by replacing segments by 1-strings.

Lemma 3 (Grounded 1-Strings = Outer 1-Strings). *A graph G can be represented as a grounded 1-string graph if and only if it can be represented as an outer 1-string graph.*

Ordered Representations. Given a graph G and a permutation π of the vertices, we say that a grounded (segment or string) representation of G is π-*ordered* if the base points of the cooresponding segments or strings are in the order of π on the grounding line, up to inversion and cyclic shifts. In the same fashion, we define π-*ordered* for outer (segment or string) representations and (downward) ray representations, where rays are ordered by their angles with the horizontal axis.

The Complexity Class $\exists\mathbb{R}$ and the Stretchability Problem. The complexity class $\exists\mathbb{R}$ is the collection of decision problems that are polynomial-time equivalent to deciding the truth of sentences in the first-order theory of the reals of the form $\exists x_1 \exists x_2 \ldots \exists x_n F(x_1, x_2, \ldots, x_n)$, where F is a quantifier-free formula involving inequalities and equalities of polynomials in the real variables x_i. This complexity class can be understood as a "real" analogue of NP. It can easily be seen to contain NP, and is known to be contained in PSPACE [4].

In recent years, this complexity class revealed itself most useful for characterizing the complexity of realizability problems in computational geometry. A standard example is the *pseudoline stretchability problem*.

Matoušek [18, p. 132] defines an *arrangement of pseudolines* as a finite collection of curves in the plane that satisfy the following two conditions: (i) each curve is x-monotone and unbounded in both directions and (ii) every two of the curves intersect in exactly one point and they cross at the intersection. In the stretchability problem, one is given the combinatorial structure of an arrangement of pseudolines in the plane as input, and is asked whether the same combinatorial structure can be realized by an arrangement of *straight lines*. If this is the case, then we say that the arrangement is *stretchable*. This structure can for instance be given in the form of a set of n *local sequences*: the left-to-right order of the intersections of each line with the $n - 1$ others. Equivalently, the input is the underlying rank-3 oriented matroid. The stretchability problem is known to be ∃ℝ-complete [26]. We refer the reader to the surveys by Schaefer [24], Matoušek [19], and Cardinal [5] for further details.

Computational Complexity Questions. Given a graph class $\mathcal{G}$, we define **Recognition**$(\mathcal{G})$ as the following decision problem:

Recognition $(\mathcal{G})$
Input: A graph $G = (V, E)$.
Question: Does G belong to the graph class $\mathcal{G}$?

We will be concerned with the problems **Recognition** (rays), **Recognition** (grounded segments) and **Recognition** (outer segments).

Potentially the recognition problem could become easier if we have some additional information. In our case it is natural to ask if a given outer 1-string representation of a graph G has an outer segment representation. The same goes for grounded 1-strings and grounded segments. Finally, we will consider outer 1-strings and rays. Formally, we define the decision problem **Stretchability** $(\mathcal{G}, \mathcal{F})$ as follows.

Stretchability $(\mathcal{G}, \mathcal{F})$
Input: graph $G = (V, E)$ and representation R that shows that G belongs to $\mathcal{F}$.
Question: Does G belong to the graph class $\mathcal{G}$?

Note that we need to assume that $\mathcal{F}$ is a graph class defined by intersections of certain objects.

3   Cycle Lemma

For some of our constructions, we would like to force that the segments or strings representing the vertices of a graph appear in a specified order on the grounding line or circle. More exactly, we would like to force the representation to be π-ordered for some given permutation π. To this end, we first study some properties of the representation of cycles, which in turn will help us to enforce this order.

Given a graph $G = (V, E)$ on n vertices $V = \{v_1, \ldots, v_n\}$ and a permutation π of the vertices of G, we define the *order forcing graph* G^π as follows. The vertices $V(G^\pi)$ are defined by $V \cup \{1, \ldots, 2n^2\}$ and the edges $E(G^\pi)$ are defined by $E \cup \{(2in, v_{\pi(i)}) \mid i = 1, \ldots, n\} \cup \{(i, i+1) \mid i = 1, \ldots, 2n^2\}$ (here we conveniently assume $2n^2 + 1 = 1$). The definition is illustrated on Fig. 3.

For the sake of simplicity, we think of π as being the identity and the vertices as being indexed in the correct way. The vertices of G are called *relevant*, and the additional vertices of G^π are called *cycle vertices*. Note that on the cycle, the distance between any two cycle vertices u, v that are adjacent to different relevant vertices is at least $2n$.

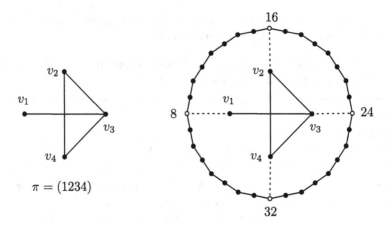

Fig. 3. Illustration of the definition of order forcing graphs.

Lemma 4 (Cycle Lemma). *Let G be a graph and π be a permutation of the vertices of G. Then there exists a π-ordered representation of G if and only if there exists a representation of G^π. This is true for the following graph classes: grounded segment graphs, ray graphs, outer segment graphs, and outer 1-string graphs.*

Due to space constraints, we omit the proof of Lemma 4. Figure 4 shows the representations of order forcing graphs in the cases of outer string and grounded segment graphs.

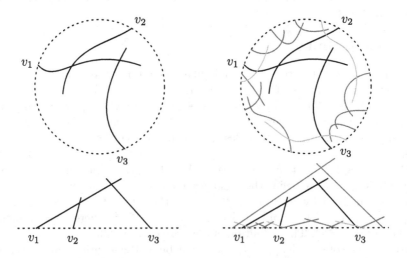

Fig. 4. Representations of order forcing graphs for ordered representations of outer string graphs and grounded segment graphs.

4   Stretchability

The main purpose of this section is to show that the recognition of the graph classes defined above is $\exists\mathbb{R}$-complete. For this we will use Lemma 4 extensively. It is likely that our techniques can be applied to other graph classes as well.

Theorem 1. *The following problems are $\exists\mathbb{R}$-complete:*

- **Recognition** *(outer segments) and* **Stretchability** *(outer segments, outer 1-strings),*
- **Recognition** *(grounded segments) and* **Stretchability** *(grounded segments, grounded 1-strings),*
- **Recognition** *(rays) and* **Stretchability** *(rays, outer 1-strings).*

Proof. We first show $\exists\mathbb{R}$-membership. Note that each of the straight-line objects we consider can be represented with at most four variables: for segments, we use two variables for each endpoint, and for rays, we use two variables for the apex and two variables for the direction. The condition that two objects intersect can be formulated with constant-degree polynomials in those variables. Hence, each of the problems can be formulated as a sentence in the first-order theory of the reals of the desired form.

Let us now turn our attention to the $\exists\mathbb{R}$-hardness. It is sufficient to show hardness for the stretchability problems, as the problems can only become easier with additional information. We will reduce from stretchability of pseudoline arrangements. Given a pseudoline arrangement $\mathcal{L}$, we will construct a graph $G_{\mathcal{L}}$ and a permutation π such that:

1. If $\mathcal{L}$ is stretchable then $G_{\mathcal{L}}$ has a π-ordered representation with *grounded segments*.
2. If $\mathcal{L}$ is not stretchable then there does not exist a π-ordered representation of $G_{\mathcal{L}}$ as an *outer segment graph*.

By Lemma 4 $G_{\mathcal{L}}$ has a π-ordered representation if and only if $G_{\mathcal{L}}^{\pi}$ has a representation. Recall that we know the following relations for the considered graph classes.

$$\text{grounded segments} \quad \subseteq \quad \text{rays} \quad \subseteq \quad \text{outer segments.}$$

Thus, item 1 implies that if $\mathcal{L}$ is stretchable then $G_{\mathcal{L}}$ has a π-ordered representation with rays or outer segments. Furthermore, item 2 implies that if $\mathcal{L}$ is not stretchable then $G_{\mathcal{L}}$ has neither a π-ordered representation with rays nor with grounded segments.

We start with the construction of $G_{\mathcal{L}}$ and π. Let $\mathcal{L}$ be an arrangement of n pseudolines. Recall that we can represent $\mathcal{L}$ by x-monotone curves. Let ℓ_1 and ℓ_2 be two vertical lines such that all the intersections of $\mathcal{L}$ lie between ℓ_1 and ℓ_2. We cut away the part outside the strip bounded by ℓ_1 and ℓ_2. This gives us a π-ordered grounded 1-string representation $R_{\mathcal{L}}$ with respect to the grounding line ℓ_1.

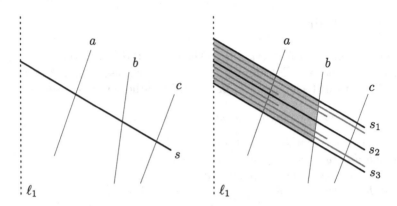

Fig. 5. Illustration of Theorem 1: Construction of $G_{\mathcal{L}}$ and its grounded 1-string representation.

Now we replace each string s representing a pseudoline in $\mathcal{L}$ by the following construction (extending π accordingly): We split s into three similar copies s_1, s_2, s_3, shifted vertically by an offset that is chosen sufficiently small so that the three copies intersect the other pseudolines (and their shifted copies) in the same order. For each successive intersection point of s with a pseudoline s' in $\mathcal{L}$, we add a pair of strings grounded on either side of the base point of s_2 and between the base points of s_1 and s_3, intersecting none of s_1, s_2 and s_3. The two strings intersect all the pseudolines of $\mathcal{L}$ that s intersects, up to and including s', in the same order as s does. All the strings for s are pairwise nonintersecting and nested around s_2; see Fig. 5. We refer to these pairs of strings as *probes*. The probes are meant to enforce the order of the intersections in all π-ordered representations.

We now prove item 1. We suppose there is a straight line representation of $\mathcal{L}$, which we denote by $\mathcal{K}$. Again let ℓ_1 and ℓ_2 be two vertical lines such that all intersections of $\mathcal{K}$ are contained in the vertical strip between them. This gives us a collection of grounded segments $R_{\mathcal{K}}$. One can check that the above construction involving probes can be implemented using straight line segments, just as illustrated in Fig. 5. Thus, $R_{\mathcal{K}}$ is a π-ordered grounded segment representation of $G_{\mathcal{L}}$, as claimed.

Next, we turn our attention to item 2 and suppose that $\mathcal{L}$ is not stretchable. Let us further suppose, for the purpose of contradiction, that we have a π-ordered outer segment representation of $G_{\mathcal{L}}$. We show that keeping only the middle copy s_2 of each segment s representing a pseudoline of $\mathcal{L}$ in our construction, we obtain a realization of $\mathcal{L}$ with straight lines. For this, we need to prove that the construction of the probes indeed forces the order of the intersections. We consider each such segment s_2 and orient it from its base point to its other endpoint. Now suppose that there exist strings a and b such that the order of intersections of s_2 with a and b with respect to this orientation does not agree with that of the pseudoline arrangement. (In Fig. 5, suppose that s_2 crosses the lines b before a in the left-to-right order.) We consider the convex region bounded by the arc of the grounding circle between the base points of s_1 and s_3, and segments from s_1, b, and s_3. This convex region is split into two convex boxes by s_2. The pair of probes corresponding to the intersection of s_2 and a is completely contained in this region, with one probe in each box. But now the line a must enter both boxes, thereby intersecting s_2 on the left of b with respect to the chosen orientation, a contradiction. Therefore, the order of the intersections is preserved, and the collection of segments s_2 is a straight line realization of $\mathcal{L}$, a contradiction to the assumption that $\mathcal{L}$ is not stretchable.   □

5 Rays and Segments

Theorem 2 (Rays $\subsetneq$ Outer Segments). *There are graphs that admit a representation as outer segment graphs but not as ray graphs.*

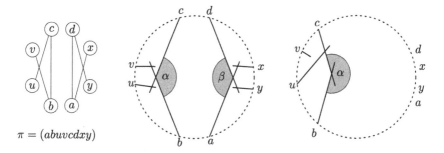

$\pi = (abuvcdxy)$

Fig. 6. Illustration of Theorem 2. On the left is a graph G together with a permutation π of the vertices displayed. In the middle is a π-ordered outer segment representation of G. The right drawing illustrates that the angles α and β must each be at most $180°$.

Proof. Consider the graph G and a permutation π as displayed in Fig. 6. We show that G has a π-ordered representation as an outer segment graph, but not as a ray graph. This implies that G^π has a representation as an outer segment graph, but not as a ray graph as well, see Lemma 4.

Given any π-ordered representation of G, we define the angle α as the angle at the intersection of b and c towards the segments d, x, y, a and we define the angle β as the angle at the intersection of a and d towards the segments b, u, v, c, as can be seen in Fig. 6.

We show that both α and β are smaller than $180°$ in any outer segment representation. As the two cases are symmetric we show it only for α. Assume $\alpha \geq 180°$ as on the right of Fig. 6. If u intersects c (as shown in the figure) then it blocks v from intersecting b, as v must not intersect u. Likewise, v intersecting b would block u from intersecting c. This shows $\alpha, \beta < 180°$.

As the angles are smaller than $180°$, we conclude that either the extensions of a and b or the extensions of c and d must meet outside of the grounding circle. Recall that we considered any representation of G. By Lemma 2 it holds for every ray graph that there exists at least one representation of G with outer segments such that *all* extensions meet *within* the grounding circle. (The lemma also holds for ordered representations.) Thus there cannot be a π-ordered ray representation of G.   □

Theorem 3 (Downward Rays $\subsetneq$ Rays). *There are graphs that admit a representation as ray graphs but not as downward ray graphs.*

Consider the graph G and the permutation π as displayed in Fig. 7 (left). Clearly, G has a π-ordered representation as a ray graph, as can be seen from the outer segment representation of G as shown in Fig. 7 (middle). We can show that G does not have a π-ordered representation as a grounded segment graph. Hence G^{π} has a representation as a ray graph, but not as a grounded segment graph or a downward ray graph; see Lemma 4 and Lemma 1. This yields the theorem, the complete proof of which is omitted.

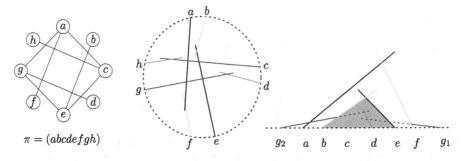

Fig. 7. Illustration of Theorem 3: A graph G together with a permutation π of the vertices (left); A π-ordered outer segment representation of G (middle); The segment g cannot enter the gray triangle without intersecting b or f (right).

Acknowledgments. This work was initiated during the Order & Geometry Workshop organized by Piotr Micek and the second author at the Gultowy Palace near Poznan, Poland, on September 14–17, 2016. We thank the organizers and attendees, who contributed to an excellent work atmosphere. Some of the problems tackled in this paper were brought to our attention during the workshop by Michal Lason. The first author also thanks Sergio Cabello for insightful discussions on these topics.

References

1. Benzer, S.: On the topology of the genetic fine structure. Proc. Nat. Acad. Sci. **45**(11), 1607–1620 (1959)

2. Cabello, S., Cardinal, J., Langerman, S.: The clique problem in ray intersection graphs. Discrete Comput. Geom. **50**(3), 771–783 (2013)
3. Cabello, S., Jejčič, M.: Refining the hierarchies of classes of geometric intersection graphs. Electron. J. Combin. **24**(1), P1.33 (2017)
4. Canny, J.F.: Some algebraic and geometric computations in PSPACE. In: Proceedings of STOC, pp. 460–467. ACM (1988)
5. Cardinal, J.: Computational geometry column 62. ACM SIGACT News **46**(4), 69–78 (2015)
6. Chalopin, J., Gonçalves, D.: Every planar graph is the intersection graph of segments in the plane: extended abstract. In: Proceedings of STOC, pp. 631–638. ACM (2009)
7. Chaplick, S., Felsner, S., Hoffmann, U., Wiechert, V.: Grid intersection graphs and order dimension. arXiv:1512.02482 (2015)
8. Chaplick, S., Hell, P., Otachi, Y., Saitoh, T., Uehara, R.: Intersection dimension of bipartite graphs. In: Gopal, T.V., Agrawal, M., Li, A., Cooper, S.B. (eds.) TAMC 2014. LNCS, vol. 8402, pp. 323–340. Springer, Cham (2014). doi:10.1007/978-3-319-06089-7_23
9. Ehrlich, G., Even, S., Tarjan, R.E.: Intersection graphs of curves in the plane. J. Comb. Theory Ser. B **21**(1), 8–20 (1976)
10. Felsner, S.: The order dimension of planar maps revisited. SIAM J. Discrete Math. **28**(3), 1093–1101 (2014)
11. Keil, J.M., Mitchell, J.S.B., Pradhan, D., Vatshelle, M.: An algorithm for the maximum weight independent set problem on outerstring graphs. Comput. Geom. **60**, 19–25 (2017)
12. Kostochka, A.V., Nešetřil, J.: Coloring relatives of intervals on the plane, I: chromatic number versus girth. Eur. J. Comb. **19**(1), 103–110 (1998)
13. Kostochka, A.V., Nešetřil, J.: Colouring relatives of intervals on the plane, II: intervals and rays in two directions. Eur. J. Comb. **23**(1), 37–41 (2002)
14. Kratochvíl, J.: String graphs. I. The number of critical nonstring graphs is infinite. J. Comb. Theory Ser. B **52**(1), 53–66 (1991)
15. Kratochvíl, J.: String graphs. II. Recognizing string graphs is NP-hard. J. Comb. Theory Ser. B **52**(1), 67–78 (1991)
16. Kratochvíl, J., Matoušek, J.: String graphs requiring exponential representations. J. Comb. Theory Ser. B **53**(1), 1–4 (1991)
17. Kratochvíl, J., Matoušek, J.: Intersection graphs of segments. J. Comb. Theory Ser. B **62**(2), 289–315 (1994)
18. Matoušek, J.: Lectures on Discrete Geometry. Graduate Texts in Mathematics, vol. 212. Springer, New York (2002). doi:10.1007/978-1-4613-0039-7
19. Matoušek, J.: Intersection graphs of segments and $\exists \mathbb{R}$. arXiv:1406.2636 (2014)
20. McKee, T.A., McMorris, F.: Topics in Intersection Graph Theory. Society for Industrial and Applied Mathematics (1999)
21. Mustaţă, I., Nishikawa, K., Takaoka, A., Tayu, S., Ueno, S.: On orthogonal ray trees. Discrete Appl. Math. **201**, 201–212 (2016)
22. Naji, W.: Reconnaissance des graphes de cordes. Discrete Math. **54**(3), 329–337 (1985)
23. Rok, A., Walczak, B.: Outerstring graphs are χ-bounded. In: Proceedings of SoCG, pp. 136–143. ACM (2014)
24. Schaefer, M.: Complexity of some geometric and topological problems. In: Eppstein, D., Gansner, E.R. (eds.) GD 2009. LNCS, vol. 5849, pp. 334–344. Springer, Heidelberg (2010). doi:10.1007/978-3-642-11805-0_32

25. Schaefer, M., Sedgwick, E., Štefankovič, D.: Recognizing string graphs in NP. J. Comput. Syst. Sci. **67**(2), 365–380 (2003)
26. Shor, P.W.: Stretchability of pseudolines is NP-hard. In: Applied Geometry and Discrete Mathematics. DIMACS Series in DMTCS, vol. 4, pp. 531–554. AMS (1990)
27. Shrestha, A.M.S., Tayu, S., Ueno, S.: On orthogonal ray graphs. Discrete Appl. Math. **158**(15), 1650–1659 (2010)
28. Sinden, F.W.: Topology of thin film RC circuits. Bell Syst. Tech. J. **45**(9), 1639–1662 (1966)
29. Soto, J.A., Telha, C.: Jump number of two-directional orthogonal ray graphs. In: Günlük, O., Woeginger, G.J. (eds.) IPCO 2011. LNCS, vol. 6655, pp. 389–403. Springer, Heidelberg (2011). doi:10.1007/978-3-642-20807-2_31
30. Wessel, W., Pöschel, R.: On circle graphs. In: Sachs, H. (ed.) Graphs, Hypergraphs and Applications, pp. 207–210. Teubner (1985)

On H-Topological Intersection Graphs

Steven Chaplick[1], Martin Töpfer[2], Jan Voborník[3], and Peter Zeman[3(✉)]

[1] Lehrstuhl für Informatik I, Universität Würzburg, Würzburg, Germany
steven.chaplick@uni-wuerzburg.de
[2] Faculty of Mathematics and Physics, Computer Science Institute,
Charles University in Prague, Prague, Czech Republic
topfer@iuuk.mff.cuni.cz
[3] Department of Applied Mathematics, Faculty of Mathematics and Physics,
Charles University in Prague, Prague, Czech Republic
{vobornik,zeman}@kam.mff.cuni.cz,
http://www1.informatik.uni-wuerzburg.de/en/staff

Abstract. Biró, Hujter, and Tuza introduced the concept of H-graphs (1992), intersection graphs of connected subgraphs of a subdivision of a graph H. They naturally generalize many important classes of graphs, e.g., interval graphs and circular-arc graphs. Our paper is the first study of the recognition and dominating set problems of this large collection of intersection classes of graphs.

We negatively answer the question of Biró, Hujter, and Tuza who asked whether H-graphs can be recognized in polynomial time, for a fixed graph H. Namely, we show that recognizing H-graphs is NP-complete if H contains the diamond graph as a minor. On the other hand, for each tree T, we give a polynomial-time algorithm for recognizing T-graphs and an $\mathcal{O}(n^4)$-time algorithm for recognizing $K_{1,d}$-graphs. For the dominating set problem (parameterized by the size of H), we give FPT- and XP-time algorithms on $K_{1,d}$-graphs and H-graphs, respectively. Our dominating set algorithm for H-graphs also provides XP-time algorithms for the independent set and independent dominating set problems on H-graphs.

1 Introduction

An intersection representation of a graph assigns a set to each vertex and uses intersections of those sets to encode its edges. More formally, an intersection representation $\mathcal{R}$ of a graph G is a collection of sets $\{R_v : v \in V(G)\}$ such that $R_u \cap R_v \neq \emptyset$ if and only if $uv \in E(G)$. Many important classes of graphs arise from restricting the sets R_v to geometric objects (e.g., intervals, convex sets).

We study H-*graphs*, intersection graphs of connected subsets of a fixed topological pattern given by a graph H, introduced by Biró et al. [1]. We obtain new

---

This paper including its full appendix is on ArXiv, see arXiv.org/pdf/1608.02389v2.pdf. Thus, for any reference to an appendix, see the ArXiv version.

P. Zeman—This author was supported by Charles University as GAUK 1334217.

H.L. Bodlaender and G.J. Woeginger (Eds.): WG 2017, LNCS 10520, pp. 167–179, 2017.
https://doi.org/10.1007/978-3-319-68705-6_13

algorithmic results on the recognition and dominating set problems. First, we discuss some closely related graph classes.

Interval graphs (INT) form one of the most studied and well-understood classes of intersection graphs. In an *interval representation*, each set R_v is a closed interval of the real line; see Fig. 1a. A primary motivation for studying interval graphs (and related classes) is the fact that many important computational problems can be solved in linear time on them; see for example [3,5,12].

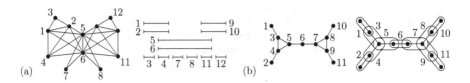

Fig. 1. (a) An interval graph and one of its interval representations. (b) A chordal graph and one of its representations as an intersection graph of subtrees of a tree.

A graph is *chordal* when it does not have an induced cycle of length at least four. Equivalently, as shown by Gavril [8], a graph is chordal if and only if it can be represented as an intersection graph of subtrees of some tree; see Fig. 1b. This immediately implies that INT is a subclass of the chordal graphs (CHOR).

While the recognition problem can be solved in linear time for CHOR [13] and such algorithms can be used to generate an intersection representation by subtrees of a tree, asking for special host trees can be more difficult. For example, when the desired tree T is part of the input (together with a graph G), deciding whether G is a T-graph is NP-complete [11]. Additionally, some other important computational problems, for example the dominating set [4] and graph isomorphism [12], are harder on chordal graphs than on interval graphs.

The *split graphs* (SPLIT) form an important subclass of chordal graphs. These are the graphs that can be partitioned into a clique and an independent set. Note that every split graph can be represented as an intersection graph of subtrees of a star S_d, where S_d is the complete bipartite graph $K_{1,d}$.

Circular-arc graphs (CARC) are a natural generalization of interval graphs. Here, each set R_v corresponds to an arc of a circle.

H-Graphs. Biró et al. [1] introduced *H-graphs*. Let H be a fixed graph. A graph G is an *intersection graph of* H if it is an intersection graph of connected subgraphs of H, i.e., the assigned subgraphs H_v and H_u of H share a vertex if and only if $uv \in E(G)$.

A *subdivision* H' of a graph H is obtained when the edges of H are replaced by internally disjoint paths of arbitrary lengths. A graph G is a *topological intersection graph of* H if G is an intersection graph of a subdivision H' of H. We say that G is an *H-graph* and the collection $\{H'_v : v \in V(G)\}$ of connected subgraphs of H' is an *H-representation* of G. The class of all *H-graphs* is denoted by H-GRAPH. We have the following relations: INT $= K_2$-GRAPH, SPLIT $\subsetneq$ $\bigcup_{d=2}^{\infty} S_d$-GRAPH, CARC $= K_3$-GRAPH, and CHOR $= \bigcup_{\text{Tree } T} T$-GRAPH.

H-graphs were introduced in the context of the (p, k)*pre-coloring extension problem* (PrColExt(p, k)). Here, one is given a graph G together with a p-coloring of $W \subseteq V(G)$, and the goal is to find a proper k-coloring of G containing this *pre-coloring*. Biró et al. [1] provide an XP (in k and $\|H\|$) algorithm to solve PrColExt(k, k) on H-GRAPH.

We have an infinite hierarchy of graph classes between interval and chordal graphs since INT $\subseteq T$-GRAPH $\subsetneq$ CHOR, for every tree T. Some important computational problems are polynomial on interval graphs and hard on chordal graphs. An interesting question is the complexity of those problems on T-graphs, for a fixed tree T.

Our Results. Biró, Hujter, and Tuza asked the following question which we answer negatively.

Problem 1. (Biró et al. [1]). Let H be an arbitrary fixed graph. Is there a polynomial algorithm testing whether a given graph G is an H-graph?

In Theorem 10, we prove that for each fixed graph H containing a diamond as a minor, it is NP-complete to recognize H-graphs. We do this by a reduction from the problem of testing if the *interval dimension* of a partial order of *height* 1 is at most 3. Moreover, we give an $\mathcal{O}(n^4)$-time and a polynomial-time algorithm for recognizing S_d-graphs and T-graphs, respectively (Theorems 7 and 9). Note that our results imply that the complexity of recognition of H-graphs is open for cactus graphs.

Conjecture 2. If H is a cactus graph, then the recognition of H-graphs can be solved in polynomial time.

Further, we solve the problem of finding a minimum dominating set on S_d-graphs (Theorem 12) in $\mathcal{O}(d \cdot n \cdot (n+m)) + (2^d(d+2^d)^{O(1)})$ time and for H-graphs (Theorem 13) in $n^{\mathcal{O}(\|H\|)}$ time. To achieve this running time, we assume that *the intersection representation is provided as a part of the input*.

Preliminaries. We assume that the reader is familiar with the following standard and parameterized computational complexity classes: NP, XP, and FPT (see, e.g., [6] for further details). Let G be an H-graph. For a subdivision H' certifying $G \in H$-GRAPH, we use H'_v to denote the subgraph of H' corresponding to $v \in V(G)$. The vertices of H and H' are called *nodes*. If H is a tree, then its degree 1 nodes are called *leaves* and its nodes of degree at least three are called *branching points*. Let a, b be two nodes of H'. By $P_{[a,b]}$ we denote the path from a to b. Further, we define $P_{(a,b]} = P_{[a,b]} - a$, and $P_{[a,b)}$, $P_{(a,b)}$ analogously. For $V_1, \ldots, V_k \subseteq V(G)$, let $G[V_1, \ldots, V_k]$ be the subgraph of G induced by $V_1 \cup \cdots \cup V_k$. For a graph G, we assume G has n vertices and m edges. In many results we implicitly use the well-known fact [7] that interval graphs are characterized by having a linear order on their maximal cliques so that for each vertex, its maximal cliques occur consecutively.

2   Recognition of T-Graphs

We solve the recognition problem for the class T-GRAPH, where T is an arbitrary fixed tree. First, we provide an $\mathcal{O}(n^4)$ algorithm which either finds the minimum d such that G is an S_d-graph, or reports that G has no such representation. Further, we give an $n^{\mathcal{O}(\|T\|^2)}$-time algorithm to test whether G is a T-graph.

We begin with a lemma that motivates our general approach. It implies that if G is a T-graph, then there exists a representation of G such that every branching point is contained in some maximal clique of G.

Lemma 3. *Let G be a T-graph and let $\mathcal{R}$ be its T-representation. Then $\mathcal{R}$ can be modified such that for every node $b \in V(T')$, we have $b \in \bigcap_{v \in C} V(T'_v)$, for some maximal clique C of G.*

Proof. For every node x of the subdivision T', let $V_x = \{u \in V(G) : x \in V(T'_u)\}$ be the set of vertices of G corresponding to the subtrees passing through x. Let b be a node such that V_b is not a maximal clique. We pick a maximal clique C with $C \supsetneq V_b$. Since $\mathcal{R}$ satisfies the Helly property, there is a node $a \in \bigcap\{V(T'_v) : v \in C\}$. Let x be the node of $P_{(b,a]}$ closest to b such that $V_x \supsetneq V_b$. Then, for each $v \in V_x \setminus V_b$, we update T'_v to be $T'_v \cup P_{[b,x]}$. Thus, we obtain a correct representation of G with $V_b = V_x$. We repeat this process until V_b is a maximal clique.   □

The General Approach. It is well-known that chordal graphs, and therefore also T-graphs, have at most n maximal cliques and that they can be listed linear time. According to Lemma 3, if G is a T-graph, then it has a representation such that every branching point of T is contained in the representation of $G[C]$, for some maximal clique C of G.

Our approach is to try all the possible mappings $f \colon \mathcal{B} \to \mathcal{C}$, where $\mathcal{B}$ is the set of branching points of T and $\mathcal{C}$ is the set of maximal cliques of G. The number of such mappings is at most n^t, where $n = |V(G)|$ and $t = |V(T)|$. For every mapping $f \colon \mathcal{B} \to \mathcal{C}$, we test whether there exists a T-representation of G with $b \in \bigcap_{v \in f(b)} V(T'_v)$, for every $b \in \mathcal{B}$. By Lemma 3, such a representation exists if and only if G is a T-graph. To find such a representation, we need to find suitable interval representations of the connected components of $G \setminus \bigcup_{b \in \mathcal{B}} f(b)$ on the segments $T' - \mathcal{B}$ such that the following conditions hold:

(i) If $X_1, \ldots, X_k$ are the connected components placed on a path $P_{(b,l]}$, where $b \in \mathcal{B}$ and l is a leaf, then the induced subgraph $G[f(b), V(X_1), \ldots, V(X_k)]$ has an interval representation with $f(b)$ being the leftmost maximal clique.

(ii) If $X_1, \ldots, X_k$ are the connected components placed on a path $P_{(b,b')}$, where $b, b' \in \mathcal{B}$, then $G[f(b), V(X_1), \ldots, V(X_k), f(b')]$ has an interval representation with $f(b)$ and $f(b')$ being the leftmost and rightmost maximal cliques, respectively.

Recognition of S$_d$-Graphs. In the case when $T = S_d$, we have $\mathcal{B} = \{b\}$ and $V(T) = \{b\} \cup \{l_1, \ldots, l_d\}$. The number of mappings $f : \{b\} \to \mathcal{C}$ is exactly the same as the number of maximal cliques of G, which is at most n (otherwise it is not an S_d-graph). For every maximal clique C of G, we try to construct a T-representation $\mathcal{R}$ such that $b \in \bigcap_{v \in C} V(T'_v)$.

Assume that G has such a T-representation, for some maximal clique C. Then the connected components of $G \setminus C$ are interval graphs and each connected component can be represented on one of the paths $P_{(b, l_i)}$, which is a subdivision of the edge bl_i; see Fig. 2a and c. However, some pairs of connected components of $G \setminus C$ cannot be placed on the same path $P_{(b, l_i)}$, since their "neighborhoods" in C are not "compatible". The goal is to define a partial order $\triangleright$ on the components of $G - C$ such that for every linear chain $X_1 \triangleright \cdots \triangleright X_k$, the induced subgraph $G[C, V(X_1), \ldots, V(X_k)]$ can be represented on some path $P_{(b, l_i)}$; see Fig. 2b.

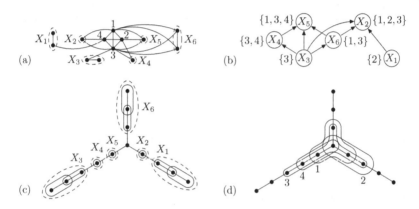

Fig. 2. (a) An example of an S_d-graph G with a maximal clique $C = \{1, 2, 3, 4\}$. (b) The partial ordering $\triangleright$ on the connected components of $G \setminus C$ with chain cover of size 3: $X_2 \triangleright X_1$, $X_5 \triangleright X_4 \triangleright X_3$, and X_6. (c) The connected components placed on the paths $P_{(b, l_1)}$, $P_{(b, l_2)}$, and $P_{(b, l_3)}$, according to the chain cover of $\triangleright$. (d) The subtrees T'_1, T'_2, T'_3, T'_4 corresponding to the vertices of the maximal clique C give an S_d-representation of G with $b \in \bigcap_{v \in C} V(T'_v)$.

We define $N_C(u)$ and $N_C(X)$ to be the *neighbourhoods* of the vertex v in C and of the components X in C, respectively. Formally,

$$N_C(u) = \{v \in C : vu \in E(G)\} \quad \text{and} \quad N_C(X) = \bigcup \{N_C(u) : u \in V(X)\}.$$

Note that, if we have two components X and X' on the same branch where $N_C(X') \subseteq N_C(u)$ for every $u \in V(X)$, then X must be closer to C than X'. We say that components X and X' are *equivalent* if there is a subset C' of C such that $N_C(u) = C'$ for every $u \in V(X)$ and $N_C(u') = C'$ for every $u' \in V(X')$. Note that equivalent components X and X' can be represented in an interval representation of $G[C, V(X), V(X')]$ in an arbitrary order and they *can be treated as one component.* We use one representative component for each equivalence class, and denote this set of non-equivalent representative components by $\mathcal{X}$. For $X, X' \in \mathcal{X}$, we let:

$$X \triangleright X' \iff \text{for every } u \in V(X), N_C(X') \subseteq N_C(u). \tag{1}$$

For a proof of the next lemmas, see Appendix A.

Lemma 4. *The relation $\triangleright$ is a partial ordering on $\mathcal{X}$.*

Lemma 5. *Let $X_1, \ldots, X_k \in \mathcal{X}$. Then the subgraph $G[C, V(X_1), \ldots, V(X_k)]$ has an interval representation with C being the leftmost clique if and only if $X_1 \triangleright \cdots \triangleright X_k$ and each $G[C, X_i]$ has an interval representation with C being the leftmost clique.*

The following lemma gives a necessary and a sufficient condition for G to be an S_d-graph having an S_d-representation with $b \in \bigcap_{v \in C} V(T'_v)$. For a proof see Appendix A.

Lemma 6. *A graph G has an S_d-representation with $b \in \bigcap_{v \in C} V(R_v)$ if and only if the following hold: (i) For every $X \in \mathcal{X}$, the induced subgraph $G[C, X]$ has an interval representation with C being the leftmost clique. (ii) The partial order $\triangleright$ on $\mathcal{X}$ has a chain cover of size at most d.*

By combining Lemmas 5 and 6 we obtain an algorithm for recognizing S_d-graphs. For a given graph G and its maximal clique C, we do the following:

1. We delete the maximal clique C and construct the partial order $\triangleright$ on the set of non-equivalent connected components $\mathcal{X}$.
2. We test whether the partial order $\triangleright$ can be covered by at most d linear chains.
3. For each linear chain $X_1^i \triangleright \cdots \triangleright X_k^i$, $1 \leq i \leq d$, we construct an interval representation $\mathcal{R}_i$ of the induced subgraph $G[C, V(X_1), \ldots, V(X_k)]$, with C being the leftmost maximal clique, on one of the paths of the subdivided S_d.
4. We complete the whole representation by placing each $\mathcal{R}_i$ on the path $P_{[b, l_i]}$ so that $b \in \bigcap_{v \in C} V(T'_v)$.

Theorem 7. *Recognition of S_d-graphs can be solved in $\mathcal{O}(n^4)$ time.*

Proof. Every chordal graph has at most n maximal cliques, where n is the number of vertices, and they can be listed in linear time [13]. For every clique C, our algorithm tries to find an S_d-representation with $b \in \bigcap_{v \in C} V(T'_v)$. Note that by forgetting the orientation in the partial order $\triangleright$ that we get a comparability graph. Every clique in the comparability induces a linear chain in $\triangleright$. For comparability graphs, a minimum clique-cover can be found in $\mathcal{O}(n^3)$ time [10]. The overall time complexity of our algorithm is therefore $\mathcal{O}(n^4)$. □

Recognition of T-graphs. The algorithm for recognizing T-graphs is a generalization of the algorithm for recognizing S_d-graphs, described above. For every mapping $f\colon \mathcal{B} \to \mathcal{C}$, we try to construct a representation $\mathcal{R}$ such that $b \in \bigcap_{v \in f(b)} V(T'_v)$, for every $b \in \mathcal{B}$. If possible, we show how to place the connected components of $G - \bigcup_{b \in \mathcal{B}} f(b)$ according to (i) and (ii). Otherwise, G is not a T-graph.

Note that if $f(b) = f(b')$, then for every branching point b'' which lies on the path from b to b', we must have $f(b) = f(b'') = f(b')$. Therefore, for $C \in f(\mathcal{B})$, the branching points in $f^{-1}(C)$, together with the paths connecting them, have to form a subtree of T'. Similarly, if G is disconnected, the branching points corresponding to maximal cliques belonging to one connected component of G, together with the paths connecting them, form a subtree of T'.

Suppose that G has a T-representation. The connected components of $G - \bigcup_{b \in \mathcal{B}} f(b)$, are interval graphs. As in the previous section, we use relationships between their sets of neighbours in the maximal cliques to find a valid placement of these components on the paths between the branching points and the paths between a branching point and a leaf. The first step of our algorithm is to find the components which *have to* be represented on a path $P_{(b,b')}$ between two branching points. The following lemma deals with this problem.

Lemma 8. *Let X be a connected component of $G - \bigcup_{b \in \mathcal{B}} f(b)$ and $b, b' \in \mathcal{B}$. If the sets $(f(b) \setminus f(b')) \cap N_{f(b)}(X)$ and $(f(b') \setminus f(b)) \cap N_{f(b')}(X)$ are nonempty, then X has to be represented on $P_{(b,b')}$.*

Proof. We put $C = f(b)$ and $C' = f(b')$. Let $v \in (C \setminus C') \cap N_C(H)$ and $u \in (C' \setminus C) \cap N_{C'}(H)$. Since $v \notin C'$, the subtree T'_v cannot pass through b'. Similarly T'_u cannot pass through b. Therefore, the only possible path where X can be represented is $P_{(b,b')}$. □

Let $X_{b,b'}$ be the disjoint union of the components satisfying the conditions of Lemma 8. If the induced interval subgraph $G[C, V(H_{b,b'}), C']$ has a representation such that the cliques C and C' are the leftmost and the rightmost, respectively, then we can represent $H_{b,b'}$ in the middle of the path $P_{(b,b')}$. (If no such representation exists, then G cannot be represented on T' for this particular $f\colon \mathcal{B} \to \mathcal{C}$.) This means that there exist nodes x, y of $P_{(b,b')}$ such that the representation of $X_{b,b'}$ is on the subpath $P_{x,y}$ of $P_{(b,b')}$. We remove the subpath $P_{(x,y)}$. We do this for each pair of neighbouring branching points.

Let $G' = G - \{X_{b,b'} : bb' \in E(T')\}$. Suppose that $b \in \mathcal{B}$. Let $l_1, \ldots, l_p$ be the leaves of T and $b_1, \ldots, b_q$ the branching points of T' that are *adjacent* to b. Let $x_1, \ldots, x_q$ and $y_1, \ldots, y_q$ be the points of the paths $P_{[b,b_1]}, \ldots, P_{[b,b_q]}$, respectively, such that X_{b,b_i} is represented on the subpath P_{x_i, y_i}. We define S^b to be the star consisting of the paths $P_{[b,l_1]}, \ldots, P_{[b,l_p]}, P_{[b,x_1)}, \ldots, P_{[b,x_q)}$. It remains to find a representation of the graph G' on disjoint disjoint union of subdivided stars. Moreover, the representation of the induced subgraph $G[f(b), V(X_{b,b_i}), f(b_i)]$ imposes restrictions on the path $P_{(b,x_i)}$. Suppose that a vertex $v \in f(b)$ is adjacent to a vertex from X_{b,b_i}. Then if we want the represent a connected component X of $G' - \bigcup_{b \in \mathcal{B}} f(b)$ on the path $P_{(b,x_i)}$, we can do this only if every vertex of X is adjacent to v.

Disjoint Stars With Restrictions. We reduced the problem of recognizing T-graphs to the following problem. On the input we have a graph G, k disjoint subdivided stars $S^{b_1}, \ldots, S^{b_k}$ with branching points $b_1, \ldots, b_k$, respectively, a mapping $f \colon \{b_1, \ldots, b_k\} \to \mathcal{C}$, and for every path from b_i to a leaf of S^{b_i} a subset of $f(b_i)$ of *restrictions*. We want to find a representation of G on $S^{b_1}, \ldots, S^{b_k}$ such that $b_i \in \bigcap_{v \in f(b_i)} V(R_v)$, and for every connected component X of $G - \bigcup f(b_i)$, the vertices $V(X)$ have to be adjacent to every vertex in the subset of restrictions corresponding to the path on which X is represented.

To solve this problem, we proceed similarly as in the recognition of S_d-graphs. We define a partial ordering on the connected components of $G - C$, where $C = \bigcup f(b_i)$. The notions $N_C(u)$ and $N_C(X)$ are defined as in the same way as in the algorithm for recognizing S_d-graphs. We get a partial ordering $\triangleright$ on the set of non-equivalent connected components $\mathcal{X}$ of $G - C$. Moreover, to each component $X \in \mathcal{X}$, we assign a list of colors $L(X)$ which correspond to the subpaths from a branching point to a leaf in the stars $S^{b_1}, \ldots, S^{b_k}$, on which they can be represented. Each list $L(H)$ has size at most $d = \sum d_i$, where d_i is the degree of b_i.

Suppose that there exists a chain cover of $\triangleright$ of size d such that for every chain $X_1 \triangleright \cdots \triangleright X_\ell$ in this cover we can pick a color belonging to every $L(X_j)$ such that no two components get the same color. Here, a representation of G satisfying the restrictions can be constructed analogously as in the proof of Theorem 6.

The partial ordering $\triangleright$ on the components $\mathcal{X}$ defines a comparability graph P with a list of colors $L(v)$ assigned to every vertex $v \in V(P)$. If we find the *list coloring* c of its complement $\overline{P}$, i.e., a coloring that for every vertex v uses only colors from its list $L(v)$, then the vertices of the same color in $\overline{P}$ correspond to a chain (clique) in P. Therefore, we have reduced our problem to list coloring co-comparability graphs with lists of bounded size.

Bounded List Coloring of Co-comparability Graphs. We showed that to solve the problem of recognizing T-graphs we need to solve the ℓ-list coloring problem for co-comparability graphs where $\ell = 2 \cdot |E(T)|$. In particular, given a co-comparability graph G, a set of colors S such that $|S| \leq \ell$, and a set $L(v) \subseteq S$ for each vertex v, we want to find a proper coloring $c \colon V(G) \to S$ such that for every vertex v, we have $c(v) \in L(v)$.

In [2] the *capacitated coloring problem* is solved on co-comparability graphs. Namely, given a graph G, an integer $s \geq 1$ of colors, and positive integers $\alpha_1^*, \ldots, \alpha_s^*$, a *capacitated s-coloring* φ of G is a proper s-coloring such that the number of vertices assigned color i is bounded by α_i^*, i.e., $|\varphi^{-1}(i)| \leq \alpha_i^*$. They prove that the capacitated coloring of co-comparability graphs can be solved in polynomial time for fixed s.

In Appendix E we modify [2] to solve the s-list coloring problem on co-comparability graphs in $\mathcal{O}(n^{s^2+1} s^3)$ time. This provides the following theorem.

Theorem 9. *Recognition of T-graphs can be solved in $n^{\mathcal{O}(\|T\|^2)}$.*

3   Recognition Hardness

In this section we negatively answer Problem 6.3 of Biró et al. [1]. Namely, we prove that testing for membership in H-GRAPH is NP-complete when the *diamond graph* D is a minor of H. The graph D is obtained by deleting an edge from a 4-vertex clique. Note that this sharply contrasts the polynomial time solvability of recognizing circular arc graphs (i.e., when H is a cycle).

Our hardness proof stems from the NP-hardness of testing whether a partial order (poset) with *height* one has *interval dimension* at most three (H1ID3) – shown by Yannakakis [14]. Note that having *height one* means that every element is either a minima or maxima of $\mathcal{P}$. Consider a collection $\mathcal{I}$ of closed intervals on the real line. A poset $\mathcal{P}_\mathcal{I} = (\mathcal{I}, <)$ can be defined on $\mathcal{I}$, by considering intervals $x, y \in \mathcal{I}$ and setting $x < y$ if and only if the right endpoint of x is strictly to the left of the left endpoint of y. A partial order $\mathcal{P}$ is called an *interval order* when there is an $\mathcal{I}$ such that $\mathcal{P} = \mathcal{P}_\mathcal{I}$. The *interval dimension* of a poset $\mathcal{P} = (P, <)$, is the minimum number of interval orders whose intersection is $\mathcal{P}$, i.e., for elements $x, y \in P$, $x < y$ if and only if x is before y in all of the interval orders. Finally, the *incomparability graph* $G_\mathcal{P}$ of a poset $\mathcal{P} = (P, <)$ is the graph with $V(G) = P$ and uv is an edge if and only if u and v are not comparable in $\mathcal{P}$. Notice that, when $\mathcal{P}$ has height one, the vertex set of $G_\mathcal{P}$ is *co-bipartite*, i.e., it naturally partitions into two cliques, one $K_{\max}$ on the maxima of $\mathcal{P}$ and one $K_{\min}$ on the minima of $\mathcal{P}$. We now prove the theorem of this section.

Theorem 10. *Testing $G \in H$-GRAPH, is* NP-*complete if D is a minor of H.*

Proof. Here, we prove the theorem for $H = D$. In Appendix C, we show how to extend this proof for every H containing D as a minor.

First, we summarize the idea behind our proof. As stated above we will encode an instance $\mathcal{P}$ of H1ID3 as an instance of membership testing in D-GRAPH. For a given height one poset $\mathcal{P}$, we construct its incomparability graph $G_\mathcal{P}$, slightly augment $G_\mathcal{P}$ to a graph G and show that G is in D-GRAPH if and only if the interval dimension of $\mathcal{P}$ is at most three. In particular, the three paths connecting the two degree 3 vertices in D will encode the three interval orders whose intersection is $\mathcal{P}$. An example is provided in Fig. 3.

To obtain our reduction we first consider H-representations of the graph T_3 where T_3 is obtained by subdividing every edge of the star S_3 exactly once. This tree is neither an interval graph nor a circular arc graph (this is well-known and easily checked). Thus: every H-representation of T_3 uses a node of degree 3 – we mark this property as $(*)$ for later reference.

Consider a height one poset $\mathcal{P} = (P, <)$ and the graph $G_\mathcal{P}$. Let $T_{\max}$ and $T_{\min}$ each be a copy of T_3. The graph G is formed by taking the disjoint union of $T_{\max}$, $T_{\min}$, and $G_\mathcal{P}$ and then making every vertex of $T_{\max}$ adjacent to every vertex of $K_{\max}$, and every vertex of $T_{\min}$ adjacent to every vertex of $K_{\min}$. We claim that $\mathcal{P}$ has interval dimension at most three if and only if G is in D-GRAPH.

For the reverse direction, consider a D'-representation $\{D'_v : v \in V(G), D'_v \subseteq V(D')\}$ of G where D' is a subdivision of D. From the observation $(*)$ (above), in

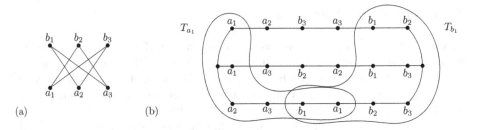

Fig. 3. (a) A Hasse diagram of a partially ordered set of interval dimension 3, but not 2. A realizer of it is given by the following three interval orders: $l_{a_1}l_{a_2}l_{a_3}r_{a_2}r_{a_3}l_{b_1}r_{a_1}l_{b_2}l_{b_3}r_{b_1}r_{b_2}r_{b_3}$, $l_{a_1}l_{a_2}l_{a_3}r_{a_1}r_{a_3}l_{b_2}r_{a_2}l_{b_1}l_{b_3}r_{b_1}r_{b_2}r_{b_3}$, and $l_{a_1}l_{a_2}l_{a_3}r_{a_1}r_{a_2}l_{b_3}r_{a_3}l_{b_1}l_{b_2}r_{b_1}r_{b_2}r_{b_3}$ (where l_x is the left endpoint of x and r_x is the right endpoint of x). (b) Illustrating part of a D-representation. The nodes are labeled by the extreme points of the intervals in the realizer. Namely, for each minima a_i, three points are labeled (one for each of its intervals). Each such point corresponds to the rightmost point of the corresponding interval, i.e., the maximum points of a_i's intervals. For example, T_{a_1} is then formed by taking the three shortest paths from the leftmost point to each point labeled a_1. Labels are placed symmetrically for the maxima.

D', one degree 3 node is contained in the representation of $T_{\min}$ and the other is contained in the representation of $T_{\max}$. We refer to the former degree 3 node of D' as $u_{\min}$ and the latter as $u_{\max}$. Since each maxima y of $\mathcal{P}$ is not adjacent to any vertex of $T_{\min}$, D'_y cannot contain $u_{\min}$, i.e., D'_y is a subtree of the tree $D'\backslash\{u_{\min}\}$. In particular, D'_y defines one (possibly empty) subpath/interval (originating in $u_{\max}$) in each of the three paths connecting $u_{\max}$ and $u_{\min}$. Similarly, for each minima x, D'_x defines one (possibly empty) subpath/interval (originating in $u_{\min}$) in each of the three paths connecting $u_{\max}$ and $u_{\min}$. It is easy to see that these intervals provide the needed three interval orders whose intersection is $\mathcal{P}$.

For the forward direction, we consider the three sets of intervals $\mathcal{I}_1, \mathcal{I}_2, \mathcal{I}_3$ where each interval in $\mathcal{I}_i$ is labelled according to its corresponding element of $\mathcal{P}$, and $\mathcal{P} = \mathcal{P}_{\mathcal{I}_1} \cap \mathcal{P}_{\mathcal{I}_2} \cap \mathcal{P}_{\mathcal{I}_3}$, i.e., certifying that $\mathcal{P}$ has interval dimension at most three. Without loss of generality we may assume that the intervals of the minima all have their left endpoints at 0 and their right endpoints as integers in the range $[0, n-1]$. Similarly, the intervals of the maxima all have their right endpoints at n and their left endpoints as integers in the range $[1, n]$. With this in mind, for each minimal element x we use x_i to denote the right endpoint of its interval in $\mathcal{I}_i$ ($i = 1, 2, 3$) and for each maximal element y, we use y_i to denote the left endpoint of its interval in $\mathcal{I}_i$.

We subdivide the diamond D so that each path between the degree 3 vertices contains $n+5$ nodes and call this new graph D'. We then label the nodes of D' as follows. The two degree 3 nodes are labelled $u_{\min}$ and $u_{\max}$, and we label the three $(u_{\min}, u_{\max})$-paths as:

- $u_{\min}, \alpha_{\min}, \alpha'_{\min}, \alpha_0, \alpha_1, \ldots, \alpha_n, \alpha'_{\max}, \alpha_{\max}, u_{\max}$;
- $u_{\min}, \beta_{\min}, \beta'_{\min}, \beta_0, \beta_1, \ldots, \beta_n, \beta'_{\max}, \beta_{\max}, u_{\max}$; and
- $u_{\min}, \gamma_{\min}, \gamma'_{\min}, \gamma_0, \gamma_1, \ldots, \gamma_n, \gamma'_{\max}, \gamma_{\max}, u_{\max}$.

It remains to describe the D'-representation of G. Each minimal element x is represented by the minimal subtree of D' which includes the nodes $u_{\min}, \alpha_{x_1}, \beta_{x_2}, \gamma_{x_3}$. Similarly, each maximal element y is represented by the minimal subtree of D' which includes the nodes $u_{\max}, \alpha_{y_1}, \beta_{y_2}, \gamma_{y_3}$. We can now see that comparable elements of $\mathcal{P}$ are represented by disjoint subgraphs of D' and the incomparable elements map to intersecting subgraphs.

For the vertices of $T_{\min}$ (the vertices of $T_{\max}$ are represented analogously):

- the tree induced by $u_{\min}, \alpha'_{\min}, \beta'_{\min}, \gamma'_{\min}$ represents the degree 3 vertex;
- the three degree 2 vertices a, b, and c are respectively represented by the three edges $\alpha'_{\min}\alpha_{\min}$, $\beta'_{\min}\beta_{\min}$, and $\gamma'_{\min}\gamma_{\min}$; and
- for each of a, b, c the corresponding degree 1 neighbor is represented by $\alpha'_{\min}, \beta'_{\min}, \gamma'_{\min}$ respectively.

Clearly, in this construction, the graphs $T_{\min}$ and $T_{\max}$ are correctly represented. Moreover, the subtree of each of the minima includes all of $u_{\min}, \alpha_{\min}, \alpha'_{\min}, \beta_{\min}, \beta'_{\min}, \gamma_{\min}, \gamma'_{\min}$, but none of the corresponding max elements. Thus, each minima is universal to $T_{\min}$ and non-adjacent to the vertices of $T_{\max}$ as needed. Symmetrically, each maxima is universal to $T_{\max}$ and non-adjacent to the vertices of $T_{\min}$. Thus, G is in D-GRAPH. See Fig. 3 for an illustration. □

4 Dominating Set

In this section, we discuss the minimum dominating set problem on H-GRAPH. This section is divided into two parts. In the first part we solve the minimum dominating set problem on S_d-GRAPH in FPT-time parameterized by d. In the second part we consider H-GRAPH (for general H), and solve the problem in XP-time parameterized by $\|H\| = |V(H)| + |E(H)|$. Based on the latter result, we also obtain XP-time algorithms for maximum independent set and independent dominating set on H-GRAPH (these are also parameterized by $\|H\|$). A useful tool for these results is Lemma 11. This easy consequence of a standard minimum dominating set algorithm [9] for interval graphs is proven in Appendix D.

Lemma 11. *Let $G = (V, E)$ be an interval graph and let $C_1, \ldots, C_k$ be the left-to-right ordering of the maximal cliques in an interval representation of G. For every $x \in C_1$, a dominating set of G which is minimum subject to including x can be found in linear time.*

Theorem 12. *For an S_d-graph G, a minimum dominating set of G can be found in $\mathcal{O}(d \cdot n \cdot (n + m)) + 2^d (d + 2^d)^{\mathcal{O}(1)}$ time.*

Proof. Let G be an S_d-graph and let S' be a subdivision of the star S_d such that G has an S-representation. Let b be the central branching point of S and let $l_1, \ldots, l_d$ be the leaves of S. Recall that, by Lemma 3, we may assume $b \in \bigcap\{S'_v : v \in C\}$, for some maximal clique C of G. Let $C_{i,1}, \ldots, C_{i,k_i}$ be the maximal cliques of G as they appear on the branch $P_{(b,l_i]}$, for $i = 1, \ldots, d$.

For each $G_i = G[C_{i,1}, \ldots, C_{i,k_i}]$, we use the standard interval graph greedy algorithm [9] to find the size d_i of a minimum dominating set in G_i. Let B_i be the set of vertices of C that can appear in a minimum dominating set of G_i. By Lemma 11, a minimum dominating set D_i^x containing a vertex $x \in C$ can be found in linear time. We have $x \in B_i$ if and only if $|D_i^x| = d_i$. Therefore, every $B_1, \ldots, B_d$ can be found in $\mathcal{O}(d \cdot n \cdot (n + m))$ time. Let $\mathcal{B} = \{B_1, \ldots, B_d\}$.

If B_i is empty, then no minimum dominating set of G_i contains a vertex from C. So for G_i, we pick an arbitrary minimum dominating set D_i. Note that D_i dominates $C \cap C_{i,1}$ regardless of the choice of D_i. Thus, if $\bigcup_{i=1}^d D_i$ dominates C, then it is a minimum dominating set of G. Otherwise, $\{x\} \cup \bigcup_{i=1}^d D_i$ is a minimum dominating set of G where x is an arbitrary vertex of C.

Let us assume now that the B_i's are nonempty (every branch with an empty B_i can be simply ignored). Let H be a subset of C such that $H \cap B_i$ is not empty, for every $i = 1, \ldots, d$, and $|H|$ is smallest possible. For every branch $P_{(b,b_i]}$, we pick a minimum dominating set D_i of G_i containing an arbitrary vertex $x_i \in H \cap B_i$. Now, the union $D_1 \cup \cdots \cup D_d$ is a minimum dominating set of G. It remains to show how to find the set H in time depending only on d.

Finding the set H can be seen as a set cover problem where $\mathcal{B}$ is the ground set. Namely, we have one set for each vertex x in C where the set of x is simply its subset of $\mathcal{B}$, and our goal is to cover $\mathcal{B}$. Note, if two vertices cover the same subset of $\mathcal{B}$ it suffices to keep just one of them for our set cover instance, i.e., giving us at most 2^d sets over a ground set of size d. Such a set cover instance can be solved in $2^d(d + 2^d)^{\mathcal{O}(1)}$ time (see Theorem 6.1 [6]).

Thus, we spend $\mathcal{O}(d \cdot n \cdot (n + m)) + 2^d(d + 2^d)^{\mathcal{O}(1)}$ time in total. $\square$

H-Graphs. We now consider the H-GRAPH for general H. Here we will solve the problem in XP-time parameterized by $\|H\|$. Recall that, when H is a cycle, H-GRAPH = CARC, i.e., minimum dominating sets can be found efficiently [5]. Thus, we assume H is not a cycle.

To introduce our main idea, we need some notation. Consider $G \in H$-GRAPH and let H' be a subdivision of H such that G has an H'-representation $\{H'_v : v \in V(G)\}$. We distinguish two important types of nodes in H'; namely, $x \in V(H')$ is called *high degree* when it has at least three neighbors and x is *low degree* otherwise. As usual, the high degree nodes play a key role. In particular, if we know the sub-solution which *dominates* the high degree nodes of H', then the remaining part of the solution must be strictly contained in the low degree part of H'. Moreover, since H is not a cycle, the subgraph $H'_{\leq 2}$ of H' induced by its low degree nodes is a collection of paths. In particular, the vertices v of G where H'_v only contains low degree nodes, induce an interval graph $G_{\leq 2}$ and, as such, we can efficiently find minimum dominating sets on them. Thus, the general idea here is to first enumerate the possible sub-solutions on the high degree nodes, then efficiently (and optimally) extend each sub-solution to a complete solution. In particular, one can show that in any minimum dominating set these sub-solutions consist of at most $2 \cdot |E(H)|$ vertices, and from this observation it is not difficult to produce the claimed $n^{\mathcal{O}(\|H\|)}$-time algorithm. Thus, we have the following theorem whose full proof is given in Appendix D.

Theorem 13. *For an H-graph G the minimum dominating set problem can be solved in $n^{O(\|H\|)}$ time.*

We remark that the above approach can also be applied to solve the maximum independent set and independent dominating set problems in $n^{O(\|H\|)}$ time. This approach is successful since these problem can be solved efficiently on interval graphs. Improving these XP-time algorithms to FPT-time remains open.

Corollary 14. *For an H-graph G, the maximum independent set problem and independent dominating set problem can both be solved in $n^{O(\|H\|)}$ time.*

References

1. Biro, M., Hujter, M., Tuza, Z.: Precoloring extension I. Interval graphs. Discret. Math. **100**(1), 267–279 (1992)
2. Bonomo, F., Mattia, S., Oriolo, G.: Bounded coloring of co-comparability graphs and the pickup and delivery tour combination problem. Theor. Comput. Sci. **412**(45), 6261–6268 (2011)
3. Booth, K.S., Lueker, G.S.: Testing for the consecutive ones property, interval graphs, and planarity using PQ-tree algorithms. J. Comput. System Sci. **13**, 335–379 (1976)
4. Booth, K.S., Johnson, J.H.: Dominating sets in chordal graphs. SIAM J. Comput. **11**(1), 191–199 (1982)
5. Chang, M.S.: Efficient algorithms for the domination problems on interval and circular-arc graphs. SIAM J. Comput. **27**(6), 1671–1694 (1998)
6. Cygan, M., Fomin, F.V., Kowalik, Ł., Lokshtanov, D., Marx, D., Pilipczuk, M., Pilipczuk, M., Saurabh, S.: Parameterized Algorithms, vol. 4. Springer, Cham (2015). doi:10.1007/978-3-319-21275-3
7. Fulkerson, D.R., Gross, O.A.: Incidence matrices and interval graphs. Pac. J. Math. **15**, 835–855 (1965)
8. Gavril, F.: The intersection graphs of subtrees in trees are exactly the chordal graphs. J. Comb. Theory, Ser. B **16**(1), 47–56 (1974)
9. Golumbic, M.C.: Algorithmic Graph Theory and Perfect Graphs, vol. 57. Elsevier, Amsterdam (2004)
10. Golumbic, M.C.: The complexity of comparability graph recognition and coloring. Computing **18**(3), 199–208 (1977)
11. Klavík, P., Kratochvíl, J., Otachi, Y., Saitoh, T.: Extending partial representations of subclasses of chordal graphs. Theor. Comput. Sci. **576**, 85–101 (2015)
12. Lueker, G.S., Booth, K.S.: A linear time algorithm for deciding interval graph isomorphism. J. ACM (JACM) **26**(2), 183–195 (1979)
13. Rose, D.J., Tarjan, R.E., Lueker, G.S.: Algorithmic aspects of vertex elimination on graphs. SIAM J. Comput. **5**(2), 266–283 (1976)
14. Yannakakis, M.: The complexity of the partial order dimension problem. SIAM J. Algebr. Discrete Methods **3**(3), 351–358 (1982)

The Hardness of Embedding Grids and Walls

Yijia Chen[1], Martin Grohe[2], and Bingkai Lin[3(✉)]

[1] School of Computer Science, Fudan University, Shanghai, China
yijiachen@fudan.edu.cn
[2] Department of Computer Science, RWTH Aachen University, Aachen, Germany
grohe@informatik.rwth-aachen.de
[3] JST, ERATO, Kawarabayashi Large Graph Project,
National Institute of Informatics, Tokyo, Japan
lin@nii.ac.jp

Abstract. The dichotomy conjecture for the parameterized embedding problem states that the problem of deciding whether a given graph G from some class $\mathcal{K}$ of "pattern graphs" can be embedded into a given graph H (that is, is isomorphic to a subgraph of H) is fixed-parameter tractable if $\mathcal{K}$ is a class of graphs of bounded tree width and W[1]-complete otherwise.

Towards this conjecture, we prove that the embedding problem is W[1]-complete if $\mathcal{K}$ is the class of all grids or the class of all walls.

1 Introduction

The *graph embedding* a.k.a *subgraph isomorphism* problem is a fundamental algorithmic problem, which, as a fairly general pattern matching problem, has numerous applications. It has received considerable attention since the early days of complexity theory (see, e.g.,[10,12,18,22]). Clearly, the embedding problem is NP-complete, because the clique problem and the Hamiltonian path or cycle problem are special cases. The embedding problem and special cases like the clique problem or the longest path problem have also played an important role in the development of fixed-parameter algorithms and parameterized complexity theory (see [15,17]). The problem is complete for the class W[1] when parameterized by the size of the pattern graph; in fact, the special case of the clique problem may be regarded as the paradigmatic W[1]-complete problem [8,9]. On the other hand, interesting special cases such as the longest path and longest cycle problems are fixed-parameter tractable [1,19]. This immediately raises the question for which pattern graphs the problem is fixed-parameter tractable.

Let us make this precise. An *embedding* from a graph G to a graph H is an injective mapping $f : V(G) \to V(H)$ such that for all edges $vw \in E(G)$ we have $f(v)f(w) \in E(H)$. For each class $\mathcal{K}$ of graphs, we consider the following parameterized problem.

Full version available at https://arxiv.org/abs/1703.06423.

© Springer International Publishing AG 2017
H.L. Bodlaender and G.J. Woeginger (Eds.): WG 2017, LNCS 10520, pp. 180–192, 2017.
https://doi.org/10.1007/978-3-319-68705-6_14

> p-EMB($\mathcal{K}$)
>> *Instance:* Graph G (the *pattern graph*) and H (the *target graph*), where $G \in \mathcal{K}$.
>> *Parameter:* $|G|$.
>> *Problem:* Decide whether there is an embedding from G to H.

Plehn and Voigt [20] proved that p-EMB($\mathcal{K}$) is fixed-parameter tractable if $\mathcal{K}$ is a class of graphs of bounded tree width. No tractable classes $\mathcal{K}$ of unbounded treewidth are known. The conjecture, which may have been stated in [13] first, is that there are no such classes.

Dichotomy Conjecture. p-EMB($\mathcal{K}$) is fixed-parameter tractable if and only if $\mathcal{K}$ is a class of bounded treewidth and W[1]-complete otherwise.[1]

Progress towards this conjecture has been slow. Even the innocent-looking case where $\mathcal{K}$ is the class of complete bipartite graphs had been open for a long time; only recently the third author of this paper proved that it is W[1]-complete [16].

Before we present our contribution, let us discuss why we expect a dichotomy in the first place. The main reason is that similar dichotomies hold for closely related problems. The first author, jointly with Thurley and Weyer [4], proved the version of the conjecture for the *strong embedding*, or *induced subgraph isomorphism* problem. Building on earlier work by Dalmau et al. [6] as well as joint work with Schwentick and Segoufin [14], the second author [13] proved that the parameterized homomorphism problem p-HOM($\mathcal{K}$) for pattern graphs from a class $\mathcal{K}$ is fixed-parameter tractable if and only if the cores of the graphs in $\mathcal{K}$ have bounded tree width and W[1]-complete otherwise.

Let us remark that there is no P vs. NP dichotomy for the classical (unparameterized) embedding problem; this can easily be proved along the lines of corresponding results for the homomorphism and strong embedding problems using techniques from [2,4,13].

Our Contribution

We make further progress towards the Dichotomy Conjecture by establishing hardness for two more natural graph classes of unbounded tree width.

Theorem 1.1. p-EMB($\mathcal{K}$) *is* W[1]-*hard for the classes* $\mathcal{K}$ *of all grids and all walls.*

See Sect. 2 and in particular Fig. 1 for the definition of grids and walls. Grids and walls are interesting in this context, because they are often viewed as the

---

[1] There is a minor issue here regarding the computability of the class $\mathcal{K}$: if we want to include classes $\mathcal{K}$ that are not recursively enumerable here then we need the nonuniform notion of fixed-parameter tractability [11].

"generic" graphs of unbounded tree width: by Robertson and Seymour's [21] Excluded Grid Theorem, a class $\mathcal{K}$ of graphs has unbounded tree width if and only if all grids (and also all walls) appear as minors of the graphs in $\mathcal{K}$.

Just like the hardness result of the embedding problem for the class of all complete bipartite graphs [16], our theorem looks simple and straightforward, but it is not. In fact, we started to work on this right after the hardness for complete bipartite graphs was proved, hoping that we could adapt the techniques to grids. This turned out to be a red herring. The proof we eventually found is closer to the proof of the dichotomy result for the homomorphism problem [13] (also see [3]). The main part of our proof is fairly generic and has nothing to do with grids or walls. We prove a general hardness result (Theorem 3.1) for p-EMB($\mathcal{K}$) under the technical condition that the graphs in $\mathcal{K}$ have "rigid skeletons" and unbounded tree width even after the removal of these skeletons. We think that this theorem may have applications beyond grids and walls.

Organization of the Paper

We introduce necessary notions and notations in Sect. 2. For some technical reason, we need a colored version p-COL-EMB of p-EMB. In Sect. 2.1 the problem p-COL-EMB is shown to be W[1]-hard on any class of graphs of unbounded treewidth. Then in Sect. 3, we set up the general framework. In particular, we explain the notion of skeletons, and prove the general hardness theorem. The classes of grids and walls are shown to satisfy the assumptions of this theorem. We leave the proofs in the full version of the paper. In the final Sect. 4 we conclude with some open problems.

2   Preliminaries

A graph G consists of a finite set of vertices $V(G)$ and a set of edges $E(G) \subseteq \binom{V}{2}$. Every edge is denoted interchangeably by $\{u, v\}$ or uv. We assume familiarity with the basic notions and terminology from graph theory, e.g., degree, path, cycle etc., which can be found in e.g., [7]. By $\text{dist}^G(u, v)$ we denote the distance between vertices u and v in a graph G, i.e., the length of a shortest path between u and v.

Let $s, t \in \mathbb{N}$. We use $G_{s,t}$ and $W_{s,t}$ to denote the generic $(s \times t)$-*grid* and $(s \times t)$-*wall*, respectively. Figure 1 gives two examples.

Let G and H be two graphs. A *homomorphism* from G to H is a mapping $h : V(G) \rightarrow V(H)$ such that for every edge $uv \in E(G)$ we have $h(u)h(v) \in E(H)$. If in addition h is injective, then h is an *embedding* from G to H. A homomorphism from G to itself is also called an *endomorphism*, and similarly an embedding from G to itself is an *automorphism*.

A *subgraph* G' of G satisfies $V(G') \subseteq V(G)$ and $E(G') \subseteq E(G)$. We say that G' is a *core* of G if there is a homomorphism from G to G', and if there is no homomorphism from G to any proper subgraph of G'. It is well known that all cores of G are isomorphic, hence we can speak of *the* core of G, written core(G).

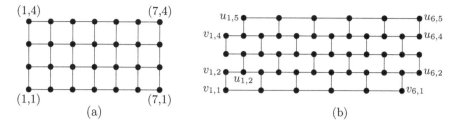

Fig. 1. (a) A (7×4)-grid. (b) A (5×4)-wall.

Sometimes, we also consider *colored* graphs in which a graph G is equipped with a coloring $\chi : V(G) \to C$ which maps every vertex to a color in the color set C. We leave it to the reader to generalize the notions of homomorphism, embedding, and core from graphs to colored graphs. One easy but important fact is that if in the colored graph (G, χ) every vertex has a distinct color then $\mathrm{core}(G) = G$.

The notions of tree decomposition and treewidth are by now standard. In particular, $\mathrm{tw}(G)$ denotes the treewidth of the graph G. For a $(s \times t)$-grid $G_{s,t}$, we have $\mathrm{tw}(G_{s,t}) = \min\{s, t\}$, and for a $(s \times t)$-wall $W_{s,t}$, we have $\mathrm{tw}(W_{s,t}) = \min\{s, t\}+1$. The treewidth of a colored graph (G, χ) is the same as the treewidth of the underlying uncolored graph G, i.e., $\mathrm{tw}(G, \chi) = \mathrm{tw}(G)$.

In a *parameterized problem* (Q, κ) every problem instance $x \in \{0,1\}^*$ has a parameter $\kappa(x) \in \mathbb{N}$ which is computable in polynomial time from x. (Q, κ) is *fixed-parameter tractable* (FPT) if we can decide for every instance $x \in \{0,1\}^*$ whether $x \in Q$ in time $f(\kappa(x)) \cdot |x|^{O(1)}$, where $f : \mathbb{N} \to \mathbb{N}$ is a computable function. Thus, FPT plays the role of P in parameterized complexity. On the other hand, the so-called class W[1] is generally considered as a parameterized analog of NP. The precise definition of W[1] is not used in our proofs, so the reader is referred to the standard textbooks, e.g., [5,8,11]. Let (Q_1, κ_1) and (Q_2, κ_2) be two parameterized problems. An fpt-reduction from (Q_1, κ_1) and (Q_2, κ_2) is a mapping $R : \{0,1\}^* \to \{0,1\}^*$ such that for every $x \in \{0,1\}^*$

- $x \in Q_1 \iff R(x) \in Q_2$,
- $R(x)$ can be computed in time $f(\kappa_1(x)) \cdot |x|^{O(1)}$, where $f : \mathbb{N} \to \mathbb{N}$ is a computable function,
- $\kappa_2(R(x)) \leq g(\kappa_1(x))$, where $g : \mathbb{N} \to \mathbb{N}$ is computable.

Now we state a version of the main result of [13] which is most appropriate for our purpose.

Theorem 2.1. *Let $\mathcal{K}$ be a recursively enumerable[2] class of colored graphs such that for every $k \in \mathbb{N}$ there is a colored graph $(G, \chi) \in \mathcal{K}$ whose core has treewidth at least k. Then p-$\mathrm{HOM}(\mathcal{K})$ is hard for W[1] (under fpt-reductions).*

---

[2] If $\mathcal{K}$ is not recursively enumerable, there is still a "non-uniform" hardness result. See [13] for a discussion.

2.1   From Homomorphism to Colored Embedding

Let $\mathcal{K}$ be a class of graphs. We consider the following colored version of the embedding problem for $\mathcal{K}$.

---

p-COL-EMB($\mathcal{K}$)

  Instance: Two graphs G and H with $G \in \mathcal{K}$, and a function $\chi : V(H) \to V(G)$.

  Parameter: $|G|$.

  Problem: Decide wether there is an embedding h from G to H such that $\chi(h(v)) = v$ for every $v \in V(G)$.

Thus, in the p-COL-EMB($\mathcal{K}$) problem we partition the vertices of H and associate one part with each vertex of G. Then we ask for an embedding where each vertex of G is mapped to its part. The following lemma can be easily deduced from Theorem 2.1.

Lemma 2.1. *Let $\mathcal{K}$ be a recursively enumerable class of graphs with unbounded treewidth. Then p-COL-EMB($\mathcal{K}$) is hard for* W[1].

3   Frames and Skeletons

In this section, we prove a general hardness result for p-EMB($\mathcal{K}$) given that the graphs in $\mathcal{K}$ have "rigid skeletons" and unbounded tree width even after the removal of these skeletons. Roughly speaking, for every $G \in \mathcal{K}$, we define a graph G^* and thus a graph class $\mathcal{K}^* = \{G^* : G \in \mathcal{K}\}$ such that: (1) There is an fpt-reduction from p-COL-EMB($\mathcal{K}^*$) to p-EMB($\mathcal{K}$); (2) $\mathcal{K}^*$ has unbounded tree width. Note that the W[1]-hardness of p-EMB($\mathcal{K}$) then follows from Lemma 2.1.

We will start with the definitions of a graph operator $/$ and the notion of skeleton. The graph G^* is then defined as follows. For $S = (F, D)$ a skeleton of G, let $G^* = (G \setminus F)/D$. To achieve (1), for every graph H and a coloring $\chi : V(H) \to V(G^*)$, we define a graph $P(G, S, H, \chi)$ such that there is an embedding from G^* to H with respect to χ iff there is an embedding from G to $P(G, S, H, \chi)$.

Let G be a graph and $D \subseteq V(G)$ such that the degree of every $v \in D$ is at most 2, i.e., $\deg^G(v) \leq 2$. For every $u, v \in V(G) \setminus D$ we say they are *close* (with respect to D) if there is a path in G between u and v whose internal vertices are all in D. We define G/D as the graph given by

$$V(G/D) := V(G) \setminus D,$$
$$E(G/D) := \{uv \mid u, v \in V(G) \setminus D, u \neq v, \text{ and they are close}\}.$$

Let $u \in D$. We say that u is *associated with a vertex* $v \in V(G/D)$ if u is on a path in G between v and a vertex $w \in D$ with $\deg^G(w) = 1$ whose internal vertices are all in D. Similarly, u is *associated* with some edge $e = vw \in E(G/D)$ if u is on a path in G between v and w whose internal vertices are all in D. It should be clear that u can only be associated with a unique vertex or a unique edge in G/D, and not both. Furthermore, some $w \in D$ might not be associated with any vertex or edge; this happens precisely to all w on a path or cycle with all vertices in D.

To simplify presentation, from now on we fix a graph G.

Definition 3.1. *A set $F \subseteq V(G)$ is a* frame *for G if every endomorphism h of G with $F \subseteq h(V(G))$ is surjective.*

Remark 3.1. Let $F \subseteq F' \subseteq V(G)$ and F be a frame for G. Then F' is also a frame for G.

Definition 3.2. *Let $F, D \subseteq V(G)$ such that*

(S1) F is a frame for G,
(S2) $F \cap D = \emptyset$,
(S3) for every $v \in D$

$$\deg^{G \setminus F}(v) = \left| \{ u \in V \mid u \notin F \text{ and } \{u, v\} \in E \} \right| \leq 2.$$

Then we call $\mathcal{S} = (F, D)$ a skeleton *of G.*

Example 3.1. Consider the grid $G_{7,8}$. It has a skeleton (F, D) with

$$F = \left\{ (i, j) \mid i \in \{1, 2, 6, 7\} \text{ or } j \in \{1, 7, 8\} \right\} \cup \left\{ (4, 2j) \mid j \in [3] \right\} \quad \text{and}$$
$$D = \left\{ (2i + 1, 2j) \mid i \in [2] \text{ and } j \in [3] \right\} \cup \left\{ (2i, 2j + 1) \mid i \in \{2\} \text{ and } j \in [2] \right\},$$

as shown in Fig. 2.

Definition 3.3. *Let $\mathcal{S} = (F, D)$ be a skeleton of G. For every graph H and every mapping*

$$\chi : V(H) \to V(G) \setminus (F \cup D),$$

we construct a product graph $P = P(G, \mathcal{S}, H, \chi)$ *as follows.*

(P1) The vertex set is $V(P) := \bigcup_{i \in [4]} V_i$ with

$$V_1 = \left\{ (u, a) \mid u \in V(G) \setminus (F \cup D) \text{ and } a \in V(H) \text{ with } \chi(a) = u \right\},$$
$$V_2 = \Big\{ (u, u) \ \Big| \ u \in F \text{ or } \big(u \in D \text{ without being associated}$$
$$\text{with any vertex or edge in } (G \setminus F)/D \big) \Big\},$$
$$V_3 = \big\{ (u, \boldsymbol{v}_{u,a}) \mid u \in D, a \in V(H), \text{ and } \chi(a) = v$$
$$\text{with } u \text{ being associated with } v \text{ in } (G \setminus F)/D \big\},$$
$$V_4 = \big\{ (u, \boldsymbol{v}_{u,e}) \mid u \in D, e = \{a, b\} \in E(H), \chi(a) = v, \text{ and } \chi(b) = w$$
$$\text{with } u \text{ being associated with } \{v, w\} \text{ in } (G \setminus F)/D \big\}.$$

Note that in the definition of V_3 and V_4 all $\boldsymbol{v}_{u,a}$ and $\boldsymbol{v}_{u,e}$ are fresh elements.

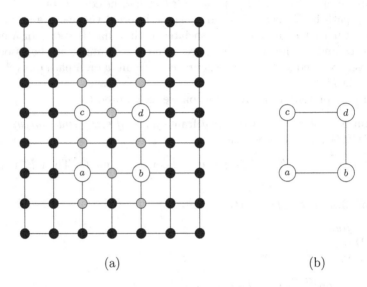

Fig. 2. (a) A skeleton for $G_{7,8}$, where F is the set of black vertices and D the set of light gray vertices. (b) The graph $(G \setminus F)/D$.

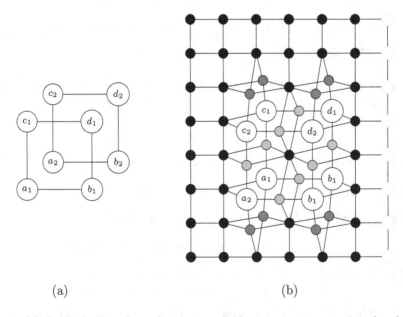

Fig. 3. (a) A graph H and a coloring χ such that every vertex a_i is colored with $\chi(a_i) := a$, every b_i with $\chi(b_i) := b$, and so on. (b) The product $P = P(G_{7,8}, \mathcal{S}, H, \chi)$.

(P2) *The edge set is* $E(P) := \bigcup_{1 \le i \le j \le 4} E_{ij}$ *with*

$$E_{11} = \{(u,a)(v,b) \mid (u,a),(v,b) \in V_1, uv \in E(G), \text{ and } ab \in E(H)\},$$
$$E_{12} = \{(u,a),(v,v) \mid (u,a) \in V_1, (v,v) \in V_2, \text{ and } uv \in E(G)\},$$
$$E_{13} = \{(u,a)(v,\boldsymbol{v}_{v,a}) \mid (u,a) \in V_1, (v,\boldsymbol{v}_{v,a}) \in V_3, \text{ and } uv \in E(G)\},$$
$$E_{14} = \{(u,a)(v,\boldsymbol{v}_{v,e}) \mid (u,a) \in V_1, (v,\boldsymbol{v}_{v,e}) \in V_4, uv \in E(G), \text{ and } a \in e\},$$
$$E_{22} = \{(u,u)(v,v) \mid (u,u),(v,v) \in V_2 \text{ and } uv \in E(G)\},$$
$$E_{23} = \{(u,u)(v,\boldsymbol{v}_{v,a}) \mid (u,u) \in V_2, (v,\boldsymbol{v}_{v,a}) \in V_3, \text{ and } uv \in E(G)\},$$
$$E_{24} = \{(u,u)(v,\boldsymbol{v}_{v,e}) \mid (u,u) \in V_2, (v,\boldsymbol{v}_{v,e}) \in V_4, \text{ and } uv \in E(G)\},$$
$$E_{33} = \{(u,\boldsymbol{v}_{u,a})(v,\boldsymbol{v}_{v,a}) \mid (u,\boldsymbol{v}_{u,a}),(v,\boldsymbol{v}_{v,a}) \in V_3 \text{ and } uv \in E(G)\},$$
$$E_{34} = \emptyset,$$
$$E_{44} = \{(u,\boldsymbol{v}_{u,e})(v,\boldsymbol{v}_{v,e}) \mid (u,\boldsymbol{v}_{u,e}),(v,\boldsymbol{v}_{v,e}) \in V_4 \text{ and } uv \in E(G)\}.$$

Example 3.2. Let H be the graph in Fig. 3(a). Moreover, we color every a_i with $\chi(a_i) := a$, every b_i with $\chi(b_i) := b$, and so on. We consider the grid $G_{7,8}$ and the skeleton $\mathcal{S} = (F,D)$ defined in Example 3.1. Then Fig. 3(b) is the product $P = P(G_{7,8}, \mathcal{S}, H, \chi)$. In particular, V_1 is the set of white vertices, V_2 is the set of black vertices, V_3 is the set of gray vertices, and V_4 is the set of remaining light gray vertices. To make the picture less cluttered, we label the vertex (a,a_i) by a_i etc. in P.

In Definition 3.3 the reader might notice that in each pair (u,a), (u,u), $(u,\boldsymbol{v}_{u,a})$, or $(u,\boldsymbol{v}_{u,e})$ the first coordinate is uniquely determined by the second coordinate. Thus:

Lemma 3.1. *Let $h : V(G) \to V(P)$ be injective, e.g., h is an embedding from G to P. Then the mapping $\pi_2 \circ h$ is injective, too. Here $\pi_2(u,z) = z$ for every $(u,z) \in V(P)$ is the projection on the second coordinate.*

Lemma 3.2. *π_1 is a homomorphism from P to G, where $\pi_1(u,z) = u$ for every $(u,z) \in V(P)$ is the projection on the first coordinate.*

Proof. Observe that by (P2) for every edge $(u,w)(v,z) \in E(P)$ we have $uv \in E(G)$. □

Lemma 3.3. *Let h be a homomorphism from G to P. Then the mapping $\pi_1 \circ h$ is an endomorphism of G. Moreover, if*

$$\{(u,u) \mid u \in F\} \subseteq h(V(G)),$$

then $\pi_1 \circ h$ is an automorphism of G.

Proof. By Lemma 3.2 $\pi_1 \circ h$ is an endomorphism of G. If $\{(u,u) \mid u \in F\} \subseteq h(V(G))$, then $F \subseteq \pi_1 \circ h(V(G))$. Since F is a frame, $\pi_1 \circ h$ has to be surjective. □

Lemma 3.4. *Let h be a homomorphism from G to P such that $\pi_1 \circ h$ is an automorphism of G. Then there is a homomorphism $\bar{h}$ from $(G \setminus F)/D$ to H such that $\chi(\bar{h}(v)) = v$ for every $v \in V(G) \setminus (F \cup D)$. Note this implies that $\bar{h}$ is an embedding.*

Proof. Let $\rho := \pi_1 \circ h$. By assumption ρ is an automorphism of G, hence so is ρ^{-1}. Thus $h \circ \rho^{-1}$ is a homomorphism from G to P with

$$\pi_1 \circ (h \circ \rho^{-1}) = (\pi_1 \circ h) \circ \rho^{-1} = (\pi_1 \circ h) \circ (\pi_1 \circ h)^{-1} = \text{id}.$$

Hence for every $u \in V(G)$ there is a w such that

$$h \circ \rho^{-1}(u) = (u, w) \tag{1}$$

Let $\bar{h} := \pi_2 \circ h \circ \rho^{-1}$. We claim that $\bar{h}$ restricted to $V(G) \setminus (F \cup D)$ is the desired homomorphism from $(G \setminus F)/D$ to H.

Let $u \in V(G) \setminus (F \cup D)$. By the definition of $V(P)$ in (P1) and (1) we have $\bar{h}(u) \in V_1$, and thus by the definition of V_1, $\bar{h}(u) \in V(H)$ with $\chi(\bar{h}(u)) = u$. Next, let $uv \in E((G \setminus F)/D)$. We have to show $\bar{h}(u)\bar{h}(v) \in E(H)$. By the definition of $(G \setminus F)/D$ there is a path $u = v_1 \to v_2 \to \cdots \to v_k = v$ in $G \setminus F$ with $k \geq 2$ and all $v_i \in D$ for $1 < i < k$. If $k = 2$, then $uv \in E(G)$. Then

$$\{(u, \bar{h}(u)), (v, \bar{h}(v))\} = \{h \circ \rho^{-1}(u), h \circ \rho^{-1}(v)\} \in E(P),$$

because $h \circ \rho^{-1}$ is a homomorphism. Then $\{\bar{h}(u), \bar{h}(v)\} \in E(H)$ follows directly from the definition of E_{11} in (P2). So assume $k > 2$. Again by (1) and (P1) for some pairwise distinct $a, b \in V(H)$ and $w_2, \ldots, w_{k-1}$

$$h \circ \rho^{-1}(u) = (u, a),$$
$$h \circ \rho^{-1}(v_2) = (v_2, w_2),$$
$$\vdots$$
$$h \circ \rho^{-1}(v_{k-1}) = (v_{k-1}, w_{k-1}),$$
$$h \circ \rho^{-1}(v) = (v, b).$$

As every v_i is associated with $\{u, v\}$, there are $e_2, \ldots, e_{k-1} \in E(H)$ with $w_i = v_{v_i, e_i}$ by the definition of V_4 in (P1). Since $h \circ \rho^{-1}$ is a homomorphism from G to P,

$$\{(u, a), (v_2, v_{v_2, e_2})\} \in E(P),$$
$$\{(v_2, v_{v_2, e_2}), (v_3, v_{v_3, e_3})\} \in E(P),$$
$$\vdots$$
$$\{(v_{k-2}, v_{v_{k-2}, e_{k-2}}), (v_{k-1}, v_{v_{k-1}, e_{k-1}})\} \in E(P),$$
$$\{(v_{k-1}, v_{v_{k-1}, e_{k-1}}), (v, b)\} \in E(P).$$

Then by the definition of E_{44} in (P2), we conclude $e_2 = \cdots = e_{k-1}$. Finally, the definition of E_{14} implies that $e_2 = \{a, b\}$, i.e., $\{\bar{h}(u), \bar{h}(v)\} \in E(H)$.   □

Lemma 3.5. *If there is an embedding $\bar{h}$ from $(G \setminus F)/D$ to H with $\chi(\bar{h}(v)) = v$ for every $v \in V(G) \setminus (F \cup D)$, then there is an embedding from G to P.*

Proof. We define a mapping $h : V(G) \to V(P)$ and show that it is an embedding.

- For $u \in V(G) \setminus (F \cup D)$ let $h(u) := (u, \bar{h}(u))$, which is well defined by $\chi(\bar{h}(u)) = u$.
- For $u \in F$ let $h(u) := (u, u)$.
- For $u \in D$ without being associated with any vertex or edge in $(G \setminus F)/D$ let $h(u) := (u, u)$.
- Let $u \in D$ be associated with a (unique) $v \in V((G \setminus F)/D)$. We set $h(u) := (u, \boldsymbol{v}_{u, \bar{h}(v)})$.
- Let $u \in D$ be associated with a (unique) $vw \in E((G \setminus F)/D)$. We set $h(u) := (u, \boldsymbol{v}_{u, \bar{h}(v)\bar{h}(w)})$.

The injectivity of h is trivial. To see that it is a homomorphism, let $uv \in E(G)$ and we need to establish $h(u)h(v) \in E(P)$.

- Assume $u, v \in V(G) \setminus (F \cup D)$. Then $uv \in E(G)$ implies $uv \in E((G \setminus F)/D)$, and as $\bar{h}$ is a homomorphism from $(G \setminus F)/D$ to H, it follows that $\bar{h}(u)\bar{h}(v) \in E(H)$. So by the definition of E_{11} in (P1) we conclude $(u, \bar{h}(u))(v, \bar{h}(v)) \in E(P)$.
- Let $u \in V(G) \setminus (F \cup D)$ and $v \in D$. Furthermore, assume that v is associated with an edge $wz \in E((G \setminus F)/D)$. Hence, $h(u) = (u, \bar{h}(u))$ and $h(v) = (v, \boldsymbol{v}_{v, \bar{h}(w)h(z)})$. Recall $uv \in E(G)$, therefore $u = w$ or $u = z$. Then, $\bar{h}(u) \in \{\bar{h}(w), \bar{h}(z)\}$, and the definition of E_{14} in (P2) implies that $h(u)h(v) \in E(P)$.
- Assume both $u, v \in D$ and they are associated with some edges $e_1, e_2 \in E((G \setminus F)/D)$. Then $e_1 = e_2$ by $uv \in E(G)$, and $h(u)h(v) \in E(P)$ follows from the definition of E_{44} in (P2).
- All the remaining cases are similar and easy. □

Definition 3.4. *A skeleton $\mathcal{S} = (F, D)$ is rigid if for every graph H, every $\chi : V(H) \to V(G) \setminus (F \cup D)$, and every embedding h from G to $P = P(G, \mathcal{S}, H, \chi)$, it holds that $\{(u, u) \mid u \in F\} \subseteq h(V(G))$.*

Proposition 3.1. *There is an algorithm which lists all rigid skeletons of an input graph G.*

Proof. Let G be a graph and $F, D \subseteq V(G)$. Clearly it is decidable whether $\mathcal{S} = (F, D)$ is a skeleton by Definition 3.2. Moreover, we observe that $\mathcal{S}$ is *not* a rigid skeleton if and only if there is graph H, a mapping $\chi : V(H) \to V(G) \setminus (F \cup D)$, and an embedding h from G to $P = P(G, \mathcal{S}, H, \chi)$ such that

$$\{(u, u) \mid u \in F\} \not\subseteq h(V(G)). \tag{2}$$

We define a set

$$X = \big\{a \in V(H) \mid (u, a) \in h(G) \text{ for some } u \in V(G) \setminus (F \cup D)\big\}$$
$$\cup \big\{a \in V(H) \mid (u, \boldsymbol{v}_{u,a}) \in h(G) \text{ for some } u \in D\big\}$$
$$\cup \big\{a, b \in V(H) \mid (u, \boldsymbol{v}_{u,ab}) \in h(G) \text{ for some } u \in D\big\}.$$

It is routine to verify that h is an embedding from G to $P' = P(G, \mathcal{S}, H[X], \chi_{\restriction X})$ such that (2) also holds. Hence, the induced subgraph $H[X]$ with the coloring $\chi_{\restriction X}$ also witnesses that $\mathcal{S}$ is not rigid. Observe that $|X| \leq 2|V(G)|$.

Therefore, to list all the rigid skeletons of G, we enumerate all pairs $\mathcal{S} = (F, D)$,

- check whether $\mathcal{S}$ is a skeleton,
- and if so, then check whether it is rigid by going through all graphs on the vertex set $[n]$ with $n \leq 2|V(G)|$.   □

Definition 3.5. *A class $\mathcal{K}$ of graphs is* rich *if for every $k \in \mathbb{N}$ there is a graph $G \in \mathcal{K}$ such that G has a rigid skeleton (F, D) with*

$$tw((G \setminus F)/D) \geq k. \tag{3}$$

Theorem 3.1. *Let $\mathcal{K}$ be a recursively enumerable and rich class of graphs. Then $p\text{-}\mathrm{EMB}(\mathcal{K})$ is hard for* W[1].

Proof. We define a sequence of graphs $G_1, G_2, \ldots$ and sets $F_i, D_i \subseteq V(G_i)$ as follows. For every $i \in \mathbb{N}$ we enumerate graphs G in the class $\mathcal{K}$ one by one. For every G we list all the rigid skeletons (F, D) of G by Proposition 3.1. Then we check whether there is such a rigid skeleton (F, D) satisfying (3). If so, we let $G_i := G$, $(F_i, D_i) := (F, D)$, and define $G_i^* := (G_i \setminus F_i)/D_i$. By our assumption, G_i will be found eventually, and G_i^* is well defined and computable from G_i. It follows that the class $\mathcal{K}^* := \left\{ G_i^* \mid i \in \mathbb{N} \right\}$ is recursively enumerable and has unbounded treewidth.

By Lemma 2.1, we conclude that $p\text{-}\mathrm{COL\text{-}EMB}(\mathcal{K}^*)$ is W[1]-hard. Hence it suffices to give an fpt-reduction from $p\text{-}\mathrm{COL\text{-}EMB}(\mathcal{K}^*)$ to $p\text{-}\mathrm{EMB}(\mathcal{K})$. Let $G_i^* \in \mathcal{K}^*$. Thus $G_i^* = (G_i \setminus F_i)/D_i$ for the rigid skeleton $\mathcal{S}_i = (F_i, D_i)$. Then for every graph H and $\chi : V(H) \mapsto V(G_i^*)$ we claim that

there is an embedding h from G_i^* to H with $\chi(h(v)) = v$ for every $v \in V(G_i^*)$

$\Longleftrightarrow$ there is an embedding from G_i to $P(G_i, \mathcal{S}_i, H, \chi)$.

The direction from left to right is by Lemma 3.5. The other direction follows from the rigidity of $\mathcal{S}_i$, Lemmas 3.3 and 3.4.   □

In the full version of this paper, we prove the richness of grids and walls. More precisely:

Proposition 3.2. *Let $\mathcal{K}$ be a class of graphs.*

(i) *If for every $k \in \mathbb{N}$ there exists a grid $\mathrm{G}_{s,t} \in \mathcal{K}$ with $\min\{s, t\} \geq k$. Then $\mathcal{K}$ is rich.*
(ii) *If for every $k \in \mathbb{N}$ there exists a wall $\mathrm{W}_{s,t} \in \mathcal{K}$ with $\min\{s, t\} \geq k$. Then $\mathcal{K}$ is rich.*

Now the following more general version of Theorem 1.1 is an immediate consequence of Theorem 3.1 and Proposition 3.2.

Theorem 3.2. *Let $\mathcal{K}$ be a recursively enumerable class of graphs. Then $p\text{-}\mathrm{EMB}(\mathcal{K})$ is $W[1]$-hard if one of the following conditions is satisfied.*

1. *For every $k \in \mathbb{N}$ there exists a grid $G_{k_1,k_2} \in \mathcal{K}$ with $\min\{k_1, k_2\} \geq k$.*
2. *For every $k \in \mathbb{N}$ there exists a wall $W_{k_1,k_2} \in \mathcal{K}$ with $\min\{k_1, k_2\} \geq k$.*

4   Conclusions

We have shown that the parameterized embedding problem on the classes of all grids and all walls is hard for $W[1]$. Our proof exploits some general structures in those graphs, i.e., frames and skeletons, thus is more generic than other known $W[1]$-hard cases. We expect that our machinery can be used to solve some other cases. However, it could be seen that the class of complete bipartite graphs is not rich. Hence the result of [16] is not a special case of our Theorem 3.1. Resolving the **Dichotomy Conjecture** for the embedding problem might require a unified understanding of the cases of biclique and grids.

A remarkable phenomenon of the homomorphism problem is that the polynomial time decidability of $\mathrm{HOM}(\mathcal{K})$ coincides with the fixed-parameter tractability of $p\text{-}\mathrm{HOM}(\mathcal{K})$ for any class $\mathcal{K}$ of graphs [13], assuming $\mathrm{FPT} \neq W[1]$. For the embedding problem this is certainly not true, as for the class $\mathcal{K}$ of all paths $\mathrm{EMB}(\mathcal{K})$ is NP-hard, yet $p\text{-}\mathrm{EMB}(\mathcal{K}) \in \mathrm{FPT}$. Thus, in the **Dichotomy Conjecture**, the tractable side is really in terms of fixed-parameter tractability. But it is still interesting and important to give a precise characterization of those $\mathcal{K}$ whose $\mathrm{EMB}(\mathcal{K})$ are solvable in polynomial time. Let $\mathcal{K}$ be a *hereditary* class of graphs, i.e., closed under taking induced subgraphs. Jansen and Marx [15] showed that $\mathrm{EMB}(\mathcal{K})$ is solvable in *randomized* polynomial time if every graph in $\mathcal{K}$ can be made into the disjoint union of isolated edges and vertices by deleting $O(1)$ vertices, and NP-hard otherwise. But for general $\mathcal{K}$, we don't even have a good conjecture.

Acknowledgement. Yijia Chen is partially supported by the Sino-German Center for Research Promotion (CDZ 996) and National Nature Science Foundation of China (Project 61373029). Bingkai Lin is partially supported by the JSPS KAKENHI Grant (JP16H07409) and JST ERATO Grant (JPMJER1201) of Japan.

References

1. Alon, N., Yuster, R., Zwick, U.: Color-coding. J. ACM **42**(4), 844–856 (1995)
2. Bodirsky, M., Grohe, M.: Non-dichotomies in constraint satisfaction complexity. In: Aceto, L., Damgård, I., Goldberg, L.A., Halldórsson, M.M., Ingólfsdóttir, A., Walukiewicz, I. (eds.) ICALP 2008. LNCS, vol. 5126, pp. 184–196. Springer, Heidelberg (2008). doi:10.1007/978-3-540-70583-3_16
3. Chen, H., Müller, M.: One hierarchy spawns another: graph deconstructions and the complexity classification of conjunctive queries. In: Proceedings of the Joint Meeting of the 23rd EACSL Annual Conference on Computer Science Logic and the 29th Annual ACM/IEEE Symposium on Logic in Computer Science, pp. 32:1–32:10 (2014)

4. Chen, Y., Thurley, M., Weyer, M.: Understanding the complexity of induced subgraph isomorphisms. In: Aceto, L., Damgård, I., Goldberg, L.A., Halldórsson, M.M., Ingólfsdóttir, A., Walukiewicz, I. (eds.) ICALP 2008. LNCS, vol. 5125, pp. 587–596. Springer, Heidelberg (2008). doi:10.1007/978-3-540-70575-8_48

5. Cygan, M., Fomin, F.V., Kowalik, L., Lokshtanov, D., Marx, D., Pilipczuk, M., Pilipczuk, M., Saurabh, S.: Parameterized Algorithms. Springer, Cham (2015). doi:10.1007/978-3-319-21275-3

6. Dalmau, V., Kolaitis, P.G., Vardi, M.Y.: Constraint satisfaction, bounded treewidth, and finite-variable logics. In: Van Hentenryck, P. (ed.) CP 2002. LNCS, vol. 2470, pp. 310–326. Springer, Heidelberg (2002). doi:10.1007/3-540-46135-3_21

7. Diestel, R.: Graph Theory. Graduate Texts in Mathematics, vol. 173, 4th edn. Springer, Heidelberg (2012). doi:10.1007/978-3-662-53622-3

8. Downey, R.G., Fellows, M.R.: Parameterized Complexity. Springer, New York (1999). doi:10.1007/978-1-4612-0515-9

9. Downey, R.G., Fellows, M.R.: Fixed-parameter tractability and completeness II: on completeness for W[1]. Theoret. Comput. Sci. **141**, 109–131 (1995)

10. Eppstein, D.: Subgraph isomorphism in planar graphs and related problems. J. Graph Algorithms Appl. **3**, 1–27 (1999)

11. Flum, J., Grohe, M.: Parameterized Complexity Theory. Springer, Heidelberg (2006). doi:10.1007/3-540-29953-X

12. Garey, M.R., Johnson, D.S.: Computers and Intractability: A Guide to the Theory of NP-Completeness. Freeman, New York (1979)

13. Grohe, M.: The complexity of homomorphism, constraint satisfaction problems seen from the other side. J. ACM **54**(1), 1:1–1:24 (2007)

14. Grohe, M., Schwentick, T., Segoufin, L.: When is the evaluation of conjunctive queries tractable. In: Proceedings on the 33rd Annual ACM Symposium on Theory of Computing, Heraklion, Crete, Greece, 6–8 July 2001, pp. 657–666 (2001)

15. Jansen, B.M.P., Marx, D.: Characterizing the easy-to-find subgraphs from the viewpoint of polynomial-time algorithms, kernels, and Turing kernels. In: Proceedings of the Twenty-Sixth Annual ACM-SIAM Symposium on Discrete Algorithms, SODA 2015, San Diego, CA, USA, 4–6 January 2015, pp. 616–629 (2015)

16. Lin, B.: The parameterized complexity of k-biclique. In: Proceedings of the Twenty-Sixth Annual ACM-SIAM Symposium on Discrete Algorithms, SODA 2015, San Diego, CA, USA, 4–6 January 2015, pp. 605–615 (2015)

17. Marx, D., Pilipczuk, M.: Everything you always wanted to know about the parameterized complexity of subgraph isomorphism (but were afraid to ask). ArXiv (CoRR), abs/1307.2187 (2013)

18. Matula, D.W.: Subtree isomorphism in $o(n^{5/2})$. In: Alspach, P.H.B., Miller, D. (eds.) Algorithmic Aspects of Combinatorics. Annals of Discrete Mathematics, vol. 2, pp. 91–106. Elsevier (1978)

19. Monien, B.: How to find longest paths efficiently. In: Analysis and Design of Algorithms for Combinatorial Problems. North Holland Mathematics Studies, vol. 109, pp. 239–254. North Holland (1985)

20. Plehn, J., Voigt, B.: Finding minimally weighted subgraphs. In: Möhring, R.H. (ed.) WG 1990. LNCS, vol. 484, pp. 18–29. Springer, Heidelberg (1991). doi:10.1007/3-540-53832-1_28

21. Robertson, N., Seymour, P.D.: Graph minors V. Excluding a planar graph. J. Comb. Theory Ser. B **41**, 92–114 (1986)

22. Ullman, J.R.: An algorithm for subgraph isomorphism. J. ACM **23**(1), 31–42 (1976)

Approximately Coloring Graphs Without Long Induced Paths

Maria Chudnovsky[1], Oliver Schaudt[2(✉)], Sophie Spirkl[1], Maya Stein[3], and Mingxian Zhong[4]

[1] Princeton University, Princeton, NJ 08544, USA
[2] RWTH Aachen University, Aachen, Germany
schaudt@mathc.rwth-aachen.de
[3] Universidad de Chile, Santiago, Chile
[4] Columbia University, New York, NY 10027, USA

Abstract. It is an open problem whether the 3-coloring problem can be solved in polynomial time in the class of graphs that do not contain an induced path on t vertices, for fixed t. We propose an algorithm that, given a 3-colorable graph without an induced path on t vertices, computes a coloring with max $\left\{5, 2\left\lceil\frac{t-1}{2}\right\rceil - 2\right\}$ many colors. If the input graph is triangle-free, we only need max $\left\{4, \left\lceil\frac{t-1}{2}\right\rceil + 1\right\}$ many colors. The running time of our algorithm is $O((3^{t-2} + t^2)m + n)$ if the input graph has n vertices and m edges.

1 Introduction

A *k-coloring* of a graph G is a function $c : V(G) \to \{1, \ldots, k\}$ so that $c(v) \neq c(u)$ for all $vu \in E(G)$. In the *k-coloring problem*, one has to decide whether a given graph admits a k-coloring or not; it is NP-complete for all $k \geq 3$, as Karp proved in his seminal paper [14].

Coloring H-free Graphs. One way of dealing with this hardness is to restrict the structure of the instances. In this paper we study *H-free graphs*, that is, graphs that do not contain a fixed graph H as an induced subgraph. It is known that the k-coloring problem is NP-hard on H-free graphs if H is any graph other than a subgraph of a chordless path [11,13,16,17]. Therefore, we further restrict our attention to P_t-free graphs, P_t being the chordless path on t vertices.

A substantial number of papers study the complexity of coloring P_t-free graphs, and most of the results are gathered in the survey paper of Golovach *et al.* [8]. Let us recall a few results that define the current state-of-the-art regarding the complexity of k-coloring in P_t-free graphs.

Theorem 1 (Bonomo *et al.* [2]). *The 3-coloring problem can be solved in polynomial time in the class of P_7-free graphs. This holds true even if each vertex comes with a subset of $\{1, 2, 3\}$ of feasible colors.*

M. Chudnovsky—Supported by NSF grant DMS-1550991.
S. Spirkl—Supported by Fondecyt grant 1140766 and Millennium Nucleus Information and Coordination in Networks.

H.L. Bodlaender and G.J. Woeginger (Eds.): WG 2017, LNCS 10520, pp. 193–205, 2017.
https://doi.org/10.1007/978-3-319-68705-6_15

It is an intriguing open question whether the 3-coloring problem is solvable in polynomial time in the class of P_t-free graphs, whenever $t > 7$ is fixed.

Going back to P_5-free graphs, an elegant algorithm of Hoàng et al. [10] shows that this class is structurally restricted enough to allow for a polynomial time algorithm solving the k-coloring problem.

Theorem 2 (Hoàng et al. [10]). *The k-coloring problem can be solved in polynomial time in the class of P_5-free graphs, for each fixed k.*

The above result is interesting also for the fact that if k is part of the input, the k-coloring problem in P_5-free graphs becomes NP-hard again [16]. Regarding negative results, the following theorem of Huang is the best known so far.

Theorem 3 (Huang [12]). *For all $k \geq 5$, the k-coloring problem is NP-complete in the class of P_6-free graphs. Moreover, the 4-coloring problem is NP-complete in the class of P_7-free graphs.*

The only cases when the complexity of k-coloring P_t-free graphs is not known is when $k = 4$, $t = 6$, or when $k = 3$ and $t \geq 8$. Our contribution is an approximation algorithm for the latter case. This line of research was first suggested to us by Chuzhoy [3].

Approximation. The hardness of approximating the k-coloring problem has been in the focus of the research on approximation algorithms. Dinur et al. [4] proved that coloring a 3-colorable graph with C colors, where C is any constant, is NP-hard assuming a variant of the Unique Games Conjecture. More precisely, the assumption is that a certain label cover problem is NP-hard (where the label cover instances are what the authors call α-shaped).

On the upside, it is known how to color a 3-colorable graph with relatively few colors in polynomial time, and there has been a long line of subsequent improvements on the number of colors needed. The current state of the art, according to our knowledge, is the following result, which combines a semidefinite programming result by Chlamtac [1] with a combinatorial algorithm for the case of large minimum degree.

Theorem 4 (Kawarabayashi and Thorup [15]). *There is a polynomial time algorithm to color a 3-colorable n-vertex graph with $O(n^{0.19996})$ colors.*

In this work we combine these two lines of research and strive to use the structure of P_t-free graphs to give an approximation algorithm for the 3-coloring problem. We are inspired by a result of Gyárfás, who proved the following.

Theorem 5 (Gyárfás [9]). *If G is a graph with no induced subgraph isomorphic to P_t, then $\chi(G) \leq (t-1)^{\omega(G)-1}$.*

Thus, for a graph with no P_t, we can check if it is $(t-1)^2$-colorable or not 3-colorable by checking whether it contains a K_4. For a connected graph, Theorem 5 also holds if the requirement of being P_t-free is weakened to the assumption that there is a vertex v in G that does not start an induced P_t in G. We use a technique similar to the proof of Theorem 5 in the proof of our key lemma, Lemma 1. We take advantage, however, from the fact that the input graph is 3-colorable. This allows us to improve the bound of $(t-1)^2$ on the number of colors given by Gyárfás' theorem. We remark that our result is not an improvement of Theorem 5, but incomparable to it.

Our contribution. We prove the following.

Theorem 6. *Let $t \in \mathbb{N}$. There is an algorithm that computes for any 3-colorable P_t-free graph G*

(a) a coloring of G with at most $\max\left\{5, 2\left\lceil\frac{t-1}{2}\right\rceil - 2\right\}$ colors, and a triangle of G, or

(b) a coloring of G with at most $\max\left\{4, \left\lceil\frac{t-1}{2}\right\rceil + 1\right\}$ colors

with running time $O((3^{t-2} + t^2)|E(G)| + |V(G)|)$.

There is a variant of this problem where we replace the requirement that G is P_t-free with the weaker restriction that G has at least one vertex which is not a starting vertex of a P_t in each connected component. We give an algorithm for this harder problem as well, with a worse approximation bound, see Lemmas 1 and 2 below. Additionally, we give a hardness result, Theorem 8, to show that Lemma 1 can probably not be improved.

We remark that our algorithm can easily be implemented so that it takes an arbitrary graph as its input. It then either refutes the graph by outputting that it contains a P_t or that it is not 3-colorable or computes a coloring as promised by Theorem 6. In the case that the graph is refuted for not being 3-colorable, the algorithm can output a certificate that is easily checked in polynomial time. If the graph is refuted because it contains an induced P_t, our algorithm outputs the path.

2   Algorithm

We start with a lemma that uses ideas from Theorem 5 to color connected graphs in which some vertex does not start a P_t. It is exact up to $t = 4$; in Sect. 3 we show that 3-coloring becomes NP-hard for $t \geq 5$, which means that our result is tight in this sense.

Lemma 1. *Let G be connected, $v \in V(G)$, and $t \in \mathbb{N}$. There is a polynomial-time algorithm that outputs*

(a) that G is not 3-colorable, or
(b) an induced path P_t starting with vertex v, or
(c) a $\max\{2, t-2\}$-coloring of G, or
(d) a $\max\{3, 2t-5\}$-coloring of G and a triangle in G.

Proof. We prove this by induction on t. For $t \leq 4$, let $Z = V(G) \setminus (\{v\} \cup N(v))$. Consider a component C of $G[Z]$. By connectivity, there is a vertex $x \in V(C)$ such that $N(x) \cap N(v) \neq \emptyset$. Since G has no P_4 starting at v, each neighbor of x in C is adjacent to all of $N(x) \cap N(v)$. Thus $N(y) \cap N(v) = N(x) \cap N(v)$ for every $y \in V(C)$. In particular, if $|C| \geq 2$, we found a triangle.

Color v with color 1, and give each vertex in a singleton component of Z color 1. For each non-singleton component C of Z, note that if C is not bipartite, then G is not 3-colorable (and we have outcome (a)). So assume C is bipartite, and color all vertices from one partition class with 1. Call G' the subgraph of G that contains all yet uncolored vertices. (So all remaining vertices of Z from singleton components of $G' - N(v)$.)

If G' has no edges, we can color $V(G')$ with color 2 to obtain a valid 2-coloring of G, and are done with outcome (c). If G' is bipartite and has an edge xy, then we can color $V(G')$ with colors 2 and 3 to obtain a valid 3-coloring of G. Observe that if $x, y \in N(v)$, then G has a triangle, and that otherwise, we can assume $x \in N(v)$ and $y \in Z$. In G, vertex y belongs to a non-trivial component of $G - N(v)$; thus, as noted above, G has a triangle containing xy. In either case, we have outcome (d).

Now assume G' is not bipartite, that is, G' has an odd cycle C_ℓ, on vertices $c_1, \ldots, c_\ell$, say. Then, for each c_i lying in Z, we know that in G, there is a vertex c_i' (from the non-trivial component of Z that c_i belongs to) which is adjacent to all three of c_{i-1}, c_i, c_{i+1} (mod ℓ). So in any valid 3-coloring of G, vertices c_{i-1} and c_{i+1} have the same color. Thus we need to use at least three colors on $V(C_\ell) \cap N(v)$, which makes it impossible to color v, unless we use a 4th color, and we can output (a). This proves the result for $t \leq 4$.

Now let $t \geq 5$, and assume that the result is true for all smaller values of t. For every component C of $Z = V(G) \setminus (\{v\} \cup N(v))$, there is a vertex w_C in $N(v)$ with neighbors in C. We apply the induction hypothesis (for $t - 1$) to $G_C := G[V(C) \cup \{w_C\}]$. If this subgraph is not 3-colorable, neither is G (and we have outcome (a)). If there is an induced P_{t-1} starting at w_C, then we can add v to this path and have found an induced P_t in G starting at v, giving outcome (b). If neither outcome (a) nor outcome (b) occured in any component, then each component C of $G[Z]$ (without w_C) can be colored with $2(t-1) - 5$ colors if the algorithm detected a triangle in G_C, and with $t - 3$ colors otherwise.

If $N(v)$ is a stable set, and no triangle was detected, then we color each component of $G[Z]$ with $t - 3$ colors (which can be repeated), and use one more color for $N(v)$, and repeat one of the colors from Z for v to obtain a $(t - 2)$-coloring of G, obtaining outcome (c).

Therefore, we may assume that the algorithm detected a triangle in $G[\{v\} \cup N(v)]$ or some G_C, and we output this triangle. If $G[N(v)]$ is not bipartite, then G is not 3-colorable. Otherwise, we color each component of $G[Z]$ with the same at most $2(t - 1) - 5$ colors, color $N(v)$ with two new colors, and repeat a color from Z for v. Then, this yields a coloring of G with $2t - 5$ colors, and we found a triangle, which is outcome (d).

In the following, we will use a slightly modified version of this lemma:

Corollary 7. *Let G be connected, $v \in V(G)$, and $t \in \mathbb{N}$. Then, there is a polynomial-time algorithm that outputs*

(a) that G is not 3-colorable, or
(b) an induced path P_t starting with vertex v, or
(c) a $\max\{1, t-2\}$-coloring of $G - v$, or
(d) a $\max\{2, 2t-5\}$-coloring of $G - v$ and a triangle in G.

Proof. This is a direct consequence of Lemma 1 unless $t \le 3$. If $t \le 3$, then we can find an induced P_t starting at v unless v is adjacent to every vertex in $G - v$. So assume v is adjacent to every other vertex. If $G - v$ is not bipartite, then G is not 3-colorable. Otherwise, $G - v$ is 2-colorable and the algorithm detects a triangle, or $G - v$ is 1-colorable.

Lemma 2. *The algorithm from Lemma 1 (and from Corollary 7) can be implemented with a running time of $O(t|E(G)|)$ for a connected input graph G.*

Proof. For $t \le 4$, we can compute $v, N(v)$ and $Z = V(G) \setminus (\{v\} \cup N(v))$ in time $\mathcal{O}(|E(G)|)$. The components of Z can be found in linear time. By going through each vertex w in $N(v)$, and for each such x, going through each component C of Z following a connected enumeration of $V(C)$, we can check that w has exactly 0 or $|V(C)|$ neighbors; if this is not true for some component C, then we have found a P_t starting at v, obtaining outcome (b).

Otherwise, color v with color 1, as well as all components in Z of size 1. If a component C contains two or more vertices, then we check if it is bipartite (in linear time); if not, then since there is a neighbor w of C in $N(v)$ and w is complete to C, we output that G is not 3-colorable for outcome (a). If C is bipartite, we choose one of the partition classes of the bipartition, and give all vertices in this class color 1.

Let G' be the remaining graph after removing all vertices colored so far. We check if G' has an edge; if not, then we can give a 2-coloring of G and output (c). If G' has an edge xy, then check if G' is bipartite. If so, we can get a valid 3-coloring of G. Moreover, xy lies in a triangle (either because $x, y \in N(v)$ or because x and y have a common neighbour in Z), and we can output (d). So assume we found that G' is not bipartite, that is we found an odd cycle C_ℓ in G', on vertices $c_1, \ldots, c_\ell$, say. For each $c_i \in Z \cap V(G')$, there is a vertex $c_i' \in Z \cap V(G)$ adjacent to all three of $c_{i-1}, c_i, c_{i+1} \pmod{\ell}$, hence we can output (a), as vertices c_i', $V(C_\ell)$ and v induce an obstruction to 3-coloring G.

Now let $t \ge 5$. We compute $N(v)$ in time $|d(v)|$, compute components of $G - (\{v\} \cup N(v))$ in linear time, check if $N(v)$ is bipartite in linear time (if not, return that G is not 3-colorable), check if $N(v)$ contains two adjacent vertices, and correspondingly 1 or 2-color $N(v)$. Then we go through $N(v)$ to find a neighbor w_C for each component C of $G - (\{v\} \cup N(v))$ and run the algorithm with vertex w_C and parameter $t - 1$ on the component C.

If the outcome in any component C is an induced P_{t-1} starting at w_c, we can add v at the start of the path and get outcome (b). If some component is not 3-colorable, then neither is G, giving outcome (a). Otherwise, we find the necessary colorings (and possibly a triangle) to output (c) or (d).

Note that no edge occurs in two components, therefore we require $O(|E(G)|)$ processing time before using recursion and a total amortized running time of at most $O((t-1)|E(G)|)$ for recursive calls of the algorithm, which implies the overall running time.

Let S be a set of vertices of G, then we let $F(S)$ denote the smallest set so that $S \subseteq F(S)$ and no vertex in $G - F(S)$ has two adjacent neighbors in $F(S)$. $F(S)$ can be computed by repeatedly adding vertices that have two adjacent neighbors in the current set. In a 3-coloring, the colors of the vertices in S uniquely determine the colors of all vertices in $F(S)$.

Lemma 3. *Let G be connected, $v \in V(G)$ and $k, t \in \mathbb{N}$. There is a polynomial-time algorithm that outputs*

(a) *that G is not 3-colorable, or*
(b) *an induced path P_t in G, or*
(c) *an induced path P_k in G starting in v, or*
(d) *a set S of size $\max\{1, k-2\}$ with $v \in S$, and a $\max\{1, \lceil\frac{t-1}{2}\rceil - 2\}$-coloring of $G - (F(S) \cup N(F(S)))$, or*
(e) *a set S of size $\max\{1, k-2\}$ with $v \in S$, and a $\max\{2, 2\lceil\frac{t-1}{2}\rceil - 5\}$-coloring of $G - (F(S) \cup N(F(S)))$, and a triangle in G.*

Proof. We prove this by induction on k. If $k \leq 3$, then this follows from Corollary 7 with input k and vertex v, by setting $S = \{v\}$ and noting that then $F(S) = S$.

Now let $k > 3$. Note that we can assume $k \leq t$, because otherwise we can run the algorithm for k set to t, and all outcomes except (c) will be valid for the original k as well, and if we do get outcome (c), we can use it as outcome (b) instead. Furthermore, if $3 \leq k \leq \lceil\frac{t-1}{2}\rceil$, then the result follows from Corollary 7 with input k and vertex v, and setting $S = \{v\}$.

Consider $Z = V(G) \setminus (N(v) \cup \{v\})$. Let $\mathcal{C} = \{C_1, \ldots, C_r\}$ be the list of components of $G[Z]$, and let $\mathcal{D} = \{D_1, \ldots, D_l\}$ be the list of components of $G[N(v)]$. We now describe a procedure where at each step, we color one of the components of $\mathcal{C}$, and then put it aside, to go on working with the remaining graph, until one component $D \in \mathcal{D}$ has neighbors in all remaining components of $\mathcal{C}$.

The details are as follows. While there is not a single component in $\mathcal{D}$ with neighbors in every component of $\mathcal{C}$, let $D, D' \in \mathcal{D}$, $C, C' \in \mathcal{C}$ so that C has a neighbor x in D but no neighbor in D', and C' has a neighbor x' in D' but no neighbor in D. To see this, choose a components $D \in \mathcal{D}$ that has neighbors in as many components of $\mathcal{C}$ as possible. If some $C' \in \mathcal{C}$ has no neighbor in D, then there is a component $D' \in \mathcal{D}$ with a neighbor in C' by connectivity. But C' has a neighbor in D' but not D, so by choice of D, D' cannot have a neighbor in all components $C \in \mathcal{C}$ in which D has a neighbor; thus let C be a component in $\mathcal{C}$ so that D has a neighbor in C and D' does not. These are the desired components.

Then, we apply Corollary 7 to $\{x\} \cup C$ (with parameter $\lceil \frac{t-1}{2} \rceil$ and vertex x) and to $\{x'\} \cup C'$ (with parameter $\lceil \frac{t-1}{2} \rceil$ and vertex x'). If either of these graphs is not 3-colorable, then G is not 3-colorable. If, in both cases, there is an induced $P_{\lceil \frac{t-1}{2} \rceil}$ starting at x and at x', respectively, then, since x has no neighbors in $C' \cup D'$ and x' has no neighbors in $D \cup C$, we can combine them, using the path xvx', to obtain an induced $P_{2\lceil \frac{t-1}{2} \rceil + 1}$ in G, which contains an induced P_t. Thus, we can assume that for at least one of the two components, we found a coloring instead. In particular, we found a coloring of C or of C' with $\max\{1, \lceil \frac{t-1}{2} \rceil - 2\}$ colors, or a triangle in G, and a coloring of C or of C' with $\max\{2, 2\lceil \frac{t-1}{2} \rceil - 5\}$ colors. We then remove the component with the coloring and continue.

Finally, we arrive at a point where there is a component $D \in \mathcal{D}$ that has neighbors in all remaining components of $\mathcal{C}$. Note that if S includes v and any vertex $x \in D$, then $F(S) \supseteq D$ and thus $N(D) \subseteq F(S) \cup N(F(S))$. Therefore, we call a remaining component of $\mathcal{C}$ *good* if it is contained in $N(D)$, and *bad* otherwise. Our goal is to find a vertex $x \in D$ with neighbors in all bad components.

While there is no vertex in D with neighbors in all bad components, we can find two bad components C, C' among the remaining components of $\mathcal{C}$ such that C has a neighbor y in D, C' has a neighbor y' in D, y has no neighbors in C' and y' has no neighbors in C. As before, we can find these components by choosing y with neighbors in as many bad components as possible, and then letting y' be a vertex with a neighbor in a bad component C' in which y does not have a neighbor. Consequently, y' has no neighbor in at least one bad component C in which y does have a neighbor.

As C and C' are bad, there exist components E and E', of $C \setminus N(D)$, and of $C' \setminus N(D)$, respectively. Let x be the first vertex on a shortest path P from E to y, and define x' and P' analogously. Apply Corollary 7 to $G[\{x\} \cup E]$ (with parameter $\lceil \frac{t-2}{2} \rceil$ and vertex x) and to $G[\{x'\} \cup E']$ (with parameter $\lceil \frac{t-2}{2} \rceil$ and vertex x'). If either of these two graphs is not 3-colorable, then G is not 3-colorable. If, in both cases, there is an induced $P_{\lceil \frac{t-2}{2} \rceil}$, say P starting at x and P' starting at x', respectively, then we can combine these paths to an induced path of length at least t by taking $xPyy'P'x'$ or $xPyvy'P'x'$ (depending on whether yy' is an edge or not). Thus, we can assume that for at least one of $G[\{x\} \cup E]$, $G[\{x'\} \cup E']$, we found a coloring instead. In particular, we found a coloring of E or of E' with $\max\{1, \lceil \frac{t-2}{2} \rceil - 2\}$ colors, or a triangle in G, and a coloring with $\max\{2, 2\lceil \frac{t-2}{2} \rceil - 5\}$ colors. We then remove the component with the coloring and continue.

When this terminates, there is a single vertex $v' \in D$ that has neighbors in all remaining components of $\mathcal{C}$, except possibly those contained in $N(D)$. Let V' be the set of vertices in those components. Then we can apply the induction hypothesis with $k-1$, t, and v' to $G[V' \cup \{v'\}]$. If this graph is not 3-colorable, neither is G. If it contains an induced path P_t, so does G. If it contains an induced path P_{k-1} starting in v', then we can add v to this path to obtain an induced path P_k starting in v. If there is a set S of size $k-3$ with $v' \in S$, and a max $\{1, \lceil \frac{t-1}{2} \rceil - 2\}$-coloring of $G - (F(S) \cup N(F(S)))$ or a max $\{2, 2\lceil \frac{t-1}{2} \rceil - 5\}$-coloring of $G - (F(S) \cup N(F(S)))$ and a triangle, then we proceed as follows. We add v to S, and now S has size $k-2$. Moreover, since both v and v' in S, we know that $D \in F(S)$, thus all vertices in $\{v\} \cup N(v) \cup D \cup N(D)$ are in $F(S) \cup N(F(S))$. We colored different components of $Z \setminus N(D)$ at different stages, but we can reuse the colors used on these components. Therefore, this leads to outcome (d) or (e), depending on if the algorithm detected a triangle at any stage.

Lemma 4. *The algorithm from Lemma 3 can be implemented with running time* $O\left(k \lceil \frac{t-1}{2} \rceil |E(G)|\right)$ *for a connected input graph G.*

Proof. For $k \leq 3$, this follows from Lemma 2. For $k > 3$, we compute $Z = V(G) \setminus (N(v) \cup \{v\})$ and the components $C_1, \ldots C_r$ of $G[Z]$ and the components $D_1, \ldots, D_r$ of $G[N(v)]$ as in the proof of Lemma 3; all this can be done in linear time.

Now, subsequently, for $j = 1, \ldots, r$, we consider D_j, and some of the C_i adjacent to it, and after possibly coloring and deleting these components C_i, we might also delete D_j. More precisely, for $j = 1, \ldots, r$, we consider those C_i that only have neighbors in D_j. For each such C_i, choose a neighbor x_{ij} in D_j and apply Corollary 7 to $G[C_i \cup \{x_{ij}\}]$. If this graph is not 3-colorable, then neither is G. If the algorithm returns a coloring (and possibly a triangle), then this is the coloring we will use in C_i, as explained in the proof of Lemma 3, so we can delete C_i. (But, if the algorithm found a triangle, we shall remember this triangle for a possible output, at least if it is the first one to be found.) Otherwise, the algorithm returns a path of length $\lceil \frac{t-1}{2} \rceil$, which we keep. After going through all C_i with neighbors only in D_j, we delete D_j if the algorithm always returned a coloring (or if there were no C_i to consider). That is, we keep D_j if and only if for some i we found a path starting from x_{ij} with interior in C_i.

The amortized time it takes to process all D_j is $O(\lceil \frac{t-1}{2} \rceil |E(G)|)$. This is so because every component C_i is used for the algorithm from Corollary 7 at most once.

In the end, if there are D_j, $D_{j'}$ that we did not delete, then there is a component C_i that only has neighbors in D_j, and another component $C_{i'}$ that only has neighbors in $D_{j'}$, and both C_i and $C_{i'}$ contain a $P_{\lceil \frac{t-1}{2} \rceil}$ in their interior, starting at x_{ij} and $x_{i'j'}$ respectively. By connecting them using the middle segment $x_{ij}vx_{i'j'}$, we find a P_t in G that we can output as outcome (b). Otherwise, there is only one D_j left at the end. Since whenever we deleted a component $D_{j'}$, we ensured that each remaining C_i has a neighbor in some D_j with $j \neq j'$, this means that D_j has neighbors in all C_i that we did not color yet.

Let Z' be the set of vertices of remaining components C_i, and let $Z'' = Z' \setminus N(D)$. Each component of $G[Z'']$ is contained in some component C_i and thus it has a neighbor in $N(D)$. Therefore, we can apply the same argument as before to components of $G[Z'']$ and components of $N(D)$ with neighbors in them.

Whenever the algorithm for Corollary 7 outputs a coloring we keep it (and we also keep the possibly found triangle, if it is the first triangle to be found), and if it outputs that a component is not 3-colorable, then G is not 3-colorable, and if there is a path $P_{\lceil\frac{t-2}{2}\rceil}$, we keep track of it. (Note that components of $N(D)$ that are not adjacent to Z'' get deleted automatically.) When this terminates, if there are still two components of $N(D)$, then there are two paths we can combine to a P_t as in Lemma 3. Otherwise, a single component D^* of $N(D)$ has a neighbor in all remaining components $C'_1, \ldots, C'_s$, so there is a vertex $v' \in D$ so that $\{v'\} \cup D^* \cup C'_1 \cup \cdots \cup C'_s$ is connected, and v' is the only neighbor of v in that subgraph. Next, we apply induction for $k-1$ with root vertex v on that set. If this finds a set S and a coloring, we add v to S. If it finds a path P_{k-1}, we add v to the path. If it finds a P_t, we output it. If it is not 3-colorable, then neither is G.

The total running time of the recursive application of the algorithm is

$$O\left((k-1)\left\lceil\frac{t-1}{2}\right\rceil |E(G)|\right),$$

and all preprocessing steps leading there can be implemented with a running time of $O\left(\left\lceil\frac{t-1}{2}\right\rceil |E(G)|\right)$, which implies the result.

We can now give the proof of our main result.

Proof (Proof of Theorem 6). We use the algorithm from Lemma 3 with $k = t$. Since only outcomes (d) and (e) can occur in this setting, it is sufficient to show that if G is 3-colorable, we can find a 3-coloring of $F(S) \cup N(F(S))$ in time $O(3^{t-2} \cdot \text{poly}(|V(G)|))$, as follows: For each vertex in the set S, we try each possible color for a total of at most 3^{t-2} possibilities. By definition of $F(S)$, this determines the color of each vertex in $F(S)$, and for each vertex in $N(F(S))$, there are at most two possible colors. Thus, we reduced to a 2-list-coloring problem, which can be solved in linear time by reduction to 2SAT [5,6,18]. Hence if G is 3-colorable, we can 3-color $F(S) \cup N(F(S))$, and add these three new colors to the coloring from Lemma 3.

The total running time follows from the running time of the algorithm in Lemma 4 in addition to an algorithm determining the connected components of G.

By combining Theorem 6 with Lemma 1, we obtain the algorithms for coloring 3-colorable P_t-free graphs with the number of colors shown in Table 1 (if there is a triangle) and Table 2 (if there is no triangle). For t larger than shown in the table, Theorem 6 uses a smaller number of colors.

Table 1. Number of colors we use for a 3-colorable P_t-free graph if there is a triangle

t	3	4	5	6	7	8	9	10	11	> 11
$\max\{3, 2t-5\}$	3	3	5	7	9	11	13	15	17	
$\max\{5, 2\lceil\frac{t-1}{2}\rceil - 2\}$	5	5	5	5	5	6	6	8	8	
Best option	3	3	3^1	5	5	6	6	8	8	$2\lceil\frac{t-1}{2}\rceil - 2$

[1]If $t = 5$, we can improve the number of colors required if there is a triangle to 3, because it cannot happen that there are two components C, C' of $G - (\{v\} \cup N(v))$ and components D, D' of $G[N(v)]$ so that C has a neighbor in D but not D', and C' has a neighbor in D' but not D', because this already yields an induced P_5. Thus, by induction, all vertices will be in $F(S) \cup N(F(S))$, where we can test for 3-colorability as described in Theorem 6.

Table 2. Number of colors we use for a 3-colorable P_t-free graph if there is no triangle

t	3	4	5	6	7	8	9	10	11	> 11
$\max\{2, t-2\}$	2	2	3	4	5	6	7	8	9	
$\max\{4, \lceil\frac{t-1}{2}\rceil + 1\}$	4	4	4	4	4	5	5	6	6	
Best option	2	2	3	4	4	5	5	6	6	$\lceil\frac{t-1}{2}\rceil + 1$

3   Hardness Result

In this section, we show that improving Lemma 1 is hard. More precisely:

Theorem 8. *Let G be a connected graph and $v \in V(G)$ so that there is no induced P_t in G starting at v. Then, deciding k-colorability on this class of graphs is NP-hard if $k \geq 4$ and $t \geq 3$ or if $k = 3$ and $t \geq 5$. It can be solved in polynomial time if $t \leq 2$ or if $k = 3$ and $t \leq 4$.*

Proof. For the polynomial time solvability, observe that if $t \leq 2$, then $|V(G)| \leq 1$. If $k = 3, t \leq 4$, then the result follows from Lemma 1.

For the hardness, first consider the case $k \geq 4$, $t \geq 3$. In this case, we can reduce the 3-coloring problem to this problem by taking any instance G and adding a clique of size $k - 3$ complete to G. Then, no vertex in this clique starts a P_3, but the resulting graph is k-colorable if and only if G is 3-colorable.

It remains to consider the case $k = 3$, $t \geq 5$. We show a reduction from the NP-complete problem NAE-3SAT [7]. An instance of NAE-3SAT is a boolean formula with variables $x_1, \ldots, x_n$ and clauses $C_1, \ldots, C_m$, where each clause contains exactly three literals (variables or their negations). It is a YES-instance if and only if there is an assignment of the variables as true or false so that for every clause, not all three literals in the clause are true, and not all three are false.

We construct a graph G as follows: G contains a vertex v, vertices labeled x_i and $\overline{x}_i$ for $1 \leq i \leq n$, and a triangle T_j for each clause C_j. The vertex v is adjacent to all vertices x_i and $\overline{x_i}$, but not to any of the triangles. For each i, x_i is adjacent to $\overline{x_i}$. For each clause C_j, we assign each literal a vertex of the triangle T_j, and connect this vertex to the literal (the vertex labeled x_i or $\overline{x_i}$). There are no other edges in G.

Then, there is no P_5 starting at v in G, because such a path would have to contain exactly one of the vertices labeled x_i and $\overline{x_i}$, and this would be the second vertex of the path. As there are no edges between the triangles T_j, all remaining vertices of the P_5 would have to be in one triangle T_j. But no triangle can contain a P_3. Therefore, G is a valid instance.

It remains to show that G is 3-colorable if and only if the instance of NAE-3SAT is a YES-instance. If G has a 3-coloring, then the neighbors of v are 2-colored (say with colors 1 and 2) and x_i never receives the same color as $\overline{x_i}$. Assign the variables so that literals colored 1 are true, and those colored 2 are false. Then, if there is a clause C_j so that all of its literals are true, this means that each vertex of T_j has a neighbor colored 1, so T_j uses only colors 2 and 3, which is impossible in a valid coloring of a triangle. For the same reason, there cannot be a clause so that all of its literals are false. Thus, G was constructed from a YES-instance.

Conversely, if the instance we started with is a YES-instance, we color v with color 3, true literals with color 1, and false literals with color 2. For each triangle T_j, one of the vertices adjacent to a true literal is colored 2, one of the vertices adjacent to a false literal is colored 1, and the remaining vertex is colored 3. This is a valid 3-coloring of G.

4   Conclusion

In this paper we showed how to color a given 3-colorable P_t-free graph with a number of colors that is t, roughly. The running time of our algorithm is of the form $O(f(t) \cdot n^{O(1)})$, when the input graph has n vertices, and thus FPT in the parameter t. (The class FPT contains the fixed parameter tractable problems, which are those that can be solved in time $f(k) \cdot |x|^{O(1)}$ for some computable function f.)

In view of this, it seems to be an intriguing question whether the 3-coloring problem is fixed-parameter tractable when parameterized by the length of the longest induced path. That is, whether there is an algorithm with running time $O(f(t) \cdot n^{O(1)})$ that decides 3-colorability in P_t-free graphs. So far, however, it is not even known whether there is an XP algorithm to decide 3-colorability in P_t-free graphs. (XP is the class of parameterized problems that can be solved in time $O(n^{f(k)})$ for some computable function f.) If such an XP-algorithm existed, this would show that the problem is in P whenever t is fixed. Therefore, attempting to prove W[1]-hardness seems to be more reasonable than trying to prove that the problem is in FPT.

Another question we addressed is k-coloring connected graphs so that some vertex is not the end vertex of an induced P_t. We showed that coloring in this case is NP-hard whenever $k = 3$ and $t \geq 5$, or $k \geq 4$ and $t \geq 3$. Lemma 1 gives a simple algorithm for an $f(t)$-approximate coloring for $k = 3$ and any t, and it would be interesting to have a complementing result proving hardness of approximation. On the other hand, any improvement of Lemma 1 would immediately yield an improvement of our main result.

Acknowledgments. We are thankful to Paul Seymour for many helpful discussions. We thank Stefan Hougardy for pointing out [15] to us.

References

1. Chlamtac, E.: Approximation algorithms using hierarchies of semidefinite programming relaxations. In: 48th Annual IEEE Symposium on Foundations of Computer Science (FOCS 2007) (2007)
2. Bonomo, F., Chudnovsky, M., Maceli, P., Schaudt, O., Stein, M., Zhong, M.: Three-coloring and list three-coloring graphs without induced paths on seven vertices (2015, preprint)
3. Chuzhoy, J.: Private communication
4. Dinur, I., Mossel, E., Regev, O.: Conditional hardness for approximate coloring. SIAM J. Comput. **39**, 843–873 (2009)
5. Edwards, K.: The complexity of colouring problems on dense graphs. Theoret. Comput. Sci. **43**, 337–343 (1986)
6. Erdős, P., Rubin, A.L., Taylor, H.: Choosability in graphs. Congr. Numer. **26**, 125–157 (1979)
7. Garey, M.R., Johnson, D.S.: A Guide to the Theory of NP-Completeness. W.H. Freemann, New York (1979)
8. Golovach, P.A., Johnson, M., Paulusma, D., Song, J.: Survey on the computational complexity of colouring graphs with forbidden subgraphs. J. Graph Theory (to appear). doi:10.1002/jgt.22028
9. Gyárfás, A.: Problems from the world surrounding perfect graphs. Applicationes Mathematicae **19**(3–4), 413–441 (1987)
10. Hoàng, C.T., Kamiński, M., Lozin, V., Sawada, J., Shu, X.: Deciding k-colorability of P_5-free graphs in polynomial time. Algorithmica **57**(1), 74–81 (2010)
11. Holyer, I.: The NP-completeness of edge-coloring. SIAM J. Comput. **10**, 718–720 (1981)
12. Huang, S.: Improved complexity results on k-coloring P_t-free graphs. Eur. J. Comb. **51**, 336–346 (2016)
13. Kamiński, M., Lozin, V.: Coloring edges and vertices of graphs without short or long cycles. Contrib. Discret. Math. **2**, 61–66 (2007)
14. Karp, R.: Reducibility among combinatorial problems. In: Miller, R., Thatcher, J. (eds.) Complexity of Computer Computations, pp. 85–103. Plenum Press, New York (1972)
15. Kawarabayashi, K.-I., Thorup, M.: Coloring 3-colorable graphs with $o(n^{1/5})$ colors. In: LIPIcs-Leibniz International Proceedings in Informatics, vol. 25. Schloss Dagstuhl-Leibniz-Zentrum fuer Informatik (2014)
16. Král', D., Kratochvíl, J., Tuza, Z., Woeginger, G.J.: Complexity of coloring graphs without forbidden induced subgraphs. In: Brandstädt, A., Le, V.B. (eds.) WG 2001. LNCS, vol. 2204, pp. 254–262. Springer, Heidelberg (2001). doi:10.1007/3-540-45477-2_23

17. Leven, D., Galil, Z.: NP-completeness of finding the chromatic index of regular graphs. J. Algorithms **4**, 35–44 (1983)
18. Vizing, V.G.: Coloring the vertices of a graph in prescribed colors. Diskret. Analiz **29**(3), 10 (1976)

New and Simple Algorithms for Stable Flow Problems

Ágnes Cseh[1(✉)] and Jannik Matuschke[2]

[1] Institute of Economics, Hungarian Academy of Sciences and Corvinus
University of Budapest, Budaörsi út 45., Budapest 1112, Hungary
cseh.agnes@krtk.mta.hu
[2] TUM School of Management and Department of Mathematics,
Technische Universität München, Arcisstraße 21, 80333 Munich, Germany
jannik.matuschke@tum.de

Abstract. Stable flows generalize the well-known concept of stable
matchings to markets in which transactions may involve several agents,
forwarding flow from one to another. An instance of the problem consists of a capacitated directed network, in which vertices express their
preferences over their incident edges. A network flow is stable if there is
no group of vertices that all could benefit from rerouting the flow along
a walk.

Fleiner [13] established that a stable flow always exists by reducing it
to the stable allocation problem. We present an augmenting-path algorithm for computing a stable flow, the first algorithm that achieves polynomial running time for this problem without using stable allocation as
a black-box subroutine. We further consider the problem of finding a stable flow such that the flow value on every edge is within a given interval.
For this problem, we present an elegant graph transformation and based
on this, we devise a simple and fast algorithm, which also can be used to
find a solution to the stable marriage problem with forced and forbidden
edges. Finally, we study the highly complex stable multicommodity flow
model by Király and Pap [24]. We present several graph-based reductions
that show equivalence to a significantly simpler model. We further show
that it is NP-complete to decide whether an integral solution exists.

1 Introduction

Stability is a well-known concept used for matching markets where the aim is to
reach a certain type of social welfare, instead of profit-maximization [29]. The
measurement of optimality is not maximum cardinality or minimum cost, but
the certainty that no two agents are willing to selfishly modify the market situation. Stable matchings were first formally defined in the seminal paper of Gale

The authors were supported by the Hungarian Academy of Sciences under its
Momentum Programme (LP2016-3/2016) and its János Bolyai Research Scholar-
ship, OTKA grant K108383 and COST Action IC1205 on Computational Social
Choice, and by the Alexander von Humboldt Foundation with funds of the German
Federal Ministry of Education and Research (BMBF).

© Springer International Publishing AG 2017
H.L. Bodlaender and G.J. Woeginger (Eds.): WG 2017, LNCS 10520, pp. 206–219, 2017.
https://doi.org/10.1007/978-3-319-68705-6_16

and Shapley [17]. They described an instance of the college admission problem and introduced the terminology based on marriage that since then became widespread. Besides resident allocation, variants of the stable matching problem are widely used in other employer allocation markets [30], university admission decisions [2,4], campus housing assignments [5,28] and bandwidth allocation [16]. A recent honor proves the currentness and importance of results in the topic: in 2012, Lloyd S. Shapley and Alvin E. Roth were awarded the Sveriges Riksbank Prize in Economic Sciences in Memory of Alfred Nobel for their outstanding results on market design and matching theory.

In the stable marriage problem, we are given a bipartite graph, where the two classes of vertices represent men and women, respectively. Each vertex has a strictly ordered preference list over his or her possible partners. A matching is *stable* if it is not *blocked* by any edge, that is, no man-woman pair exists who are mutually inclined to abandon their partners and marry each other [17].

In practice, the stable matching problem is mostly used in one of its capacitated variants, which are the stable many-to-one matching, many-to-many matching and allocation problems. The *stable flow* problem can be seen as a high-level generalization of all these settings. To the best of our knowledge, it is the most complex graph-theoretical generalization of the stable marriage model, and thus plays a crucial role in the theoretical understanding of the power and limitations of the stable marriage concept. From a practical point of view, stable flows can be used to model markets in which interactions between agents can involve chains of participants, e.g., supply chain networks involving multiple independent companies.

In the stable flow problem, a directed network with preferences models a market situation. Vertices are vendors dealing with some goods, while edges connecting them represent possible deals. Through his preference list, each vendor specifies how desirable a trade would be to him. Sources and sinks model suppliers and end-consumers. A feasible network flow is stable, if there is no set of vendors who mutually agree to modify the flow in the same manner. A blocking walk represents a set of vendors and a set of possible deals so that all of these vendors would benefit from rerouting some flow along the blocking walk.

Literature Review. The notion of stability was extended to so-called "vertical networks" by Ostrovsky in 2008 [26]. Even though the author proves the existence of a stable solution and presents an extension of the Gale-Shapley algorithm, his model is restricted to unit-capacity acyclic graphs. Stable flows in the more general setting were defined by Fleiner [13], who reduced the stable flow problem to the stable allocation problem.

The best currently known computation time for finding a stable flow is $\mathcal{O}(|E|\log|V|)$ in a network with vertex set V and edge set E. This bound is due to Fleiner's reduction to the stable allocation problem and its fastest solution described by Dean and Munshi [9]. Since the reduction takes $\mathcal{O}(|V|)$ time and does not change the instance size significantly and the weighted stable allocation problem can be solved in $\mathcal{O}(|E|^2\log|V|)$ time [9], the same holds for the maximum weight stable flow problem. The Gale-Shapley algorithm can also be extended for stable flows [8], but its straightforward implementation requires exponential running time, just like in the stable allocation problem.

It is sometimes desirable to compute stable solutions using certain forced edges or avoiding a set of forbidden edges. This setting has been an actively researched topic for decades [6,10,14,20,25]. This problem is known to be solvable in polynomial time in the one-to-one matching case, even in non-bipartite graphs [14]. Though Knuth presented a combinatorial method that finds a stable matching in a bipartite graph with a given set of forced edges or reports that none exists [25], all known methods for finding a stable matching with both forced and forbidden edges exploit a somewhat involved machinery, such as rotations [20], LP techniques [11,12,21] or reduction to other advanced problems in stability [10,14].

In many flow-based applications, various goods are exchanged. Such problems are usually modeled by multicommodity flows [22]. A maximum multicommodity flow can be computed in strongly polynomial time [31], but even when capacities are integer, all optimal solutions might be fractional, and finding a maximum integer multicommodity flow is NP-hard [19]. Király and Pap [24] introduced the concept of stable multicommodity flows, in which edges have preferences over which commodities they like to transport and the preference lists at the vertices may depend on the commodity. They show that a stable solution always exists, but it is PPAD-hard to find one.

Our Contribution and Structure. In this paper we discuss new and simplified algorithms and complexity results for three differently complex variants of the stable flow problem. Section 2 contains preliminaries on stable flows.

- In Sect. 3 we present a *polynomial algorithm for stable flows*. To derive a fast, elegant, and direct solution method, we combine the well-known pseudo-polynomial Gale-Shapley algorithm and the proposal-refusal pointer machinery known from stable allocations into an augmenting-path algorithm for computing a stable flow.
- Then, in Sect. 4 *stable flows with restricted intervals* are discussed. We provide a simple combinatorial algorithm to find a flow with flow value within a pre-given interval for each edge. Surprisingly, our algorithm directly translates into a very simple new algorithm for the problem of stable matchings with forced and forbidden edges in the classical stable marriage case. Unlike the previously known methods, our result relies solely on elementary graph transformations.
- Finally, in Sect. 5 we study *stable multicommodity flows*. First, we provide tools to simplify stable multicommodity flow instances to a great extent by showing that it is without loss of generality to assume that no commodity-specific preferences at the vertices and no commodity-specific capacities on the edges exist. Then, we reduce 3-SAT to the integral stable multicommodity flow problem and show that it is NP-complete to decide whether an integral solution exists even if the network in the input has integral capacities only.

2 Preliminaries

A *network* (D, c) consists of a directed graph $D = (V, E)$ and a capacity function $c : E \to \mathbb{R}_{\geq 0}$ on its edges. The vertex set of D has two distinct elements, also called *terminal vertices*: a source s, which has outgoing edges only and a sink t, which has incoming edges only.

Definition 1 (flow). *Function $f : E \to \mathbb{R}_{\geq 0}$ is a* flow *if it fulfills both of the following requirements:*

1. *capacity constraints: $f(uv) \leq c(uv)$ for every $uv \in E$;*
2. *flow conservation: $\sum_{uv \in E} f(uv) = \sum_{vw \in E} f(vw)$ for all $v \in V \setminus \{s, t\}$.*

A stable flow instance is a triple $\mathcal{I} = (D, c, O)$. It comprises a network (D, c) and O, the preference ordering of vertices on their incident edges. Each non-terminal vertex ranks its incoming and also its outgoing edges strictly and separately. If v prefers edge vw to vz, then we write $r_v(vw) < r_v(vz)$. Terminals do not rank their edges, because their preferences are irrelevant with respect to the following definition.

Definition 2 (blocking walk, stable flow). *A* blocking walk *of flow f is a directed walk $W = \langle v_1, v_2, ..., v_k \rangle$ such that all of the following properties hold:*

1. *$f(v_i v_{i+1}) < c(v_i v_{i+1})$, for each edge $v_i v_{i+1}$, $i = 1, ..., k-1$;*
2. *$v_1 = s$ or there is an edge $v_1 u$ such that $f(v_1 u) > 0$ and $r_{v_1}(v_1 v_2) < r_{v_1}(v_1 u)$;*
3. *$v_k = t$ or there is an edge wv_k such that $f(wv_k) > 0$ and $r_{v_k}(v_{k-1}v_k) < r_{v_k}(wv_k)$.*

A flow is stable, *if there is no blocking walk with respect to it in the graph.*

Unsaturated walks fulfilling point 2 are said to *dominate f at start*, while walks fulfilling point 3 *dominate f at the end*. We can say that a walk blocks f if it dominates f at both ends.

Throughout the paper, we will assume that the digraph D does not contain loops or parallel edges, the source s only has outgoing edges, the sink t only has incoming edges, and that no isolated vertices exist. All these assumptions are without loss of generality and only for notational convenience.

Problem 1. SF
Input: $\mathcal{I} = (D, c, O)$; a directed network (D, c) and O, the preference ordering of vertices.
Question: Is there a flow f so that no walk blocks f?

Theorem 1 (Fleiner [13]). SF *always has a stable solution and it can be found in polynomial time. Moreover, for a fixed SF instance, each edge incident to s or t has the same value in every stable flow.*

This result is based on a reduction to the stable allocation problem. The second half of Theorem 1 can be seen as the flow generalization of the so-called *Rural Hospitals Theorem*, known for stable matching instances in general graphs. Part of this theorem states that if a vertex is unmatched in one stable matching, then all stable solutions leave it unmatched [18].

Algorithm 1. Augmenting path algorithm for stable flows

Initialize π, ρ.
while $\pi[s] \neq \emptyset$ **do**
    Let W be an s-t-path or cycle in $H_{\pi,\rho}$.
    Let $\Delta := \min_{e \in W} c_f(e)$.
    Augment f by Δ along W.
    while $\exists\, uv \in E_{H_{\pi,\rho}}$ with $c_f(uv) = 0$ **do**
        UpdatePointers (u)

---

3 A Polynomial-Time Augmenting Path Algorithm for Stable Flows

Using Fleiner's construction [13], it is possible to find a stable flow efficiently by computing a stable allocation instead. Also the popular Gale-Shapley algorithm can be extended to SF. As described in [8], this yields a preflow-push type algorithm, in which vertices forward or reject excessive flow according to their preference lists. While this algorithm has the advantage of operating directly on the network without transforming it to a stable allocation instance, it requires pseudo-polynomial running time.

In the following, we describe a polynomial time algorithm to produce a stable flow that operates directly on the network D. Our method is based on the well-known augmenting path algorithm of Ford and Fulkerson [15], also used by Baïou and Balinski [1] and Dean and Munshi [9] for stability problems. The main idea is to introduce proposal and refusal pointers to keep track of possible Gale-Shapley steps and execute them in bulk. Each such iteration corresponds to augmenting flow along an s-t-path or cycle in a restricted residual network.

In the algorithm, every vertex $v \in V \setminus \{t\}$ is associated with two pointers, the *proposal pointer* $\pi[v]$ and the *refusal pointer* $\rho[v]$. Initially $\pi[v]$ points to the first-choice outgoing edge on v's preference list, whereas $\rho[v]$ is inactive. Throughout the algorithm $\pi[v]$ traverses the outgoing edges of v in order of increasing rank on v's preference list (for the source s, we assume an arbitrary preference order) until it gets advanced beyond the final outgoing edge. Then $\rho[v]$ becomes active and traverses the incoming edges of v in order of decreasing rank on v's preference list.

With any state of the pointers π, ρ, we associate a helper graph $H_{\pi,\rho}$, which contains the edges pointed at by the proposal pointers and the reversals of the edges pointed at by the refusal pointer. A recursive update procedure for advancing the pointers along their lists ensures that throughout the algorithm, $H_{\pi,\rho}$ contains an s-t-path or a cycle, with all edges having positive residual capacity c_f with respect to the current flow. The algorithm then augments the flow along this path or cycle and updates the pointers of saturated edges. Once $\pi[s]$ has traversed all outgoing edges of s, the algorithm has found a stable flow. As in each iteration, at least one pointer is advanced, the running time of the algorithm is polynomial in the size of the graph. See the full version of the paper [7] for a listing of the subroutine `UpdatePointers` and a complete analysis.

4   Stable Flows with Restricted Intervals

Various stable matching problems have been tackled under the assumption that restricted edges are present in the graph [10,14]. A restricted edge can be *forced* or *forbidden*, and the aim is to find a stable matching that contains all forced edges, while it avoids all forbidden edges. Such edges correspond to transactions that are particularly desirable or undesirable from a social welfare perspective, but it is undesirable or impossible to push the participating agents directly to use or avoid the edges. We thus look for a stable solution in which the edge restrictions are met voluntarily.

A natural way to generalize the notion of a restricted edge to the stable flow setting is to require the flow value on any given edge to be within a certain interval. To this end, we introduce a *lower capacity* function $\mathfrak{l} : E \to \mathbb{R}_{\geq 0}$ and an *upper capacity* function $\mathfrak{u} : E \to \mathbb{R}_{\geq 0}$.

Problem 1. SF RESTRICTED
Input: $\mathcal{I} = (D, c, O, \mathfrak{l}, \mathfrak{u})$; an SF instance (D, c, O), a lower capacity function $\mathfrak{l} : E \to \mathbb{R}_{\geq 0}$ and an upper capacity function $\mathfrak{u} : E \to \mathbb{R}_{\geq 0}$.
Question: Is there a stable flow f so that $\mathfrak{l}(uv) \leq f(uv) \leq \mathfrak{u}(uv)$ for all $uv \in E$?

Note that in the above definition, the upper bound $\mathfrak{u}$ does not affect blocking walks, i.e., a blocking walk can use edge uv, even if $f(uv) = \mathfrak{u}(uv) < c(uv)$ holds.

SF RESTRICTED generalizes the natural notion of requiring flow to use an edge to its full capacity (by setting $\mathfrak{l}(uv) = c(uv)$) and of requiring flow not to use an edge at all (by setting $\mathfrak{u}(uv) = 0$), which corresponds to the traditional cases of forced and forbidden edges. It turns out that any given instance of SF RESTRICTED can be transformed to an equivalent instance in which $\mathfrak{l}(uv), \mathfrak{u}(uv) \in \{0, c(uv)\}$ for all $uv \in E$. We describe the corresponding reduction in the full version [7]. Henceforth, we will assume that our instances are of this form and use the notation $Q := \{uv \in E : \mathfrak{l}(uv) = c(uv)\}$ and $P := \{uv \in E : \mathfrak{u}(uv) = 0\}$ for the sets of forced and forbidden edges, respectively.

In the following, we describe a polynomial algorithm that finds a stable flow with restricted intervals or proves its nonexistence. We show that restricted intervals can be handled by small modifications of the network that reduce the problem to the unrestricted version of SF. We show this separately for the case that only forced edges occur, which we call SF FORCED, in Sect. 4.1 and for the case that only forbidden edges occur, called SF FORBIDDEN, in Sect. 4.2. It is straightforward to see that both results can be combined to solve the general version of SF RESTRICTED. All missing proofs can be found in the full version [7].

We mention that it is also possible to solve SF RESTRICTED by transforming the instance first into a weighted SF instance, and then into a weighted stable allocation instance, both solvable in $\mathcal{O}(|E|^2 \log |V|)$ time [9]. The advantages of our method are that it can be applied directly to the SF RESTRICTED instance and

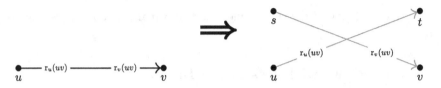

Fig. 1. Substituting forced edge uv by edges sv and ut in D'.

it also gives us insights to solving the stable roommate problem with restricted edges directly, as pointed out at the end of Sects. 4.1 and 4.2. Moreover, our running time is $\mathcal{O}(|P||E|\log|V|)$, where P is the set of edges with $u(uv) < c(uv)$.

4.1 Forced Edges

Let us first consider a single forbidden edge uv. We modify graph D to derive a graph D'. The modification consists of deleting the forced edge uv and introducing two new edges sv and ut to substitute it. Both new edges have capacity $c(uv)$ and take over uv's rank on u's and on v's preference lists, respectively, as shown in Fig. 1.

Lemma 1. *Let f be a flow in D with $f(uv) = c(uv)$. Let f' be the flow in D' derived by setting $f'(sv) = f'(ut) = f(uv)$ and $f'(e) = f(e)$ for all $e \in E \setminus \{uv\}$. Then f is stable if and only if f' is stable.*

Proof. We first observe that the set of edges not saturated by f in D is the same as the set of edges not saturated by f' in D'. This is because uv is saturated by f and ut, sv are saturated by f'. Now let $u'v'$ be such an unsaturated edge. Note that there is an edge $u'w'$ with $r_{u'}(u'w) > r_{u'}(u'v')$ and $f(u'w) > 0$ if and only if there is an edge $u'w'$ with $r_{u'}(u'w') > r_{u'}(u'v')$ and $f'(u'w') > 0$. The same is true for incoming edges at v' dominated by $u'v'$. In other words, the dominance situation at all vertices is the same for f and f'. This implies that any blocking walk for f in D is a blocking walk for f' in D' and vice versa.   □

Repeated application of Lemma 1 in conjunction with the Rural Hospital Theorem (Theorem 1) allows us to transform any instance of SF FORCED into a standard stable flow instance.

Theorem 2. SF FORCED *can be solved in time* $\mathcal{O}(|E|\log|V|)$.

Stable matchings with forced edges. The technique described above also provides a fairly simple method for the stable matching problem with forced edges even in non-bipartite graphs, because the Rural Hospitals Theorem holds for that case as well. After deleting each forced edge $uw \in Q$ from the graph, we add uw_s and u_tw edges to each of the pairs, where w_s and u_t are newly introduced vertices. These edges take over the rank of uw. Unlike in SF, here we need to introduce two separate dummy vertices to each forced edge, simply due to the matching constraints. There is a stable matching containing all forced

Fig. 2. Adding edges sv in D^+ and ut in D^- to forbidden edge uv.

edges if and only if an arbitrary stable matching covers all of these new vertices w_s and u_t. The running time of this algorithm is $\mathcal{O}(|E|)$, since it is sufficient to construct a single stable solution in an instance with at most $2|V|$ vertices. More vertices cannot occur, because in a matching problem more than one forced edge incident to a vertex immediately implies infeasibility.

4.2 Forbidden Edges

In order to handle SF FORBIDDEN, we present here an argumentation of the same structure as in the previous section. First, we show how to solve the problem of stable flows with a single forbidden edge by solving two instances on two different extended networks. Then we show how these constructions can be used to obtain an algorithm for the general case.

Notation. For $e = uv \in P$, we define edges $e^+ = sv$ and $e^- = ut$. We set $c(e^+) = \varepsilon > 0$ and set $r_v(e^+) = r_v(e) - 0.5$, i.e., e^+ occurs on v's preference list exactly before e. Likewise, we set $c(e^-) = \varepsilon$ and $r_u(e^-) = r_u(e) - 0.5$, i.e., e^- occurs on u's preference list exactly before e. For $F \subseteq P$ we define $E^+(F) := \{e^+ : e \in F\}$ and $E^-(F) := \{e^- : e \in F\}$.

A single forbidden edge. Assume that $P = \{e_0\}$ for a single edge e_0. First we present two modified instances that will come handy when solving SF FORBIDDEN. The first is the graph D^+, which we obtain from D by adding an edge e_0^+ to E. Similarly, we obtain the graph D^- by adding e_0^- to E. Both graphs are illustrated in Fig. 2.

In the following, we characterize SF FORBIDDEN instances with the help of D^+ and D^-. Our claim is that SF FORBIDDEN in D has a solution if and only if there is a stable flow f^+ in D^+ with $f^+(e^+) = 0$ or there is a stable flow f^- in D^- with $f^-(e^-) = 0$. These existence problems can be solved easily in polynomial time, since all stable flows have the same value on edges incident to terminal vertices by Theorem 1.

We will use the following straightforward observation. It follows from the fact that deletion of an edge that does not carry any flow in a stable flow neither affects flow conversation nor can create blocking walks.

Observation 1. If $f(e) = 0$ for an edge $e \in E$ and stable flow f in D, then f remains stable in $D \setminus e$ as well.

Lemma 2. *Let f be a flow in $D = (V, E)$ with $f(e_0) = 0$. Then f is a stable flow in D if and only if at least one of the following conditions hold:*

Property 1: The flow f^+ with $f^+(e) = f(e)$ for all $e \in E$ and $f^+(e_0^+) = 0$ is stable in $(V, E \cup \{e_0^+\})$.
Property 2: The flow f^- with $f^-(e) = f(e)$ for all $e \in E$ and $f^-(e_0^-) = 0$ is stable in $(V, E \cup \{e_0^-\})$.

Proof. Sufficiency of any of the two properties follows immediately from Observation 1 by deletion of e_0^+ or e_0^-, respectively.

To see necessity, assume that f is a stable flow in D. By contradiction assume that neither f^+ nor f^- is stable. Then there is a blocking walk W^+ for f^+ and a blocking walk W^- for f^-. Since W^+ is not a blocking walk for f in D, it must start with e_0^+. Since W^- is not a blocking walk for f in D, it must end with e_0^-. Let $W'^+ := W^+ \setminus \{e_0^+\}$ and $W'^- := W^- \setminus \{e_0^-\}$. Consider the concatenation $W := W'^- \circ e_0 \circ W'^+$. Note that W is an unsaturated walk in D. If $W'^- \neq \emptyset$, then W starts with the same edge as W^- and thus dominates f at the start. If $W'^- = \emptyset$, then W starts with e_0, which dominates any flow-carrying edge dominated by ut, and hence it dominates f at the start also in this case. By analogous arguments it follows that W also dominates f at the end. Hence W is a blocking walk, contradicting the stability of f. We conclude that at least one of Properties 1 or 2 must be true if f is stable. □

General case. The method described above can be used to solve SF FORBIDDEN if $|P| = 1$: We simply compute stable flows f^+ in D^+ and f^- in D^-. If $f^+(e^+) = 0$ or $f^-(e^-) = 0$, we have found a stable flow in f avoiding the forbidden edge e_0. More generally, for $|P| > 1$, Lemma 2 guarantees that we can add either e^+ or e^- for each forbidden edge $e \in P$ without destroying any stable flow avoiding the forbidden edges. However, it is not straightforward to decide for which forbidden edges to add e^+ and for which to add e^-. Simply applying our method greedily for each forbidden edge does not lead to correct results, since the steps can impact each other, as shown in the full version of the paper [7]. In the following, we describe how to resolve this issue and obtain a polynomial time algorithm for SF FORBIDDEN.

For any $A, \subseteq E$, let us denote by $D[A|B]$ the network with vertices V and edges $E \cup E^+(A) \cup E^-(B)$. Our algorithm maintains a partition of the forbidden edges in two groups P^+ and P^-. Initially $P^+ = P$ and $P^- = \emptyset$. In every iteration, we compute a stable flow f in $D[P^+|P^-]$. If $f(e^+) > 0$ for some $e \in P^+$, we move e from P^+ to P^- and repeat. If $f(e^+) = 0$ for all $e \in P^+$ but $f(e^-) > 0$ for some $e \in P^-$, we will show that no stable flow avoiding all forbidden edges exists in D. Finally, if we reach a flow f where neither of these two things happens, f is a stable flow in D avoiding all forbidden edges.

For the analysis of Algorithm 2, the following consequence of the augmenting path algorithm presented earlier (Algorithm 1) is helpful. It allows us to prove an important invariant of the Algorithm 2.

Algorithm 2. Stable flow with forbidden edges

Initialize $P^+ = P$ and $P^- = \emptyset$.

repeat

  | Compute a stable flow f in $D[P^+|P^-]$.

  | **if** $\exists\, e \in P^+$ *with* $f(e^+) > 0$ **then**

  | $\lfloor$ $P^+ := P^+ \setminus \{e\}$ and $P^- := P^- \cup \{e\}$

until $f(e^+) = 0$ *for all* $e \in P^+$;

if $\exists\, e \in P^-$ *with* $f(e^-) > 0$ **then**

  $\lfloor$ **return** $\emptyset$

else

  $\lfloor$ **return** f

Lemma 3. *Let f be a stable flow in D. Let f' be a stable flow in $D' = D - e'$ for some edge $e' \in \delta^+(s)$. Then $f'(e) \geq f(e)$ for all $e \in \delta^+(s) \setminus \{e'\}$.*

Lemma 4. *Algorithm 2 maintains the following invariant: There is a stable flow in D avoiding P iff there is a stable flow in $D[\emptyset|P^-]$ avoiding $P^+ \cup E^-(P^-)$.*

The correctness of Algorithm 2 follows immediately from the above invariant. The running time of this algorithm is bounded by $\mathcal{O}(|P||E|\log|V|)$, as each stable flow f can be computed in $\mathcal{O}(|E|\log|V|)$ time and in each round either $|P^+|$ decreases by one or the algorithm terminates. Note that both forced and forbidden edges in the same instance can be handled by our two algorithms, applying them one after the other. Finally, we can conclude the following result:

Theorem 3. SF RESTRICTED *can be solved in $\mathcal{O}(|P||E|\log|V|)$ time.*

Stable matchings with forbidden edges. As before, we finish this part with the direct interpretation of our results in SR and SM instances. To each forbidden edge $uw \in P$ we introduce edges uw_s or u_tw. According to the preference lists, they are slightly better than uw itself. A stable matching with forbidden edges exists, if there is a suitable set of these uw_s and u_tw edges such that all w_s and u_t are unmatched. Our algorithm for several forbidden edges runs in $\mathcal{O}(|P||E|)$ time, because computing stable solutions in each of the $|P|$ or less rounds takes only $\mathcal{O}(|E|)$ time in SM. With this running time, it is somewhat slower than the best known method [10] that requires only $\mathcal{O}(|E|)$ time, but it is a reasonable assumption that the number of forbidden edges is small.

5   Stable Multicommodity Flows

A *multicommodity network* $(D, c^i, c), 1 \leq i \leq n$ consists of a directed graph $D = (V, E)$, non-negative commodity capacity functions $c^i : E \rightarrow \mathbb{R}_{\geq 0}$ for all the n commodities and a non-negative cumulative capacity function $c : E \rightarrow \mathbb{R}_{\geq 0}$ on E. For every commodity i, there is a *source* $s^i \in V$ and a *sink* $t^i \in V$, also referred to as the *terminals of commodity i*.

Definition 3 (multicommodity flow). *A set of functions* $f^i : E \to \mathbb{R}_{\geq 0}$, $1 \leq i \leq n$ *is a* multicommodity flow *if it fulfills all of the following requirements:*

1. *capacity constraints for commodities:*
   $f^i(uv) \leq c^i(uv)$ *for all* $uv \in E$ *and commodity* i;
2. *cumulative capacity constraints:*
   $f(uv) = \sum_{1 \leq i \leq n} f^i(uv) \leq c(uv)$ *for all* $uv \in E$;
3. *flow conservation:*
   $\sum_{uv \in E} f^i(uv) = \sum_{vw \in E} f^i(vw)$ *for all* $v \in V \setminus \{s^i, t^i\}$.

The concept of stability was extended to multicommodity flows by Király and Pap [24]. A stable multicommodity flow instance $\mathcal{I} = (D, c^i, c, O_E, O_V^i), 1 \leq i \leq n$ comprises a network $(D, c^i, c), 1 \leq i \leq n$, *edge preferences* O_E over commodities, and *vertex preferences* $O_V^i, 1 \leq i \leq n$ over incident edges for commodity i. Each edge uv ranks all commodities in a strict order of preference. Separately for every commodity i, each non-terminal vertex ranks its incoming and also its outgoing edges strictly with respect to commodity i. Note that these preference orderings of v can be different for different commodities and they do not depend on the edge preferences O_E over the commodities. If edge uv prefers commodity i to commodity j, then we write $\mathrm{r}_{uv}(i) < \mathrm{r}_{uv}(j)$. Analogously, if vertex v prefers edge vw to vz with respect to commodity i, then we write $\mathrm{r}_v^i(vw) < \mathrm{r}_v^i(vz)$. We denote the flow value with respect to commodity i by $f^i = \sum_{u \in V} f^i(s^i u)$.

Definition 4 (stable multicommodity flow). *A directed walk* $W = \langle v_1, v_2, ..., v_k \rangle$ blocks *flow* f *with respect to commodity* i *if all of the following properties hold:*

1. $f^i(v_j v_{j+1}) < c^i(v_j v_{j+1})$ *for each edge* $v_j v_{j+1}$, $j = 1, ..., k-1$;
2. $v_1 = s^i$ *or there is an edge* $v_1 u$ *such that* $f^i(v_1 u) > 0$ *and* $\mathrm{r}_{v_1}^i(v_1 v_2) < \mathrm{r}_{v_1}^i(v_1 u)$;
3. $v_k = t^i$ *or there is an edge* wv_k *such that* $f^i(wv_k) > 0$ *and* $\mathrm{r}_{v_k}^i(v_{k-1} v_k) < \mathrm{r}_{v_k}^i(wv_k)$;
4. *if* $f(v_j v_{j+1}) = c(v_j v_{j+1})$, *then there is a commodity* i' *such that* $f^{i'}(v_j v_{j+1}) > 0$ *and* $\mathrm{r}_{v_j v_{j+1}}(i) < \mathrm{r}_{v_j v_{j+1}}(i')$.

A multicommodity flow is stable, *if there is no blocking walk with respect to any commodity.*

Note that due to point 4, this definition allows saturated edges to occur in a blocking walk with respect to commodity i, provided that these edges are inclined to trade in some of their forwarded commodities for more flow of commodity i. On the other hand, the role of edge preferences is limited: blocking walks still must start at vertices who are willing to reroute or send extra flow along the first edge of the walk according to their vertex preferences with respect to commodity i.

Problem 2. SMF
Input: $\mathcal{I} = (D, c^i, c, O_E, O_V^i)$, $1 \leq i \leq n$; a directed multicommodity network (D, c^i, c), $1 \leq i \leq n$, edge preferences over commodities O_E and vertex preferences over incident edges $O_V^i, 1 \leq i \leq n$.
Question: Is there a multicommodity flow f so that no walk blocks f with respect to any commodity?

Theorem 4 (Király, Pap [24]**).** *A stable multicommodity flow exists for any instance, but it is* PPAD-*hard to find.*

PPAD-hardness is a somewhat weaker evidence of intractability than NP-hardness [27]. Király and Pap use a polyhedral version of Sperner's lemma [23] to prove this existence result. Note that SMF is one of the very few problems in stability [3] where a stable solution exists, but no extension of the Gale-Shapley algorithm is known to solve it – not even a variant with exponential running time.

5.1 Problem Simplification

The definition of SMF given above involves many distinct components and constraints. It is natural to investigate how far the model can be simplified without losing any of its generality. It turns out that the majority of the commodity-specific input data can be dropped, as shown by Theorem 5, which we prove in the full version of the paper [7]. This result not only simplifies the instance, but it also sheds light to the most important characteristic of the problem, which seems to be the preference ordering of edges over commodities.

Theorem 5. *There is a polynomial-time transformation that, given an instance $\mathcal{I}$ of* SMF*, constructs an instance $\mathcal{I}'$ of* SMF *with the following properties:*

1. *all commodities have the same source and sink,*
2. *at each vertex, the preference lists are identical for all commodities,*
3. *there are no commodity-specific edge capacities,*

and there is a polynomially computable bijection between the stable multicommodity flows of $\mathcal{I}$ and the stable multicommodity flows of $\mathcal{I}'$.

5.2 Integral Multicommodity Stable Flows

Finally, we discuss the problem of integer stable multicommodity flows, introduced in [24]. The proof of our result can be found in the full version [7].

Theorem 6. *It is* NP-*complete to decide whether there is an integer stable multicommodity flow in a given network. This holds even if all commodities share the same set of terminal vertices and all vertices have the same preferences with respect to all commodities (but edges might have different capacities with respect to different commodities).*

Acknowledgment. We thank Tamás Fleiner for discussions on Lemma 1.

References

1. Baïou, M., Balinski, M.: Many-to-many matching: stable polyandrous polygamy (or polygamous polyandry). Discrete Appl. Math. **101**, 1–12 (2000)

2. Balinski, M., Sönmez, T.: A tale of two mechanisms: student placement. J. Econ. Theory **84**, 73–94 (1999)
3. Biró, P., Kern, W., Paulusma, D., Wojuteczky, P.: The stable fixtures problem with payments. Games Econ. Behav. (2017). doi:10.1016/j.geb.2017.02.002. ISSN 0899-8256
4. Braun, S., Dwenger, N., Kübler, D.: Telling the truth may not pay off: an empirical study of centralized university admissions in Germany. B.E. J. Econ. Anal. Policy **10**, article 22 (2010)
5. Chen, Y., Sönmez, T.: Improving efficiency of on-campus housing: an experimental study. Am. Econ. Rev. **92**, 1669–1686 (2002)
6. Cseh, Á., Manlove, D.F.: Stable marriage and roommates problems with restricted edges: complexity and approximability. Discrete Optim. **20**, 62–89 (2016)
7. Cseh, Á., Matuschke, J.: New and simple algorithms for stable flow problems. *CoRR*, abs/1309.3701 (2017)
8. Cseh, Á., Matuschke, J., Skutella, M.: Stable flows over time. Algorithms **6**, 532–545 (2013)
9. Dean, B.C., Munshi, S.: Faster algorithms for stable allocation problems. Algorithmica **58**, 59–81 (2010)
10. Dias, V.M.F., da Fonseca, G.D., de Figueiredo, C.M.H., Szwarcfiter, J.L.: The stable marriage problem with restricted pairs. Theoret. Comput. Sci. **306**, 391–405 (2003)
11. Feder, T.: A new fixed point approach for stable networks and stable marriages. J. Comput. Syst. Sci. **45**, 233–284 (1992)
12. Feder, T.: Network flow and 2-satisfiability. Algorithmica **11**, 291–319 (1994)
13. Fleiner, T.: On stable matchings and flows. Algorithms **7**, 1–14 (2014)
14. Fleiner, T., Irving, R.W., Manlove, D.F.: Efficient algorithms for generalised stable marriage and roommates problems. Theoret. Comput. Sci. **381**, 162–176 (2007)
15. Ford, L.R., Fulkerson, D.R.: Flows in Networks. Princeton University Press, Princeton (1962)
16. Gai, A.-T., Lebedev, D., Mathieu, F., de Montgolfier, F., Reynier, J., Viennot, L.: Acyclic preference systems in P2P networks. In: Kermarrec, A.-M., Bougé, L., Priol, T. (eds.) Euro-Par 2007. LNCS, vol. 4641, pp. 825–834. Springer, Heidelberg (2007). doi:10.1007/978-3-540-74466-5_88
17. Gale, D., Shapley, L.S.: College admissions and the stability of marriage. Am. Math. Mon. **69**, 9–15 (1962)
18. Gale, D., Sotomayor, M.: Some remarks on the stable matching problem. Discrete Appl. Math. **11**, 223–232 (1985)
19. Garey, M.R., Johnson, D.S.: Computers and Intractability. Freeman, San Francisco (1979)
20. Gusfield, D., Irving, R.W.: The Stable Marriage Problem: Structure and Algorithms. MIT Press, Cambridge (1989)
21. Irving, R.W., Leather, P., Gusfield, D.: An efficient algorithm for the "optimal" stable marriage. J. ACM **34**, 532–543 (1987)
22. Jewell, W.S.: Multi-commodity network solutions. Operations Research Center, University of California (1966)
23. Király, T., Pap, J.: A note on kernels and Sperner's Lemma. Discrete Appl. Math. **157**, 3327–3331 (2009)
24. Király, T., Pap, J.: Stable multicommodity flows. Algorithms **6**, 161–168 (2013)
25. Knuth, D.: Mariages Stables. Les Presses de L'Université de Montréal (1976). English translation in Stable Marriage and its Relation to Other Combinatorial Problems. CRM Proceedings and Lecture Notes, vol. 10. American Mathematical Society (1997)

26. Ostrovsky, M.: Stability in supply chain networks. Am. Econ. Rev. **98**, 897–923 (2008)
27. Papadimitriou, C.H.: On the complexity of the parity argument and other inefficient proofs of existence. J. Comput. Syst. Sci. **48**, 498–532 (1994)
28. Perach, N., Polak, J., Rothblum, U.G.: A stable matching model with an entrance criterion applied to the assignment of students to dormitories at the Technion. Int. J. Game Theory **36**, 519–535 (2008)
29. Roth, A.E.: The evolution of the labor market for medical interns and residents: a case study in game theory. J. Polit. Econ. **92**, 991–1016 (1984)
30. Roth, A.E., Sotomayor, M.A.O.: Two-Sided Matching: A Study in Game-Theoretic Modeling and Analysis. Econometric Society Monographs, vol. 18. Cambridge University Press (1990)
31. Tardos, É.: A strongly polynomial algorithm to solve combinatorial linear programs. Oper. Res. **34**(2), 250–256 (1986)

Clique-Width and Well-Quasi-Ordering
of Triangle-Free Graph Classes

Konrad K. Dabrowski[1]([✉]), Vadim V. Lozin[2], and Daniël Paulusma[1]

[1] Department of Computer Science, Durham University,
South Road, Durham DH1 3LE, UK
{konrad.dabrowski,daniel.paulusma}@durham.ac.uk
[2] Mathematics Institute, University of Warwick, Coventry CV4 7AL, UK
v.lozin@warwick.ac.uk

Abstract. Daligault, Rao and Thomassé asked whether every hereditary graph class that is well-quasi-ordered by the induced subgraph relation has bounded clique-width. Lozin, Razgon and Zamaraev (WG 2015) gave a negative answer to this question, but their counterexample is a class that can only be characterised by infinitely many forbidden induced subgraphs. This raises the issue of whether their question has a positive answer for finitely defined hereditary graph classes. Apart from two stubborn cases, this has been confirmed when at most two induced subgraphs H_1, H_2 are forbidden. We confirm it for one of the two stubborn cases, namely for the case $(H_1, H_2) = $ (triangle, $P_2 + P_4$) by proving that the class of (triangle, $P_2 + P_4$)-free graphs has bounded clique-width and is well-quasi-ordered. Our technique is based on a special decomposition of 3-partite graphs. We also use this technique to completely determine which classes of (triangle, H)-free graphs are well-quasi-ordered.

1 Introduction

A graph class $\mathcal{G}$ is well-quasi-ordered by some containment relation if for any infinite sequence $G_0, G_1, \ldots$ of graphs in $\mathcal{G}$, there is a pair i, j with $i < j$ such that G_i is contained in G_j. A graph class $\mathcal{G}$ has bounded clique-width if there exists a constant c such that every graph in $\mathcal{G}$ has clique-width at most c. Both being well-quasi-ordered and having bounded clique-width are highly desirable properties of graph classes in the area of theoretical computer science. To illustrate this, let us mention the seminal project of Robertson and Seymour on graph minors that culminated in 2004 in the proof of Wagner's conjecture, which states that the set of all finite graphs is well-quasi-ordered by the minor relation. As an algorithmic consequence, given a minor-closed graph class, it is possible to test in cubic time whether a given graph belongs to this class. The algorithmic importance of having bounded clique-width follows from the fact that many well-known NP-hard problems, such as GRAPH COLOURING and HAMILTON CYCLE, become

This paper received support from EPSRC (EP/K025090/1 and EP/L020408/1) and Leverhulme Trust RPG-2016-258.

H.L. Bodlaender and G.J. Woeginger (Eds.): WG 2017, LNCS 10520, pp. 220–233, 2017.
https://doi.org/10.1007/978-3-319-68705-6_17

polynomial-time solvable for graph classes of bounded clique-width (this follows from combining results from several papers [4,13,16,22] with a result of Oum and Seymour [21]).

Courcelle [3] proved that the class of graphs obtained from graphs of clique-width 3 via one or more edge contractions has unbounded clique-width. Hence the clique-width of a graph can be much smaller than the clique-width of its minors. On the other hand, the clique-width of a graph is at least the clique-width of any of its induced subgraphs (see, for example, [5]). We therefore focus on *hereditary* classes, that is, on graph classes that are closed under taking induced subgraphs. Our goal is to increase our understanding of the relation between well-quasi-orders and clique-width of hereditary graph classes.

It is readily seen that a class of graphs is hereditary if and only if it can be characterised by a unique set $\mathcal{F}$ of minimal forbidden induced subgraphs, which due to their minimality form an antichain, that is, no graph in $\mathcal{F}$ is an induced subgraph of another graph in $\mathcal{F}$. Note that the class of cycles is not well-quasi-ordered by the induced subgraph relation. As every cycle has clique-width at most 4, having bounded clique-width does not imply being well-quasi-ordered by the induced subgraph relation. In 2010, Daligault et al. [10] asked about the reverse implication: does every hereditary graph class that is well-quasi-ordered by the induced subgraph relation have bounded clique-width? In 2015, Lozin et al. [20] gave a negative answer. As the set $\mathcal{F}$ in their counter-example is infinite, the question of Daligault et al. [10] remains open for *finitely defined* hereditary graph classes, that is, hereditary graph classes for which $\mathcal{F}$ is finite.

Conjecture 1 [20]. If a finitely defined hereditary class of graphs $\mathcal{G}$ is well-quasi-ordered by the induced subgraph relation, then $\mathcal{G}$ has bounded clique-width.

If Conjecture 1 is true, then for finitely defined hereditary graph classes the aforementioned algorithmic consequences of having bounded clique-width also hold for the property of being well-quasi-ordered by the induced subgraph relation. A hereditary graph class defined by a single forbidden induced subgraph H is well-quasi-ordered by the induced subgraph relation if and only if it has bounded clique-width if and only if H is an induced subgraph of P_4 (see, for instance, [9,11,18]). Hence Conjecture 1 holds when $\mathcal{F}$ has size 1. We consider the case when $\mathcal{F}$ has size 2, say $\mathcal{F} = \{H_1, H_2\}$. Such graph classes are called *bigenic* or (H_1, H_2)-*free* graph classes. In this case Conjecture 1 is also known to be true except for two stubborn open cases, namely $(H_1, H_2) = (K_3, P_2 + P_4)$ and $(H_1, H_2) = (\overline{P_1 + P_4}, P_2 + P_3)$; see [7].

Our Results. We prove that the class of $(K_3, P_2 + P_4)$-free graphs has bounded clique-width and is well-quasi-ordered by the induced subgraph relation. We do this by using a general technique explained in Sect. 3. This technique is based on extending (a labelled version of) well-quasi-orderability or boundedness of clique-width of the bipartite graphs in a hereditary graph class X to a special subclass of 3-partite graphs in X. The crucial property of these 3-partite graphs is that no three vertices from the three different partition classes form a clique or independent set. We call such 3-partite graphs *curious*. A more restricted version of this concept was used to prove that $(K_3, P_1 + P_5)$-free graphs have bounded clique-width [6]. In Sect. 4 we show how to generalise results for curious $(K_3, P_2 + P_4)$-free graphs to the whole class of $(K_3, P_2 + P_4)$-free graphs and that our technique can also be applied to prove that $(K_3, P_1 + P_5)$-free graphs are well-quasi-ordered.

Consequences of Our Results. Previously, well-quasi-orderability was known for (K_3, P_6)-free graphs [1], $(P_2 + P_4)$-free bipartite graphs [17] and $(P_1 + P_5)$-free bipartite graphs [17]. It has also been shown that H-free bipartite graphs are not well-quasi-ordered if H contains an induced $3P_1 + P_2$ [18], $3P_2$ [12] or $2P_3$ [17]. This leads to the following dichotomy.

Theorem 1. *Let H be a graph. The class of H-free bipartite graphs is well-quasi-ordered by the induced subgraph relation if and only if $H = sP_1$ for some $s \geq 1$ or H is an induced subgraph of $P_1 + P_5$, $P_2 + P_4$ or P_6.*

Now combining the aforementioned known results for (K_3, H)-free graphs and H-free bipartite graphs with our new results yields exactly the same dichotomy for (K_3, H)-free graphs as the one in Theorem 1.

Theorem 2. *Let H be a graph. The class of (K_3, H)-free graphs is well-quasi-ordered by the induced subgraph relation if and only if $H = sP_1$ for some $s \geq 1$ or H is an induced subgraph of $P_1 + P_5$, $P_2 + P_4$, or P_6.*

Future Work. The class of $(\overline{P_1 + P_4}, P_2 + P_3)$-free graphs is the only bigenic graph class left for which Conjecture 1 still needs to be verified. After updating the summaries in [7] with our new results, this class is also one of the six remaining bigenic graph classes for which well-quasi-orderability is still open. And it is one of the six remaining bigenic graph classes for which we do not know if their clique-width is bounded [2]. Hence, a new approach is required to solve this case.

   Besides our technique based on curious graphs, we also expect that Theorem 2 will itself be a useful ingredient for showing results for other graph classes, just as Theorem 1 has already proven to be useful (see e.g. [17]).

   For clique-width the following dichotomy is known for H-free bipartite graphs.

Theorem 3 [8]. *Let H be a graph. The class of H-free bipartite graphs has bounded clique-width if and only if $H = sP_1$ for some $s \geq 1$ or H is an induced subgraph of $K_{1,3} + 3P_1$, $K_{1,3} + P_2$, $P_1 + S_{1,1,3}$ or $S_{1,2,3}$.*

It would be interesting to determine whether (K_3, H)-free graphs allow the same dichotomy with respect to the boundedness of their clique-width. The evidence so far is affirmative, but in order to answer this question two remaining cases need to be solved, namely $(H_1, H_2) = (K_3, P_1 + S_{1,1,3})$ and $(H_1, H_2) = (K_3, S_{1,2,3})$; see Sect. 2 for the definition of the graph $S_{h,i,j}$. Both cases turn out to be highly non-trivial; in particular, the class of $(K_3, P_1 + S_{1,1,3})$-free graphs contains the class of $(K_3, P_1 + P_5)$-free graphs, and the class of $(K_3, S_{1,2,3})$-free graphs contains both the classes of $(K_3, P_1 + P_5)$-free and $(K_3, P_2 + P_4)$-free graphs.

2   Preliminaries

We consider only finite, undirected graphs without multiple edges or self-loops. The *disjoint union* $(V(G) \cup V(H), E(G) \cup E(H))$ of two vertex-disjoint graphs G and H is denoted by $G + H$ and the disjoint union of r copies of a graph G is denoted by rG. The *complement* $\overline{G}$ of a graph G has vertex set $V(\overline{G}) = V(G)$ and an edge between two distinct vertices u, v if and only if $uv \notin E(G)$. For a subset $S \subseteq V(G)$, we let $G[S]$ denote the subgraph of G *induced* by S. If $S = \{s_1, \ldots, s_r\}$, we may also write $G[s_1, \ldots, s_r]$. We write $G' \subseteq_i G$ to indicate that G' is an induced subgraph of G.

The graphs C_r, K_r, $K_{1,r-1}$ and P_r denote the cycle, complete graph, star and path on r vertices, respectively. The graphs K_3 and $K_{1,3}$ are also called the *triangle* and *claw*, respectively. The graph $S_{h,i,j}$, for $1 \le h \le i \le j$, denotes the *subdivided claw*, that is, the tree that has only one vertex x of degree 3 and exactly three leaves, which are of distance h, i and j from x, respectively. Observe that $S_{1,1,1} = K_{1,3}$. We let $\mathcal{S}$ denote the class of graphs, each connected component of which is either a subdivided claw or a path. For a set of graphs $\{H_1, \ldots, H_p\}$, a graph G is $(H_1, \ldots, H_p)$-*free* if it has no induced subgraph isomorphic to a graph in $\{H_1, \ldots, H_p\}$; if $p = 1$, we may write H_1-free instead of (H_1)-free.

For a graph $G = (V, E)$, the set $N(u) = \{v \in V \mid uv \in E\}$ denotes the *neighbourhood* of $u \in V$. A graph is k-*partite* if its vertex can be partitioned into k (possibly empty) independent sets; 2-partite graphs are also known as *bipartite* graphs.

Let X be a set of vertices in a graph $G = (V, E)$. A vertex $y \in V \setminus X$ is *complete* to X if it is adjacent to every vertex of X and *anti-complete* to X if it is adjacent to no vertex of X. A set of vertices $Y \subseteq V \setminus X$ is *complete* (resp. *anti-complete*) to X if every vertex in Y is complete (resp. anti-complete) to X. A vertex $y \in V \setminus X$ *distinguishes* X if y has both a neighbour and a non-neighbour in X. The set X is a *module* of G if no vertex in $V \setminus X$ distinguishes X. A module X is *non-trivial* if $1 < |X| < |V|$, otherwise it is *trivial*. A graph is *prime* if it has only trivial modules. Two (non-adjacent) vertices are *false twins* if they share the same neighbours. Prime graphs on at least three vertices contain no false twins, as any such pair of vertices would form a non-trivial module.

The *clique-width* $\mathrm{cw}(G)$ of a graph G is the minimum number of labels needed to construct G by using the following four operations:

1. $i(v)$: creating a new graph consisting of a single vertex v with label i;
2. $G_1 \oplus G_2$: taking the disjoint union of two labelled graphs G_1 and G_2;
3. $\eta_{i,j}$: joining each vertex with label i to each vertex with label j $(i \ne j)$;
4. $\rho_{i \to j}$: renaming label i to j.

A class of graphs $\mathcal{G}$ has *bounded* clique-width if there is a constant c such that the clique-width of every graph in $\mathcal{G}$ is at most c; otherwise the clique-width is *unbounded*. For an induced subgraph G' of a graph G, the *subgraph complementation* operation replaces every edge present in G' by a non-edge, and vice versa. For two disjoint vertex subsets S and T in G, the *bipartite complementation* operation replaces every edge with one end-vertex in S and the other one in T

by a non-edge and vice versa. Let $k \geq 0$ be a constant and let γ be some graph operation. A class $\mathcal{G}'$ is (k, γ)-*obtained* from a class $\mathcal{G}$ if:

1. every graph in $\mathcal{G}'$ is obtained from a graph in $\mathcal{G}$ by performing γ at most k times, and
2. for every $G \in \mathcal{G}$ there exists at least one graph in $\mathcal{G}'$ obtained from G by performing γ at most k times.

We say that γ *preserves* boundedness of clique-width if for any finite constant k and any graph class $\mathcal{G}$, any graph class $\mathcal{G}'$ that is (k, γ)-obtained from $\mathcal{G}$ has bounded clique-width if and only if $\mathcal{G}$ has bounded clique-width.

Fact 1. Vertex deletion preserves boundedness of clique-width [19].

Fact 2. Subgraph complementation preserves boundedness of clique-width [15].

Fact 3. Bipartite complementation preserves boundedness of clique-width [15].

Lemma 1 [5]. *Let G be a graph and let $\mathcal{P}$ be the set of all induced subgraphs of G that are prime. Then $\mathrm{cw}(G) = \max_{H \in \mathcal{P}} \mathrm{cw}(H)$.*

Lemma 2 [6]. *Let G be a connected $(K_3, C_5, S_{1,2,3})$-free graph that does not contain a pair of false twins. Then G is either bipartite or a cycle.*

A *quasi order* $\leq$ on a set X is a reflexive, transitive binary relation. Two elements $x, y \in X$ in this quasi-order are *comparable* if $x \leq y$ or $y \leq x$, otherwise they are *incomparable*. A set of elements in a quasi-order is a *chain* if every pair of elements is comparable and it is an *antichain* if every pair of elements is incomparable. The quasi-order $\leq$ is a *well-quasi-order* if any infinite sequence of elements $x_1, x_2, x_3, \ldots$ in X contains a pair (x_i, x_j) with $x_i \leq x_j$ and $i < j$. Equivalently, a quasi-order is a well-quasi-order if and only if it has no infinite strictly decreasing sequence $x_1 \geq x_2 \geq x_3 \geq \cdots$ and no infinite antichain. For an arbitrary set M, let M^* denote the set of finite sequences of elements of M. A quasi-order $\leq$ on M defines a quasi-order $\leq^*$ on M^* as follows: $(a_1, \ldots, a_m) \leq^* (b_1, \ldots, b_n)$ if and only if there is a sequence of integers $i_1, \ldots, i_m$ with $1 \leq i_1 < \cdots < i_m \leq n$ such that $a_j \leq b_{i_j}$ for $j \in \{1, \ldots, m\}$. We call $\leq^*$ the *subsequence relation*.

Lemma 3 (Higman's Lemma [14]). *If $(M, \leq)$ is a well-quasi-order then $(M^*, \leq^*)$ is a well-quasi-order.*

For a quasi-order $(W, \leq)$, a graph G is a *labelled* graph if each vertex v of G is equipped with an element $l_G(v) \in W$ (the *label* of v). Given two labelled graphs G and H, we say that G is a *labelled induced subgraph* of H if G is isomorphic to an induced subgraph of H and there is an isomorphism that maps each vertex v of G to a vertex w of H with $l_G(v) \leq l_H(w)$. Clearly, if $(W, \leq)$ is a well-quasi-order, then a class of graphs X cannot contain an infinite sequence of labelled graphs that is strictly-decreasing with respect to the labelled induced subgraph relation. We therefore say that a graph class X is well-quasi-ordered by the *labelled* induced subgraph relation if it contains no infinite antichains of labelled graphs whenever $(W, \leq)$ is a *well*-quasi-order. Such a class is readily seen to also be well-quasi-ordered by the induced subgraph relation. Similarly to the notion of preserving boundedness of clique-width, we say that a graph operation γ *preserves* well-quasi-orderability by the labelled induced subgraph relation if for any finite constant k and any graph class $\mathcal{G}$, any graph class $\mathcal{G}'$ that is (k, γ)-obtained from $\mathcal{G}$ is well-quasi-ordered by this relation if and only if $\mathcal{G}$ is.

Lemma 4 [7]. *Subgraph and bipartite complementations and vertex deletion preserve well-quasi-orderability by the labelled induced subgraph relation.*

Lemma 5 [1]. *A hereditary class X of graphs is well-quasi-ordered by the labelled induced subgraph relation if and only if the set of prime graphs in X is. In particular, X is well-quasi-ordered by the labelled induced subgraph relation if and only if the set of connected graphs in X is.*

Lemma 6 [1,17]. *$(P_7, S_{1,2,3})$-free bipartite graphs are well-quasi-ordered by the labelled induced subgraph relation.*

Let $(L_1, \leq_1)$ and $(L_2, \leq_2)$ be well-quasi-orders. We define the *Cartesian Product* $(L_1, \leq_1) \times (L_2, \leq_2)$ of these well-quasi-orders as the order $(L, \leq_L)$ on the set $L := L_1 \times L_2$ where $(l_1, l_2) \leq_L (l'_1, l'_2)$ if and only if $l_1 \leq_1 l'_1$ and $l_2 \leq_2 l'_2$. Lemma 3 implies that $(L, \leq_L)$ is also a well-quasi-order. If G has a labelling with elements of L_1 and of L_2, say $l_1 : V(G) \to L_1$ and $l_2 : V(G) \to L_2$, we can construct the *combined labelling* in $(L_1, \leq_1) \times (L_2, \leq_2)$ that labels each vertex v of G with the label $(l_1(v), l_2(v))$. We omit the proof of the next lemma.

Lemma 7. *Fix a well-quasi-order $(L_1, \leq_1)$ that has at least one element. Let X be a class of graphs. For each $G \in X$ fix a labelling $l_G^1 : V(G) \to L_1$. Then X is well-quasi-ordered by the labelled induced subgraph relation if and only if for every well-quasi-order $(L_2, \leq_2)$ and every labelling of the graphs in X by this order, the combined labelling in $(L_1, \leq_1) \times (L_2, \leq_2)$ obtained from these labellings also results in a well-quasi-ordered set of labelled graphs.*

For an integer $k \geq 1$, a graph G is *k-uniform* if there is a symmetric square $0, 1$ matrix K of order k and a graph F_k on vertices $1, 2, \ldots, k$ such that $G \in \mathcal{P}(K, F_k)$, where $\mathcal{P}(K, F_k)$ is a graph class defined as follows. Let H be the disjoint union of infinitely many copies of F_k. For $i = 1, \ldots, k$, let V_i be the subset of $V(H)$ containing vertex i from each copy of F_k. Construct from H an infinite graph $H(K)$ on the same vertex set by applying a subgraph complementation to V_i if and only if $K(i, i) = 1$ and by applying a bipartite complementation to a pair V_i, V_j if and only if $K(i, j) = 1$. Thus, two vertices $u \in V_i$ and $v \in V_j$ are adjacent in $H(K)$ if and only if $uv \in E(H)$ and $K(i, j) = 0$ or $uv \notin E(H)$ and $K(i, j) = 1$. Then, $\mathcal{P}(K, F_k)$ is the hereditary class consisting of all the finite induced subgraphs of $H(K)$. The minimum k such that G is k-uniform is the *uniformicity* of G. The second of the next two lemmas follows directly from the above definitions.

Lemma 8 [18]. *Any class of graphs of bounded uniformicity is well-quasi-ordered by the labelled induced subgraph relation.*

Lemma 9. *Every k-uniform graph has clique-width at most $2k$.*

3   Partitioning 3-Partite Graphs

Let G be a 3-partite graph given with a partition of its vertex set into three independent sets V_1, V_2 and V_3. Suppose each V_i can be partitioned into sets $V_i^0, \ldots, V_i^\ell$ such that, taking subscripts modulo 3: for $i \in \{1,2,3\}$ if $j < k$ then V_i^j is complete to V_{i+1}^k and anti-complete to V_{i+2}^k. For $i \in \{0,\ldots,\ell\}$ let $G^i = G[V_1^i \cup V_2^i \cup V_3^i]$. Then the graphs G^i are the *slices* of G. If the slices belong to some class X, then G can be *partitioned into slices from X*; see Fig. 1 for an example.

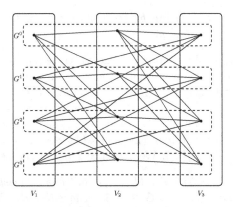

Fig. 1. A 3-partite graph that is partitioned into slices $G^0, \ldots, G^3$ isomorphic to P_3.

Lemma 10. *If G is a 3-partite graph that can be partitioned into slices of clique-width at most k then G has clique-width at most $\max(3k, 6)$.*

Proof. Since every slice G^j of G has clique-width at most k, it can be constructed using the labels $1, \ldots, k$. Applying relabelling operations if necessary, we may assume that at the end of this construction, every vertex receives the label 1. We can modify this construction so that we use the labels $1_1, \ldots, k_1, 1_2, \ldots, k_2, 1_3, \ldots, k_3$ instead, in such a way that at all points in the construction, for each $i \in \{1,2,3\}$ every constructed vertex in V_i has a label in $\{1_i, \ldots, k_i\}$. To do this we replace:

- creation operations $i(v)$ by $i_j(v)$ if $v \in V_j$,
- relabel operations $\rho_{j \to k}()$ by $\rho_{j_1 \to k_1}(\rho_{j_2 \to k_2}(\rho_{j_3 \to k_3}()))$ and
- join operations $\eta_{j,k}()$ by
$$\eta_{j_1,k_1}(\eta_{j_1,k_2}(\eta_{j_1,k_3}(\eta_{j_2,k_1}(\eta_{j_2,k_2}(\eta_{j_2,k_3}(\eta_{j_3,k_1}(\eta_{j_3,k_2}(\eta_{j_3,k_3}()))))))))).$$

This modified construction uses $3k$ labels and at the end of it, every vertex in V_i is labelled with label 1_i. We may do this for every slice G^j of G independently. We now show how to use these constructed slices to construct $G[V(G^0) \cup \cdots \cup V(G^j)]$ using six labels in such a way that every vertex in V_i is labelled with label 1_i. We do this by induction. If $j = 0$ then $G[V(G^0)] = G^0$, so we are done. If $j > 0$ then by the induction hypothesis, we can construct $G[V(G^0) \cup \cdots \cup V(G^{j-1})]$ in this way. Consider the copy of G^j constructed earlier and relabel its vertices using the operations $\rho_{1_1 \to 2_1}, \rho_{1_2 \to 2_2}$ and $\rho_{1_3 \to 2_3}$ so that in this copy of G^j, every vertex in V_i is labelled 2_i. Next take the disjoint union of the obtained graph with the constructed $G[V(G^0) \cup \cdots \cup V(G^{j-1})]$. Then, apply join operations $\eta_{1_1,2_2}, \eta_{1_2,2_3}$ and $\eta_{1_3,2_1}$. Finally, apply the relabelling operations $\rho_{2_1 \to 1_1}, \rho_{2_2 \to 1_2}$ and $\rho_{2_3 \to 1_3}$. This constructs $G[V(G^0) \cup \cdots \cup V(G^j)]$ in such a way that every vertex in V_i is labelled with 1_i. By induction, G has clique-width at most $\max(3k,6)$. □

Lemma 11. *Let X be a hereditary graph class containing a class Z. Let Y be the set of 3-partite graphs in X that can be partitioned into slices from Z. If Z is well-quasi-ordered by the labelled induced subgraph relation then so is Y.*

Proof. For each graph G in Y, we may fix a partition into independent sets (V_1, V_2, V_3) with respect to which the graph can be partitioned into slices from Z. Let $(L_1, \leq_1)$ be the well-quasi-order with $L_1 = \{1,2,3\}$ in which every pair of distinct elements is incomparable. By Lemma 7, we need only consider labellings of graphs in G of the form $(i, l(v))$ where $v \in V_i$ and $l(v)$ belongs to an arbitrary well-quasi-order L. Suppose G can be partitioned into slices $G^1, \ldots, G^k$, with vertices labelled as in G. The slices along with the labellings completely describe the edges in G. Suppose H is another such graph, partitioned into slices $H^1, \ldots, H^k$. If $(H^1, \ldots, H^\ell)$ is smaller than $(G^1, \ldots, G^k)$ under the subsequence relation, then H is an induced subgraph of G. The result follows by Lemma 3. □

We will now introduce curious graphs. Let G be a 3-partite graph given together with a partition of its vertex set into three independent sets V_1, V_2 and V_3. An induced K_3 or $3P_1$ in G is *rainbow* if it has exactly one vertex in each set V_i. We say that G is *curious with respect to the partition* (V_1, V_2, V_3) if it contains no rainbow K_3 or $3P_1$ when its vertex set is partitioned in this way. We say that G is *curious* if there is a partition (V_1, V_2, V_3) with respect to which G is curious. We will prove that given a hereditary class X, if the bipartite graphs in X are well-quasi-ordered by the labelled induced subgraph relation or have bounded clique-width, then the same is true for the curious graphs in X. A linear order $(x_1, x_2, \ldots, x_k)$ of the vertices of an independent set I is

- *increasing* if $i < j$ implies $N(x_i) \subseteq N(x_j)$,
- *decreasing* if $i < j$ implies $N(x_i) \supseteq N(x_j)$,
- *monotone* if it is either increasing or decreasing.

Bipartite graphs that are $2P_2$-free are also known as bipartite *chain* graphs. It is readily seen that a bipartite graph G is $2P_2$-free if and only if the vertices in each independent set of the bipartition admit a monotone ordering. Suppose G

is a curious graph with respect to some partition (V_1, V_2, V_3). We say that (with respect to this partition) the graph G is a curious graph of *type t* if exactly t of the graphs $G[V_1 \cup V_2]$, $G[V_1 \cup V_3]$ and $G[V_2 \cup V_3]$ contain an induced $2P_2$. If G is a curious graph of type 0 or 1 with respect to the partition (V_1, V_2, V_3) then without loss of generality, we may assume that $G[V_1 \cup V_2]$ and $G[V_1 \cup V_3]$ are both $2P_2$-free. We omit the proof of the next lemma.

Lemma 12. *Let G be a curious graph with respect to (V_1, V_2, V_3), such that $G[V_1 \cup V_2]$ and $G[V_1 \cup V_3]$ are both $2P_2$-free. Then the vertices of V_1 admit a linear ordering which is decreasing in $G[V_1 \cup V_2]$ and increasing in $G[V_1 \cup V_3]$.*

Lemma 13. *If G is a curious graph of type 0 or 1 with respect to a partition (V_1, V_2, V_3) then G can be partitioned into slices that are bipartite.*

Proof. Let $x_1, \ldots, x_\ell$ be a linear order on V_1 satisfying Lemma 12. Let $V_1^0 = \emptyset$ and for $i \in \{1, \ldots, \ell\}$, let $V_1^i = \{x_i\}$. We partition V_2 and V_3 as follows. For $i \in \{0, \ldots, \ell\}$, let $V_2^i = \{y \in V_2 \mid x_j y \in E(G)$ if and only if $j \leq i\}$. For $i \in \{0, \ldots, \ell\}$, let $V_3^i = \{z \in V_3 \mid x_j z \notin E(G)$ if and only if $j \leq i\}$. In particular, note that the vertices of $V_2^\ell \cup V_3^0$ and $V_2^0 \cup V_3^\ell$ are complete and anti-complete to V_1, respectively. The following properties hold: if $j < k$ then V_1^j is complete to V_2^k and anti-complete to V_3^k, and if $j > k$ then V_1^j is anti-complete to V_2^k and complete to V_3^k. If $j < k$ and $y \in V_2^j$ is non-adjacent to $z \in V_3^k$ then $G[x_k, y, z]$ is a rainbow $3P_1$, a contradiction. If $j > k$ and $y \in V_2^j$ is adjacent to $z \in V_3^k$ then $G[x_j, y, z]$ is a rainbow K_3, a contradiction. It follows that: if $j < k$ then V_2^j is complete to V_3^k and if $j > k$ then V_2^j is anti-complete to V_3^k.

For $i \in \{0, \ldots, \ell\}$, let $G^i = G[V_1^i \cup V_2^i \cup V_3^i]$. The above properties about the edges between the sets V_j^i show that G can be partitioned into the slices $G^0, \ldots, G^\ell$. Now, for each $i \in \{0, \ldots, \ell\}$, V_1^i is anti-complete to V_3^i, so every slice G^i is bipartite. This completes the proof. □

Lemma 14. *Fix $t \in \{2, 3\}$. If G is a curious graph of type t with respect to a partition (V_1, V_2, V_3) then G can be partitioned into slices of type at most $t - 1$.*

Proof Sketch. Fix $t \in \{2, 3\}$ and let G be a curious graph of type t with respect to a partition (V_1, V_2, V_3). We may assume that $G[V_1 \cup V_2]$ contains an induced $2P_2$.

We start with the following claim (we omit the proof).

Claim 1. Given a $2P_2$ in $G[V_1 \cup V_2]$, every vertex of V_3 has exactly two neighbours in the $2P_2$ and these neighbours either both lie in V_1 or both lie in V_2.

Consider a maximal set $\{H^1, \ldots, H^q\}$ of vertex-disjoint sets that induce copies of $2P_2$ in $G[V_1 \cup V_2]$. We say that a vertex of V_3 *distinguishes* two graphs $G[H^i]$ and $G[H^j]$ if its neighbours in H^i and H^j do not belong to the same set V_k. We group these sets H^i into *blocks* $B^1, \ldots, B^p$ that are not distinguished by any vertex of V_3. In other words, every B^i contains at least one $2P_2$ and every vertex of V_3 is complete to one of the sets $B^i \cap V_1$ and $B^i \cap V_2$ and anti-complete to the other. For $j \in \{1, 2\}$, let $B_j^i = B^i \cap V_j$. We define a relation $<_B$ on the blocks as follows: $B^i <_B B^j$ holds if B_1^i is complete to B_2^j, while B_2^i is anti-complete to B_1^j. For distinct blocks B^i, B^j at most one of $B^i <_B B^j$ and $B^j <_B B^i$ can hold.

We need the following two claims (we omit their proofs).

Claim 2. Let B^i and B^j be distinct blocks. There is a vertex $z \in V_3$ that differentiates B^i and B^j. If z is complete to $B_2^i \cup B_1^j$ and anti-complete to $B_1^i \cup B_2^j$ then $B^i <_B B^j$ (see also Fig. 2). If z is complete to $B_2^j \cup B_1^i$ and anti-complete to $B_1^j \cup B_2^i$ then $B^j <_B B^i$.

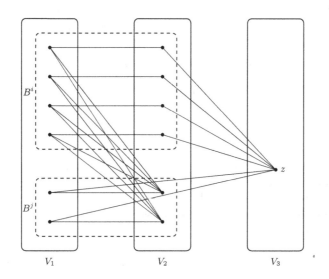

Fig. 2. Two blocks B^i and B^j with $B^i <_B B^j$ and a vertex $z \in V_3$ differentiating them.

Claim 3. The relation $<_B$ is transitive.

Combining Claims 1–3, we find that $<_B$ is a linear order on the blocks. We obtain the following conclusion, which we call the *chain property*.

Claim 4. The set of blocks has a linear order $B^1 <_B B^2 <_B \cdots <_B B^p$ so that

(i) if $i < j$ then B_1^i is complete to B_2^j, while B_2^i is anti-complete to B_1^j and
(ii) for each $z \in V_3$ there exists an $i \in \{0, \ldots, p\}$ such that if $j \le i$ then z is complete to B_2^j and anti-complete to B_1^j and if $j > i$ then z is anti-complete to B_2^j and complete to B_1^j.

We now consider the set of vertices in $V_1 \cup V_2$ that do not belong to any set B^i. Let R denote this set and note that $G[R]$ is $2P_2$-free by maximality of the set $\{H^1, \ldots, H^q\}$. For $i \in \{1,2\}$ let $R_i = R \cap V_i$. We make the following claim (we omit its proof).

Claim 5. If $x \in R_1$ has a neighbour in B_2^i, then x is complete to B_2^{i+1}, and if x has a non-neighbour in B_2^i, then x is anti-complete to B_2^{i-1}. If $x \in R_2$ has a non-neighbour in B_1^i, then x is anti-complete to B_1^{i+1}, and if x has a neighbour in B_1^i, then x is complete to B_1^{i-1}.

Claim 5 allows us to update the sequence of blocks as follows:

Update Procedure. *For $i \in \{1,2\}$, if R_i contains a vertex x that has both a neighbour y and a non-neighbour y' in B_{3-i}^j for some j, we add x to the sets B_i^j and B^j and remove it from R_i.*

We make the following claim (we omit its proof).

Claim 6. Applying the Update Procedure preserves the chain property of the blocks B^i.

By Claim 6 we may apply the Update Procedure exhaustively, after which the chain property will continue to hold. Once this procedure is complete, every remaining vertex of R_1 will be either complete or anti-complete to each set B_2^j. In fact, by Claim 5, we know that for every vertex $x \in R_1$, there is an $i \in \{0, \ldots, p\}$ such that x has a neighbour in all B_2^j with $j > i$ (if such a j exists) and x has a non-neighbour in all B_2^j with $j \leq i$ (if any such j exists). Since x is complete or anti-complete to each set B_2^j, we obtain the following conclusion:

- for every vertex $x \in R_1$, there is an $i \in \{0, \ldots, p\}$ such that x is complete to all B_2^j with $j > i$ (if such a j exists) and x is anti-complete to all B_2^j with $j \leq i$ (if any such j exists). We denote the corresponding subset of R_1 by Y_1^i.

By symmetry, we also obtain the following:

- for every vertex $x \in R_2$, there is an $i \in \{0, \ldots, p\}$ such that x is complete to all B_1^j with $j \leq i$ (if such a j exists) and x is anti-complete to all B_1^j with $j > i$ (if any such j exists). We denote the corresponding subset of R_2 by Y_2^i.

We also partition the vertices of V_3 into $p+1$ subsets $V_3^0, \ldots, V_3^p$ such that the vertices of V_3^j are complete to B_2^i and anti-complete to B_1^i for $i \leq j$ and complete to B_1^i and anti-complete to B_2^i for $i > j$. (So V_3^0 is complete to B_i^1 for all i and V_3^p is complete to B_i^2 for all i).

Claim 7. For each i, if $j < i$ then V_3^i is anti-complete to Y_1^j and complete to Y_2^j, and if $j > i$ then V_3^i is complete to Y_1^j and anti-complete to Y_2^j.

Suppose that $z \in V_3^i$ and $x \in Y_1^j$ and $y \in Y_2^j$ (note that such vertices x and y do not exist if Y_1^j or Y_2^j, respectively, is empty). First suppose that $j < i$ and choose arbitrary vertices $x' \in B_1^i$, $y' \in B_2^i$. Note that x and z are both complete to B_2^i and y and z are both anti-complete to B_1^i. Then z cannot be adjacent to x otherwise $G[x, y', z]$ would be a rainbow K_3 and z must be adjacent to y, otherwise $G[x', y, z]$ would be a rainbow $3P_1$. Now suppose $i < j$ and choose arbitrary vertices $x' \in B_1^{i+1}$, $y' \in B_2^{i+1}$. Note that x and z are both anti-complete to B_2^{i+1} and y and z are both complete to B_1^{i+1}. Then z must be adjacent to x otherwise $G[x, y', z]$ would be a rainbow $3P_1$ and z must be non-adjacent to y, otherwise $G[x', y, z]$ would be a rainbow K_3. This completes the proof of Claim 7.

Let G^i denote the subgraph of G induced by $Y_1^i \cap Y_2^i \cap V_3^i$. By Claims 4, 6 and 7 the graph G can be partitioned into slices: $G^0, G[B^1], G^1, G[B^2], \ldots, G[B^p], G^p$. Recall that the graph G is of type t and $G[V_1 \cup V_2]$ contains an induced $2P_2$. Since $G[Y_1^i \cup Y_2^i]$ is $2P_2$-free (by construction, since the original sequence $H^1, H^2, \ldots, H^q$ of $2P_2$s was maximal), it follows that each G^i is of type at most $t - 1$. Furthermore, since each $G[B_i]$ is bipartite, it forms a curious graph in which the set V_3 is empty, so it has type at most 1. This completes the proof. $\square$

We are now ready to state the main result of this section.

Theorem 4. *Let X be a hereditary class of graphs. If the set of bipartite graphs in X is well-quasi-ordered by the labelled induced subgraph relation or has bounded clique-width, then this property also holds for the set of curious graphs in X.*

Proof. Suppose that the class of bipartite graphs in X is well-quasi-ordered by the labelled induced subgraph relation or has bounded clique-width. By Lemmas 10, 11 and 13, the curious graphs of type at most 1 also have this property. Applying Lemmas 10, 11 and 14 once, we obtain the same conclusion for curious graphs of type at most 2. Applying Lemmas 10, 11 and 14 again, we obtain the same conclusion for curious graphs of type at most 3, that is, all curious graphs. $\square$

4   Applications of Our Technique

We start with two lemmas. The first is implicit in the proofs of Lemma 9 and Theorem 3 in [6]; we omit the proof of the second.

Lemma 15 [6]. *There is a constant c, such that given any $(K_3, P_1 + P_5)$-free graph G that contains an induced C_5, we can apply at most c vertex deletions and at most c bipartite complementation operations to obtain a graph H that is the disjoint union of $(K_3, P_1 + P_5)$-free curious graphs.*

Lemma 16. *There is a constant c, such that given any prime $(K_3, P_2 + P_4)$-free graph G that contains an induced C_5, we can apply at most c vertex deletions and at most c bipartite complementation operations to obtain a graph H that is the disjoint union of $(K_3, P_2 + P_4)$-free curious graphs and 3-uniform graphs.*

We can now prove the following theorem.[1]

Theorem 5. *For $H \in \{P_2+P_4, P_1+P_5\}$ the class of (K_3, H)-free graphs is well-quasi-ordered by the labelled induced subgraph relation and has bounded clique-width.*

Proof. Let $H \in \{P_2 + P_4, P_1 + P_5\}$. By Lemmas 1 and 5, we need only consider prime graphs in this class. Recall that a prime graph on at least three vertices cannot contain two vertices that are false twins, otherwise these two vertices would form a non-trivial module. Therefore, by Lemma 2, and since $H \subseteq_i S_{1,2,3}$, the classes of prime (K_3, H)-free graphs containing an induced C_7 is precisely the graph C_7. We may therefore restrict ourselves to C_7-free graphs. Since the graphs in the class are H-free, it follows they contain no induced cycles on eight or more vertices. We may therefore restrict ourselves to prime (K_3, C_7, H)-free graphs that either contain an induced C_5 or are bipartite. By Lemmas 15 or 16, given any prime (K_3, C_7, H)-free that contains an induced C_5, we can apply at most a constant number of vertex deletions and bipartite complementation operations to obtain a graph that is a disjoint union of (K_3, H)-free curious graphs and (in the $H = P_2 + P_4$ case) 3-uniform graphs. By Lemmas 4, 8 and 9, Facts 1 and 3, and Theorem 4, it is sufficient to only consider bipartite (K_3, C_7, H)-free graphs. These graphs are H-free bipartite graphs. Furthermore, they form a subclass of the class of $(P_7, S_{1,2,3})$-free bipartite graphs, since $H \subseteq_i P_7, S_{1,2,3}$. $(P_7, S_{1,2,3})$-free bipartite graphs are well-quasi-ordered by the labelled induced subgraph relation by Lemma 6 and have bounded clique-width by Theorem 3. $\square$

References

1. Atminas, A., Lozin, V.V.: Labelled induced subgraphs and well-quasi-ordering. Order **32**(3), 313–328 (2015)
2. Blanché, A., Dabrowski, K.K., Johnson, M., Lozin, V.V., Paulusma, D., Zamaraev, V.: Clique-width for graph classes closed under complementation. In: Proceeding of MFCS 2017. LIPIcs, vol. 83, pp. 73:1–73:14 (2017)
3. Courcelle, B.: Clique-width and edge contraction. Inf. Process. Lett. **114**(1–2), 42–44 (2014)
4. Courcelle, B., Makowsky, J.A., Rotics, U.: Linear time solvable optimization problems on graphs of bounded clique-width. Theory Comput. Syst. **33**(2), 125–150 (2000)
5. Courcelle, B., Olariu, S.: Upper bounds to the clique width of graphs. Discrete Appl. Math. **101**(1–3), 77–114 (2000)
6. Dabrowski, K.K., Dross, F., Paulusma, D.: Colouring diamond-free graphs. J. Comput. Syst. Sci. (in press)
7. Dabrowski, K.K., Lozin, V.V., Paulusma, D.: Well-quasi-ordering versus clique-width: New results on bigenic classes. Order (in press)
8. Dabrowski, K.K., Paulusma, D.: Classifying the clique-width of H-free bipartite graphs. Discret. Appl. Math. **200**, 43–51 (2016)

---

[1] It was already known [6] that the class of $(K_3, P_1 + P_5)$-free graphs has bounded clique-width but it was not known that it is well-quasi-ordered.

9. Dabrowski, K.K., Paulusma, D.: Clique-width of graph classes defined by two forbidden induced subgraphs. Comput. J. **59**(5), 650–666 (2016)

10. Daligault, J., Rao, M., Thomassé, S.: Well-quasi-order of relabel functions. Order **27**(3), 301–315 (2010)

11. Damaschke, P.: Induced subgraphs and well-quasi-ordering. J. Graph Theory **14**(4), 427–435 (1990)

12. Ding, G.: Subgraphs and well-quasi-ordering. J. Graph Theory **16**(5), 489–502 (1992)

13. Espelage, W., Gurski, F., Wanke, E.: How to solve NP-hard graph problems on clique-width bounded graphs in polynomial time. In: Brandstädt, A., Le, V.B. (eds.) WG 2001. LNCS, vol. 2204, pp. 117–128. Springer, Heidelberg (2001). doi:10.1007/3-540-45477-2_12

14. Higman, G.: Ordering by divisibility in abstract algebras. Proc. London Math. Soc. **s3–2**(1), 326–336 (1952)

15. Kamiński, M., Lozin, V.V., Milanič, M.: Recent developments on graphs of bounded clique-width. Discret. Appl. Math. **157**(12), 2747–2761 (2009)

16. Kobler, D., Rotics, U.: Edge dominating set and colorings on graphs with fixed clique-width. Discrete Appl. Math. **126**(2–3), 197–221 (2003)

17. Korpelainen, N., Lozin, V.V.: Bipartite induced subgraphs and well-quasi-ordering. J. Graph Theory **67**(3), 235–249 (2011)

18. Korpelainen, N., Lozin, V.V.: Two forbidden induced subgraphs and well-quasi-ordering. Discret. Math. **311**(16), 1813–1822 (2011)

19. Lozin, V.V., Rautenbach, D.: On the band-, tree-, and clique-width of graphs with bounded vertex degree. SIAM J. Discret Math. **18**(1), 195–206 (2004)

20. Lozin, V.V., Razgon, I., Zamaraev, V.: Well-quasi-ordering does not imply bounded clique-width. In: Mayr, E.W. (ed.) WG 2015. LNCS, vol. 9224, pp. 351–359. Springer, Heidelberg (2016). doi:10.1007/978-3-662-53174-7_25

21. Oum, S.-I., Seymour, P.D.: Approximating clique-width and branch-width. J. Comb. Theory, Ser. B **96**(4), 514–528 (2006)

22. Rao, M.: MSOL partitioning problems on graphs of bounded treewidth and clique-width. Theor. Comput. Sci. **377**(1–3), 260–267 (2007)

Finding Cut-Vertices in the Square Roots of a Graph

Guillaume Ducoffe[1,2(✉)]

[1] Université Côte d'Azur, Inria, CNRS, I3S, Sophia Antipolis, France
guillaume.ducoffe@inria.fr, guillaume.ducoffe@ici.ro
[2] National Institute for Research and Development in Informatics,
Bucharest, Romania

Abstract. The square of a given graph $H = (V, E)$ is obtained from H by adding an edge between every two vertices at distance two in H. Given a graph class $\mathcal{H}$, the $\mathcal{H}$-SQUARE ROOT PROBLEM asks for the recognition of the squares of graphs in $\mathcal{H}$. In this paper, we answer positively to an open question of [Golovach et al. IWOCA 2016] by showing that the squares of *cactus-block graphs* can be recognized in polynomial time. Our proof is based on new relationships between the decomposition of a graph by cut-vertices and the decomposition of its square by clique cutsets. More precisely, we prove that the closed neighbourhoods of cut-vertices in H induce maximal subgraphs of $G = H^2$ with no clique-cutset. Furthermore, based on this relationship, we can compute from a given graph G the block-cut tree of a desired square root (if any). Although the latter tree is not uniquely defined, we show surprisingly that it can only differ marginally between two different roots. Our approach not only gives the first polynomial-time algorithm for the $\mathcal{H}$-SQUARE ROOT PROBLEM for several graph classes $\mathcal{H}$, but it also provides a unifying framework for the recognition of the squares of trees, block graphs and cactus graphs—among others.

1 Introduction

This paper deals with the well-known concepts of *square* and *square root* in graph theory. Roughly, the square of a given graph is obtained by adding an edge between the pairs of vertices at distance two (technical definitions are postponed to Sect. 2). A square root of a given graph G has G as its square. The reason for this terminology is that when encoding a graph as an adjacency matrix A (with $1's$ on the diagonal), its square has for adjacency matrix A^2–obtained from A using Boolean matrix multiplication. The squares of graphs appear, somewhat naturally, in the study of coloring problems: when it comes about modelling interferences at a bounded distance in a radio network [46]. Unsurprisingly, there is

This work is partially supported by ANR project Stint under reference ANR-13-BS02-0007 and ANR program "Investments for the Future" under reference ANR-11-LABX-0031-01.

H.L. Bodlaender and G.J. Woeginger (Eds.): WG 2017, LNCS 10520, pp. 234–248, 2017.
https://doi.org/10.1007/978-3-319-68705-6_18

an important literature on the topic, with nice structural properties of square graphs being undercovered [2,6,15,30,33,35]. In particular, an elegant characterization of the squares of graphs has been given in [37]. However, this does not lead to an efficient (polynomial-time) algorithm for recognizing square graphs. Our main focus in the paper is on the existence of such algorithms. They are, in fact, unlikely to exist since the problem has been proved NP-complete [36]. In light of this negative result, there has been a growing literature trying to identify the cases where the recognition of the squares of graphs remains tractable [10,22,25–27,32,38]. We are interested in the variant where the desired square root (if any) must belong to some specified graph class.

1.1 Related Work

There is a complete dichotomy result for the problem when it is parameterized by the *girth* of a square root. More precisely, the squares of graphs with girth at least six can be recognized in polynomial time, and it is NP-complete to decide whether a graph has a square root with girth at most five [1,13,14]. One first motivation for our work was to obtain similar dichotomy results based on the *separators* in a square root. We are thus more interested in graph classes with nice separability properties, such as chordal graphs. Recognizing the squares of chordal graphs is already NP-complete [26]. However, it can be done in polynomial time for many subclasses [26–28,34,39,43].

The most relevant examples to explain our approach are the classes of trees [43], block graphs [28] and cacti [19]. The squares of all these graphs can be recognized in polynomial time. Perhaps surprisingly, whereas the case of trees is a well-known success story for which many algorithmic improvements have been proposed over the years [9,28,32,43], the polynomial-time recognition of the squares of cactus graphs has been proved only very recently. A common point to these three above classes of graphs is that they can be decomposed into very simple subgraphs by using *cut-vertices* (respectively, in edges for trees, in complete graphs for block graphs and in cycles for cactus graphs). This fact is exploited in the polynomial-time recognition algorithms for the squares of these graphs. We observe that more generally, cut-vertices play a discrete, but important role, in the complexity of the recognition of squares, even for general graphs. As an example, most hardness results rely on a gadget called a "tail", that is a particular case of cut-vertices in the square roots [14,36]. Interestingly, this tail construction imposes for some vertex in the square to be a cut-vertex with the same closed neighbourhood in any square root. It is thus natural to ask whether more general considerations on the cut-vertices can help to derive additional constraints on the closed neighbourhoods in these roots. Our results prove that it is the case.

As stated before, we are not the first to study the properties of cut-vertices in the square roots. In this respect, the work in [19] has been a major source of inspiration for this paper. However, most of the results so far obtained are specific to some graph classes and they hardly generalize to more general graphs [19,28]. Evidence of this fact is that whereas both the squares of block graphs and the squares of cacti can be recognized in polynomial time, the techniques involved in these two cases do not apply to the slightly more general class of *cactus-block graphs* (graphs that can be decomposed by cut-vertices into cycles and complete graphs) [19]. In the end, the characterization of the cut-vertices in these roots is only partial – even for cactus roots –, with most of the technical work for the recognition algorithm being rather focused on the notion of *tree decompositions* (*e.g.*, clique-trees for chordal squares, or decomposition of the square into bounded-treewidth graphs). Roughly, tree decompositions [42] aim at decomposing graphs into pieces, called *bags*, organized in a tree-like manner. The decomposition of a square root of a graph by its cut-vertices leads to a specific type of tree decompositions for this graph that are called "H-tree decompositions" [18]. Note that it is not known whether a H-tree decomposition can be computed in polynomial time. In contrast, we use in this work another type of tree decompositions, called an *atom tree*, that generalizes the notion of clique-trees for every graph. It can be computed in polynomial time [4].

1.2   Our Contributions

Our work is based on new relationships between the cut-vertices in a given graph and the *clique-cutsets* of its square (separators being a clique). These results are presented in Sect. 3. In particular, we obtain a complete characterization of the *atoms* of a graph (maximal subgraphs with no clique cutset) based on the *blocks* of its square roots (maximal subgraphs with no cut-vertices).

The most difficult part is to show how to "reverse" these relationships: from the square back to a square root. We prove in Sect. 4 that it can be done to some extent. More precisely, in Sect. 4.1 we show that the "essential" cut-vertices of the square roots, with at least two connected components not fully contained in their closed neighbourhoods, are in some sense unique (independent of the root) and that they can be computed in polynomial time, along with their closed neighbourhood in any square root. Indeed, structural properties of these vertices allow to reinterpret them as the cut-vertices of some incidence graphs that can be locally constructed from the intersection of the atoms in an atom tree (tree decomposition whose bags are exactly the atoms). Proving a similar characterization for non essential cut-vertices remains to be done. We give sufficient conditions and a complete characterization of the closed neighbourhoods of the non essential cut-vertices for a large class of graphs in Sect. 4.2.

Then, inspired from these above results, we introduce a novel framework in Sect. 5 for the recognition of squares[1]. Assuming a square root exists, we can push further some ideas of Sect. 4 in order to compute, for every block in this root, a graph that is isomorphic to its square. This way, a square root can be computed for each square of a block separately. However, we need to impose additional constraints on these roots in order to be able to reconstruct from them a square root for the original graph. We thus reduce the recognition of

---

[1] Sufficient conditions for the framework to be applied are rather technical. They will be properly stated in a journal version.

the squares to a stronger variant of the problem for the squares of biconnected graphs. Let us point out that this approach can be particularly beneficial when the blocks of a root are assumed to be part of a well-structured graph class.

In Sect. 6, we finally answer positively to an open question of [19] by proving that the squares of cactus-block graphs can be recognized in polynomial time. Our result is actually much more general, as it gives a unifying algorithm for many graph classes already known to be tractable (*e.g.*, trees, block graphs and cacti) and it provides the first polynomial time recognition algorithm for the squares of related graph classes – such as Gallai trees [16]. In its full generality, the result applies to "*j-cactus-block graphs*": a generalization of cactus-block graphs where each block is either a complete graph or the kth-power of a cycle, for some $1 \leq k \leq j$. As expected this last result is obtained by using our framework. This application is not straightforward. Indeed, we need to show the existence of a j-cactus-block root with some "good" properties in order for the framework to be applied. We also need to show that a stronger variant of the recognition of squares (discussed in Sect. 5) can be solved in polynomial time for j-cactus-block graphs when j is a fixed constant. We do so by introducing classical techniques from the study of circular-arc graphs [45].

Although we keep the focus on square roots, we think that our approach could be generalized in order to compute the cut-vertices in the *p-th roots* of a graph (*e.g.*, see [9] for related work on p-th tree roots). This is left for future work. Due to lack of space, most proofs are only sketched or postponed to our technical report [11]. Definitions and preliminary results are given in Sect. 2. We conclude this paper in Sect. 7 with some open questions.

2   Preliminaries

We use standard graph terminology from [7]. All graphs in this study are finite, unweighted and simple (hence with neither loops nor multiple edges), unless stated otherwise. Given a graph $G = (V, E)$ and a set $S \subseteq V$, we will denote by $G[S]$ the subgraph of G that is induced by S. The open neighbourhood of S, denoted by $N_G(S)$, is the set of all vertices in $G[V \setminus S]$ that are adjacent to at least one vertex in S. Similarly, the closed neighbourhood of S is denoted by $N_G[S] = N_G(S) \cup S$. For every $u, v \in V$, vertex v is *dominated* by u if $N_G[v] \subseteq N_G[u]$. In particular, if $N_G[u] = N_G[v]$ then we say u and v are *true twins*. If even more strongly, we have $N_G[w] \subseteq N_G[u]$ for every $w \in N_G[v]$, then u is a *maximum neighbour* of v.

2.1   Squares and Powers of Graphs

For every connected graph G and for every $u, v \in V$, the distance between u and v in G, denoted by $dist_G(u, v)$, is equal to the minimum length (number of edges) of a uv-path in G. The jth-power of G is the graph $G^j = (V, E_j)$ with same vertex-set as G and an edge between every two distinct vertices at distance at most j in G. In particular, the *square* of a graph $G = (V, E)$ is the graph $G^2 = (V, E_2)$ with same vertex-set V as G and an edge between every two distinct vertices $u, v \in V$ such that $N_G[u] \cap N_G[v] \neq \emptyset$. Conversely, if there exists a graph H such that G is isomorphic to H^2 then H is called a *square root* of G. On the one hand it is easy to see that not all graphs have a square root. For example, if G is a tree with at least three vertices then it does not have any square root. On the other hand, note that a graph can have more than one square root. As an example, the complete graph K_n with n-vertices is the square of any diameter two n-vertex graph.

In what follows, we will focus on the following recognition problem:

Problem 1 (𝓗-SQUARE ROOT).

Input: A graph $G = (V, E)$.
Question: Is G the square of a graph in $\mathcal{H}$?

Our proofs will make use of the notions of subgraphs, induced subgraphs and *isometric subgraphs*, the latter denoting a subgraph H of a connected graph G such that $dist_H(x, y) = dist_G(x, y)$ for every $x, y \in V(H)$. Furthermore, let H be a square root of a given graph $G = (V, E)$. Given a walk $W = (x_0, x_1, \ldots, x_l)$ in G, an *H-extension* of W is any walk W' of H that is obtained from W by adding, for every i such that x_i and x_{i+1} are nonadjacent in H, a common neighbour $y_i \in N_H(x_i) \cap N_H(x_{i+1})$ between x_i and x_{i+1}.

2.2   Graph Decompositions

A set $S \subseteq V$ is a *separator* in a graph $G = (V, E)$ if its removal increases the number of connected components. A *full component* in $G[V \setminus S]$ is any connected component C in $G[V \setminus S]$ satisfying that $N_G(C) = S$ (note that a full component might fail to exist). The set S is called a *minimal separator* in G if it is a separator and there are at least two full components in $G[V \setminus S]$. Minimal separators are closely related to the notion of Robertson and Seymour's *tree decompositions* (*e.g.*, see [8, 20, 23, 40]). Formally, a *tree-decomposition* $(T, \mathcal{X})$ of G is a pair consisting of a tree T and of a family $\mathcal{X} = (X_t)_{t \in V(T)}$ of subsets of V indexed by the nodes of T and satisfying:

- $\bigcup_{t \in V(T)} X_t = V$;
- for any edge $e = \{u, v\} \in E$, there exists $t \in V(T)$ such that $u, v \in X_t$;
- for any $v \in V$, $\{t \in V(T) \mid v \in X_t\}$ induces a subtree, denoted by T_v, of T.

The sets X_t are called *the bags* of the decomposition.

In what follows, we will consider two main types of minimal separators.

Cut-Vertices. If $S = \{v\}$ is a separator then it is a minimal one and we call it a *cut-vertex* of G. Following the terminology of [19], we name v an *essential* cut-vertex if there are at least two components C_1, C_2 of $G \setminus v$ such that $C_1 \not\subseteq N_G(v)$

and similarly $C_2 \not\subseteq N_G(v)$; otherwise, v is called a *non essential* cut-vertex[2]. A graph $G = (V, E)$ is *biconnected* if it is connected and it does not have a cut-vertex. Examples of biconnected graphs are cycles and complete graphs. Furthermore, the *blocks* of G are the maximal biconnected subgraphs of G. For every connected graph G there is a tree whose nodes are the blocks and the cut-vertices of G, sometimes called the *block-cut tree*, that is obtained by adding an edge between every block B and every cut-vertex v such that $v \in B$. The block-cut tree of a given connected graph G can be computed in linear time [24].

It has been observed that every graph with a square root is biconnected [15]. We often use this fact in what follows.

Clique Cutsets. More generally, if S is a minimal separator inducing a complete subgraph of $G = (V, E)$ then we call it a *clique cutset* of G. A connected graph $G = (V, E)$ is *prime* if it does not have a clique cutset. Cycles and complete graphs are again examples of prime graphs, and it can be observed more generally that every prime graph is biconnected. The *atoms* of G are the maximal prime subgraphs of G. They can be computed in polynomial time [29, 44]. A *clique-atom* is an atom inducing a complete subgraph. Furthermore, a *simplicial vertex* is a vertex $v \in V$ such that $N_G[v]$ is a clique. If the atoms of G are given, then the clique-atoms and the simplicial vertices of G can be computed in linear time [12]. Finally, it has been proved in [4] that the atoms of G are the bags of a tree decomposition of G, sometimes called an *atom tree*. An atom tree can be computed in $\mathcal{O}(nm)$-time, and it is not necessarily unique [4].

3   Basic Properties of the Atoms in a Square

We start presenting relationships between the block-cut tree of a given graph and the decomposition of its square by clique cutsets (Theorem 1). These relationships are compared after the proof to some existing results in the literature for the $\mathcal{H}$-SQUARE ROOT problem. More precisely, our approach in this paper is based on the following relationship between the clique cutsets in a graph G and the cut-vertices in its square-roots (if any).

Proposition 1. *Let $H = (V, E)$ be a graph. The closed neighbourhood of any cut-vertex in H is a clique-atom of $G = H^2$.*

Proof. Let $v \in V$ be a cut-vertex of H and let $A_v = N_H[v]$. It is clear that A_v is a clique of G and so, this set induces a prime subgraph of G. In particular, A_v must be contained in an atom A of G. Suppose for the sake of contradiction that $A \neq A_v$. Let $u \in A \setminus A_v$. This vertex u is contained in some connected component C_u of $H \setminus v$. Furthermore since v is a cut-vertex of H, there exists $w \in N_H(v) \setminus C_u$. We claim that $S = (C_u \cap N_H(v)) \cup \{v\}$ is an uw-clique separator of G.

---

[2] The authors in [19] have rather focused on the stronger notion of *important* cut-vertices, that requires the existence of an additional third component C_3 of $G \setminus v$ such that $C_3 \not\subseteq N_G(v)$. We do not use this notion in our paper.

Indeed, let us consider any uw-path $\mathcal{P}$ in G. We name $\mathcal{Q} = (x_0 = u, x_1, \ldots, x_l = w)$ an arbitrary H-extension of $\mathcal{P}$. Since $\mathcal{Q}$ is an uw-walk in H, and u and w are in different connected components of $H \backslash v$, there exists an i such that $x_i \in C_u$, $x_{i+1} = v$. In particular, $x_i \in C_u \cap N_H(v) = S \backslash v$. Furthermore, by construction, for every two consecutive vertices x_i, x_{i+1} in the H-extension $\mathcal{Q}$, at least one of x_i or x_{i+1} belongs to $\mathcal{P}$. As a result, every uw-path in G intersects S, that proves the claim and so, that contradicts the fact that A is an atom of G. Therefore, $A = A_v$. Since A_v is a clique it is indeed a clique-atom of G.   □

The above Proposition 1 unifies and generalizes some previous results that have been found only for specific graph classes [19,28]. For example, it has been proved in [28] that for every block-graph H, the closed neighbourhoods of its cut-vertices are maximal cliques of its square. Our result shows that it holds for *any* square root H (not only block-graphs). Indeed, a clique-atom is always a maximal clique. Furthermore, our purpose with Theorem 1 is to give a partial characterization of the remaining atoms of the square. Ideally, we would have liked them to correspond to the blocks of its square roots. It turns out that this is not always the case. However, there are strong ties between the two.

Theorem 1. *Let H be a square root of a given graph $G = (V, E)$. Then, the atoms of G are exactly:*

- *the cliques $A_v = N_H[v]$, for every cut-vertex v of H;*
- *and for every block B of H, the atoms A' of $H[B]^2$ that are not dominated in H by a cut-vertex.*

4   Computation of the Cut-Vertices from the Square

Given a square graph $G = (V, E)$, we aim at computing all the cut-vertices in some square root H of G. More precisely, given two square roots H_1 and H_2 of G, we say that H_1 is "finer" than H_2 if the blocks of H_1 are contained in the blocks of H_2. The latter defines a partial ordering over the square roots of G, of which we call *maxblock square roots* its minimal elements. This notion is related to, but different than, the notion of minimal square root studied in [19][3]. The following section is based on Proposition 1, that gives a necessary condition for a vertex to be a cut-vertex in any maxblock square root $H_{\max}$ of G. Indeed, it follows from this Proposition 1 that there is a mapping from the cut-vertices of $H_{\max}$ to the clique-atoms of its square $G = H_{\max}^2$. This mapping is injective but in general it is not surjective. In what follows, we present sufficient conditions for a clique-atom of G to be the closed neighbourhood of a cut-vertex in *any* maxblock square root of G. In particular, we obtain a complete characterization for the essential cut-vertices.

---

[3] Let $\mathcal{H}$ be closed under edge deletion. If G has a square root in $\mathcal{H}$ then there exists a *finest* square root $H \in \mathcal{H}$ such that H is a minimal square root of G.

4.1   Recognition of the Essential Cut-Vertices

We recall that a cut-vertex v of $H_{\max}$ is called essential if there are two vertices in different connected components of $H_{\max}\backslash v$ that are both at distance two from v in $H_{\max}$. The remaining of the section is devoted to prove the following result.

Theorem 2. *Let $G = (V, E)$ be a square graph. Every maxblock square root of G has the same set $\mathcal{C}$ of essential cut-vertices. Furthermore, every vertex $v \in \mathcal{C}$ has the same neighbourhood A_v in any maxblock square root of G. All the vertices $v \in \mathcal{C}$ and their neighbourhood A_v can be computed in $\mathcal{O}(n + m)$-time if an atom tree of G is given.*

Algorithm 1. Computation of the essential cut-vertices

Require: A graph $G = (V, E)$; an atom tree $(T_G, \mathcal{A})$ of G.
Ensure: Returns (if G is a square) the set $\mathcal{C}$ of essential cut-vertices, and for every
    $v \in \mathcal{C}$ its neighbourhood A_v, in any maxblock square root of G.
  1: $\mathcal{C} \leftarrow \emptyset$.
  2: **for all** clique-atom $A \in \mathcal{A}$ **do**
  3:     Compute the incidence graph $I_A = Inc(\Omega(A), A)$, with $\Omega(A)$ being the multiset
      of neighbourhoods of the connected components of $G\backslash A$.
  4:     **if** $\bigcap\limits_{S \in \Omega(A)} S = \{v\}$ **and** v is a cut-vertex of I_A **then**
  5:        $\mathcal{C} \leftarrow \mathcal{C} \cup \{v\}$; $A_v \leftarrow A$.

The proof of Theorem 2 mainly follows from the correctness proof and the complexity analysis of Algorithm 1. Its basic idea is that the essential cut-vertices in any maxblock square root of G are exactly the cut-vertices in some "incidence graphs", that are locally constructed from the neighbourhoods of each clique-atom in the atom tree. Formally, for every clique-atom A of G, let $\Omega(A)$ contain $N_G(C)$ for every connected component C of $G\backslash A$ (note that $\Omega(A)$ is a multiset, with its cardinality being equal to the number of connected components in $G\backslash A$). The incidence graph $I_A = Inc(\Omega(A), A)$ is the bipartite graph with respective sides $\Omega(A)$ and A and an edge between every $S \in \Omega(A)$ and every $u \in S$.

We first need to observe that for every $v \in A$, v is a cut-vertex of I_A if and only if there is a bipartition P, Q of the connected components of $G\backslash A$ such that $N_G(P) \cap N_G(Q) = \{v\}$. Then, we subdivide the correctness proof of Algorithm 1 in two lemmas.

Lemma 1. *Let H be a square root of a given graph $G = (V, E)$, let $v \in V$ be an essential cut-vertex of H and let $A_v = N_H[v]$. Then, v has a neighbour in G in every connected component of $G\backslash A_v$. Furthermore, there is a bipartition P, Q of the connected components of $G\backslash A_v$ such that $N_G(P) \cap N_G(Q) = \{v\}$.*

Proof. First, observe that for every connected component D of $G\backslash A_v$, we have that $N_H(D) \cap A_v \neq \emptyset$. Since $A_v = N_H[v]$, it follows that $v \in N_G(D)$. In particular, v has a neighbour in G in every connected component of $G\backslash A_v$. Second, let $C_1, C_2, \ldots, C_k$ be all the connected components of $H\backslash v$ such that $C_i \not\subseteq A_v$. Note that $k \geq 2$ by the hypothesis. Furthermore, since for every $i \neq j$ and for every $u_i \in C_i \backslash A_v$, $u_j \in C_j \backslash A_v$, we have $dist_H(u_i, u_j) = dist_H(u_i, v) + dist_H(u_j, v) \geq 4$, there can be no edge between $C_i \backslash A_v$ and $C_j \backslash A_v$ in G. It implies that for every component D of $G \backslash A_v$, there is an $1 \leq i \leq k$ such that $D \subseteq C_i \backslash A_v$. So, let us group the components of $G \backslash A_v$ in order to obtain the sets $C_i \backslash A_v$, $1 \leq i \leq k$. For every $1 \leq i \leq k$, we have $\{v\} \subseteq N_G(C_i \backslash A_v) \subseteq (N_H(v) \cap C_i) \cup \{v\}$. In particular, for every $i \neq j$, we obtain $N_G(C_i \backslash A_v) \cap N_G(C_j \backslash A_v) = \{v\}$. Hence, let us bipartition the sets $C_i \backslash A_v$ into two nonempty supersets P and Q; by construction we have $N_G(P) \cap N_G(Q) = \{v\}$.   □

It turns out that conversely, Lemma 1 also provides a sufficient condition for a vertex v to be an essential cut-vertex in some square root of G (and in particular, in any maxblock square root). We formalize this next.

Lemma 2. *Let $H_{\max}$ be a maxblock square root of a given graph $G = (V, E)$, and let $v \in V$. Suppose there is a clique-atom A_v of G and a bipartition P, Q of the connected components of $G \backslash A_v$ such that $N_G(P) \cap N_G(Q) = \{v\}$. Then, for every square root H of G, we have $N_H(P) \cup N_H(Q) \subseteq N_H(v) \subseteq A_v$. In particular, v is an essential cut-vertex of $H_{\max}$ and $N_{H_{\max}}[v] = A_v$.*

Correctness of Algorithm 1 follows from Lemmas 1 and 2. In order to obtain a linear-time implementation, we replace the incidence graph I_A with a "reduced version" I_A^*, where we only consider the adhesion sets in an atom tree of G (intersection of A with the adjacent atoms in the atom tree). Indeed, doing so we simply discard the neighbourhoods of some components that are strictly contained in the neighbourhood of another component. Using the fact that G is biconnected, it can be shown that this does not affect the outcome. This allows us to achieve a time complexity that is linear in the size of the atom tree, and so, linear in the size of the input graph G.

4.2   Sufficient Conditions for Non Essential Cut-Vertices

We let open whether a good characterization of non essential cut vertices can be found. The remaining of this section is devoted to partial results in this direction. In general, not all the maxblock square roots of a graph have the same set of non essential cut-vertices. Our main result in this section is a complete characterization of the closed neighbourhoods of such vertices in any *finest* square root with some prescribed properties being satisfied by its blocks (Theorem 3).

Non essential cut-vertices are strongly related to simplicial vertices in the square. In general, if a clique-atom of G contains a simplicial vertex then it may not necessarily represent the closed neighbourhood of such a cut-vertex. However, we can prove it is always the case if the vertex is *simple*, *i.e.*, it is simplicial and the closed neighbourhoods of its neighbours can be linearly ordered by inclusion.

Lemma 3. *Let $H_{\max}$ be a maxblock square root of a graph $G = (V, E)$. If there exists a simple vertex u in G then it has a neighbour $v \in N_G(u)$ that is a non essential cut-vertex of $H_{\max}$. Furthermore, $N_{H_{\max}}[v] = N_G[u]$.*

Before concluding this section, we now state its main result.

Theorem 3. *Let $G = (V, E)$ be a connected graph that is not a complete graph. Furthermore let $H_{\max}$ be a finest square root of G with the property that, for every block B of $H_{\max}$, we have: $H_{\max}[B]$ has no dominated vertex, unless B is a clique[4]; and $H_{\max}[B]^2$ is prime. Then, a clique-atom A of G is the closed neighbourhood of a non essential cut-vertex in $H_{\max}$ if and only if it is a leaf in some atom tree of G.*

Sketch Proof. Let H be any square root of G with its blocks satisfying the two assumptions of the theorem. By analogy between the block-cut tree of H and an atom tree of G, it can be shown that the closed neighbourhood of a non essential cut-vertex in H satisfies the condition of the theorem. Conversely, if a clique-atom of G is a leaf in some atom tree, then either it is the closed neighbourhood of some (non essential) cut-vertex, or it is the square of a block B of H with diameter two. In the latter case, we deduce from the hypothesis – that there can be no dominated vertex in B – that B must contain a single cut-vertex v of H. Let us pairwise connect all the neighbours of v in B. Then, let us make of all the remaining vertices in $B \backslash N_H[v]$ a set of pending vertices adjacent to an arbitrary neighbour $u \in N_H(v) \cap B$. In doing so, we keep the property to be a square root of G and we strictly increase the number of blocks. $\square$

5   Reconstructing the Block-Cut Tree of a Square Root

Given a graph $G = (V, E)$, we propose a generic approach in order to compute the block-cut tree of one of its square-roots (if any). More precisely, we remind that a square root $H_{\max}$ of G is called a *maxblock square root* if there does not exist any other square root $H \neq H_{\max}$ of G with all its blocks being contained in the blocks of $H_{\max}$. We suppose we are given the closed neighbourhoods of all the cut-vertices in some maxblock square root $H_{\max}$ of G (the cut-vertices may not be part of the input). Based on this information, we show how to compute for every block of $H_{\max}$ a graph that is isomorphic to its square.

Theorem 4. *Let $H_{\max}$ be a maxblock square root of a graph $G = (V, E)$, and let $A_1, A_2, \ldots, A_k$ be the closed neighbourhoods of every cut-vertex in $H_{\max}$. For every block B of $H_{\max}$, we can compute a graph G_B that is isomorphic to its square. Furthermore, if B is not isomorphic to K_2 then we can also compute the mapping from $V(G_B)$ to B. It can be done in $\mathcal{O}(n + m)$-time in total if an atom tree of G is given.*

Sketch Proof. This part reuses the same techniques as Sect. 4.1. Given a clique-atom A and its incidence graph I_A, we can compute the blocks of I_A. Then, let us define the following equivalence relation over the connected components of $G \backslash A$: $C \sim_A C'$ if and only if $N_G(C)$ and $N_G(C')$ (taken as elements of $\Omega(A)$) are in the same block of I_A. The latter relation naturally extends to

---

[4] This first assumption on the blocks may look a bit artificial. However, we emphasize that it holds for every regular graph [3].

(a) Square root H.   (b) Square $G = H^2$.   (c) Incidence graph.   (d) Block-cut tree.

Fig. 1. Computation of the connected components in a square root.

an equivalence relation over $V \backslash A$: for every two components C, C' of $G \backslash A$ and for every $u \in C$, $u' \in C'$, $u \equiv_A u'$ if and only if $C \sim_A C'$. In doing so, the equivalence classes of $\equiv_A$ partition the set $V \backslash A$. We refer to Fig. 1 for an illustration of the procedure. Furthermore, it can be proved that when A is the closed-neighbourhood of a cut-vertex v in $H_{\max}$, the equivalence classes of $\equiv_A$ are exactly the sets $C_i \backslash A$, $1 \le i \le l$, with $C_1, \ldots, C_l$ being the connected components of $H_{\max}$. Applying this procedure sequentially to all the clique-atoms that represent the closed neighbourhood of a cut-vertex in $H_{\max}$, we can compute the squares of each block of $H_{\max} \backslash v$. This can be done in total $\mathcal{O}(n + m)$-time by carefully using the adhesion sets in an atom tree of G.   □

Then, we wish to solve the $\mathcal{H}$-SQUARE ROOT problem for each square of a block separately. However, doing so, we may not be able to reconstruct a square root for the original graph. Indeed, the closed neighbourhoods of cut-vertices are imposed, and these additional constraints may be violated by the partial solutions. We thus need to solve the following stronger version of the problem.

Problem 2 (*$\mathcal{H}$-SQUARE ROOT WITH NEIGHBOURS CONSTRAINTS*).

Input: A graph $G = (V, E)$; a list $\mathcal{N}_F$ of pairs $\langle v, N_v \rangle$ with $v \in V$, $N_v \subseteq V$; a list $\mathcal{N}_A$ of subsets $N_i \subseteq V$, $1 \le i \le k$.
Question: Are there a graph $H \in \mathcal{H}$ and a sequence $v_1, v_2, \ldots, v_k \in V$ of pairwise distinct vertices such that H is a square root of G, and:
  – $\forall \langle v, N_v \rangle \in \mathcal{N}_F$, we have $N_H[v] = N_v$
  – $\forall 1 \le i \le k$, we have $N_H[v_i] = N_i$; furthermore, $\langle v_i, N_i \rangle \notin \mathcal{N}_F$?

To our best knowledge, this variant has not been studied before in the literature. We show how to solve it for some graph classes in the next section. Intuitively, the list $\mathcal{N}_F$ represents the *essential* cut-vertices and their closed neighbourhoods in the block. The list $\mathcal{N}_A$ represents the closed neighbourhoods of non essential cut-vertices. Furthermore, non essential cut-vertices correspond to the vertices $v_1, \ldots, v_k$ to be computed. Notice that we need to ensure that all the v_i's are distinct in case there may be true twins in a square root. We also need to ensure that $\langle v_i, N_i \rangle \notin \mathcal{N}_F$ for the same reason.

6   Application to Trees of Cycle-Powers

A cycle-power graph is any j^{th}-power C_n^j of the n-node cycle C_n, for some $j, n \ge 1$. A *tree of cycle-powers* is a graph whose blocks are cycle-power graphs. In particular, a j-cactus-block graph is a graph whose blocks are complete graphs or k^{th}-powers of cycles, for any $1 \leq k \leq j$. This above class generalizes the classes of trees, block graphs and cacti: where all the blocks are edges, complete subgraphs and cycles, respectively. Other relevant examples are the class of *cactus-block graphs* (*a.k.a.*, 1-cactus-block graphs with our terminology): where all the blocks are either cycles or complete subgraphs [41]; and the *Gallai trees*, that are the cactus-block graphs with no block being isomorphic to an even cycle [16]. Our main result in this section is that the squares of these graphs can be recognized in polynomial time:

Theorem 5. *For every fixed $j \geq 1$, the squares of j-cactus-block graphs can be recognized in $\mathcal{O}(nm)$-time.*

Up to simple changes, the proof of Theorem 5 applies to all the subclasses mentioned above. This solves for the first time the complexity of the $\mathcal{H}$-SQUARE ROOT problem for the cactus-block graphs and Gallai trees:

Theorem 6. *Squares of cactus-block graphs, resp. squares of Gallai trees, can be recognized in $\mathcal{O}(nm)$-time.*

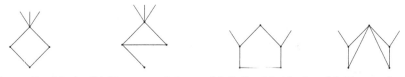

(a) A pending block. (b) Non-essential cut. (c) Splittable block. (d) Essential cut.

Fig. 2. Local modifications of the blocks.

The proof of Theorem 5 is twofold. We seek for a square root H of G that is a tree of cycle-powers and maximizes its number of blocks. First we show that the cut-vertices in this square root are exactly those characterized by Theorems 2 and 3. We do so by adapting the respective techniques from Lemma 2 and Theorem 3 in order to increase the number of cut-vertices. An illustration is provided with Fig. 2. Then, we need to show that $\mathcal{H}$-SQUARE ROOT WITH NEIGHBOURS CONSTRAINTS can be solved in linear time for j-cactus-block graphs. This is done by exploiting the fact that cycle-power graphs are *circular-arc graphs* (intersection graphs of intervals on the cycle) with a unique circular-arc model [21,31].

7   Conclusion

We intend the framework introduced in this paper to be applied for solving the $\mathcal{H}$-SQUARE ROOT problem in other graph classes – *e.g.*, graphs with *special treewidth* at most two [5]. Furthermore, we leave the existence of a full characterization of non essential cut-vertices in the square roots as an interesting open question. More generally, we aim at better understanding the relationships between small-size separators in a graph and small-diameter separators in its square. As an example, we believe that by studying the relationships between edge-separators in a graph and quasi-clique cutsets in its square (clique with one edge removed), we could improve the recognition of the squares of outerplanar graphs [17]. Let us mention that the complexity of recognizing the squares of planar graphs is still open.

References

1. Adamaszek, A., Adamaszek, M.: Uniqueness of graph square roots of girth six. Electron. J. Comb. **18**(P139), 1 (2011)
2. Agnarsson, G., Halldórsson, M.M.: Coloring powers of planar graphs. SIAM J. Discrete Math. **16**(4), 651–662 (2003)
3. Aigner, M., Fromme, M.: A game of cops and robbers. Discrete Appl. Math. **8**(1), 1–12 (1984)
4. Berry, A., Pogorelcnik, R., Simonet, G.: Organizing the atoms of the clique separator decomposition into an atom tree. Discrete Appl. Math. **177**, 1–13 (2014)
5. Bodlaender, H.-L., Kratsch, S., Kreuzen, V., Kwon, O.-J., Ok, S.: Characterizing width two for variants of treewidth. Discrete Appl. Math. **216**(Part 1), 29–46 (2017)
6. Bonamy, M., Lévêque, B., Pinlou, A.: 2-distance coloring of sparse graphs. J. Graph Theory **77**(3), 190–218 (2014)
7. Bondy, J.A., Murty, U.S.R.: Graph Theory. Graduate Texts in Mathematics. Springer, London (2008)
8. Bouchitté, V., Todinca, I.: Treewidth and minimum fill-in: grouping the minimal separators. SIAM J. Comput. **31**(1), 212–232 (2001)
9. Chang, M.-S., Ko, M.-T., Lu, H.-I.: Linear-time algorithms for tree root problems. In: Arge, L., Freivalds, R. (eds.) SWAT 2006. LNCS, vol. 4059, pp. 411–422. Springer, Heidelberg (2006). doi:10.1007/11785293_38
10. Cochefert, M., Couturier, J.-F., Golovach, P.A., Kratsch, D., Paulusma, D.: Parameterized algorithms for finding square roots. Algorithmica **74**(2), 602–629 (2016)
11. Ducoffe, G.: Finding cut-vertices in the square roots of a graph. Technical report hal-01477981, UCA, Inria, CNRS, I3S, France (2017). https://hal.archives-ouvertes.fr/hal-01477981
12. Ducoffe, G., Coudert, D.: Clique-decomposition revisited. In: Revision (Research Report on HAL, hal-01266147) (2017)
13. Farzad, B., Karimi, M.: Square-root finding problem in graphs, a complete dichotomy theorem. Technical report, arXiv arXiv:1210.7684 (2012)
14. Farzad, B., Lau, L.C., Tuy, N.N.: Complexity of finding graph roots with girth conditions. Algorithmica **62**(1–2), 38–53 (2012)
15. Fleischner, H.: The square of every two-connected graph is Hamiltonian. J. Comb. Theory Ser. B **16**(1), 29–34 (1974)
16. Gallai, T.: Graphen mit triangulierbaren ungeraden Vielecken. Magyar Tud. Akad. Mat. Kutató Int. Közl **7**, 3–36 (1962)
17. Golovach, P., Heggernes, P., Kratsch, D., Lima, P., Paulusma, D.: Algorithms for outerplanar graph roots and graph roots of pathwidth at most 2. In: Bodlaender, H.L., Woeginger, G.J. (eds.) WG 2017. LNCS, vol. 10520, pp. 275–288. Springer, Cham (2017). doi:10.1007/978-3-319-68705-6_z. arXiv:1703.05102

18. Golovach, P., Kratsch, D., Paulusma, D., Stewart, A.: A linear kernel for finding square roots of almost planar graphs. In: SWAT, pp. 4:1–4:14 (2016)
19. Golovach, P.A., Kratsch, D., Paulusma, D., Stewart, A.: Finding cactus roots in polynomial time. In: Mäkinen, V., Puglisi, S.J., Salmela, L. (eds.) IWOCA 2016. LNCS, vol. 9843, pp. 361–372. Springer, Cham (2016). doi:10.1007/978-3-319-44543-4_28
20. Golumbic, M.C.: Algorithmic Graph Theory and Perfect Graphs, vol. 57. Elsevier, Amsterdam (2004)
21. Golumbic, M.C., Hammer, P.L.: Stability in circular arc graphs. J. Algorithms 9(3), 314–320 (1988)
22. Harary, F., Karp, R.M., Tutte, W.T.: A criterion for planarity of the square of a graph. J. Comb. Theory 2(4), 395–405 (1967)
23. Heggernes, P.: Minimal triangulations of graphs: a survey. Discrete Math. 306(3), 297–317 (2006)
24. Hopcroft, J., Tarjan, R.: Algorithm 447: efficient algorithms for graph manipulation. Commun. ACM 16(6), 372–378 (1973)
25. Lau, L.C.: Bipartite roots of graphs. ACM Trans. Algorithms (TALG) 2(2), 178–208 (2006)
26. Lau, L.C., Corneil, D.G.: Recognizing powers of proper interval, split, and chordal graphs. SIAM J. Discrete Math. 18(1), 83–102 (2004)
27. Le, V.B., Nguyen, N.T.: A good characterization of squares of strongly chordal split graphs. Inf. Process. Lett. 111(3), 120–123 (2011)
28. Le, V.B., Tuy, N.N.: The square of a block graph. Discrete Math. 310(4), 734–741 (2010)
29. Leimer, H.-G.: Optimal decomposition by clique separators. Discrete Math. 113(1–3), 99–123 (1993)
30. Lih, K.-W., Wang, W.-F., Zhu, X.: Coloring the square of a K_4-minor free graph. Discrete Math. 269(1), 303–309 (2003)
31. Lin, M.C., Rautenbach, D., Soulignac, F.J., Szwarcfiter, J.L.: Powers of cycles, powers of paths, and distance graphs. Discrete Appl. Math. 159(7), 621–627 (2011)
32. Lin, Y.-L., Skiena, S.S.: Algorithms for square roots of graphs. SIAM J. Discrete Math. 8(1), 99–118 (1995)
33. Lloyd, E., Ramanathan, S.: On the complexity of distance-2 coloring. In: ICCI, pp. 71–74. IEEE (1992)
34. Milanič, M., Schaudt, O.: Computing square roots of trivially perfect and threshold graphs. Discrete Appl. Math. 161(10), 1538–1545 (2013)
35. Molloy, M., Salavatipour, M.R.: A bound on the chromatic number of the square of a planar graph. J. Comb. Theory Ser. B 94(2), 189–213 (2005)
36. Motwani, R., Sudan, M.: Computing roots of graphs is hard. Discrete Appl. Math. 54(1), 81–88 (1994)
37. Mukhopadhyay, A.: The square root of a graph. J. Comb. Theory 2(3), 290–295 (1967)
38. Nestoridis, N.V., Thilikos, D.M.: Square roots of minor closed graph classes. Discrete Appl. Math. 168, 34–39 (2014)
39. Le, V.B., Oversberg, A., Schaudt, O.: Polynomial time recognition of squares of ptolemaic graphs and 3-sun-free split graphs. In: Kratsch, D., Todinca, I. (eds.) WG 2014. LNCS, vol. 8747, pp. 360–371. Springer, Cham (2014). doi:10.1007/978-3-319-12340-0_30
40. Parra, A., Scheffler, P.: Characterizations and algorithmic applications of chordal graph embeddings. Discrete Appl. Math. 79(1–3), 171–188 (1997)

41. Randerath, B., Volkmann, L.: A characterization of well covered block-cactus graphs. Australas. J. Comb. **9**, 307–314 (1994)
42. Robertson, N., Seymour, P.: Graph minors. II. algorithmic aspects of tree-width. J. Algorithms **7**(3), 309–322 (1986)
43. Ross, I.C., Harary, F.: The square of a tree. Bell Syst. Tech. J. **39**(3), 641–647 (1960)
44. Tarjan, R.E.: Decomposition by clique separators. Discrete Math. **55**(2), 221–232 (1985)
45. Tucker, A.: Characterizing circular-arc graphs. Bull. Am. Math. Soc. **76**(6), 1257–1260 (1970)
46. Wegner, G.: Graphs with given diameter and a coloring problem. University of Dortmund, Technical report (1977)

The Minimum Shared Edges Problem on Grid-Like Graphs

Till Fluschnik[✉], Meike Hatzel, Steffen Härtlein, Hendrik Molter,
and Henning Seidler

Institut für Softwaretechnik und Theoretische Informatik,
TU Berlin, Berlin, Germany
{till.fluschnik,meike.hatzel,hendrik.molter}@tu-berlin.de,
{haertlein,henning.seidler}@campus.tu-berlin.de

Abstract. We study the NP-hard *Minimum Shared Edges* (MSE) problem on graphs: decide whether it is possible to route p paths from a start vertex to a target vertex in a given graph while using at most k edges more than once. We show that MSE can be decided on bounded (i.e. finite) grids in linear time when both dimensions are either small or large compared to the number p of paths. On the contrary, we show that MSE remains NP-hard on subgraphs of bounded grids.

Finally, we study MSE from a parametrised complexity point of view. It is known that MSE is fixed-parameter tractable with respect to the number p of paths. We show that, under standard complexity-theoretical assumptions, the problem parametrised by the combined parameter k, p, maximum degree, diameter, and treewidth does not admit a polynomial-size problem kernel, even when restricted to planar graphs.

1 Introduction

Routing in street-like networks is a frequent task. Graphs modelling street networks are often (almost) planar, that is, they can be drawn in the plane with (almost) no edge crossings. As a special case, a graph modelling the street network in Manhattan is similar to a grid graph. We study the following problem, originally introduced by Omran et al. [17], from a computational (parametrised) complexity perspective on planar and grid-like graphs:

MINIMUM SHARED EDGES (MSE)
Input: An undirected graph $G = (V, E)$, two distinct vertices $s, t \in V$, and two integers $k, p \in \mathbb{N}$.
Question: Are there p paths from s to t in G such that at most k edges appear in more than one of the p paths?

A full version is available at https://arxiv.org/abs/1703.02332.

T. Fluschnik—Supported by the DFG, project DAMM (NI 369/13-2).

H. Molter—Partially supported by the DFG, project DAPA (NI 369/12).

H.L. Bodlaender and G.J. Woeginger (Eds.): WG 2017, LNCS 10520, pp. 249–262, 2017.
https://doi.org/10.1007/978-3-319-68705-6_19

Note that Omran et al. [17] originally defined the problem on directed graphs (we refer to this as DIRECTED MINIMUM SHARED EDGES or DMSE). While Omran et al. motivate MSE by applications in security management, the problem can further appear in the following scenario. A network company wants to upgrade their network since it still uses old copper cables. To improve the throughput, some of these cables shall be replaced by modern optical fibre cables. The network routes information from a source location to a target location and the company wants to achieve a certain minimal throughput. Since digging up the conduits for the cables is much more expensive than the actual cables, we can neglect the cost of the cables and upgrade them to arbitrary bandwidth, because once open, we can lay as many cables as necessary into a conduit. The company wants to find the minimum number of conduits that have to be dug up in order to achieve the desired bandwidth.

Related Work. Omran et al. [17] showed that DMSE is NP-complete on acyclic digraphs. The problems MSE and DMSE were both shown to be NP-complete even if the input graph is planar [11]. Moreover, MSE is solvable in linear time on unbounded (i.e. infinite) grid graphs [8]. DMSE is $\lfloor k/2 \rfloor$-approximable [3], but there is no polynomial-time approximation of factor $2^{(\log(n))^{1-\epsilon}}$ for any $\epsilon > 0$ unless NP $\subseteq$ DTIME($n^{\text{polylog}(n)}$) [17].

Analysing its parametrised complexity, Fluschnik et al. [9] showed that MSE is fixed-parameter tractable when parametrised by the number p of paths but does not admit a polynomial-size problem kernel unless NP $\subseteq$ coNP/poly, MSE is W[1]-hard when parametrised by tw $+ k$, where tw denotes the treewidth of the input graph, and W[2]-hard when parametrised by the number k of shared edges. On graphs of bounded treewidth, MSE is solvable in polynomial time [2,19].

Our Contribution. We give both positive and negative results for MSE on grid-like graphs. On the positive side, we show that if the dimensions of the grid are smaller than the number p of paths, then MSE is trivially decidable, and if the dimensions of the grid are at least the number p of paths, then we provide an arithmetic criterion to decide MSE in linear-time (Sect. 3.1). On the negative side, we prove that the situation changes when subgraphs of bounded grids (which we refer to as *holey grids*) are considered, that is, we prove that MSE on subgraphs of bounded grids is NP-hard (Sect. 3.2). Similarly, we prove that DMSE is NP-hard for acyclic subgraphs of directed bounded grids (Sect. 3.3). Our NP-hardness results improve upon the known hardness results [11] as the graphs we consider are more restricted. Moreover, we show that MSE parametrised by $k + p + \Delta + \text{diam} + \text{tw}$, where Δ and diam denote the maximum degree and diameter, respectively, does not admit a polynomial-size problem kernel, unless NP $\subseteq$ coNP/poly, even on planar graphs (Sect. 4), improving an existing kernelization lower bound [9]. Due to space constraints, several proofs and details are deferred to a full version (this is indicated by a ($\star$)).

2   Preliminaries

We use basic notation from graph theory [7] and parametrised complexity [6]. By $\mathbb{N}$ we denote the natural numbers containing zero.

Graph Theory. Unless stated otherwise, we assume that all graphs are finite, undirected, simple and without self-loops. We refer with $V(G)$ and $E(G)$ to the vertex set and edge set, respectively, of a graph G. An edge set $P \subseteq E$ is called a *path* if we have $P = \{\{v_{i-1}, v_i\} \mid 0 < i \leq n\}$ for some pairwise distinct vertices $v_0, \ldots, v_n$. In this case we say P is a v_0-v_n-path of length n. The distance $\mathrm{dist}_G(u, v)$ between two vertices $u, v \in V(G)$ is defined as the length of a shortest u-v-path (we set $\mathrm{dist}_G(u, v) = \infty$ if there is no u-v path in G).

Grids. For $n, m \in \mathbb{N}$, let $G_{n \times m}$ be the (bounded) $n \times m$-grid, that is, the undirected graph (V, E) with the set of vertices $V := \{(x, y) \in \mathbb{N} \times \mathbb{N} \mid x < n, y < m\}$ and the set of edges $E := \{\{(v, w), (x, y)\} \mid |v - x| + |w - y| = 1\}$. The *coordinates* of a vertex are denoted by $v := (v_x, v_y)$. We call the vertices of degree less than four the *rim* of the grid. For a given vertex $v \in V$ we define $\partial_x v := v_x$ and $\partial_y v := v_y$, $\overline{\partial}_x v := n - 1 - v_x$, and $\overline{\partial}_y v := m - 1 - v_y$. We also use $\partial v := \partial_x v + \partial_y v$ and $\overline{\partial} v := \overline{\partial}_x v + \overline{\partial}_y v$.

Parametrised Complexity. A pair $Q = (P, \kappa)$ with $P \subseteq \Sigma^*$ and $\kappa : \Sigma^* \to \mathbb{N}$ is called a *parametrised problem*. A parametrised problem $Q = (P, \kappa)$ admits a *problem kernel* (or is *kernelisable*) if there is a polynomial-time algorithm transforming any instance $\mathcal{I}$ of Q into an instance $\mathcal{I}'$ such that (i) $\mathcal{I} \in Q \Leftrightarrow \mathcal{I}' \in Q$, and (ii) the size of $\mathcal{I}'$ (the *kernel*) is bounded by a computable function $f(\kappa(\mathcal{I}))$. If f is a polynomial, then the problem is said to admit a polynomial (problem) kernel. A parametrised problem is *fixed-parameter tractable* (or in FPT) if each instance (x, κ) can be decided in $f(\kappa(x)) \cdot |x|^{O(1)}$ time, where f is a computable function. A (decidable) parametrised problem is in FPT if and only if it is kernelisable. A parametrised problem that is W-hard is presumably not in FPT.

Further Notation. Let $\mathcal{I} = (G, s, t, p, k)$ be an instance of MSE. If (i) $\mathfrak{P}$ is a multiset of p s-t-paths $\{P_1, \ldots, P_p\}$, and (ii) $|\{e \in E \mid \exists i < j : e \in P_i \cap P_j\}| \leq k$, then we say $\mathfrak{P}$ is a *solution* for $\mathcal{I}$. We say that $\mathfrak{P}$ is a *trivial solution* if $P_i = P_j$ for all $i, j \in [p]$. An edge is called *shared* if it occurs in at least two paths of $\mathfrak{P}$.

3   On Bounded and Holey Grids

The class of grid graphs appeared frequently in the literature: There is work on grid graphs and related graphs with respect to finding paths [14,15], routing [4], or structural properties [1,13]. In this section we study the complexity of MSE on bounded grids and their subgraphs. We show that MSE is solvable in linear time on bounded grids when both dimensions are either small or large compared to the number p of paths (Sect. 3.1) and becomes NP-hard for subgraphs of bounded grids (Sects. 3.2 and 3.3). We remark that MSE is solvable in linear-time on the class of unbounded grids [8].

3.1   Bounded Grids

We fix some instance $\mathcal{I} := (G = G_{n \times m}, s, t, p, k)$ for the remainder of the section. Since the problem is invariant under symmetry and swapping s and t, we may assume s lies left and below of t and $\partial_x s \leq \partial_y s$. To show optimality of the constructions we regard edge cuts of size less than p. Assume $\mathcal{I}$ has a solution $\mathfrak{P}$. We know [8] that after contraction of the shared edges, the graph must allow an s-t-flow of value at least p. Therefore, every cut smaller than p has to be eliminated by a contraction, that is, it must contain a shared edge.

We distinguish the following different cases depending on the dimensions of the grid in relation to the number p of paths: p-*small* grid ($p > \max\{n, m\}$), p-*large* grid ($p \leq \min\{n, m\}$), and p-*narrow* grids (neither p-small nor p-large). We leave open whether MSE is solvable in polynomial-time on p-narrow grids. However, ongoing work indicates that the question can be answered positively.

On p-small Grids. If $p > \max\{n, m\}$, then every set of horizontal edges with endpoints having the same coordinates in the grid forms an s-t-cut of size smaller than p (analogously for every set of horizontal edges). Hence, intuitively, any set of p s-t-paths share an edge for each horizontal or vertical level they cross. Indeed, we prove that every instance on p-small grids is a yes instance if and only if it admits the trivial solution.

Lemma 1 ($\star$). *If $m < p$ and $n < p$, then we have a solution if and only if* $\text{dist}_G(s, t) \leq k$.

On p-large Grids. Compared to the situation on p-small grids, p-large grids allow for non-trivial solutions. Nevertheless, we prove that the existence of such non-trivial solutions is expressed by arithmetic conditions which can be checked in linear time. These arithmetic conditions basically relate p, k, and the positions of s and t relative to the rim of the grid. If s lies sufficiently far away from the corner formed by the left and lower rim, then only every second path in our construction introduces a new shared edge at this part. However, if s lies close to the corner (or if p is large enough), there is a critical number of paths after which every additional path introduces at least one new shared edge. The same happens at the side of t. Thus we obtain the following cases.

Lemma 2. *Let $p \leq m$ and $p \leq n$. Then there is a non-trivial solution if and only if either*

- $p \leq 2(\partial s + 2) - \deg(s)$ *and* $k \geq \left\lceil \frac{1}{2}(p - \deg(s)) \right\rceil + \left\lceil \frac{1}{2}(p - \deg(t)) \right\rceil$, *or*
- $2(\partial s + 2) - \deg(s) < p \leq 2(\overline{\partial} t + 2) - \deg(t)$
  and $k \geq p - (\partial s + 2) + \left\lceil \frac{1}{2}(p - \deg(t)) \right\rceil$, *or*
- $p > 2(\overline{\partial} t + 2) - \deg(t)$ *and* $k \geq 2p - (\partial s + \overline{\partial} t + 4)$.

For simplification, we introduce the following arrow notation. For $(x, y) \in V$ we define $(x, y) \rightarrow (x + \ell, y) := \{\{(x + i, y), (x + i + 1, y)\} \in E \mid 0 \leq i < \ell\}$. Analogously we define $\uparrow, \downarrow$ and $\leftarrow$. We also use the concatenation of these expressions such that e.g. $u \rightarrow v \uparrow w := (u \rightarrow v) \cup (v \uparrow w)$.

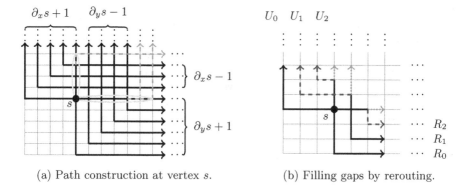

(a) Path construction at vertex s.   (b) Filling gaps by rerouting.

Fig. 1. Sketched aspects of the construction described in the proof of Lemma 2. (a) Path construction at vertex s; dashed: one shared edge; black: only a shared edge every two paths; orange ellipses enclose shared edges. (b) Filling gaps by rerouting in the construction; dotted: paths before reconstruction, dashed: rerouted paths for filling the gaps. (Colour figure online)

Proof. From s we construct path fragments (cf. Fig. 1) going upwards:

$$U_i := s \uparrow (s_x, s_y + i) \leftarrow (i, s_y + i) \uparrow (i, m - 1 - i), \qquad 0 \leq i < \partial_x s,$$
$$U_i := s \rightarrow (i, s_y) \uparrow (i, m - 1 - i), \qquad \partial_x s \leq i.$$

Next, we construct path fragments going to the right:

$$R_i := s \rightarrow (s_x + i, s_y) \downarrow (s_x + i, i) \rightarrow (n - 1 - i, i), \qquad 0 \leq i < \partial_y s,$$
$$R_i := s \uparrow (s_x, i) \rightarrow (n - 1 - i, i), \qquad \partial_y s \leq i.$$

To obtain the solution we add the path fragments in the following order. This process is illustrated in Fig. 1a.

(A) We start with U_0, $U_{\partial_x s}$, R_0, and $R_{\partial_y s}$ which have no shared edge. This yields $\deg(s)$ paths, since some of these are identical if s lies on the rim. Then for $i = 1, \ldots, \partial_y s - 1$ we add R_i and $U_{\partial_x s + i}$, where R_i new adds a shared edge (the other common edges were already shared before). Afterwards we continue adding U_i and $R_{\partial_y s + i}$ for $i = 1, \ldots, \partial_x s - 1$, where U_i adds a shared edge. Thus every other path fragment adds a shared edge. We stop as soon as we have constructed p paths.

(B) Continue adding R_i for $i = \partial s + 1, \ldots, \partial t$ and U_j for $j = \partial s + 1, \ldots, p - \partial t$ until we reach p paths. Here each single fragment adds another shared edge.

For $p \leq 2\partial s$ we add the following modifications. If $p \leq 2\partial_y s + 2$, that is, $R_{\partial_y s}$ is the last right-going fragment, extend this fragment downwards such that the endpoints of the R_i form a consecutive line. If the U_i leave a gap, that is, U_j is not part of the construction for some $0 < j < \partial_x s$, we take the rightmost up-going fragment and route it leftwards along the first free row, and then continue as U_j. In the end, if necessary, we extend $U_{\partial_x s}$ leftwards, like we did with $R_{\partial_y s}$. Thus the endpoints of the up-going fragments form a consecutive line as well. These steps do not introduce further shared edges. See Fig. 1b for an illustration.

So in the end we may assume we have constructed fragments $U_0, \ldots, U_{u-1}$ and $R_0, \ldots, R_{r-1}$ for some $u, r \in \mathbb{N}$ with $u + r = p$. At t we proceed analogously, simply mirrored. Hence we have r down-going fragments D_i and u left-going fragments L_i. Then we obtain the solution

$$\mathfrak{P} := \{U_i \cup L_i \mid i = 0, \ldots, u\} \cup \{R_i \cup D_i \mid i = 0, \ldots, r\}.$$

Feasibility. Furthermore the R_i only use the lower r rows of the grid whereas the L_i use the upper u rows. Since $r + u = p \leq m$, these do not intersect, that is, we do not get further shared edges. The same holds for the U_i and D_i, since $p \leq n$.

Let k_s and k_t denote the number of shared edges used to construct the path fragments at s, and at t respectively. Thus, we have a solution if $k \geq k_s + k_t$.

If $p \leq 2(\partial s + 2) - \deg(s)$, then we only use part A. From the $\deg(s)$-th path to the p-th path, every other path adds a new shared edge, so $k_s = \lceil \frac{1}{2}(p - \deg(s)) \rceil$. Furthermore, $\partial s \leq \overline{\partial} t$, so at t we also only use part A. Therefore we get $k_t = \lceil \frac{1}{2}(p - \deg(t)) \rceil$. Hence $k \geq k_s + k_t = \lceil \frac{1}{2}(p - \deg(s)) \rceil + \lceil \frac{1}{2}(p - \deg(t)) \rceil$.

If $2(\partial s + 2) - \deg(s) < p \leq 2(\overline{\partial} t + 2) - \deg(t)$, we still only use part A at t getting $k_t = \lceil \frac{1}{2}(p - \deg(t)) \rceil$. But at s we also use part B. Assume that s lies in the interior of the grid. Then, when completely executing part A, we use $\partial s - 2$ shared edges to construct $2\partial s$ paths. This leaves $k_s - (\partial s - 2)$ shared edges for part B. Each of those allows for another path. So we obtain the condition $p = \partial s + 2 + k_s$. If s lies on the rim or in the corner, then the argument differs slightly, but the condition is the same. So overall we get the condition $k \geq k_s + k_t = p - (\partial s + 2) + \lceil \frac{1}{2}(p - \deg(t)) \rceil$.

Finally, if $p > 2(\overline{\partial} t + 2) - \deg(t)$, then we use part B at both s and t. Thus we have $p = \partial s + 2 + k_s = \overline{\partial} t + 2 + k_t$. By adding these equalities we obtain $k_s + k_t = 2p - (\partial s + \overline{\partial} t + 4) \leq k$. So the solution is feasible.

Optimality. We only give a lower bound for the number k_s of shared edges at s. The bound for k_t follows analogously, which then gives the desired bound for k.

During part (A) of the construction, each contraction may increase the degree of s by at most 2. Hence $p \leq \deg(s) + 2k_s$, which shows $k_s \geq \lceil \frac{1}{2}(p - \deg(s)) \rceil$.

For $p \geq 2(\partial s + 2)$ we present a number of cuts of size $p - 1$. We use rectangles containing s, whose right upper corners move along a diagonal. Formally, these are $\mathrm{cut}_i := \{\{(i, y), (i+1, y)\} \mid y \leq p - 3 - i\} \cup \{\{(x, p-3-i), (x, p-2-i)\} \mid x \leq i\}$ for $i = s_x, \ldots, p - 3 - s_y$. Assume that t lies inside one of those rectangles. Then cut$_i$ for $s_x \leq i < t_x$ and $p - 3 - t_y < i \leq p - 3 - s_y$ are s-t-cuts of size $p - 1$, and these are dist$_G(s,t)$ many. In this case we need $k \geq$ dist$_G(s,t)$ which only allows the trivial solution. So we may assume that t lies outside all of these rectangles. Thus there are $p - 2 - \partial s$ many of these cuts and they separate s and t. Furthermore they are pairwise disjoint. So we get $k_s \geq p - 2 - \partial s$.

Altogether, our construction is optimal.   □

3.2   Holey Grids

In the previous section we proved that MSE is solvable in linear time on small and large (compared to the number p of paths) bounded grids. In this section we study the complexity of MSE on subgraphs of bounded grids, which we call *holey grids* and show that the problem is NP-hard on this graph class. To this end we reduce from the well known VERTEX COVER problem which is, given a graph G and a natural number $k \in \mathbb{N}$, to decide whether there exists $U \subseteq V$ with $|U| \leq k$ such that $\forall e \in E : e \cap U \neq \emptyset$. More precisely, we use that VERTEX COVER remains NP-complete on graphs with maximum degree three [12]. Note that our reduction adapts the idea of a reduction used in previous work [11].

Theorem 1. *MSE on holey grids is* NP-*hard.*

Proof. Given a VERTEX COVER-instance $\mathcal{I}_{VC} := (G = (V,E), k)$ with $\Delta(G) \leq 3$, we compute an equivalent MSE-instance $\mathcal{I}_{MSE} := (G' = (V', E'), s, t, p, k')$ on holey grids in polynomial time. We assume that $|V|$ is a power of two (otherwise we add isolated vertices until it is).

Figure 2d illustrates the graph obtained by applying the following transformation to the graph shown in Fig. 2a. The main part of the construction is a structure we refer to as *meta-grid*. The meta-grid encodes the vertex-edge incidence matrix of the original graph. We assume that the obtained graph to be embedded as shown in Fig. 2d, which serves as a reference when we use the terms "*left*", "*right*", "*up*", and "*down*". For construction purposes, we refer to paths with $\ell + 1$ vertices as ℓ-*chains* or chains of length ℓ. Whenever a chain is added in the construction, all vertices except the two end-vertices are new.

The main component in the construction is a gadget called *rainbow* (cf. [11]), see Fig. 2b. Figure 2b also shows that this gadget is a subgraph of a bounded grid. We use rainbow gadgets where the number of vertices in each band in the spectrum of the rainbow is larger than the number of allowed shared edges. This allows the rainbow gadget to restrict the number of paths that can be routed through it to at most the number of bands in the spectrum. Note that in any rainbow that is satiated with M paths $2M - 2$ edges are shared. We call the number of shared edges in a rainbow the *rainbow-offset*.

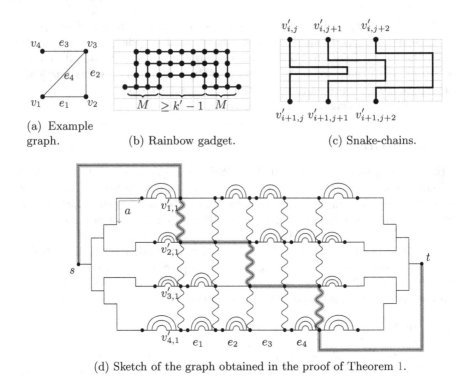

(a) Example graph.

(b) Rainbow gadget.

(c) Snake-chains.

(d) Sketch of the graph obtained in the proof of Theorem 1.

Fig. 2. An exemplified illustration of the construction in the proof of Theorem 1. (a) A graph representing an example instance of VERTEX COVER. (b) An illustration of the rainbow gadget. (c) An illustration of snake-chains. (d) Sketch of the holey grid constructed in the proof of Theorem 1, exemplified with the instance represented by (a). The highlighted path indicates the validation path.

We define M and a few other values we need in order to build the graph G' in the following:

$$M := 2(|E| + 1) + 2; \quad \text{trees} := 2 \cdot (|V| \cdot \log_2(|V|) - 2 + 2k);$$
$$c := 10; \quad c' := 2|V| + |E| \cdot |V| - 2|E|; \quad b := 2 \cdot M \cdot c' + 1;$$
$$a_0 := \frac{|V| - 1}{2}(M + c - 2) - \log |V|; \quad a := \max(a_0, |E|^3, b^2).$$

Here, c' is the number of rainbow gadgets we construct. The values a, a_0, and b are chosen to ensure certain constraints when routing paths and sharing edges while c can be understood as a scaling constant used to avoid intersections. Why the values are chosen in this way will become clear later in the proof. Next we set $p := k \cdot M + (|V| - k) + 1$ for the number of paths in $\mathcal{I}_{\text{MSE}}$ and $k' := k \cdot (2a + b|E|) + \text{trees} + c'(2M - 2)$ for the number of shared edges.

In the following we describe the construction of the meta-grid. First, we create a grid of vertices, without any edges, that has $|V|$ rows and $|E| + 1$ columns. We fix an ordering $v_1, \ldots, v_{|V|}$ on the set V of vertices and use it to identify each row of the grid with a vertex from G. Analogously, we fix an order $e_1, \ldots, e_{|E|}$ on the edge set E and use it to identify each space between two consecutive columns of the grid with an edge in G. From here on we will refer to these spaces as columns.

The first vertex in row i is denoted $v'_{i,1} \in V'$, see Fig. 2d, the second one is denoted $v'_{i,2}$, and so on. If vertex v_i is incident to edge e_j in G, then vertices $v'_{i,j}$ and $v'_{i,j+1}$ are connected by a chain of length b. If v_i is not incident to e_j in G, then vertices $v'_{i,j}$ and $v'_{i,j+1}$ are connected by a chain of length b followed by a rainbow. This completes the construction of the rows.

We embed the structure we just created in a grid such that the first vertices are vertically aligned and have vertical distance of $M + c$. Now we connect each vertex $v'_{i,j}$ with $i < |V|$ and $j < |E|$ with its respective lower neighbour $v'_{i+1,j}$ by so-called *snake-chains* of length at least $k' + 1$ (illustrated as wavy vertical lines in Fig. 2d). Note that these vertices do not necessarily lie above each other. The snake-chains are constructed as follows (also see Fig. 2c).

In every row except the lowest one, we start with the left most snake-chain. We first route it four steps down, then k' steps to the right, one down, left again until we are above its end-vertex which we then join it to by a vertical path. Then every further snake-chain is routed the following way: down by the maximum possible number of steps (at most four) such that no previous snake-chain is intersected, then k' to the right, then the minimum necessary number of steps down, such that the snake-chain can be extended to the left without intersecting a previous snake-chain until it can be routed downwards until it meets its end-vertex.

Note that the above description implies that we reduce the number of steps that a snake-chains is routed downwards every time the previous column did not contain a rainbow. After a rainbow is encountered we start with four steps down again. Since G has a maximum degree of three, there are at most three columns in every row without a rainbow, so after at most four consecutive snake-chains we encounter a rainbow in the next column. This way the snake-chains do not intersect or touch each other and the constant $c > 2 \cdot 4 + 1$ ensures that they also do not intersect any rainbows from the next row.

Now we add a source vertex s to the left of the meta-grid and construct a complete, binary tree of height $\log_2 |V|$ with s as its root and with $|V|$ leaves pointing in direction of the grid. We construct this tree in such a way that all vertices of the same level lie in the same column of the grid and from one leaf to the next we have distance two in the grid. This is possible since the number of vertices in G is a power of two. To make this tree embeddable into a grid we replace every edge by a chain of the minimal required length running along the grid structure. We connect the uppermost leaf to the first row of the meta-grid in a way such that the vertical distance between this leaf and $v'_{1,1}$ is exactly a_0. More specifically, we add a chain up and to the right until it has length a, then add a rainbow of sufficient length and connect it to $v'_{1,1}$. Each leaf of the tree is connected by a chain of length a and a following rainbow to one of the vertices in the first column of the meta-grid such that the order of the leaves and the vertices is the same. The length a is chosen such that all the chains have the same length. To avoid intersections in the a-chains these go right first: the chain leading to the row corresponding to v_i is routed $i-1$ steps to the right if $i < \frac{|V|}{2}$ and $|V| - i$ otherwise. Then the chains go up/down to their row and then right until they have length a. Note that this tree is symmetrical in the end since we work on an even number of vertices.

The same is done on the right side: we add a vertex t and a binary tree to its left with root t and the leaves are connected to the vertices in the last column of the meta-grid by a chain of length a and a rainbow. If the construction of the snake-paths causes some of the snake-paths to "stick out" to the right, then we extend the paths in the rainbows at the leaves of t as far as necessary to ensure that nothing intersects. The length of these rainbows is also used to align the leaves of the tree on this side.

Finally, we add chains of length at least $k' + 1$, the *outer-grid chains*, one connecting s to $v'_{1,1}$ and the other connecting $v'_{|V|,|E|+1}$ to t.

Intuitively, the correctness is shown as follows. We know that we can route at most M paths through a rainbow. Recall that $p := k \cdot M + (|V| - k) + 1$, so we have to do that k times. So we can pick k of the $|V|$ rows and route M paths through each. We route a single path through each of the remaining $|V| - k$ rows. Now we have to route one additional path, which has to use the outer-grid chains and the snake-chains. This path will verify that the k rows we chose to route M paths through correspond to vertices of G that constitute a vertex cover. Then each column corresponding to an edge of G has at least one row where we have a fully shared chain and no rainbow. So the remaining path can be routed through those chains and use the snake-chains to switch between rows. Of course, k' is chosen in a way that we are forced to use the described approach and that there is no solution if G does not have a vertex cover of size k. The formal proof of the correctness of the presented construction is deferred to a full version (⋆). □

3.3 Manhattan-Like Acyclic Digraphs

In the previous section, we proved that MSE is NP-hard on holey grids, i.e. subgraphs of a bounded grid. Along the line, in this section we prove that the *directed* version, DMSE, is NP-hard on the graph class of acyclic directed holey grids (we refer to this class by *Manhattan DAGs*). We remark that inspired by the street design of Manhattan, New York City, directed bounded grids (referred to as *Manhattan street networks*) are considered in the literature, also in the context of routing [16, 18].

Observe that MSE reduces to DMSE by replacing each edge $\{u, v\}$ by anti-parallel arcs $(u, v), (v, u)$. The correctness is due to the following.

Lemma 3 (⋆). *Let (G, s, t, p, k) be an instance of DMSE. If $\mathfrak{P}$ is a solution for this instance where two paths P_A and P_B use $e = (u, v) \in E$ and its inverted arc $e' = (v, u) \in E$, then we can find a solution $\mathfrak{P}'$ for the same instance that does not use both of these arcs.*

However, the directed graph obtained in the reduction is not acyclic. We show next that DMSE remains hard even on acyclic directed holey grids. On a high level, we adapt the construction presented in the proof of Theorem 1. We then direct the edges from left to right, from s towards t. Finally, we duplicate the horizontal chains (snake chains) and direct one upwards and one downwards.

Theorem 2 (⋆). *DMSE on Manhattan DAGs is* NP*-hard.*

4   The Nonexistence of Polynomial Kernels

In this section, we consider MSE from a parametrised complexity point of view. MSE is kernelisable but does not admit a polynomial problem kernel when it is parametrised by the number p of paths, unless NP $\subseteq$ coNP/poly [9]. We strengthen the latter result and complement the intractability of MSE on planar graphs by showing the following.

Theorem 3. *MSE with $\kappa(G, s, t, p, k) := p + k + \Delta(G) + \text{diam}_G + \text{tw}(G)$ as parameter does not admit a polynomial kernel, even on planar graphs, unless* NP $\subseteq$ coNP/poly.

In order to prove Theorem 3, we use a so-called OR-cross-composition due to Bodlaender et al. [5]. Therein, one uses a *polynomial equivalence relation* $\mathcal{R}$ which is an equivalence relation that is decidable in polynomial time and for each finite set S, the number of equivalence classes with respect to $\mathcal{R}$, that is, $|\{[s]_{\mathcal{R}} \mid s \in S\}|$, is polynomially bounded in the size of the largest element in S.

Definition 1 (OR-cross-composition [5]**).** *Let $L \subseteq \Sigma^*$ be some problem and $Q = (P, \kappa)$ with $P \subseteq \Sigma^*$ and $\kappa : \Sigma^* \to \mathbb{N}$ be some parametrised problem. Furthermore, let $\mathcal{R}$ be a polynomial equivalence relation on Σ^*. An* OR-cross-composition *is an algorithm that gets instances $\mathcal{I}_1, \ldots, \mathcal{I}_q$ of L as input, all of them belonging to the same equivalence class of $\mathcal{R}$, and outputs an instance $\mathcal{I}$ of Q such that*

- *$\mathcal{I} \in P$ if and only if there is at least one i such that $\mathcal{I}_i \in L$ and*
- *$\kappa(\mathcal{I})$ is polynomially bounded in $\max\{|\mathcal{I}_i| \mid i = 1, \ldots, q\} + \log q$.*

If there is an OR-cross-composition from an NP-hard problem L to some parametrised problem Q, then, unless NP $\subseteq$ coNP/poly, Q does not admit a polynomial-size kernel [5]. Using this result, we give an OR-cross-composition to prove Theorem 3. Our construction contains binary trees and we use the following structural result on binary trees with respect to MSE.

Lemma 4 (⋆). *Let T be a balanced, binary and complete tree of height h with root s, where additionally all leaves are identified with the target t. Then the only solutions for an MSE-instance (T, s, t, p, k) with $p \geq h + 3$ and $k \leq h$ are to share a complete path from s to some leaf, which is only possible for $k = h$.*

Proof (of Theorem 3). We apply the OR-cross-composition framework with MSE on planar graphs where s and t lie on the outer face as input problem. The NP-hardness of this problem is shown in Theorem 1.

We say an instance (G, s, t, p, k) of MSE is *malformed* if $\text{dist}_G(s, t) \leq k$ (trivial yes-instances), if s and t are not connected, if $p \geq 2 \cdot |E(G)|$ and $k < \text{dist}_G(s, t)$ (trivial no-instances), or if $p \leq 2$. Note that in the last case we can decide the instance in polynomial time, since MSE is fixed-parameter tractable with respect to p [9]. Hence we can decide each malformed instance in polynomial time.

We define the equivalence relation $\mathcal{R}$ as follows: two instances (G, s, t, p, k) and (G', s', t', p', k') are $\mathcal{R}$-equivalent if both are malformed or if $p = p'$ and $k = k'$. Observe that $\mathcal{R}$ is a polynomial equivalence relation.

Let $\mathcal{I}_i = (G_i, s_i, t_i, p, k)_{1 \leq i \leq q}$ be non-malformed $\mathcal{R}$-equivalent instances of MSE. We assume q to be a power of 2 (as otherwise we duplicate instances until it is). We first construct a complete binary tree T_s rooted in s of depth $\log(q)$ such that the s_i are the leaves of T_s, occurring in their canonical order. Conversely, we construct a tree T_t with root t and leaves t_i. We subdivide each edge in T_s and T_t to obtain paths of length $k+1$. In this way we obtain a new graph $G = (V, E)$ with $V := V(T_s) \cup V(T_t) \cup \bigcup_{i=1}^{q} V(G_i)$ and $E := E(T_s) \cup E(T_t) \cup \bigcup_{i=1}^{q} E(G_i)$. Also, we define the new parameters $k' := 2\log(q) \cdot (k+1) + k$ and $p' := p + \log(q)$ and get the instance $\mathcal{I} := (G, s, t, p', k')$, see Fig. 3. Now we claim that $\mathcal{I}$ is a yes-instance if and only if there is an $\mathcal{I}_y$ with $1 \leq y \leq q$ that is a yes-instance. The correctness proof is deferred to a full version $(\star)$. □

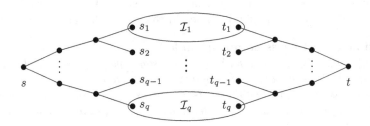

Fig. 3. Construction of $\mathcal{I}$ via an OR-cross-composition. The instances are connected by complete binary trees with roots s and t, respectively.

Recall that DMSE is NP-hard on planar acyclic digraphs with s and t lying on the outerface (Theorem 2). Hence, replacing the input instances by instances from DMSE on the aforementioned graphs, and directing the remaining edges in the trees away from s and towards t allows us to also exclude polynomial kernels for DMSE parametrised[1] by $p + k + \Delta_{\text{in}}(G) + \Delta_{\text{out}}(G)$.

Corollary 1 *Unless* NP $\subseteq$ coNP/poly, *DMSE on planar acyclic digraphs with parameter* $\kappa(G, s, t, p, k) := p + k + \Delta_{\text{in}}(G) + \Delta_{\text{out}}(G)$ *does not admit a polynomial kernel.*

---

[1] DMSE is in FPT when parametrised by $p + k$ since the search tree algorithm solving MSE in $O((p-1)^k \cdot (|V| + |E|)^2)$ time [9] can easily be adapted to the directed case.

5   Conclusion

On the positive side, we proved that MINIMUM SHARED EDGES on bounded grids is solvable in linear time when both dimensions are either small or large compared to the number p of paths. On the negative side, we proved that MSE becomes NP-hard on subgraphs of the bounded grid, even if the subgraph is directed and acyclic, and that it does not allow for polynomial kernels on planar graphs when parametrised by a combined parameter $k + p + \Delta + \mathsf{diam} + \mathsf{tw}$, unless $\mathsf{NP} \subseteq \mathsf{coNP/poly}$.

We conjecture that MSE on p-narrow grids is solvable in polynomial time. In particular, we find it interesting whether an arithmetic criterion similar to the p-large case (cf. Lemma 2) exists. Furthermore, in our reduction from VERTEX COVER, the construction yields a grid with a large amount of edges removed by taking a subgraph. Is MSE parametrised by the number of edges removed from the grid in FPT (or even admits a polynomial-size problem kernel)?

We consider it as interesting to study DMSE on *Manhattan street networks* (cf. [16]). Recently, MSE is considered with an additional time-aspect [10]. Herein, on a high level, an edge is shared if it appears in at least two paths at the same time. Another future research direction could be to study MSE with the additional time-aspect on grid-like graphs.

References

1. Alanko, S., Crevals, S., Isopoussu, A., Östergård, P.R.J., Pettersson, V.: Computing the domination number of grid graphs. Electron. J. Comb. **18**(1), 18 p. (2011). http://www.combinatorics.org/ojs/index.php/eljc/article/view/v18i1p141
2. Aoki, Y., Halldórsson, B.V., Halldórsson, M.M., Ito, T., Konrad, C., Zhou, X.: The minimum vulnerability problem on specific graph classes. J. Comb. Optim. **32**(4), 1288–1304 (2016)
3. Assadi, S., Emamjomeh-Zadeh, E., Norouzi-Fard, A., Yazdanbod, S., Zarrabi-Zadeh, H.: The minimum vulnerability problem. Algorithmica **70**(4), 718–731 (2014)
4. Bhatia, D., Leighton, T., Makedon, F., Norton, C.H.: Improved algorithms for routing on two-dimensional grids. In: Mayr, E.W. (ed.) WG 1992. LNCS, vol. 657, pp. 114–122. Springer, Heidelberg (1993). doi:10.1007/3-540-56402-0_41
5. Bodlaender, H.L., Jansen, B.M., Kratsch, S.: Kernelization lower bounds by cross-composition. SIAM J. Discrete Math. **28**(1), 277–305 (2014)
6. Cygan, M., Fomin, F.V., Kowalik, L., Lokshtanov, D., Marx, D., Pilipczuk, M., Pilipczuk, M., Saurabh, S.: Parameterized Algorithms. Springer, Cham (2015). doi:10.1007/978-3-319-21275-3
7. Diestel, R.: Graph Theory. GTM, 4th edn. Springer, Heidelberg (2010). doi:10.1007/978-3-662-53622-3. vol. 173
8. Fluschnik, T.: The parameterized complexity of finding paths with shared edges. Master's thesis, TU Berlin, March 2015. http://fpt.akt.tu-berlin.de/publications/theses/MA-till-fluschnik.pdf

9. Fluschnik, T., Kratsch, S., Niedermeier, R., Sorge, M.: The parameterized complexity of the minimum shared edges problem. In: Proceedings of the 35th IARCS Annual Conference on Foundation of Software Technology and Theoretical Computer Science, FSTTCS' 2015. LIPIcs, vol. 45, pp. 448–462. Schloss Dagstuhl - Leibniz-Zentrum fuer Informatik (2015)
10. Fluschnik, T., Morik, M., Sorge, M.: The complexity of routing with few collisions. In: Klasing, R., Zeitoun, M. (eds.) FCT 2017. LNCS, vol. 10472, pp. 257–270. Springer, Berlin (2017). doi:10.1007/978-3-662-55751-8_21
11. Fluschnik, T., Sorge, M.: The minimum shared edges problem on planar graphs. arXiv preprint arXiv:1602.01385 (2016)
12. Garey, M.R., Johnson, D.S.: Computers and Intractability: A Guide to the Theory of NP-Completeness. W. H. Freeman, New York (1979)
13. Jelínek, V.: The rank-width of the square grid. Discrete Appl. Math. **158**(7), 841–850 (2010)
14. Kanchanasut, K.: A shortest-path algorithm for Manhattan graphs. Inf. Process. Lett. **49**(1), 21–25 (1994)
15. Kanté, M.M., Moataz, F.Z., Momège, B., Nisse, N.: Finding paths in grids with forbidden transitions. In: Mayr, E.W. (ed.) WG 2015. LNCS, vol. 9224, pp. 154–168. Springer, Heidelberg (2016). doi:10.1007/978-3-662-53174-7_12
16. Maxemchuk, N.: Routing in the Manhattan street network. IEEE Trans. Commun. **35**(5), 503–512 (1987)
17. Omran, M.T., Sack, J.R., Zarrabi-Zadeh, H.: Finding paths with minimum shared edges. J. Comb. Optim. **26**(4), 709–722 (2013)
18. Varvarigos, E.A.: Optimal communication algorithms for Manhattan street networks. Discrete Appl. Math. **83**(1–3), 303–326 (1998)
19. Ye, Z.Q., Li, Y.M., Lu, H.Q., Zhou, X.: Finding paths with minimum shared edges in graphs with bounded treewidth. In: Proceedings of the 9th International Conference on Foundations of Computer Science, FCS 2013, pp. 40–46 (2013)

Linearly χ-Bounding (P_6, C_4)-Free Graphs

Serge Gaspers and Shenwei Huang$^{(\boxtimes)}$

School of Computer Science and Engineering, University of New South Wales,
Sydney, NSW 2052, Australia
dynamichuang@gmail.com

Abstract. Given two graphs H_1 and H_2, a graph G is (H_1, H_2)-free if it contains no subgraph isomorphic to H_1 or H_2. Let P_t and C_s be the path on t vertices and the cycle on s vertices, respectively. In this paper we show that for any (P_6, C_4)-free graph G it holds that $\chi(G) \leq \frac{3}{2}\omega(G)$, where $\chi(G)$ and $\omega(G)$ are the chromatic number and clique number of G, respectively. Our bound is attained by C_5 and the Petersen graph. The new result unifies previously known results on the existence of linear χ-binding functions for several graph classes. Our proof is based on a novel structure theorem on (P_6, C_4)-free graphs that do not contain clique cutsets. Using this structure theorem we also design a polynomial time 3/2-approximation algorithm for coloring (P_6, C_4)-free graphs. Our algorithm computes a coloring with $\frac{3}{2}\omega(G)$ colors for any (P_6, C_4)-free graph G in $O(n^2 m)$ time.

1 Introduction

All graphs in this paper are finite and simple. We say that a graph G *contains* a graph H if H is isomorphic to an induced subgraph of G. A graph G is H-free if it does not contain H. For a family of graphs $\mathcal{H}$, G is $\mathcal{H}$-free if G is H-free for every $H \in \mathcal{H}$. In case that $\mathcal{H}$ consists of two graphs, we simply write (H_1, H_2)-free instead of $\{H_1, H_2\}$-free. As usual, let P_t and C_s denote the path on t vertices and the cycle on s vertices, respectively. The complete graph on n vertices is denoted by K_n. For two graphs G and H, we use $G + H$ to denote the *disjoint union* of G and H. The *join* of G and H, denoted by $G \vee H$, is the graph obtained by taking the disjoint union of G and H and adding an edge between every vertex in G and every vertex in H. For a positive integer r, we use rG to denote the disjoint union of r copies of G. The *complement* of G is denoted by $\overline{G}$. The *girth* of G is the length of the shortest cycle in G. A q-coloring of a graph G is a function $\phi : V(G) \longrightarrow \{1, \ldots, q\}$ such that $\phi(u) \neq \phi(v)$ whenever u and v are adjacent in G. The *chromatic number* of a graph G, denoted by $\chi(G)$, is the minimum number q for which there exists a q-coloring of G. The *clique number* of G, denoted by $\omega(G)$, is the size of a largest clique in G. Obviously, $\chi(G) \geq \omega(G)$ for any graph G.

A family $\mathcal{G}$ of graphs is said to be χ-bounded if there exists a function f such that for every graph $G \in \mathcal{G}$ and every induced subgraph H of G it holds that $\chi(H) \leq f(\omega(H))$. The function f is called a χ-binding function for $\mathcal{G}$. The class

© Springer International Publishing AG 2017
H.L. Bodlaender and G.J. Woeginger (Eds.): WG 2017, LNCS 10520, pp. 263–274, 2017.
https://doi.org/10.1007/978-3-319-68705-6_20

of perfect graphs (a graph G is *perfect* if for every induced subgraph H of G it holds that $\chi(H) = \omega(H)$), for instance, is a χ-bounded family with χ-binding function $f(x) = x$. Therefore, χ-boundedness is a generalization of perfection. The notion of χ-bounded families was introduced by Gyárfás [15] who posed the following two meta problems:

- Does there exist a χ-binding function f for a given family $\mathcal{G}$ of graphs?
- Does there exist a *linear* χ-binding function f for $\mathcal{G}$?

The two problems have received considerable attention for hereditary classes. Hereditary classes are exactly those classes that can be characterized by *forbidden induced subgraphs*. What choices of forbidden induced subgraphs guarantee that a family of graphs is χ-bounded? Since there are graphs with arbitrarily large chromatic number and girth [11], at least one forbidden subgraph has to be acyclic. Gyárfás [14] conjectured that this necessary condition is also a sufficient condition for a hereditary class to be χ-bounded.

Conjecture 1 (Gyárfás) [14]). For every forest T, the class of T-free graphs is χ-bounded.

Gyárfás [15] proved the conjecture for $T = P_t$: every P_t-free graph G has $\chi(G) \leq (t-1)^{\omega(G)-1}$. Note that this χ-binding function is exponential in $\omega(G)$. Therefore, it is natural to ask whether there exists a linear χ-binding function for P_t-free graphs. Unfortunately, unless $t \leq 4$ in which case every P_t-free graph is perfect and hence has $\chi(G) = \omega(G)$, no linear χ-binding function exists for P_t-free graphs when $t \geq 5$ [12]. In fact, as observed in [17], the class of H-free graphs admits a linear χ-binding function if and only if H is contained in a P_4.

However, if an additional graph is forbidden, then the class could become linearly χ-bounded again. Choudum et al. [7] derived a linear χ-binding function for $(P_6, P_4 \vee P_1)$-free graphs, $(P_5, P_4 \vee P_1)$-free graphs and $(P_5, C_4 \vee P_1)$-free graphs. In the same paper, they also obtained the optimal χ-binding function $f(x) = \lceil \frac{5}{4}x \rceil$ for (P_5, C_4)-free graphs, improving a result in [12]. Later on, the same set of authors [8] obtained linear χ-binding functions for certain subclasses of $3P_1$-free graphs (thus subclasses of P_5-free graphs). In particular, they showed that the class of $(3P_1, K_4 + P_1)$-free graphs has a linear χ-binding function $f(x) = 2x$. Henning et al. [16] obtained an improved χ-binding function $f(x) = \frac{3}{2}x$ for $(3P_1, K_4 + P_1)$-free graphs.

An important subclass of P_5-free graphs is the class of $2P_2$-free graphs. It was known that for any $2P_2$-free graph it holds that $\chi \leq \binom{\omega+1}{2}$ [18]. For a slightly larger class, namely $P_2 + P_3$-free graphs, Bharathi and Choudum [1] gave an $O(\omega^3)$ bound on χ. Brause et al. [5] recently showed that $(P_5, butterfly)$-free graphs and $(P_5, hammer)$-free graphs, both of which are superclasses of $2P_2$-free graphs due to a recent structural result [9], admit cubic and quadratic χ-binding functions, respectively, where a *butterfly* is a graph isomorphic to $2P_2 \vee P_1$ and a *hammer* is a graph on five vertices $\{a, b, c, d, e\}$ where a, b, c, d in this order induces a P_4 and e is adjacent to a and b. It is not known whether any of these χ-binding functions can be improved to linear. Very recently, a linear χ-binding function has been shown to exist for $(2P_2, H)$-free graphs when H is one of $(P_1 + P_2) \vee P_1$ (usually referred to as *paw*), $P_4 \vee P_1$ (usually referred to as *gem*) or $\overline{P_5}$ (usually referred to as *house*) [5]. When H is isomorphic to C_4, it was known [2] that every such graph has $\chi \leq \omega + 1$; when H is $P_2 \vee 2P_1$ (usually referred to as *diamond*), it was known that $\chi \leq \omega + 3$. This bound in fact holds for $(P_2 + P_3, diamond)$-free graphs [1]. For more results on χ-binding functions, we refer to a survey by Randerath and Schiermeyer [17].

Our Contributions. In this paper, we prove that $f(x) = \frac{3}{2}x$ is a χ-binding function for (P_6, C_4)-free graphs. This unifies several previous results on the existence of linear χ-binding functions for, e.g., $(2P_2, C_4)$-free graphs [2], (P_5, C_4)-free graphs [7] and $(P_3 + P_2, C_4)$-free graphs [6]. On the other hand, there is an active research on classifying the complexity of coloring (H_1, H_2)-free graphs. Despite much effort, the classification is far from being complete, see [13] for a summary of partial results. The class of (P_6, C_4)-free graphs is one of the unknown cases. Here we develop an $O(n^2 m)$ 3/2-approximation algorithm for coloring (P_6, C_4)-free graphs. This is the first approximation algorithm for coloring these graphs and could be viewed as a first step towards a possible polynomial time algorithm for optimally coloring these graphs.

The remainder of the paper is organized as follows. We present some preliminaries in Sect. 2 and prove a novel structure theorem for (P_6, C_4)-free graphs without clique cutsets in Sect. 3. Using this theorem we show in Sect. 4 that every (P_6, C_4)-free graph has chromatic number at most 3/2 its clique number. Finally, we turn our proof into a 3/2-approximation algorithm in Sect. 5.

2   Preliminaries

For general graph theory notation we follow [3]. Let $G = (V, E)$ be a graph. The *neighborhood* of a vertex v, denoted by $N_G(v)$, is the set of neighbors of v. For a set $X \subseteq V(G)$, let $N_G(X) = \bigcup_{v \in X} N_G(v) \setminus X$. The *degree* of v, denoted by $d_G(v)$, is equal to $|N_G(v)|$. For $x \in V$ and $S \subseteq V$, we denote by $N_S(x)$ the set of neighbors of x that are in S, i.e., $N_S(x) = N_G(x) \cap S$. For $X, Y \subseteq V$, we say that X is *complete* (resp. *anti-complete*) to Y if every vertex in X is adjacent (resp. non-adjacent) to every vertex in Y. A vertex subset $K \subseteq V$ is a *clique cutset* if $G - K$ has more components than G and K induces a clique. A vertex is *universal* in G if it is adjacent to all other vertices. For $S \subseteq V$, the subgraph *induced* by S, is denoted by $G[S]$. A subset $M \subseteq V$ is a *dominating set* if every vertex not in M has a neighbor in M. We say that M is a *module* if every vertex not in M is either complete or anti-complete to M.

Let $u, v \in V$. We say that u and v are *twins* if u and v are adjacent and they have the same set of neighbors in $V \setminus \{u, v\}$. Note that the binary relation of being twins on V is an equivalence relation and so V can be partitioned into equivalence classes $T_1, \ldots, T_r$ of twins. The *skeleton* of G is the subgraph induced by a set of r vertices, one from each of $T_1, \ldots, T_r$. A *blow-up* of a graph G is a graph G' obtained by replacing each vertex v of G with a clique K_v of size at least 1 such that K_v and K_u are complete in G' if u and v are adjacent in G,

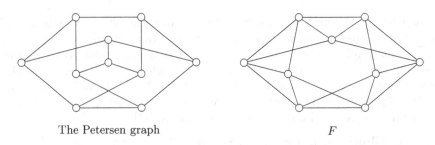

The Petersen graph   F

Fig. 1. Two smallest (P_6, C_4)-free atoms that do not contain any small vertex.

and anti-complete otherwise. Since each equivalence class of twins is a clique and any two equivalence classes are either complete or anti-complete, every graph is a blow-up of its skeleton.

A graph is *chordal* if it does not contain any induced cycle of length at least four. The following structure of (P_6, C_4)-free graphs discovered by Brandstädt and Hoàng [4] is of particular importance in our proofs below.

Lemma 1 (Brandstädt and Hoàng [4]). *Let G be a (P_6, C_4)-free graph without clique cutsets. Then the following statements hold: (i) every induced C_5 is dominating; (ii) If G contains an induced C_6 which is not dominating, G is the join of a blow-up of the Petersen graph (Fig. 1) and a (possibly empty) clique.*

3   The Structure of (P_6, C_4)-Free Atoms

A graph without clique cutsets is called an *atom*. We say that a vertex v in G is *small* if $d_G(v) \leq \frac{3}{2}\omega(G) - 1$. Our main result in this section is the following.

Theorem 1. *Every (P_6, C_4)-free atom either contains a small vertex or is the join of a blow-up of the Petersen graph or F (see Fig. 1) and a (possibly empty) clique.*

To prove the above theorem, we shall prove a number of lemmas below. Let $G = (V, E)$ be a graph and H be an induced subgraph of G. We partition $V \setminus V(H)$ into subsets with respect to H as follows: for any $X \subseteq V(H)$, we denote by $S(X)$ the set of vertices in $V \setminus V(H)$ that have X as their neighborhood among $V(H)$, i.e.,

$$S(X) = \{v \in V \setminus V(H) : N_{V(H)}(v) = X\}.$$

For $0 \leq j \leq |V(H)|$, we denote by S_j the set of vertices in $V \setminus V(H)$ that have exactly j neighbors among $V(H)$. Note that $S_j = \bigcup_{X \subseteq V(H):|X|=j} S(X)$. We say that a vertex in S_j is a *j-vertex*.

The idea is that we assume the occurrence of some induced subgraph H in G and then argue that the theorem holds in this case. Afterwards, we can assume that G is H-free in addition to being (P_6, C_4)-free. We then pick a

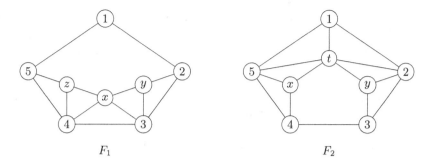

Fig. 2. Two special graphs F_1 and F_2.

different induced subgraph as H and repeat. In the end, we are able to show that the theorem holds if G contains a C_5 or C_6 (Lemmas 3 and 5). Therefore, the remaining case is that G is chordal. In this case, the theorem follows from a well-known fact [10] that every chordal graph has a *simplicial* vertex, that is, a vertex whose neighborhood induces a clique. As straightforward as the approach sounds, the difficulty is that in order to eliminate C_5 and C_6 we have to eliminate two more special graphs F_1 and F_2 (Lemmas 2 and 4) and do it in the 'right' order. We start with F_1.

Lemma 2. *If a* (P_6, C_4)*-free atom* G *contains* F_1 *(see Fig. 2), then* G *contains a small vertex.*

Proof. Let G be a (P_6, C_4)-free atom that contains an induced subgraph H that is isomorphic to F_1 with $V(H) = \{1, 2, 3, 4, 5, x, y, z\}$ where $1, 2, 3, 4, 5, 1$ induces the *underlying* five-cycle C of F_1 and x is adjacent to 3 and 4, y is adjacent to 2 and 3, z is adjacent to 4 and 5, and x is adjacent to y and z, see Fig. 2. We partition $V(G)$ with respect to C. We choose H such that $|S_2|$ maximized. Note that $x \in S(3, 4)$, $y \in S(2, 3)$ and $z \in S(4, 5)$. All indices below are modulo 5. Since G is an atom, it follows from Lemma 1 that $S_0 = \emptyset$. Moreover, it follows immediately from the (P_6, C_4)-freeness of G that $V(G) = C \cup S_1 \cup \bigcup_{i=1}^{5} S(i, i+1) \cup \bigcup_{i=1}^{5} S(i-1, i, i+1) \cup S_5$. Observe that if v is a small vertex in the graph obtained from G by deleting all universal vertices, then v is also small in G. Therefore, we may also assume that G has no universal vertices.

(1) $S_5 \cup S(i-1, i, i+1)$ *is a clique.*
    Suppose not. Let u and v be two non-adjacent vertices in $S_5 \cup S(i-1, i, i+1)$. Then $\{u, i-1, v, i+1\}$ induces a C_4, a contradiction.   ∎

(2) $S(i-1, i, i+1)$ *and* $S(i+1, i+2, i+3)$ *are anti-complete.*
    By symmetry, it suffices to prove (2) for $i = 1$. Suppose that $u \in S(5, 1, 2)$ is adjacent to $v \in S(2, 3, 4)$. Then $\{5, 4, v, u\}$ induces a C_4, and this is a contradiction.   ∎

(3) $S(i, i+1)$ *and* $S(i+1, i+2)$ *are complete, and* $S(i, i+1)$ *and* $S(i+3, i+4)$ *are anti-complete. Moreover, if both* $S(i, i+1)$ *and* $S(i+1, i+2)$ *are not empty, then both sets are cliques.*

It suffices to prove (3) for $i = 1$. Suppose that $u \in S(1,2)$ is not adjacent to $v \in S(2,3)$. Then $u, 1, 5, 4, 3, v$ induces a P_6, a contradiction. If $S(1,2)$ and $S(2,3)$ are not empty, then it follows from the C_4-freeness of G that both sets are cliques. Similarly, if $u \in S(1,2)$ is adjacent to $w \in S(3,4)$, then $\{2, 3, w, u\}$ induces a C_4. ∎

(4) $S(i)$ and $S(i+1)$ are anti-complete, and $S(i)$ and $S(i+2)$ are complete.

It suffices to prove the statement for $i = 1$. If $u \in S(1)$ is adjacent to $v \in S(2)$, then $\{1, 2, v, u\}$ induces a C_4, a contradiction. Similarly, if $u \in S(1)$ is not adjacent to $w \in S(3)$, then $u, 1, 5, 4, 3, w$ induces a P_6. ∎

Remark. (1)–(4) holds whenever we partition $V(G)$ with respect to a C_5. They will be used in the proof of Lemmas 4 and 5.

(5) $S_2 = S(2,3) \cup S(3,4) \cup S(4,5)$

Recall that $x \in S(3,4)$, $y \in S(2,3)$ and $z \in S(4,5)$. By symmetry, it suffices to show that $S(1,2) = \emptyset$. Suppose that $S(1,2)$ contains one vertex, say s. Then s is not adjacent to x and z, and x and z are adjacent by (3). This implies that $5, z, x, 3, 2, s$ induces a P_6, a contradiction. ∎

(6) $S_1 = \emptyset$.

We first observe that one of $S(i)$ and $S(i+1, i+2)$ is empty for each i. By symmetry, it suffices to show this for $i = 1$. Suppose that $u \in S(1)$ and $v \in S(2,3)$. Then either $u, 1, 5, 4, 3, v$ induces a P_6 or $\{u, 1, 2, v\}$ induces a C_4, depending on whether u and v are adjacent. This is a contradiction. Then (6) follows from the fact that $S(i, i+1) \neq \emptyset$ for $i = 2, 3, 4$. ∎

(7) S_5 and S_2 are anti-complete

Let $u \in S_5$ be an arbitrary vertex. Note that any $x' \in S(3,4)$ and any $z' \in S(4,5)$ are adjacent by (3). Consider the induced six-cycle $C' = z', 5, 1, 2, 3, x', z'$. Since u is adjacent to $5, 1, 2, 3$, it follows from the C_4-freeness of G that u is either complete or anti-complete to $\{x', z'\}$. Similarly, u is either complete or anti-complete to $\{x', y'\}$ for any $x' \in S(3,4)$ and $y' \in S(2,3)$. This implies that u is either complete or anti-complete to S_2. If u is complete to S_2, then u is a universal in G by (1) and (6), which contradicts our assumption that G has no universal vertices. Therefore, the claim follows. ∎

(8) The following statements hold between subsets of S_3 and subsets of S_2.

(8i) $S(5,1,2)$ and S_2 are anti-complete.

Let $t \in S(5,1,2)$ be an arbitrary vertex. Suppose that t is adjacent to some vertex $x' \in S(3,4)$. Then $\{4, 5, t, x'\}$ induces a C_4, a contradiction. This shows that $S(5,1,2)$ and $S(3,4)$ are anti-complete. Suppose that t has a neighbor in $S(2,3) \cup S(4,5)$, say $y' \in S(2,3)$. Then either $1, t, y', 3, 4, z$ induces a P_6 or $\{t, z, x, y'\}$ induces a C_4, depending on whether t and z are adjacent. Therefore, (8i) follows. ∎

(8ii) $S(4,5,1)$ and $S(4,5)$ are complete. By symmetry, $S(1,2,3)$ and $S(2,3)$ are complete.

Let $t \in S(4,5,1)$ and $z' \in S(4,5)$ be two arbitrary vertices. Suppose that t and z' are not adjacent. Then either $t, 5, z', x, y, 2$ induces a P_6 or $\{t, 5, z', x\}$ induces a C_4, depending on whether t and x are adjacent. ∎

(8iii) $S(2,3,4)$ *and* $S(3,4)$ *are complete. By symmetry,* $S(3,4,5)$ *and* $S(3,4)$ *are complete.*

Let $t \in S(2,3,4)$ and $x' \in S(3,4)$ be two arbitrary vertices. Suppose that t and x' are not adjacent. Then either $t, 3, x', z, 5, 1$ induces a P_6 or $\{t, 3, x', z\}$ induces a C_4, depending on whether t and z are adjacent. ∎

(8iv) $S(2,3,4)$ *and* $S(2,3)$ *are complete. By symmetry,* $S(3,4,5)$ *and* $S(4,5)$ *are complete.*

Let $t \in S(2,3,4)$ and $y' \in S(2,3)$ be two arbitrary vertices. By (3) and (8iii), x is adjacent to both t and y'. So, t and y' are adjacent, for otherwise $\{t, x, y', 2\}$ induces a C_4. ∎

(9) *The following statements hold among subsets of* S_3.

 (9i) $S(5,1,2)$ *is complete to* $S(1,2,3)$ *and* $S(4,5,1)$.

By symmetry, it suffices to show that $S(5,1,2)$ is complete to $S(4,5,1)$. Suppose that $s \in S(5,1,2)$ is not adjacent to $t \in S(4,5,1)$. Note that s is not adjacent to y by (8i). Then $s, 1, t, 4, 3, y$ induces a P_6, and this is a contradiction. ∎

(9ii) *Let* $s \in S(3,4,5)$ *and* $t \in S(4,5,1)$ *such that* s *and* t *are not adjacent. Then* t *is anti-complete to* $S(3,4)$ *and* s *is complete to* $S(2,3)$.

Let $x' \in S(3,4)$ be an arbitrary vertex. First, x' and s are adjacent by (8iii). Moreover, x' and t are not adjacent, for otherwise $\{5, t, x', s\}$ induces a C_4. This proves the first part of (9ii). Now let $y' \in S(2,3)$ be an arbitrary vertex. By (3), y' is adjacent to x. If s and y' are not adjacent, then $t, 5, s, x, y', 2$ induces a P_6, a contradiction. This shows that s is complete to $S(2,3)$. ∎

(9iii) *Let* $s \in S(2,3,4)$ *and* $t \in S(3,4,5)$ *such that* s *and* t *are not adjacent. Then* s *(respectively* t*) is anti-complete to* $S(4,5)$ *(respectively* $S(2,3)$*).*

Let $z' \in S(4,5)$ be an arbitrary vertex. By (8iv), t is adjacent to z'. If s and z' are adjacent, then $\{s, z', t, 3\}$ induces a C_4, a contradiction. This proves that s is anti-complete to $S(4,5)$. By symmetry, t is anti-complete to $S(2,3)$. ∎

We distinguish two cases depending on whether S_5 is empty.

Case 1. S_5 contains a vertex u. We prove some additional properties of the graph with the existence of u.

(a) $S(3,4,5)$ *and* $S(2,3)$ *are anti-complete. By symmetry,* $S(2,3,4)$ *and* $S(4,5)$ *are anti-complete.*

Let $t \in S(3,4,5)$ and $y' \in S(2,3)$ be two arbitrary vertices. Suppose that t and y' are adjacent. By (1) and (7), u is adjacent to t but not adjacent to y'. Then $\{t, u, 2, y'\}$ induces a C_4, a contradiction. This proves the claim. ∎

(b) $S(4,5,1)$ *and* $S(3,4)$ *are anti-complete. By symmetry,* $S(1,2,3)$ *and* $S(3,4)$ *are anti-complete.*

Let $t \in S(4,5,1)$ and $x' \in S(3,4)$ be two arbitrary vertices. By (1) and (7), u is adjacent to t but not adjacent to x'. If t and x' are adjacent, then $\{t, u, 3, x'\}$ induces a C_4, a contradiction. ∎

(c) $S(2,3,4)$ *and* $S(3,4,5)$ *are complete.*

Let $s \in S(2,3,4)$ and $t \in S(3,4,5)$ be two arbitrary vertices. Then x is adjacent to both s and t by (8iii). By (1) and (7), u is adjacent to s and t but not adjacent to x. If s and t are not adjacent, then $\{x, s, u, t\}$ induces a C_4. ∎

(d) $S(4,5,1)$ *and* $S(3,4,5)$ *are complete. By symmetry,* $S(1,2,3)$ *and* $S(2,3,4)$ *are complete.*

This follows directly from (a) and (9ii). ∎

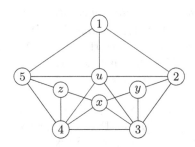

Fig. 3. The graph F_3.

It follows from (1)–(9) and (a)–(d) that G is a blow-up of a special graph F_3 (see Fig. 3). We denote by Q_v the clique that $v \in V(F_3)$ is blown into. Suppose first that $|Q_2| \leq \omega(G)/2$. Note that $N_G(y) = (Q_y \setminus \{y\}) \cup Q_x \cup Q_2 \cup Q_3$. Since $(N_G(y) \setminus Q_2) \cup \{y\}$ is a clique, it follows that $|N_G(y) \setminus Q_2| \leq \omega(G) - 1$. Therefore, $d_G(y) \leq \omega(G) - 1 + |Q_2| \leq \omega(G) - 1 + \omega(G)/2 = \frac{3}{2}\omega(G) - 1$. Now suppose that $|Q_2| > \omega(G)/2$. This implies that $|Q_1 \cup Q_u| < \omega(G)/2$. Note that $(N_G(5) \cup \{5\}) \setminus (Q_1 \cup Q_u)$ is a clique. Therefore, $d_G(5) \leq \omega(G) - 1 + |Q_1 \cup Q_u| \leq \frac{3}{2}\omega(G) - 1$.

Case 2. S_5 is empty.

(a) $S(1,2,3)$ *and* $S(2,3,4)$ *are complete. By symmetry,* $S(4,5,1)$ *and* $S(3,4,5)$ *are complete.*

Suppose that $s \in S(2,3,4)$ and $r \in S(1,2,3)$ are not adjacent. By (9ii) and (8ii), r is complete to $S(2,3)$ and anti-complete to $S(3,4)$. Note that $V(H) \setminus \{2\} \cup \{r\}$ also induces a subgraph H' that is isomorphic to F_1 whose underlying five-cycle is $C' = C \setminus \{2\} \cup \{r\}$. Clearly, s is adjacent to exactly two vertices on C'. Therefore, the number of 2-vertices with respect to C' is more than that with respect to C, and this contradicts the choice of H. ∎

(b) $S(2,3,4)$ *and* $S(3,4,5)$ *are complete.*

Suppose that $s \in S(2,3,4)$ and $r \in S(3,4,5)$ are not adjacent. By (8iii) and (8iv), s is complete to $S(2,3) \cup S(3,4)$. By (9iii), s is anti-complete to $S(4,5)$. Note that $V(H) \setminus \{3\} \cup \{s\}$ also induces a subgraph H' that is isomorphic to F_1 whose underlying five-cycle is $C' = C \setminus \{3\} \cup \{s\}$. Clearly, r

is adjacent to exactly two vertices in C'. Therefore, the number of 2-vertices with respect to C' is more than that with respect to C, and this contradicts the choice of H.   ∎

By (a), (b), (8) and (9), $S(i-1, i, i+1)$ and $S(i, i+1, i+2)$ are complete, $S(2, 3)$ is complete to $S(1, 2, 3) \cup S(2, 3, 4)$ and anti-complete to $S(4, 5, 1) \cup S(5, 1, 2)$, $S(4, 5)$ is complete to $S(3, 4, 5) \cup S(4, 5, 1)$ and anti-complete to $S(5, 1, 2) \cup S(1, 2, 3)$, and $S(3, 4)$ is complete to $S(2, 3, 4) \cup S(3, 4, 5)$ and anti-complete to $S(5, 1, 2)$, see Fig. 4.

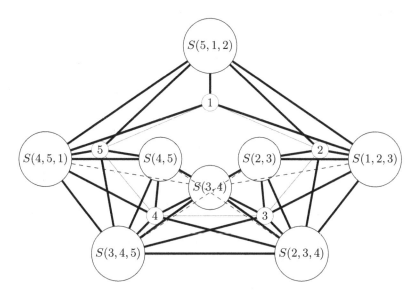

Fig. 4. The structure of G. A thick line between two sets represents that the two sets are complete, and a dotted line represents that the edges between the two sets can be arbitrary. Two sets are anti-complete if there is no line between them.

Let $Q_i = S(i-1, i, i+1) \cup \{i\}$ for $i \in \{1, 2, 3, 4, 5\}$. Suppose first that $|Q_2| \le \omega(G)/2$. Then $d_G(1) = |S(5, 1, 2) \cup Q_5| + |Q_2| \le \omega(G) - 1 + \omega(G)/2 = \frac{3}{2}\omega(G) - 1$. Now suppose that $|Q_2| > \omega(G)/2$. This implies that $|Q_1| < \omega(G)/2$. Therefore, $d_G(5) = |S(4, 5, 1) \cup Q_4 \cup S(4, 5)| + |Q_2| \le \omega(G) - 1 + \omega(G)/2 = \frac{3}{2}\omega(G) - 1$. This completes the proof of Case 2.   □

Due to page limits we omit the proof of the following lemmas.

Lemma 3. *If a (P_6, C_4, F_1)-free atom G contains a C_6, then G either contains a small vertex or is the join of a blow-up of the Petersen graph or F and a (possibly empty) clique (see Fig. 1 for the Petersen graph or F).*

Lemma 4. *If a (P_6, C_4, F_1, C_6)-free atom G contains an F_2 (see Fig. 2), then G contains a small vertex.*

Lemma 5. *If a $(P_6, C_4, F_1, C_6, F_2)$-free atom G contains a C_5, then G contains a small vertex.*

We are now ready to prove Theorem 1.

Proof (of Theorem 1). Let G be a (P_6, C_4)-free atom. It follows from Lemmas 2–5 that we can assume that G is also (C_6, C_5)-free. Therefore, G is chordal. It is well-known [10] that every chordal graph contains a vertex of degree at most $\omega(G) - 1$ and so this vertex is small. This completes the proof.

4 χ-Bounding (P_6, C_4)-Free Graphs

In this section, we shall prove the main result of this paper, that is, every (P_6, C_4)-free graph has $\chi \leq \frac{3}{2}\omega$. For that purpose, we need one additional lemma.

Lemma 6. *Let G be a graph and let H be the skeleton of G. If $\chi(H) \leq 3$, then $\chi(G) \leq \frac{3}{2}\omega(G)$.*

Proof. We prove this by induction on $|V(G)|$. The base case is that G is its own skeleton. Our assumption implies that $\chi(G) \leq 3$. If $\omega(G) \geq 2$, then it follows that $\chi(G) \leq \frac{3}{2}\omega(G)$. Otherwise, $\omega(G) = 1$, i.e., G is an independent set. So, $\chi(G) = 1 < \frac{3}{2}\omega(G)$. Now suppose that the lemma is true for any graph G' with $|V(G')| < |V(G)|$. If G contains an isolated vertex, then applying the inductive hypothesis to $G - v$ shows that $\chi(G) \leq \frac{3}{2}\omega(G)$. So, G does not contain any isolated vertex. Therefore, H does not contain any isolated vertex either. So, any vertex of H lies in a maximal clique of H with size at least 2. This implies that $\omega(G') \leq \omega(G) - 2$, where $G' = G - H$. Note that the skeleton H' of G' is an induced subgraph of H and so $\chi(H') \leq 3$. By the inductive hypothesis it follows that $\chi(G') \leq \frac{3}{2}\omega(G')$. Finally,

$$\chi(G) \leq \chi(G') + \chi(H) \leq \frac{3}{2}\omega(G') + \chi(H) \leq \frac{3}{2}(\omega(G) - 2) + 3 = \frac{3}{2}\omega(G).$$

This completes our proof.   □

Now we are ready to prove the main result of this paper.

Theorem 2. *Every (P_6, C_4)-free graph G has $\chi(G) \leq \frac{3}{2}\omega(G)$.*

Proof. We use induction on $|V(G)|$. We may assume that G is connected, for otherwise we apply the inductive hypothesis on each connected component. If G contains a clique cutset K that disconnects H_1 from H_2, let $G_i = G[H_i \cup K]$ for $i = 1, 2$. Then the inductive hypothesis implies that $\chi(G_i) \leq \frac{3}{2}\omega(G_i)$ for $i = 1, 2$. Note that $\chi(G) = \max\{\chi(G_1), \chi(G_2)\}$ and so $\chi(G) \leq \frac{3}{2}\omega(G)$. Now G is an atom. If G contains a universal vertex u, then applying the inductive hypothesis to $G - u$ implies that $\chi(G - u) \leq \frac{3}{2}\omega(G - u)$. Since $\chi(G) = \chi(G - u) + 1$ and $\omega(G) = \omega(G - u) + 1$, it follows that $\chi(G) \leq \frac{3}{2}(\omega(G) - 1) + 1 < \frac{3}{2}\omega(G)$. So, G is an atom without universal vertices. It follows then from Theorem 1 that G is either contains a small vertex v or is a blow-up of the Petersen graph or F_3. If G contains a small vertex v, then applying the inductive hypothesis to $G - v$ gives us that $\chi(G - v) \leq \frac{3}{2}\omega(G - v) \leq \frac{3}{2}\omega(G)$. Since v has degree at most $\frac{3}{2}\omega(G) - 1$, it follows that $\chi(G) = \chi(G') \leq \frac{3}{2}\omega(G)$. So, we assume that G is a blow-up of the Petersen graph or F. In other words, the skeleton of G is the Petersen graph or F. It is straightforward to check that both graphs have chromatic number 3. Therefore, $\chi(G) \leq \frac{3}{2}\omega(G)$ by Lemma 6. This completes our proof. $\square$

5   A 3/2-Approximation Algorithm

In this section, we give a polynomial time 3/2-approximation algorithm for coloring (P_6, C_4)-free graphs. The general idea is to decompose the input graph G in the following way to obtain a decomposition tree $T(G)$ where the leaves of $T(G)$ are 'basic' graphs which we know how to color and the internal nodes of $T(G)$ are subgraphs of G that are decomposed via either clique cutsets or small vertices. Since both clique cutsets and small vertices 'preserve' the colorability of graphs, a bottom-up approach on $T(G)$ will give us a coloring of G using at most $\frac{3}{2}\omega(G)$ colors. Formally, let G be a connected (P_6, C_4)-free graph. If G has a clique cutset K, then $G - K$ is a disjoint union of two subgraphs H_1 and H_2 of G. We let $G_i = H_i \cup K$ for $i = 1, 2$ and decompose G into G_1 and G_2. On the other hand, if G does not contain any clique cutset but contains a small vertex v that is not universal, then we decompose G into $G - v$. We then further decompose G_1 and G_2 or $G - v$ in the same way until either the graph has no clique cutsets and no small vertices or the graph is a clique. We refer to these subgraphs that are not further decomposed as *strong atoms*. The decomposition procedure can be represented by a binary tree $T(G)$ whose root is G, and G may have two children G_1 and G_2 or only one child $G - v$, depending on the way G is decomposed. Each leaf in $T(G)$ corresponds to a strong atom. Note that if a strong atom is not a clique, then it is the join of a blow-up of the Petersen graph or F and a clique by Theorem 1. The following lemma, whose proof we omit, says that there are only polynomially many nodes in $T(G)$ and $T(G)$ can be found in polynomial time.

Lemma 7. *$T(G)$ has $O(n^2)$ nodes and can be found in $O(mn^2)$ time.*

Theorem 3. *There is an $O(n^2m)$ algorithm to find a coloring of G that uses at most $\frac{3}{2}\omega(G)$ colors.*

Proof. The algorithm works as follows: (i) we first find $T(G)$; (ii) color each leaf X of $T(G)$ using at most $\frac{3}{2}\omega(X)$ colors; (iii) for an internal node Y, if Y has only one child $Y - y$, then color y with a color that is not used on its neighbors in Y; if Y has two children Y_1 and Y_2, then combine the colorings of Y_1 and Y_2 on the clique cutset that decomposes Y. The correctness follows from Lemma 6 and Theorem 2. It takes $O(mn^2)$ time to find $T(G)$ by Lemma 7. Moreover, it is easy to see that one can color a leaf X of $T(G)$ in time $O(n + m)$ time, and combining the coloring for a single decomposition step takes $O(n)$ time. Therefore, it takes $(O(m + n) + O(n))O(n^2) = O(n^2m)$ time to obtain a desired coloring of G. $\square$

References

1. Bharathi, A.P., Choudum, S.A.: Colouring of $(P_3 \cup P_2)$-free graphs. arXiv:1610.07177v1 [cs.DM] (2016)
2. Blázsik, Z., Hujter, M., Pluhár, A., Tuza, Z.: Graphs with no induced C_4 and $2K_2$. Discrete Math. **115**, 51–55 (1993)
3. Bondy, J.A., Murty, U.S.R.: Graph Theory. Graduate Texts in Mathematics, vol. 244. Springer, New York (2008)
4. Brandstädt, A., Hoàng, C.T.: On clique separators, nearly chordal graphs, and the maximum weight stable set problem. Theor. Comput. Sci. **389**, 295–306 (2007)
5. Brause, C., Randerath, B., Schiermeyer, I., Vumar, E.: On the chromatic number of $2K_2$-free graphs. In: Bordeaux Graph Workshop (2016)
6. Choudum, S.A., Karthick, T.: Maximal cliques in $\{P_2 \cup P_3, C_4\}$-free graphs. Discrete Math. **310**, 3398–3403 (2010)
7. Choudum, S.A., Karthick, T., Shalu, M.A.: Perfectly coloring and linearly χ-bound P_6-free graphs. J. Graph Theory **54**, 293–306 (2007)
8. Choudum, S.A., Karthick, T., Shalu, M.A.: Linear chromatic bounds for a subfamily of $3K_1$-free graphs. Graphs Comb. **24**, 413–428 (2008)
9. Dhanalakshmi, S., Sadagopan, N., Manogna, V.: On $2K_2$-free graphs-structural and combinatorial view. arXiv:1602.03802v2 [math.CO] (2016)
10. Dirac, G.A.: On rigid circuit graphs. Abhandlungen aus dem Mathematischen Seminar der Universität Hamburg **25**, 71–76 (1961)
11. Erdős, P.: Graph theory and probability. Canad. J. Math. **11**, 34–38 (1959)
12. Fouquet, J.L., Giakoumakis, V., Maire, F., Thuillier, H.: On graphs without P_5 and $\overline{P_5}$. Discrete Math. **146**, 33–44 (1995)
13. Golovach, P.A., Johnson, M., Paulusma, D., Song, J.: A survey on the computational complexity of coloring graphs with forbidden subgraphs. J. Graph Theory (to appear)
14. Gyárfás, A.: On Ramsey covering numbers. Coll. Math. Soc. János Bolyai **10**, 801–816 (1973)
15. Gyárfás, A.: Problems from the world surrounding perfect graphs. Zastosow. Mat. **19**, 413–431 (1987)
16. Henning, M.A., Löwenstein, C., Rautenbach, D.: Independent sets and matchings in subcubic graphs. Discrete Math. **312**, 1900–1910 (2012)
17. Randerath, B., Schiermeyer, I.: Vertex colouring and forbidden subgraphs-a survey. Graphs Comb. **20**, 1–40 (2004)
18. Wagon, S.: A bound on the chromatic number of graphs without certain induced subgraphs. J. Combin. Theory Ser. B **29**, 345–346 (1980)

Algorithms for Outerplanar Graph Roots and Graph Roots of Pathwidth at Most 2

Petr A. Golovach[1(⊠)], Pinar Heggernes[1], Dieter Kratsch[2], Paloma T. Lima[1], and Daniël Paulusma[3]

[1] Department of Informatics, University of Bergen, PB 7803, 5020 Bergen, Norway
{petr.golovach,pinar.heggernes,paloma.lima}@uib.no
[2] Laboratoire d'Informatique Théorique et Appliquée, Université de Lorraine, 57045 Metz Cedex 01, France
dieter.kratsch@univ-lorraine.fr
[3] School of Engineering and Computing Sciences, Durham University, Durham DH1 3LE, UK
daniel.paulusma@durham.ac.uk

Abstract. Deciding whether a given graph has a square root is a classical problem that has been studied extensively both from graph theoretic and from algorithmic perspectives. The problem is NP-complete in general, and consequently substantial effort has been dedicated to deciding whether a given graph has a square root that belongs to a particular graph class. There are both polynomial-time solvable and NP-complete cases, depending on the graph class. We contribute with new results in this direction. Given an arbitrary input graph G, we give polynomial-time algorithms to decide whether G has an outerplanar square root, and whether G has a square root that is of pathwidth at most 2.

1 Introduction

Squares and square roots of graphs form a classical and well-studied topic in graph theory, which has also attracted significant attention from the algorithms community. A graph G is the *square* of a graph H if G and H have the same vertex set, and two vertices are adjacent in G if and only if the distance between them is at most 2 in H. This situation is denoted by $G = H^2$, and H is called a *square root* of G. A square root of a graph need not be unique; it might even not exist. There are graphs without square roots, graphs with a unique square root, and graphs with several different square roots. Characterizing and recognizing graphs with square roots has therefore been an intriguing and important problem both in graph theory and in algorithms for decades.

Already in 1967, Mukhopadhyay [26] proved that a graph G on vertex set $\{v_1, \ldots, v_n\}$ has a square root if and only if G contains complete subgraphs $\{K^1, \ldots, K^n\}$, such that each K^i contains v_i, and vertex v_j belongs to K^i if and

Supported by the Research Council of Norway via the project "CLASSIS" and the Leverhulme Trust (RPG-2016-258).

H.L. Bodlaender and G.J. Woeginger (Eds.): WG 2017, LNCS 10520, pp. 275–288, 2017.
https://doi.org/10.1007/978-3-319-68705-6_21

only if v_i belongs to K^j. Unfortunately, this characterization does not yield a polynomial-time algorithm for deciding whether G has a square root. Let us formally call SQUARE ROOT the problem of deciding whether an input graph G has a square root. In 1994, it was shown by Motwani and Sudan [25] that SQUARE ROOT is NP-complete. Motivated by its computational hardness, special cases of the problem have been studied, where the input graph G belongs to a particular graph class. According to these results, SQUARE ROOT is polynomial-time solvable on planar graphs [22], and more generally, on every non-trivial minor-closed graph class [27]. Polynomial-time algorithms exist also when the input graph G belongs to one of the following graph classes: block graphs [20], line graphs [23], trivially perfect graphs [24], threshold graphs [24], graphs of maximum degree 6 [3], graphs of maximum average degree smaller than $\frac{46}{11}$ [13], graphs with clique number at most 3 [14], and graphs with bounded clique number and no long induced path [14]. On the negative side, it has been shown that SQUARE ROOT is NP-complete on chordal graphs [17]. A number of parameterized complexity results exist for the problem [3,4,13].

More interesting from our perspective, the intractability of the problem has also been attacked by restricting the properties of the square root that we are looking for. In this case, the input graph G is arbitrary, and the question is whether G has a square root that belongs to some graph class $\mathcal{H}$ specified in advance. We denote this problem by $\mathcal{H}$-SQUARE ROOT, and this is exactly the problem variant that we focus on in this paper.

Significant advances have been made also in this direction. Previous results show that $\mathcal{H}$-SQUARE ROOT is polynomial-time solvable for the following graph classes $\mathcal{H}$: trees [22], proper interval graphs [17], bipartite graphs [16], block graphs [20], strongly chordal split graphs [21], ptolemaic graphs [18], 3-sun-free split graphs [18], cactus graphs [12], cactus block graphs [8] and graphs with girth at least g for any fixed $g \geq 6$ [10]. The result for 3-sun-free split graphs has been extended to a number of other subclasses of split graphs in [19]. Observe that if $\mathcal{H}$-SQUARE ROOT is polynomial-time solvable for some class $\mathcal{H}$, then this does not automatically imply that $\mathcal{H}'$-SQUARE ROOT is polynomial-time solvable for a subclass $\mathcal{H}'$ of $\mathcal{H}$.

On the negative side, $\mathcal{H}$-SQUARE ROOT remains NP-complete for each of the following graph classes $\mathcal{H}$: graphs of girth at least 5 [9], graphs of girth at least 4 [10], split graphs [17], and chordal graphs [17]. All known NP-hardness constructions involve dense graphs [9,10,17,25], and the square roots that occur in these constructions are dense as well. This, in combination with the listed polynomial-time cases, naturally leads to the question whether $\mathcal{H}$-SQUARE ROOT is polynomial-time solvable if the class $\mathcal{H}$ is "sparse" in some sense.

Motivated by the above, in this paper we study $\mathcal{H}$-SQUARE ROOT when $\mathcal{H}$ is the class of outerplanar graphs, and when $\mathcal{H}$ is the class of graphs of pathwidth at most 2. In both cases, we show that $\mathcal{H}$-SQUARE ROOT can be solved in polynomial time. In particular, we prove that OUTERPLANAR (SQUARE) ROOT can be solved in time $O(n^4)$ and (SQUARE) ROOT OF PATHWIDTH ≤ 2 in time $O(n^6)$. Our approach for outerplanar graphs can in fact be directly applied to every subclass of outerplanar graphs that is closed under edge deletion and that can be expressed in monadic second-order logic, including cactus graphs, for which a polynomial-time algorithm is already known [12]. Due to space restrictions, some proofs are omitted; see [11] for the full version of our paper.

2 Preliminaries

We consider only finite undirected graphs without loops and multiple edges. We refer to the textbook by Diestel [7] for any undefined graph terminology.

Let G be a graph. We denote the vertex set of G by V_G and the edge set by E_G. The subgraph of G induced by a subset $U \subseteq V_G$ is denoted by $G[U]$. The graph $G - U$ is the graph obtained from G after removing the vertices of U. If $U = \{u\}$, we also write $G - u$. Similarly, we denote the graph obtained from G by deleting a set of edges S, or a single edge e, by $G - S$ and $G - e$, respectively.

The *distance* $\text{dist}_G(u, v)$ between a pair of vertices u and v of G is the number of edges of a shortest path between them. The *open neighborhood* of a vertex $u \in V_G$ is defined as $N_G(u) = \{v \mid uv \in E_G\}$, and its *closed neighborhood* is defined as $N_G[u] = N_G(u) \cup \{u\}$. For $S \subseteq V_G$, $N_G(S) = (\bigcup_{v \in S} N_G(v)) \setminus S$. Two (adjacent) vertices u, v are said to be *true twins* if $N_G[u] = N_G[v]$. A vertex v is *simplicial* if $N_G[v]$ is a clique, that is, if there is an edge between any two vertices of $N_G[v]$. The *degree* of a vertex $u \in V_G$ is defined as $d_G(u) = |N_G(u)|$. The maximum degree of G is $\Delta(G) = \max\{d_G(v) \mid v \in V_G\}$. A vertex of degree 1 is said to be a *pendant* vertex.

A *connected component* of G is a maximal connected subgraph. A vertex u is a *cut vertex* of a graph G with at least two vertices if $G - u$ has more components than G. A connected graph without cut vertices is said to be *biconnected*. An inclusion-maximal induced biconnected subgraph of G is called a *block*.

For a positive integer k, the *k-th power* of a graph H is the graph $G = H^k$ with vertex set $V_G = V_H$ such that every pair of distinct vertices u and v of G are adjacent if and only if $\text{dist}_H(u, v) \leq k$. For the particular case $k = 2$, H^2 is a *square* of H, and H is a *square root* of G if $G = H^2$.

The *contraction* of an edge uv of a graph G is the operation that deletes the vertices u and v and replaces them by a vertex w adjacent to $(N_G(u) \cup N_G(v)) \setminus \{u, v\}$. A graph G' is a contraction of a graph G if G' can be obtained from G by a series of edge contraction. A graph G' is a *minor* of G if it can be obtained from G by vertex deletions, edge deletions and edge contractions.

A graph G is *planar* if it admits an embedding on the plane such that there are no edges crossing (except in endpoints). A planar graph G is *outerplanar* if it admits a crossing-free embedding on the plane in such a way that all its vertices are on the boundary of the same (external) face. For a considered outerplanar graph, we always assume that its embedding on the plane is given. If G is a planar biconnected graph different from K_2, then for any of its embeddings, the boundary of each face is a cycle (see, e.g., [7]). If G is a biconnected outerplanar graph distinct from K_2, then the cycle C forming the boundary of the external face is unique and we call it the *boundary cycle*. By definition, all vertices of G

are laying on C, and every edge is either an edge of C or a *chord* of C, that is, its endpoints are vertices of C that are non-adjacent in C. Clearly, these chords are not intersecting in the embedding. For a vertex u, we define the *clockwise ordering with respect to* u as a clockwise ordering of the vertices on C starting from u. For a subset of vertices X, the *clockwise ordering of* X *with respect to* u is the ordering induced by the clockwise ordering of the vertices of C. See Fig. 1(a) for some examples. In our paper, we use these terms for blocks of an outerplanar graph that are distinct from K_2. Outerplanar graphs can also be characterized via forbidden minors as shown by Sysło [29].

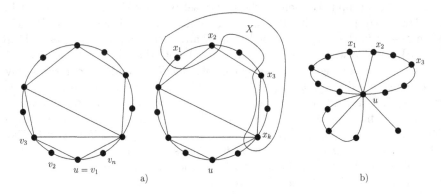

Fig. 1. (a) Clockwise orderings with respect to u of a biconnected outerplanar graph with vertex set $V_G = \{v_1, \ldots, v_n\}$ and a set $X = \{x_1, \ldots, x_k\}$. (b) Example of a set $X = \{x_1, x_2, x_3\}$ that is consecutive with respect to u; notice that the vertices x_1 and x_3 are not consecutive.

Lemma 1 [29]. *A graph G is outerplanar if and only if it does not contain K_4 and $K_{2,3}$ as minors.*

A *tree decomposition* of a graph G is a pair (T, X) where T is a tree and $X = \{X_i \mid i \in V_T\}$ is a collection of subsets (called *bags*) of V_G such that the following three conditions hold:

(i) $\bigcup_{i \in V_T} X_i = V_G$,
(ii) for each edge $xy \in E_G$, $x, y \in X_i$ for some $i \in V_T$, and
(iii) for each $x \in V_G$ the set $\{i \mid x \in X_i\}$ induces a connected subtree of T.

The *width* of a tree decomposition $(\{X_i \mid i \in V_T\}, T)$ is $\max_{i \in V_T} \{|X_i| - 1\}$. The *treewidth* $\mathbf{tw}(G)$ of a graph G is the minimum width over all tree decompositions of G. If T is restricted to be a path, then we say that (X, T) is a *path decomposition* of G. The *pathwidth* $\mathbf{pw}(G)$ of G is the minimum width over all path decompositions of G. Notice that a path decomposition of G can be seen as a sequence $(X_1, \ldots, X_r)$ of bags. We always assume that the bags $(X_1, \ldots, X_r)$ are distinct and inclusion incomparable, that is, there are no bags X_i and X_j such that $X_i \subset X_j$. The following fundamental results are due to Bodlaender [1], and Bodlaender and Kloks [2].

Lemma 2 [1,2]. *For every fixed constant c, it is possible to decide in linear time whether the treewidth or the pathwidth of a graph is at most c.*

We need the following three folklore observations about treewidth.

Observation 1. *If H is a minor (contraction) of G, then $\mathbf{tw}(H) \leq \mathbf{tw}(G)$ and $\mathbf{pw}(H) \leq \mathbf{pw}(G)$.*

Observation 2. *For an outerplanar graph G, $\mathbf{tw}(G) \leq 2$.*

Observation 3. *For a graph G and a positive integer k,*

$$\mathbf{tw}(G^k) \leq (\mathbf{tw}(G) + 1)\Delta(G)^{\lfloor k/2 \rfloor + 1}$$

and

$$\mathbf{pw}(G^k) \leq (\mathbf{pw}(G) + 1)\Delta(G)^{\lfloor k/2 \rfloor + 1}.$$

Let H be a square root of a graph G. We say that H is a *minimal* square root of G if $H^2 = G$, and no proper subgraph of H is a square root of G. We need the following simple observations.

Observation 4. *Let $\mathcal{H}$ be a graph class closed under edge deletion. If a graph G has a square root $H \in \mathcal{H}$, then G has a minimal square root that belongs to $\mathcal{H}$.*

Observation 5. *Let H be a minimal square root of a graph G that contains three vertices u, v, w that are pairwise adjacent in H. Then v or w has a neighbor $x \neq u$ in H such that x is not adjacent to u in H.*

We conclude this section by a lemma that is implicit in [12], which enables us to identify some edges that are not included in any square root.

Lemma 3. *Let x, y be distinct neighbors of a vertex u in a graph G such that x and y are at distance at least 3 in $G - u$. Then $xu, yu \notin E_H$ for any square root H of G.*

3   Outerplanar Roots

In this section, we show that it can be decided in polynomial time whether a graph has an outerplanar square root. We say that a square root H of G is an *outerplanar root* if H is outerplanar. We define the following problem:

OUTERPLANAR ROOT
Input: a graph G.
Question: is there an outerplanar graph H such that $H^2 = G$?

The main result of this section is the following.

Theorem 1. OUTERPLANAR ROOT *can be solved in time $O(n^4)$, where n is the number of vertices of the input graph.*

The remaining part of this section is devoted to the proof of Theorem 1. In Sect. 3.1 we obtain several structural results we need to construct an algorithm for OUTERPLANAR ROOT. Then, in Sect. 3.2, we construct a polynomial-time algorithm for OUTERPLANAR ROOT.

3.1   Structural Lemmas

In this section, we give several structural results about outerplanar square roots. Due to space restriction we omit the proofs.

Let H be an outerplanar root of a graph G and let $u \in V_G$. We say that two distinct vertices $x, y \in N_H(u)$ are *consecutive with respect to* u if x and y are in the same block F of H and there are no vertices of $N_H(u)$ between x and y in the clockwise ordering of the vertices of the boundary cycle of F with respect to u. For a set of vertices $X \subseteq N_H(u)$, we say that the vertices X are *consecutive with respect to* u if the vertices of X are in the same block of H and any two vertices of X consecutive in the clockwise ordering of elements of X with respect to u are consecutive with respect to u; a single-vertex set is assumed to be consecutive (see Fig. 1 (b) for an example).

As every subgraph of an outerplanar graph is outerplanar, by Observation 4, we may restrict ourselves to minimal outerplanar roots. Let H be a minimal outerplanar root of a graph G and let $u \in V_G$. Denote by $S(G, H, u)$ a collection of all subsets X of $N_H(u)$ such that $X = N_G(x) \cap N_H(u)$ for some $x \in N_G(u) \setminus N_H(u)$. We can use $S(G, H, u)$ to find edges with both endpoints in $N_H(u)$ that are not included in a square root.

Lemma 4. *Let H be a minimal outerplanar root of a graph G, and let $u \in V_G$. Then for each $X \in S(G, H, u)$, X is consecutive with respect to u.*

Lemma 5. *Let H be a minimal outerplanar root of a graph G, and let $u \in V_G$. If for two distinct vertices $x, y \in N_H(u)$ there is no set $X \in S(G, H, u)$ such that $x, y \in X$, then $xy \notin E_H$.*

We also need the following two lemmas.

Lemma 6. *Let H be a minimal outerplanar root of a graph G, and let $u \in V_G$. If $x \in N_H(u)$ is not a pendant vertex of H, then there is a vertex $y \in N_G(u) \setminus N_H(u)$ that is adjacent to x in G.*

Lemma 7. *Let H be a minimal outerplanar root of a graph G, and let $u \in V_G$. Then any $X \in S(G, H, u)$ has size at most 4.*

By combining Lemmas 4 and 7 we obtain the following lemma.

Lemma 8. *Let H be a minimal outerplanar root of a graph G, and let $u \in V_G$. Then the following holds.*

(i) *If $x, y \in N_H(u)$ do not belong to the same block of H, then for any $X \in S(G, H, u)$, $x \notin X$ or $y \notin X$.*

(ii) *If F is a block of H containing u and vertices $x_1, \ldots, x_k \in N_H(u)$ ordered in the clockwise order with respect to u in the boundary cycle of F, then for any $X \in S(G, H, u)$, $x_i \notin X$ or $x_j \notin X$ if $i, j \in \{1, \ldots, k\}$ and $|i - j| \geq 4$.*

We now state some structural results that help to decide whether an edge incident to a vertex is in an outerplanar root or not. Suppose that u and v are pendant vertices of a square root H of G and that u and v are adjacent to the same vertex of $H - \{u, v\}$. Then, in G, u and v are simplicial vertices and true twins. We use this observation in the proof of the following lemma that allows to find some pendant vertices.

Lemma 9. *Let H be a minimal outerplanar root of a graph G. If G contains at least 7 simplicial vertices that are pairwise true twins, then at least one of these vertices is a pendant vertex of H.*

We apply Lemma 3 to identify the edges incident to a vertex of sufficiently high degree in an outerplanar root using the following two lemmas.

Lemma 10. *Let G be a graph having a minimal outerplanar root H. Let also $u \in V_G$ be such that there are three distinct vertices $v_1, v_2, v_3 \in N_G(u)$ that are pairwise at distance at least 3 in $G - u$. Then for $x \in N_G(u)$, $xu \notin E_H$ if and only if there is $i \in \{1, 2, 3\}$ such that $\text{dist}_{G-u}(x, v_i) \geq 3$.*

Lemma 11. *Let G be a graph having a minimal outerplanar root H such that any vertex of H has at most 7 pendant neighbors. Let also $u \in V_G$ with $d_H(u) \geq 22$. Then there are distinct $v_1, v_2, v_3 \in N_G(u)$ that are pairwise at distance at least 3 in $G - u$.*

Notice that v_1, v_2 and v_3 are in distinct components of $H - u$. We obtain that v_3 is at distance at least 3 from v_1 and v_2 in $G - u$.

The next lemma is crucial for our algorithm. To state it, we need some additional notations. Let H be a minimal outerplanar root of a graph G such that each vertex of H is adjacent to at most 7 pendant vertices. Let U be a set of vertices of H that contains all vertices of degree at least 22. For every $u \in U$ and every block F of H containing u, we do the following. Consider the set $X = N_H(u) \cap V_F$ and denote the vertices of X by $x_1, \ldots, x_k$, where these vertices are numbered in the clockwise order with respect to u. Then

- for $i, j \in \{1, \ldots, k\}$, delete the edge $x_i x_j$ from G if $|i - j| \geq 4$.
- for $i \in \{1, \ldots, k\}$, delete the edges $x_i y$ from G for $y \in N_H(u) \setminus V_F$.

Denote by $G(H, U)$ the graph obtained in the end.

Lemma 12. *There is a constant c that depends neither on G nor on H such that*

$$\mathbf{tw}(G(H, U)) \leq c.$$

3.2 The Algorithm

In this section, we construct an algorithm for OUTERPLANAR ROOT with running time $O(n^4)$. Let G be the input graph. Clearly, it is sufficient to solve OUTERPLANAR ROOT for connected graphs. Hence, we assume that G is connected and has $n \geq 2$ vertices.

First, we preprocess G using Lemma 9 to reduce the number of pendant vertices adjacent to the same vertex in a (potential) outerplanar root of G. To do so, we exhaustively apply the following rule.

Pendants reduction. If G has a set X of simplicial true twins of size at least 8, then delete an arbitrary $u \in X$ from G.

The following lemma shows that this rule is safe.

Lemma 13. *If $G' = G - u$ is obtained from G by the application of* **Pendant reduction,** *then G has an outerplanar root if and only if G' has an outerplanar root.*

Proof. Suppose that H is a minimal outerplanar root of G. By Lemma 9, H has a pendant vertex $u \in X$. It is easy to verify that $H' = H - u$ is an outerplanar root of G'. Assume now that H' is a minimal outerplanar root of G'. By Lemma 9, H has a pendant vertex $w \in X \setminus \{u\}$, since the vertices of $X \setminus \{u\}$ are simplicial true twins of G' and $|X \setminus \{u\}| \geq 7$. Let v be the unique neighbor of w in H'. We construct H from H' by adding u and making it adjacent to v. It is readily seen that H is an outerplanar root of G. This completes the proof. □

For simplicity, we call the graph obtained by exhaustive application of the pendants reduction rule G again. The following property immediately follows from the observation that any two pendant vertices of a square root H of G adjacent to the same vertex in H are true twins of G.

Lemma 14. *Every outerplanar root of G has at most 7 pendant vertices adjacent to the same vertex.*

In the next stage of our algorithm we label some edges of G red or *blue* in such a way that the edges labeled red are included in every minimal outerplanar root and the blue edges are not included in any minimal outerplanar root. We denote by R the set of red edges and by B the set of blue edges. We also construct a set of vertices U of G such that for every $u \in U$, the edges incident to u are labeled red or blue.

Labeling. Set $U = \emptyset$, $R = \emptyset$ and $B = \emptyset$. For each $u \in V_G$ such that there are three distinct vertices $v_1, v_2, v_3 \in N_G(u)$ that are at distance at least 3 from each other in $G - u$, do the following:

(i) set $U = U \cup \{u\}$,
(ii) set $B' = \{ux \in E_G \mid \text{there is } 1 \leq i \leq 3 \text{ s.t. } \text{dist}_{G-u}(x, v_i) \geq 3\}$,
(iii) set $R' = \{ux \mid x \in N_G(u)\} \setminus B'$,
(iv) set $R = R \cup R'$ and $B = B \cup B'$,
(v) if $R \cap B \neq \emptyset$, then return a no-answer and stop.

Lemmas 10 and 11 imply the following statement.

Lemma 15. *If G has an outerplanar root, then* **Labeling** *does not stop in Step (v), and if H is a minimal outerplanar root of G, then $R \subseteq E_H$ and $B \cap E_H = \emptyset$. Moreover, every vertex $u \in V_G$ with $d_H(u) \geq 22$ is included in U.*

Next, we find the set of edges xy with $xu, yu \in R$ for some u in R that are not included in a minimal outerplanar root.

Finding irrelevant edges. Set $S = \emptyset$. For each $u \in U$ and each pair of distinct $x, y \in N_G(u)$ such that $ux, uy \in R$ do the following.

(i) If $xy \notin E_G$, then return a no-answer and stop.
(ii) If for x and y, there is no $v \in N_G(u)$ such that $vu \in B$ and $x, y \in N_G(v)$, then include xy in S.
(iii) If $R \cap S \neq \emptyset$, then return a no-answer and stop.

Combining Lemmas 5 and 15, we obtain the following claim.

Lemma 16. *If G has an outerplanar root, then* **Finding irrelevant edges** *does not stop in Steps* (i) *and* (iii), *and if H is a minimal outerplanar root of G, then $S \cap E_H = \emptyset$.*

Assume that we did not stop during the execution of **Finding irrelevant edges**. Let $G' = G - S$. We show the following.

Lemma 17. *The graph G has an outerplanar root if and only if there is a set $L \subseteq E_{G'}$ such that*

(i) $R \subseteq L$, $B \cap L = \emptyset$,
(ii) for any $xy \in E_{G'}$, $xy \in L$ or there is $z \in V_{G'}$ such that $xz, yz \in L$,
(iii) for any pair of distinct edges $xz, yz \in L$, $xy \in E_{G'}$ or there is $u \in U$ such that $xu, yu \in R$,
(iv) the graph $H = (V_G, L)$ is outerplanar.

Proof. Let H be a minimal outerplanar root of G. By Lemma 16, $E_H \cap S = \emptyset$, i.e., $E_H \subseteq E_{G'}$. Let $L = E_H$. It is straightforward to verify that (i)–(iv) are fulfilled. Assume now that there is $L \subseteq E_{G'}$ such that (i)–(iv) hold. Then we have that $H = (V_G, L)$ is an outerplanar root of G.   □

To complete the description of the algorithm, it remains to show how to check the existence of a set of edges L satisfying (i)–(iv) of Lemma 17 for given G', R and B. Notice that, if G has a minimal outerplanar root H, then G' is a subgraph of the graph $G(H, U)$ constructed in Sect. 3.1 by Lemma 8. By Lemma 12, there is a constant c that depends neither on G nor on H such that $\mathbf{tw}(G(H, U)) \leq c$. Therefore, $\mathbf{tw}(G') \leq c$ for a yes-instance. We use Lemma 2 to verify whether it holds. If we obtain that $\mathbf{tw}(G') > c$, we conclude that we have a no-instance and stop. Otherwise, we use the celebrated theorem of Courcelle [5], which states that any problem that can be expressed in monadic second-order logic can be solved in linear time on a graph of bounded treewidth. It is straightforward to see that properties (i)–(iv) can be expressed in this logic. In particular, to express outerplanarity in (iv), we can use Lemma 1 and the well-known fact that the property that G contains F as a minor can be expressed in monadic second-order logic if F is fixed (see, e.g., the book of Courcelle and Engelfriet [6]). It immediately implies that we can decide in linear time whether

L exists or not. Notice that we can modify these arguments such that we do not only check the existence of L but also find it. To do this, we can construct a dynamic programming algorithm for graphs of bounded treewidth that finds L.

Now we evaluate the running time of our algorithm. Since it can be verified in time $O(n)$ whether two vertices of G are true twins, the classes of true twins can be constructed in time $O(n^3)$. Then we can check whether each class contains simplicial vertices in time $O(n^2)$. Therefore, **Pendant reduction** can be done in time $O(n^3)$. For every vertex u, we can compute the distances between the vertices of $N_G(u)$ in $G - u$ in time $O(n^3)$. This implies that **Labeling** can be done in time $O(n^4)$. **Finding irrelevant edges** also can be done in time $O(n^4)$ by checking $O(n^2)$ pairs of vertices x and y. Then G' can be constructed in linear time. Finally, checking whether $\mathbf{tw}(G') \leq c$ and deciding whether there is a set of edges L satisfying the required properties can be done in linear time by Lemma 2 and Courcelle's theorem [5] respectively.

Notice that we can use the same arguments to decide whether a graph G has a square root H that belongs to some subclass $\mathcal{H}$ of the class of outerplanar graphs. To be able to apply our structural lemmas, we only need the property that $\mathcal{H}$ should be closed under edge deletions. Observe also that if the properties defining $\mathcal{H}$ could be expressed in monadic second-order logic, then we can apply Courcelle's theorem [5]. It gives us the following corollary.

Corollary 1. *For every subclass $\mathcal{C}$ of the class of outerplanar graphs that is closed under edge deletions and can be expressed in monadic second-order logic, it can be decided in time $O(n^4)$ whether an n-vertex graph G has a square root $H \in \mathcal{C}$.*

4   Roots of Pathwidth at Most Two

Our main approach for solving OUTERPLANAR ROOT is general in the sense that it can be adapted to find also square roots belonging to some other graph classes. In this section, we show that there is an algorithm to decide in polynomial time whether a graph has a square root of pathwidth at most 2. Notice that graphs of pathwidth 1 are caterpillars, and that it can be decided in polynomial time whether a graph G has a square root that is a caterpillar by an easy adaptation of algorithms for finding square roots that are trees [22,28].

We define the following problem:

ROOT OF PATHWIDTH ≤ 2
Input: a graph G.
Question: is there a graph H such that $\mathbf{pw}(H) \leq 2$ and $H^2 = G$?

The main difference between our algorithm for ROOT OF PATHWIDTH ≤ 2 and our algorithm for OUTERPLANAR ROOT lies in the way properties of the involved graph classes are used. To show the structural results needed for this algorithm, we use the property that a potential square root has a path decomposition of width at most 2, instead of the existence of an outerplanar embedding used in the previous section.

We briefly sketch the proof of the following theorem.

Theorem 2. ROOT OF PATHWIDTH ≤ 2 *can be solved in time* $O(n^6)$, *where* n *is the number of vertices of the input graph.*

Proof (Sketch). Let G be the input graph. It is sufficient to solve ROOT OF PATHWIDTH ≤ 2 for connected graphs. Hence, we assume that G is connected and has $n \geq 2$ vertices. Notice that the class of graphs of pathwidth at most 2 is closed under edge deletions. Therefore, by Observation 4, we can consider only minimal square roots.

First, we preprocess G to reduce the number of true twins that a given vertex of V_G might have. To do so, we show that there is a constant c_1 such that if W is a set of true twins of G of size at least c_1, then for any minimal square root H of G with $\mathbf{pw}(H) \leq 2$, either W has a vertex that is pendant in H or W has distinct nonadjacent vertices x, y, z with $d_H(x) = d_H(y) = d_H(y) = 2$. It allows us to show that if G has a set of true twins W of size at least $c_1 + 1$, then by the deletion of an arbitrary $u \in W$ from G, we obtain an equivalent instance of ROOT OF PATHWIDTH ≤ 2. From now on, we can assume that any set of true twins of G has size at most c_1. We need this to obtain forthcoming structural properties.

In the next stage of our algorithm, we label some edges of G *red* or *blue* in such a way that the edges labeled red are included in every minimal square root of pathwidth at most 2 and the blue edges are not included in any minimal square root of pathwidth at most 2. We denote by R the set of red edges and by B the set of blue edges. We also construct a set of vertices U of G such that for every $u \in U$, the edges incident to u are labeled red or blue.

The labeling is based on the following structural property. If there is $u \in V_G$ such that there are five distinct vertices $v_1, \ldots, v_5$ in $N_G(u)$ that are at distance at least 3 from each other in $G - u$, then for any square root H with $\mathbf{pw}(H) \leq 2$, $ux \notin E_H$ for $x \in N_G(u)$ if and only if there is $i \in \{1, \ldots, 5\}$ such that $\text{dist}_{G-u}(x, v_i) \geq 3$. Respectively, if we find $u \in V_G$ with the aforementioned property that there are five distinct vertices $v_1, \ldots, v_5$ in $N_G(u)$ that are at distance at least 3 from each other in $G - u$, then we include u in U and for $x \in N_G(u)$, we label ux blue if there is $i \in \{1, \ldots, 5\}$ such that $\text{dist}_{G-u}(x, v_i) \geq 3$ and we label ux red otherwise. If we get inconsistent labelings, that is, some edge should be labeled red and blue, then we stop and report that there is no square root of pathwidth at most 2.

We show that there is a constant c_2 such that, for a square root H of G with $\mathbf{pw}(H) \leq 2$, if $d_H(u) \geq c_2$, then $u \in U$ and, therefore, all the edges of G incident to u are labeled red or blue. It means that if u is a vertex of H of sufficiently high degree, then for each edge of G incident to u, we distinguish whether this edge is in a square root or not.

Next, we find the set of edges xy with $xu, yu \in R$ which for some u in U are not included in a minimal square root of pathwidth at most 2. To do it, we use Observation 5 to show that if there is no $z \in N_G(u)$ with $uz \in B$ such that $xz, yz \in E_G$, then $xy \notin E_H$ for a minimal square root H of pathwidth at most 2. Respectively, we label such edges xy blue. Again, if we get inconsistent labelings, then we stop and report that there is no square root of pathwidth at most 2.

Denote by S the set of edges labeled blue in this stage of the algorithm and let $G' = G - S$. We prove that if G has a square root of pathwidth at most 2, then there is a constant c_4 such that $\mathbf{pw}(G') \leq c_4$. The proof is based on the property that every vertex of degree at least c_2 in a (potential) square root of pathwidth at most 2 is included in U. We can verify whether $\mathbf{pw}(G') \leq c_4$ in linear time using Lemma 2. If $\mathbf{pw}(G') > c_4$, then we stop and report that there is no square root of pathwidth at most 2. Otherwise, we obtain a path decomposition of G' of width at most c_4.

Then, similarly to the proof of Theorem 1, we obtain that G has a square root of pathwidth at most 2 if and only if there is a set $L \subseteq E_{G'}$ such that

(i) $R \subseteq L$, $B \cap L = \emptyset$,
(ii) for any $xy \in E_{G'}$, $xy \in L$ or there is $z \in V_{G'}$ such that $xz, yz \in L$,
(iii) for any distinct edges $xz, yz \in L$, $xy \in E_{G'}$ or $xy \in S$,
(iv) the graph $H = (V_G, L)$ is such that $pw(H) \leq 2$.

Notice that the properties (i)–(iv) can be expressed in monadic second-order logic. In particular, (iv) can be expressed using the property that the class of graphs of pathwidth at most 2 is defined by the set of forbidden minors given by Kinnersley and Langston in [15]. Then we use Courcelle's theorem [5] to decide in linear time whether L exists or not.

To evaluate the running time, observe that to construct U, we consider each vertex $u \in V_G$ and check whether there are 5 distinct vertices in $N_G(u)$ that are at distance at least 3 from each other in $G - u$. This can be done in time $O(n^6)$ and implies that the total running time is also $O(n^6)$. □

5   Conclusions

We proved that $\mathcal{H}$-SQUARE ROOT is polynomial-time solvable when $\mathcal{H}$ is the class of outerplanar graphs or the class of graphs of pathwidth at most 2. The same result holds if $\mathcal{H}$ is any subclass of the class of outerplanar graphs that is closed under edge deletion and that can be expressed in monadic second-order logic (for instance, if $\mathcal{H}$ is the class of cactus graphs). We conclude by posing two questions:

- Is $\mathcal{H}$-SQUARE ROOT polynomial-time solvable for every class $\mathcal{H}$ of graphs of bounded pathwidth?
- Is $\mathcal{H}$-SQUARE ROOT polynomial-time solvable if $\mathcal{H}$ is the class of planar graphs?

References

1. Bodlaender, H.L.: A linear-time algorithm for finding tree-decompositions of small treewidth. SIAM J. Comput. **25**, 305–1317 (1996)
2. Bodlaender, H.L., Kloks, T.: Efficient and constructive algorithms for the pathwidth and treewidth of graphs. J. Algorithms **21**(2), 358–402 (1996)

3. Cochefert, M., Couturier, J.-F., Golovach, P.A., Kratsch, D., Paulusma, D.: Sparse square roots. In: Brandstädt, A., Jansen, K., Reischuk, R. (eds.) WG 2013. LNCS, vol. 8165, pp. 177–188. Springer, Heidelberg (2013). doi:10.1007/978-3-642-45043-3_16

4. Cochefert, M., Couturier, J., Golovach, P.A., Kratsch, D., Paulusma, D.: Parameterized algorithms for finding square roots. Algorithmica **74**, 602–629 (2016)

5. Courcelle, B.: The monadic second-order logic of graphs III: tree-decompositions, minor and complexity issues. Informatique Théorique Appl. **26**, 257–286 (1992)

6. Courcelle, B., Engelfriet, J.: Graph Structure and Monadic Second-Order Logic - A Language-Theoretic Approach, Encyclopedia of Mathematics and its Applications, vol. 138. Cambridge University Press, Cambridge (2012)

7. Diestel, R.: Graph Theory. Graduate Texts in Mathematics. Springer, Heidelberg (2012)

8. Ducoffe G., Finding cut-vertices in the square roots of a graph. In: Proceedings of the WG 2017. LNCS (to Appear)

9. Farzad, B., Karimi, M.: Square-root finding problem in graphs, a complete dichotomy theorem. CoRR abs/1210.7684 (2012)

10. Farzad, B., Lau, L.C., Le, V.B., Tuy, N.N.: Complexity of finding graph roots with girth conditions. Algorithmica **62**, 38–53 (2012)

11. Golovach, P.A., Heggernes, P., Kratsch, D., Lima, P.T., Paulusma, D.: Algorithms for outerplanar graph roots and graph roots of pathwidth at most 2. CoRR abs/1703.05102 (2017)

12. Golovach, P.A., Kratsch, D., Paulusma, D., Stewart, A.: Finding cactus roots in polynomial time. In: Mäkinen, V., Puglisi, S.J., Salmela, L. (eds.) IWOCA 2016. LNCS, vol. 9843, pp. 361–372. Springer, Cham (2016). doi:10.1007/978-3-319-44543-4_28

13. Golovach, P.A., Kratsch, D., Paulusma, D., Stewart, A.: A linear kernel for finding square roots of almost planar graphs. In: Proceedings of the 15th Scandinavian Symposium and Workshops on Algorithm Theory, SWAT 2016, vol. 53, pp. 4:1–4:14. Leibniz International Proceedings in Informatics (2016)

14. Golovach, P.A., Kratsch, D., Paulusma, D., Stewart, A.: Squares of low clique number. Electron. Notes Discrete Math. **55**, 195–198 (2016). 14th Cologne Twente Workshop 2016, CTW 2016

15. Kinnersley, N.G., Langston, M.A.: Obstruction set isolation for the gate matrix layout problem. Discrete Appl. Math. **54**(2–3), 169–213 (1994)

16. Lau, L.C.: Bipartite roots of graphs. ACM Trans. Algorithms **2**, 178–208 (2006)

17. Lau, L.C., Corneil, D.G.: Recognizing powers of proper interval, split, and chordal graphs. SIAM J. Discrete Math. **18**, 83–102 (2004)

18. Le, V.B., Oversberg, A., Schaudt, O.: Polynomial time recognition of squares of ptolemaic graphs and 3-sun-free split graphs. Theoret. Comput. Sci. **602**, 39–49 (2015)

19. Le, V.B., Oversberg, A., Schaudt, O.: A unified approach for recognizing squares of split graphs. Theoret. Comput. Sci. **648**, 26–33 (2016)

20. Le, V.B., Tuy, N.N.: The square of a block graph. Discrete Math. **310**, 734–741 (2010)

21. Le, V.B., Tuy, N.N.: A good characterization of squares of strongly chordal split graphs. Inf. Process. Lett. **111**, 120–123 (2011)

22. Lin, Y.-L., Skiena, S.S.: Algorithms for square roots of graphs. In: Hsu, W.-L., Lee, R.C.T. (eds.) ISA 1991. LNCS, vol. 557, pp. 12–21. Springer, Heidelberg (1991). doi:10.1007/3-540-54945-5_44

23. Milanic, M., Oversberg, A., Schaudt, O.: A characterization of line graphs that are squares of graphs. Discrete Appl. Math. **173**, 83–91 (2014)
24. Milanic, M., Schaudt, O.: Computing square roots of trivially perfect and threshold graphs. Discrete Appl. Math. **161**, 1538–1545 (2013)
25. Motwani, R., Sudan, M.: Computing roots of graphs is hard. Discrete Appl. Math. **54**, 81–88 (1994)
26. Mukhopadhyay, A.: The square root of a graph. J. Comb. Theory **2**, 290–295 (1967)
27. Nestoridis, N.V., Thilikos, D.M.: Square roots of minor closed graph classes. Discrete Appl. Math. **168**, 34–39 (2014)
28. Ross, I.C., Harary, F.: The square of a tree. Bell Syst. Tech. J. **39**, 641–647 (1960)
29. Sysło, M.M.: Characterizations of outerplanar graphs. Discrete Math. **26**(1), 47–53 (1979)

Enumeration and Maximum Number of Maximal Irredundant Sets for Chordal Graphs

Petr A. Golovach[1($\boxtimes$)], Dieter Kratsch[2], Mathieu Liedloff[3],
and Mohamed Yosri Sayadi[2]

[1] Department of Informatics, University of Bergen, Bergen, Norway
petr.golovach@ii.uib.no
[2] LITA, Université de Lorraine, Metz, France
{dieter.kratsch,yosri.sayadi}@univ-lorraine.fr
[3] INSA Centre Val de Loire, LIFO EA 4022, Université d'Orléans,
45067 Orléans, France
mathieu.liedloff@univ-orleans.fr

Abstract. In this paper we provide exponential-time algorithms to enumerate the maximal irredundant sets of chordal graphs and two of their subclasses. We show that the maximum number of maximal irredundant sets of a chordal graph is at most 1.7549^n, and these can be enumerated in time $O(1.7549^n)$. For interval graphs, we achieve the better upper bound of 1.6957^n for the number of maximal irredundant sets and we show that they can be enumerated in time $O(1.6957^n)$. Finally, we show that forests have at most 1.6181^n maximal irredundant sets that can be enumerated in time $O(1.6181^n)$. We complement the latter result by providing a family of forests having at least 1.5292^n maximal irredundant sets.

1 Introduction

Many NP-complete graph problems have been described in the well-known monograph of Garey and Johnson [15]. Such problems are often studied and solved as optimization problems. In the context of our paper it is worth mentioning the approach in which for any graph class it is studied whether the problem remains NP-complete or becomes polynomial-time solvable, see, e.g., [5,19]. In the last decades various new approaches to solve NP-complete problems exactly have attracted a lot of attention, among them parameterized algorithms and exact exponential-time algorithms. Let us refer the interested reader to the books [9,14]. Instead of optimization problems our paper studies enumeration problems which require to list all wanted objects of the input graph. The classical approach, called output-sensitive, measures running time in dependence of input and output length, and asks for output-polynomial algorithms and algorithms of polynomial delay. This approach has been studied since a long time and

This work is supported by the Research Council of Norway by the CLASSIS project and the French National Research Agency by the ANR project GraphEn (ANR-15-CE40-0009).

© Springer International Publishing AG 2017
H.L. Bodlaender and G.J. Woeginger (Eds.): WG 2017, LNCS 10520, pp. 289–302, 2017.
https://doi.org/10.1007/978-3-319-68705-6_22

has produced its own important open questions, see, e.g., [11,25,26]. Our approach, called input-sensitive, measures the running time in dependence the input length, and thus it typically produces exact exponential algorithms to enumerate objects. Branching algorithms are a major tool to design such algorithms. This also relates to so-called lower and upper combinatorial bounds on the maximum number of objects in an n-vertex graph which can be achieved by the described algorithmic approach but also by the use of combinatorial (non-algorithmic) means. We mention that lower bounds for this number provide lower bounds for the running time of any corresponding input-sensitive enumeration algorithm. Various papers following this second approach have been published in the last years, see e.g., [7,8,16].

MINIMUM DOMINATING SET is a classical NP-complete graph problem with a large number of algorithmic and practical applications. There are two monographs on domination in graphs [21,22]. Enumerating all minimal dominating sets of a graph, is a benchmark problem in input-sensitive enumeration. In 2008 Fomin et al. showed in a cornerstone work of input-sensitive enumeration that an n-vertex graph has at most 1.7159^n minimal dominating sets and that they can be enumerated in time $O(1.7159^n)$ [13]. In the last years enumerating all minimal dominating sets has been studied for many graph classes, as e.g. chordal, interval and split graphs [7,8].

There is a huge number of problems related to domination, a prominent one is based on the notion of *irredundancy*, see [21,22]. A set of vertices D of a graph G is *irredundant* if any $u \in D$ has a *private vertex* v in the closed neighborhood of u, that is, v is not dominated by any other vertex of D (see Sect. 2 for formal definitions). It is straightforward to see that any (inclusion) minimal dominating set is an (inclusion) maximal irredundant set but not the other way around. Still, is some graph classes, these notions can coincide. For example, it could be shown that on split graphs every maximal irredundant set is a minimal dominating set (see [25, Corollary 20]). Combined with corresponding results for minimal dominating sets in split graphs [8] this implies that n-vertex split graphs have at most $3^{n/3}$ maximal irredundant sets which can be enumerated in time $O^*(3^{n/3})$. But for the graph classes where not all maximal irredundant sets are dominating, it seems to be impossible to use the results about enumerations of minimal dominating sets to enumerate the maximal irredundant sets. The reason is that all known enumeration algorithms for minimal dominating sets heavily rely on the *local* property that for every vertex, there is a vertex in its closed neighborhood that is included in a minimal dominating set.

Optimization problems asking for minimum or maximum size of a maximal irredundant set have been studied in various algorithmic settings and with the restrictions of these problems to various graph classes [2,6,10,12,23,24]. In particular, exact exponential-time algorithms were given by Binkele-Raible et al. in [3].

We study enumeration and maximum number of inclusion maximal irredundant sets in n-vertex chordal graphs. We also provide corresponding results for two subclasses of chordal graphs, namely interval graphs and forests. We show that the maximum number of maximal irredundant sets of a chordal graph is at most 1.7549^n, and these can be enumerated in time $O(1.7549^n)$. For the class of interval graphs, we prove that the number of maximal irredundant sets is at most 1.6957^n and that these sets can be enumerated in time $O(1.6957^n)$. Also we show that forests have at most 1.6181^n maximal irredudant sets, and these can be enumerated in time $O(1.6168^n)$. Complementing these upper bounds, we provide a lower bound by constructing a family of forests on n vertices having at least 1.5292^n maximal irredundant sets. Due to space restrictions, we only sketch the proofs of these results in this extended abstract.

Let us mention that for paths, Golovach et al. [17] recently proved that the number of maximal irredundant sets is $\Theta(1.4696\ldots^n)$. Clearly, this result provides also a lower bound for interval graphs. Also very recently Golovach et al. [18] proved that an n-vertex claw-free graph has $O(1.9341^n)$ maximal irredundant sets that can be enumerated in the same time and showed the lower bound 1.5848^n for the maximum number of maximal irredundant sets for this graph class.

2 Preliminaries

We consider finite undirected graphs without loops or multiple edges. Throughout the paper we denote by $n = |V(G)|$ and $m = |E(G)|$ the numbers of vertices and edges of the input graph G respectively. For a graph G and a subset $U \subseteq V(G)$ of vertices, we write $G[U]$ to denote the subgraph of G induced by U. We write $G - U$ to denote the subgraph of G induced by $V(G) \setminus U$, and we write $G - u$ instead of $G - \{u\}$ for a single element set. For a vertex v, we denote by $N_G(v)$ the *(open) neighborhood* of v, i.e., the set of vertices that are adjacent to v in G. The *closed neighborhood* $N_G[v] = N_G(v) \cup \{v\}$. For a set of vertices $U \subseteq V(G)$, $N_G[U] = \cup_{v \in U} N_G[v]$. The *degree* of a vertex v is $d_G(v) = |N_G(v)|$. We call a vertex of degree one *pendant*. A cycle C is *induced* if it has no *chord*, i.e., there is no edge of G incident to any two vertices of C that are not adjacent in C.

A vertex v of a graph G *dominates* a vertex u if $u \in N_G[v]$; similarly v dominates a set of vertices U if $U \subseteq N_G[v]$. For two sets $D, U \subseteq V(G)$, the set D dominates U if $U \subseteq N_G[D]$. Let $D \subseteq V(G)$. A vertex v is a *private vertex* (or, simply, a *private*) for $u \in D$ w.r.t. D (or, simply, for $u \in D$), if v is dominated by u but v is not dominated by any other vertex of D. Notice that a vertex $v \in D$ is private for itself iff v is an isolated vertex of $G[D]$. A set of vertices D is an *irredundant set* of G if every vertex $v \in D$ has a private. An irredundant set D is *(inclusion) maximal* if D is irredundant but any proper superset of D does not have this property.

A graph is *chordal* if it has no induced cycle on at least 4 vertices. A graph is a *forest* if it does not contain a cycle; and a *tree* is a forest being connected. As it is standard for forests, we say that a vertex of degree one is a *leaf*. An *interval* graph is a graph that has an *interval representation* in which each vertex corresponds to an interval of the real line, and two vertices are adjacent if and only if their corresponding intervals have non empty intersection. By the classical result of Lekkerkerker and Boland [28], interval graphs can be recognized and an interval representation can be found in polynomial (in fact, it can be done in linear time [20,27]). Therefore, throughout the paper we assume that an interval graph is given together with its interval representation. We refer to the monographs by Brandstädt et al. [5] and Golumbic [19] for more properties and characterizations of these graph classes.

A vertex u of a graph G is *simplicial* if $N_G[u]$ is a clique. It is well known (see, e.g. [19]) that every chordal graph has a simplicial vertex.

We conclude this section by observing that there is a family of forests with at least 1.5292^n maximal irredundant sets.

Proposition 1. *For every $k \geq 1$, there is a forest with $n = 11k$ vertices with at least 1.5292^n maximal irredundant sets.*

3   Enumeration of Maximal Irredundant Sets for Chordal Graphs

In this section we construct a branching algorithm enumerating all maximal irredundant sets of a chordal graph and obtain an upper bound for the number of maximal irredundant sets in an n-vertex chordal graph. Our algorithm uses structural results for chordal graphs that are very similar to the results recently obtained by Abu-Khzam and Heggernes [1] and already proved to be useful for solving enumeration problems on chordal graphs. We state this structural lemma in a way slightly different from [1].

Let G be a graph and $S(G)$ the set of all simplicial vertices of G. For a vertex u, denote by $S_G(u)$ the set of all simplicial vertices v of G such that $v \in N_G(u)$. We say that u is a *semi-simplicial* vertex if $S_G(u) \neq \emptyset$ and u is a simplicial vertex of $G - S_G(u)$.

Lemma 1. *If G is a connected chordal graph with at least two vertices, then it has a semi-simplicial vertex that can be found in polynomial time.*

Now we are ready to state the main result of the section.

Theorem 1. *A chordal graph has at most 1.7549^n maximal irredundant sets, and these can be enumerated in time $O(1.7549^n)$.*

Proof. Due to space restrictions, we only sketch the proof.

Let G be a chordal graph. We consider the following branching recursive algorithm $\textsc{EnumIS}(S, F, X)$, where $S, X, F \subseteq V(G)$ are disjoint. The algorithm enumerates the maximal irredundant sets D of G such that (a) $S \subseteq D$, (b) $D \setminus S \subseteq X$, and (c) for each $v \in D \setminus S$, v has a private in $F \cup X$. To enumerate all maximal irredundant sets of G, we call $\textsc{EnumIS}(\emptyset, \emptyset, V(G))$.

In our algorithm we denote by $H = G[F \cup X]$. Notice that we require that $F \cap D = \emptyset$. Respectively, we say that the vertices of F are *forbidden* as they cannot be included in an irredundant set and the vertices of X are *free*. We also say that a vertex of H is *dominated* if it is dominated in G by a vertex of S. In each step of the algorithm, we either reduce the considered instance or branch. In every step, if we decide to include a vertex $u \in X$ in a (potential) irredundant set, we make sure that it gets a private v in $F \cup X$ that is not dominated by other vertices in the recursive calls. To do it, we declare all the neighbors of v except u forbidden. As it is standard for such algorithms, each step is applied if its conditions are fulfilled and the previous steps are not applicable.

We say that a maximal irredundant set D is *compatible* with a triple (S, F, X) of disjoint subsets of $V(G)$ if the following is fulfilled:

(a) $S \subseteq D$,
(b) $D \setminus S \subseteq X$,
(c) for each $v \in D \setminus S$, v has a private in $F \cup X$,
(d) for each $u \in S$, u has a private $v \in V(G) \setminus (F \cup X)$ with respect to D such that $N_G(v) \cap X = \emptyset$.

To show the correctness of ENUMIS(S, F, X), we prove for every step of the algorithm that if it can be applied for the instance (S, F, X) and the previous steps are not applicable, then the following holds: if D is a maximal irredundant set of G compatible with (S, F, X), then either $D = S$ and the algorithm outputs D or the algorithm produces an instance (S', F', X') such that D is compatible with it or, in the case of a branching step, ENUMIS is called in one of the branches for an instance (S', F', X') such that D is compatible with it.

1. If $X = \emptyset$, then check whether S is a maximal irredundant set of G and output it if it holds; then stop.

In the next 3 steps we reduce the input instance. It is straightforward to verify that if a maximal irredundant set D of G is compatible with (S, F, X), then these steps produce instances (S', F', X') that are compatible with D.

2. If H has a forbidden and dominated vertex x, then call ENUMIS$(S, F \setminus \{x\}, X)$.
3. If there is $x \in V(H)$ such that all the vertices of $N_H[x]$ are dominated, then call ENUMIS$(S, F, X \setminus \{x\})$.
4. If there is $x \in V(H)$ such that all the vertices of $N_H[x]$ are forbidden, then call ENUMIS$(S, F \setminus \{x\}, X)$.

Notice that from now we can assume that each dominated vertex is free and every forbidden vertex is not dominated.

From now on we branch.

5. If H has a component H' that is a complete graph, then for each $x \in V(H') \cap X$, call ENUMIS$(S \cup \{x\}, F \setminus V(H'), X \setminus V(H'))$.

If a maximal irredundant D is compatible with (S, F, X), then it can be observed that $|V(H') \cap D| = 1$. We use this property to show that in one of the branches we call our algorithm for the instance that is compatible with D.

In the next 8 steps we use simplicial vertices to organize the branching. Let D be a maximal irredundant set of G compatible with (S, F, X) and let x be a simplicial vertex of G. Note that if $x \in D$, then $N_H(x) \cap D = \emptyset$ and the vertices of $N_H[x]$ can be private vertices only for x. Suppose that $x \in D$ and is dominated. Then x has its private $v \in N_H(x)$ and D is compatible with $(S \cup \{x\}, F \setminus N_H[v], X \setminus N_H[x])$. If $x \notin D$ and is dominated, then x is not a private for any vertex of D. These observations allow us to deal with dominated simplicial vertices. If x is not dominated, we need some additional observations. If $x \in D$ and is not dominated, then x is a private for itself and D is compatible with $(S \cup \{x\}, F \setminus N_H[x], X \setminus N_H[x])$. Suppose that $x \notin D$. Assume that $D \cap N_H(x) = \{v\}$ for some $v \in X$. Then x is a private for v, because x is not dominated. Also the vertices of $N_H[x]$ cannot be privates for any vertex of $D \setminus (S \cup \{v\})$. It implies that D is compatible with $(S \cup \{v\}, F \setminus N_H[x], X \setminus N_H(x))$. If $|N_H(x) \cap D| \neq 1$, then x cannot be private for any vertex of $(D \setminus S)$ and D is compatible with $(S, F \setminus \{x\}, X \setminus \{x\})$. Using these observations, we show that Steps 6–13 produce instances compatible with D.

6. If there is a dominated simplicial vertex x of H, then branch:
   (i) for each not dominated $v \in N_H(x)$, call ENUMIS$(S \cup \{x\}, (F \cup N_H(v)) \setminus N_H[x], X \setminus N_H[v])$,
   (ii) call ENUMIS$(S, F, X \setminus \{x\})$.

Observe that from now we can assume that all simplicial vertices of H are not dominated.

7. If there is a simplicial vertex x of H with $d_H(x) \geq 3$, then branch:
   (i) for each $v \in N_H[x] \cap X$, call ENUMIS$(S \cup \{v\}, F \setminus N_H[x], X \setminus N_H[x])$,
   (ii) call ENUMIS$(S, F \setminus \{x\}, X \setminus \{x\})$.

From now we can assume that each simplicial vertex has degree 1 or 2.

8. If there is a forbidden simplicial vertex x, then branch:
   (i) for each $v \in N_H(x) \cap X$, call ENUMIS$(S \cup \{v\}, F \setminus N_H[x], X \setminus N_H[x])$,
   (ii) call ENUMIS$(S, F \setminus \{x\}, X)$.

From now we have that all simplicial vertices are free.

9. If there are adjacent simplicial vertices x and y, then branch:
   (i) call ENUMIS$(S \cup \{x\}, (F \cup N_H(x)) \setminus \{x, y\}, X \setminus N_H[x])$,
   (ii) call ENUMIS$(S, F, X \setminus \{x\})$.

To see that this step produces instances compatible with a maximal irredundant set D if D is compatible with (S, F, X), we additionally observe the following. If x and y are adjacent simplicial nondominated vertices, $x \notin D$ and is a private for some $u \in D \setminus S$, then y is a private for u as well.

From now we can assume that simplicial vertices of H are pairwise nonadjacent.

10. If there is a simplicial vertex x such that $N_H(x) \subseteq F$, then branch:
    (i) call ENUMIS$(S \cup \{x\}, F \setminus N_H[x], X \setminus \{x\})$,
    (ii) call ENUMIS$(S, F, X \setminus \{x\})$.

11. If there is a simplicial vertex x such that $N_H(x) \cap F \neq \emptyset$, then let $y \in N_H(x) \cap X$ and branch:
    (i) call ENUMIS$(S \cup \{x\}, F \setminus N_H[x], X \setminus N_H[x])$,
    (ii) call ENUMIS$(S \cup \{y\}, F \setminus N_H[x], X \setminus N_H[x])$,
    (iii) call ENUMIS$(S, F, X \setminus \{x\})$.

After Steps 10 and 11 we can assume that all the neighbors of simplicial vertices are free.

To construct the next rule, we observe that if x and y are distinct simplicial nonadjacent vertices that are not dominated and free, then either $x, y \in D$ or $x, y \notin D$ for any maximal irredundant set D compatible with (S, F, X).

12. If there are simplicial vertices x and y such that $N_H(x) = N_H(y)$, then branch:
    (i) call ENUMIS$(S \cup \{x, y\}, F, X \setminus (N_H[x] \cup \{y\}))$,
    (ii) call ENUMIS$(S, F \cup \{y\}, X \setminus \{x, y\})$.

To deal with the case $N_H(y) \subset N_H(x)$ for two simplicial vertices x and y, we note the following. Suppose that $N_H(x) = \{u, z\}$, $N_H(y) = \{z\}$ and x, y, z are free and not dominated. Let D be a maximal irredundant set of G. If either $x \in D$ or it holds that $x, y \notin D$ and $u \in D$, then we can show that $y \in D$. It gives us the next step.

13. If there are simplicial vertices x and y such that $N_H(y) \subset N_H(x)$, then denote by z the common neighbor of x, y and by u the neighbor of x that is not adjacent to y and branch:
    (i) call ENUMIS$(S \cup \{x, y\}, F, X \setminus (N_H[x] \cup \{y\}))$,
    (ii) call ENUMIS$(S \cup \{u, y\}, F, X \setminus (N_H[x] \cup \{y\}))$,
    (iii) call ENUMIS$(S \cup \{z\}, F, X \setminus (N_H[x] \cup \{y\}))$,
    (iii) call ENUMIS$(S, F, X \setminus \{x\})$.

In the remaining steps of our algorithm we use semi-simplicial vertices to organize branching. Before we start, observe that summarizing the properties of simplicial vertices that were used for branching in the previous steps, we can assume from now that every simplicial vertex of H is not dominated, free, has degrees 1 or 2 and all their neighbors are free. Also any two simplicial vertices are not adjacent and their neighborhoods are not comparable by inclusion. To show that if D is a maximal irredundant set compatible with (S, F, X), then these branching steps produce instances compatible with D, we essentially use the same observations that were used for Steps 6–13. In particular, some of our new steps are, in fact, combinations of two previous steps. The main additional observation is that if a semi-simplicial vertex $y \in D$, then $S_H(y) \cap D = \emptyset$ and one of the vertices of $S_H(y)$ is a private for y.

14. If there is a dominated semi-simplicial vertex y of H, then branch:
    (i) for each $x \in S_H(y)$, call ENUMIS$(S \cup \{y\}, F \setminus (S_H(y) \cup N_H(x)), X \setminus (S_H(y) \cup N_H(x)))$,
    (ii) call ENUMIS$(S, F, X \setminus \{y\})$.

Notice that from now we can assume that all semi-simplicial vertices of H are not dominated.

15. If there is a semi-simplicial vertex y of H such that $|S_H(y)| \geq 3$ and there is $x \in S_H(y)$ with $d_H(x) = 2$, then denote by u the neighbor of x distinct from y and branch:
    (i) call ENUMIS$(S \cup \{x\}, F, X \setminus \{x, u, y\})$,
    (ii) call ENUMIS$(S \cup \{y\}, F, X \setminus (S_H(y) \cup \{u\}))$,
    (iii) call ENUMIS$(S \cup \{u\}, F, X \setminus \{x, u, y\})$,
    (iv) call ENUMIS$(S, F, X \setminus \{x\})$.

From now we have that for every semi-simplicial vertex y of H, $|S_H(y)| \leq 2$. Next, we analyze the cases when there is a semi-simplicial y with $|S_H(y)| = 1$.

16. If there is a semi-simplicial vertex y of H such that $|S_H(y)| = 1$ and $x \in S_H(y)$ is a pendant, then branch:
    (i) call ENUMIS$(S \cup \{x\}, F, X \setminus \{x, y\})$,
    (ii) call ENUMIS$(S \cup \{y\}, F, X \setminus \{x, y\})$,
    (iii) for every $z \in (N_H(y) \setminus \{x\}) \cap X$, call ENUMIS$(S \cup \{z\}, F \setminus N_H[y], X \setminus N_H[y])$.

17. If there is a semi-simplicial vertex y of H such that $|S_H(y)| = 1$ and for $x \in S_H(y)$, $d_G(x) = 2$, then denote by z the neighbor of x in H distinct from y and branch:
    (i) call ENUMIS$(S \cup \{x\}, F, X \setminus \{x, y, z\})$,
    (ii) call ENUMIS$(S \cup \{y\}, F, X \setminus \{x, y, z\})$,
    (iii) call ENUMIS$(S \cup \{z\}, F, X \setminus \{x, y, z\})$,
    (iv) for each $v \in (N_H(y) \setminus \{x, z\}) \cap X$, call ENUMIS$(S \cup \{v\}, F \setminus N_H[y], X \setminus N_H[y])$,
    (v) call ENUMIS$(S, F \cup \{z\}, X \setminus \{x, y, z\})$.

Finally, we deal with semi-simplicial vertices y with $|S_H(y)| = 2$. Notice that two vertices of $S_H(y)$ are not adjacent in this case and have incomparable neighborhoods.

18. If there is a semi-simplicial vertex y of H such that $|S_H(y)| = 2$, $N_H(x) \neq N_H(z)$ for distinct $x, z \in S_H(y)$ and $d_H(x) = d_H(z) = 2$, then denote by u and v the neighbors of x and z respectively that are distinct from y and branch:
    (i) call ENUMIS$(S \cup \{x\}, F, X \setminus \{x, u, y\})$,
    (ii) call ENUMIS$(S \cup \{u, v\}, F, X \setminus \{x, u, y, v, z\})$,
    (iii) call ENUMIS$(S \cup \{u, z\}, F, X \setminus \{x, u, y, v, z\})$,
    (iv) call ENUMIS$(S \cup \{y\}, F, X \setminus \{x, u, y, z\})$,
    (v) call ENUMIS$(S, F, X \setminus \{x\})$.

This completes the description of our algorithm. Now we explain how we prove its correctness.

Observe that in every step of $EnumIS(S, F, X)$, we either output S or call $\text{ENUMIS}(S', F', X')$ for $|F'| + |X'| < |F| + |X|$. It implies that the algorithm is finite.

Recall that to enumerate all maximal irredundant sets of a chordal graph G we call $\text{ENUMIS}(\emptyset, \emptyset, V(G))$. The algorithm outputs sets only in Step 1. Because we check whether S is a maximal irredundant set of G and output S only if it holds, the algorithm outputs only maximal irredundant sets. We show that $\text{ENUMIS}(\emptyset, \emptyset, V(G))$ outputs every maximal irredundant set. Let D be a maximal irredundant set of G. It is straightforward to see that D is compatible with $(\emptyset, \emptyset, V(G))$. We claim that if D is compatible with a triple (S, F, X) of disjoint subsets of $V(G)$, then $\text{ENUMIS}(S, F, X)$ outputs D.

The proof is by induction on the measure $|F| + |X|$ of an instance. If $|X| = \emptyset$, then the claim is straightforward, because $D = S$ and the algorithm outputs D in Step 1. Suppose that $|X| \geq 1$. We claim that the case analysis in the algorithm is exhaustive, that is, at least one step of $\text{ENUMIS}(S, F, X)$ can be applied for an instance (S, F, X). To show it notice that, as we already underlined in the description of the algorithm, if Steps 1–13 of the algorithm could not be applied, we have that every simplicial vertex of H is not dominated, free, has degrees 1 or 2 and all their neighbors are free. Also any two simplicial vertices are not adjacent and their neigborhoods are incomparable. Notice also that because of Step 5, we can assume that every component of H has at least 2 vertices. Then by Lemma 1, each component of H has a semi-simplicial vertex. It immediately follows that one of Steps 14–18 can be applied. It is already observed in the description of the algorithm that if we apply a step, then in this step we call $EnumIS(S', F', X')$ for an instance (S', F', X') such that D is compatible with it and $|F'| + |X'| < |F| + |X|$. It follows that $\text{ENUMIS}(S', F', X')$ outputs D by the inductive assumption.

Finally, we evaluate the running time of the algorithm and obtain our upper bound for the number of maximal irredundant sets. To do it, we compute branching vectors and branching numbers for Steps 5–18 (we refer to [14] for formal definitions) with respect to the instance measure $|X| + |F|$. We show that the maximum value of the branching number is less than 1.7549. We use this to show that the number of maximal irredundant sets of an n-vertex chordal graph is at most 1.7549^n and the running time of our algorithm is $O(1.7549^n)$.   □

We conclude the section by showing the enumeration algorithm and the upper bound for the number of maximal irredundant sets for chordal graphs can be improved in the case of forests, and thus also for trees. Recall that a graph is a forest if it does not contain a cycle; and a tree is a forest being connected. In fact our algorithm for forests is a simplified version of the algorithm ENUMIS for chordal graphs. Clearly, a forest is a chordal graph. Recall that in ENUMIS we use simplicial and semi-simplicial vertices for branching. For a forest, any simplicial vertex is either a leaf or an isolated vertex. Hence, we can exclude from ENUMIS all the steps where we analyze simplicial vertices of degree at least 2. It is easy to see that a vertex u is semi-simplicial if it has at least one adjacent leaf of a forest and at most one nonleaf neighbor. It gives us the following theorem whose proof is omitted.

Theorem 2. *A forest has at most* 1.6181^n *maximal irredundant sets, and these can be enumerated in time* $O(1.6181^n)$.

4   Enumeration of Maximal Irredundant Sets for Interval Graphs

In this section we show that for the class of interval graphs, we can improve the bound obtained for chordal graphs. We need some additional notations. Recall that we assume that for an interval graph G, we are given its interval representation, that is, a set of closed intervals of the real line corresponding to the vertices such that two intervals intersects if and only if the corresponding vertices are adjacent. Slightly abusing notation, we do not distinguish a vertex and the corresponding interval in the representation. For a vertex v, we denote by ℓ_v and r_v the left end-point and the right end-point respectively of the interval corresponding to v in the representation.

Theorem 3. *An interval graph has at most* 1.6957^n *maximal irredundant sets, and these can be enumerated in time* $O(1.6957^n)$.

Proof. Again, we only sketch the proof.

Similarly to the chordal graph case, we construct a branching recursive algorithm to solve the enumeration problem. Let G be an interval graph given together with its interval representation. The algorithm ENUMIS-I(S, F, X) takes as an input disjoint subsets $S, X, F \subseteq V(G)$ such that $\ell_x \leq r_y$ for $x \in S$ and $y \in F \cup X$ and enumerates the maximal irredundant sets D of G such that (a) $S \subseteq D$, (b) $D \setminus S \subseteq X$, and (c) for each $v \in D \setminus S$, v has a private belonging to $F \cup X$. In our algorithm we denote by $H = G[F \cup X]$. In the same way as for chordal graphs, we say that the vertices of F are *forbidden* as they cannot be included in an irredundant set and the vertices of X are called *free*. We also say that a vertex of H is *dominated* if it is dominated in G by a vertex of S. In our algorithm, we either reduce the considered instance or branch. In every step, if we decide to include a vertex $u \in X$ in a (potential) irredundant set, we make sure that it gets a private v in $F \cup X$ that is not dominated by other vertices in the recursive calls. To guarantee this, we declare all the neighbors of v except u forbidden.

ENUMIS-I(S, F, X)

1. If $X = \emptyset$, then check whether S is a maximal irredundant set of G and output it if it holds; then stop.
2. If H has a forbidden and dominated vertex x, then call ENUMIS-I$(S, F \setminus \{x\}, X)$.

3. If there is $x \in V(H)$ such that all the vertices of $N_H[x]$ are dominated, then call ENUMIS-I$(S, F \setminus \{x\}, X \setminus \{x\})$.

4. If there is $x \in V(H)$ such that all the vertices of $N_H[x]$ are forbidden, then call ENUMIS-I$(S, F \setminus \{x\}, X \setminus \{x\})$.

5. Find a vertex u of H with the minimum right end-point r_u, and let $N_H(u) = \{v_1, \ldots, v_k\}$, where the vertices $v_1, \ldots, v_k$ are ordered by increasing value of their right end-points.

6. If u is dominated then find the minimum index $i \in \{1, \ldots, k\}$ such that v_i is not dominated and branch:
   (i) call ENUMIS-I$(S \cup \{u\}, F \setminus N_H[v_i], X \setminus N_H[u])$,
   (ii) call ENUMIS-I$(S, F \setminus \{u\}, X \setminus \{u\})$.

7. If u is forbidden, then branch:
   (i) for each $i = 1, \ldots, k$, if v_i is free, then call ENUMIS-I$(S \cup \{v_i\}, F \setminus N_H[u], X \setminus N_H[u])$,
   (ii) call ENUMIS-I$(S, (F \setminus \{u\}) \cup N_H(u), X \setminus N_H[u])$.

8. If $N_H(u) = \emptyset$, then call ENUMIS-I$(S \cup \{u\}, F \setminus \{u\}, X \setminus \{u\})$.

9. Let $W_0 = \emptyset$, and for $i = 1, \ldots, k$, set

$$W_i = \left(\{w \in N_H(v_i) \setminus N_H[u] \mid r_w > r_{v_i}\} \cap X\right) \setminus \left(\bigcup_{j=0}^{i-1} W_j\right).$$

10. If $W_i = \emptyset$ for $i \in \{1, \ldots, k\}$, then branch:
    (i) call ENUMIS-I$(S \cup \{u\}, F \setminus N_H[u], X \setminus N_H[u])$,
    (ii) for each $i = 1, \ldots, k$, if v_i free, then call ENUMIS-I$(S \cup \{v_i\}, F \setminus N_H[u], X \setminus N_H[u])$.

11. Otherwise, branch:
    (i) call ENUMIS-I$(S \cup \{u\}, F \setminus N_H[u], X \setminus N_H[u])$,
    (ii) for each $i = 1, \ldots, k$, if v_i free, then call ENUMIS-I$(S \cup \{v_i\}, F \setminus N_H[u], X \setminus N_H[u])$,
    (iii) for each $i = 1, \ldots, k$, if $W_i \neq \emptyset$, then for each $w \in W_i$, call ENUMIS-I$(S \cup \{w\}, F \setminus N_H[v_i], X \setminus N_H[v_i])$.

To enumerate the maximal irredundant sets of G, we call ENUMIS-I$(\emptyset, \emptyset, V(G))$.

In the same way as for chordal graphs, it is easy to see that all the sets generated by the algorithm are maximal irredundant sets, because at Step 1, where we output sets, we verify whether a generated set is a maximal irredundant set. To show that every maximal irredundant set D of G is generated by the algorithm, we prove the following claim.

Claim A. *Let D be a maximal irredundant set of an interval graph G, and assume that S, F, X are disjoint subset of $V(G)$ such that the following holds:*

(a) $\ell_x \leq r_y$ for $x \in S$ and $y \in F \cup X$,
(b) $S \subseteq D$,
(c) $D \setminus S \subseteq X$,

(d) for each $v \in D \setminus S$, v has a private in $F \cup X$,

(e) for each $u \in S$, u has a private $v \in V(G) \setminus (F \cup X)$ with respect to D such that $N_G(v) \cap X = \emptyset$.

Then either $D = S$ and the algorithm outputs D or there is a recursive call of $\textsc{EnumIS-II}(S', F', X')$ in the call of $\textsc{EnumIS-I}(S, F, X)$ such that the sets S', F', X' satisfy (a)–(e).

To get our upper bound for the number of maximal irredundant sets and evaluate the running time, we compute branching vectors and branching numbers for branching steps with respect to the measure $|X| + |F|$. We show that the maximum branching number is less than 1.6957. This implies that G has at most 1.6957^n maximal irredundant sets, and the algorithm runs in time $O(1.6957^n)$.   □

5   Conclusions

We presented enumeration algorithms for maximal irredundant sets and obtained upper bounds for the number of such sets for chordal graphs, interval graphs and forests. We proved that a chordal graph has at most 1.7549^n maximal irredundant sets, and these can be enumerated in time $O(1.7549^n)$. For interval graphs, we prove that the number of maximal irredundant sets is at most 1.6957^n and they can be enumerated in time $O(1.6957^n)$. We also show that a forest has at most 1.6181^n maximal irredundant sets and these sets can be enumerated in time $O(1.6181^n)$. We complement these results by showing a lower bound of 1.5292^n that holds for trees and forests.

It is natural to ask whether it is possible to reduce the gap between upper and lower bounds by constructing better algorithms and/or giving better lower bounds. It could also be interesting to consider other subclasses of chordal graphs, e.g., strongly chordal graphs. Besides this, what can be said about general graphs? It is worth mentioning that there is a large interest in the output-sensitive enumeration of maximal irredundant sets and minimal redundant sets that has been mentioned as an open problem at the Lorentz Workshop "Enumeration Algorithms using Structure" [4].

References

1. Abu-Khzam, F.N., Heggernes, P.: Enumerating minimal dominating sets in chordal graphs. Inf. Process. Lett. **116**(12), 739–743 (2016)
2. Bertossi, A.A., Gori, A.: Total domination and irredundance in weighted interval graphs. SIAM J. Discrete Math. **1**(3), 317–327 (1988)
3. Binkele-Raible, D., Brankovic, L., Cygan, M., Fernau, H., Kneis, J., Kratsch, D., Langer, A., Liedloff, M., Pilipczuk, M., Rossmanith, P., Wojtaszczyk, J.O.: Breaking the 2^n-barrier for irredundance: two lines of attack. J. Discrete Algorithms **9**(3), 214–230 (2011)
4. Bodlaender, H., Boros, E., Heggernes, P., Kratsch, D.: Open problems of the Lorentz workshop, "Enumeration Algorithms using Structure". Technical report Series UU-CS-2015-016, Utrecht University (2016)

5. Brandstädt, A., Le, V.B., Spinrad, J.P.: Graph classes: a survey. In: SIAM Mono-graphs on Discrete Mathematics and Applications. Society for Industrial and Applied Mathematics (SIAM), Philadelphia (1999)
6. Cockayne, E.J., Favaron, O., Payan, C., Thomason, A.G.: Contributions to the theory of domination, independence and irredundance in graphs. Discrete Math. **33**(3), 249–258 (1981)
7. Couturier, J., Heggernes, P., van't Hof, P., Kratsch, D.: Minimal dominating sets in graph classes: combinatorial bounds and enumeration. Theor. Comput. Sci. **487**, 82–94 (2013)
8. Couturier, J., Letourneur, R., Liedloff, M.: On the number of minimal dominating sets on some graph classes. Theor. Comput. Sci. **562**, 634–642 (2015)
9. Cygan, M., Fomin, F.V., Kowalik, L., Lokshtanov, D., Marx, D., Pilipczuk, M., Pilipczuk, M., Saurabh, S.: Parameterized Algorithms. Springer, Cham (2015). doi:10.1007/978-3-319-21275-3
10. Downey, R.G., Fellows, M.R., Raman, V.: The complexity of irredundant sets parameterized by size. Discrete Appl. Math. **100**(3), 155–167 (2000)
11. Eiter, T., Gottlob, G.: Identifying the minimal transversals of a hypergraph and related problems. SIAM J. Comput. **24**(6), 1278–1304 (1995)
12. Fellows, M.R., Fricke, G., Hedetniemi, S.T., Jacobs, D.P.: The private neighbor cube. SIAM J. Discrete Math. **7**(1), 41–47 (1994)
13. Fomin, F.V., Grandoni, F., Pyatkin, A.V., Stepanov, A.A.: Combinatorial bounds via measure and conquer: bounding minimal dominating sets and applications. ACM Trans. Algorithms **5**(1), 9 (2008)
14. Fomin, F.V., Kratsch, D.: Exact Exponential Algorithms. Texts in Theoretical Computer Science. An EATCS Series. Springer, Heidelberg (2010). doi:10.1007/978-3-642-16533-7
15. Garey, M.R., Johnson, D.S.: Computers and Intractability: A Guide to the Theory of NP-Completeness. W. H. Freeman & Co., New York (1979)
16. Golovach, P.A., Heggernes, P., Kratsch, D.: Enumeration and maximum number of minimal connected vertex covers in graphs. In: Lipták, Z., Smyth, W.F. (eds.) IWOCA 2015. LNCS, vol. 9538, pp. 235–247. Springer, Cham (2016). doi:10.1007/978-3-319-29516-9_20
17. Golovach, P.A., Kratsch, D., Liedloff, M., Rao, M., Sayadi, M.Y.: On maximal irredundant sets and (σ, ϱ)-dominating sets in paths. Manuscript (2016)
18. Golovach, P.A., Kratsch, D., Sayadi, M.Y.: Enumeration of maximal irredundant sets for claw-free graphs. In: Fotakis, D., Pagourtzis, A., Paschos, V.T. (eds.) CIAC 2017. LNCS, vol. 10236, pp. 297–309. Springer, Cham (2017). doi:10.1007/978-3-319-57586-5_25
19. Golumbic, M.C.: Algorithmic Graph Theory and Perfect Graphs, Annals of Dis-crete Mathematics, vol. 57, 2nd edn. Elsevier Science B.V., Amsterdam (2004)
20. Habib, M., McConnell, R.M., Paul, C., Viennot, L.: Lex-BFS and partition refine-ment, with applications to transitive orientation, interval graph recognition and consecutive ones testing. Theor. Comput. Sci. **234**(1–2), 59–84 (2000)
21. Haynes, T.W., Hedetniemi, S.T., Slater, P.J.: Domination in Graphs: Advanced Topics. Marcel Dekker Inc., New York (1998)
22. Haynes, T.W., Hedetniemi, S.T., Slater, P.J.: Fundamentals of Domination in Graphs, Monographs and Textbooks in Pure and Applied Mathematics, vol. 208. Marcel Dekker Inc., New York (1998)
23. Hedetniemi, S.T., Laskar, R., Pfaff, J.: Irredundance in graphs: a survey. In: Pro-ceedings of the Sixteenth Southeastern International Conference on Combinatorics, Graph Theory and Computing, Boca Raton, FL, vol. 48, pp. 183–193 (1985)

24. Jacobson, M.S., Peters, K.: Chordal graphs and upper irredundance, upper domination and independence. Discrete Math **86**(1–3), 59–69 (1990)
25. Kanté, M.M., Limouzy, V., Mary, A., Nourine, L.: On the enumeration of minimal dominating sets and related notions. SIAM J. Discrete Math. **28**(4), 1916–1929 (2014)
26. Khachiyan, L., Boros, E., Elbassioni, K.M., Gurvich, V.: On the dualization of hypergraphs with bounded edge-intersections and other related classes of hypergraphs. Theor. Comput. Sci. **382**(2), 139–150 (2007)
27. Korte, N., Möhring, R.H.: An incremental linear-time algorithm for recognizing interval graphs. SIAM J. Comput. **18**(1), 68–81 (1989)
28. Lekkerkerker, C.G., Boland, J.C.: Representation of a finite graph by a set of intervals on the real line. Fund. Math. **51**, 45–64 (1962/1963)

The Minimum Conflict-Free Row Split Problem Revisited

Ademir Hujdurović[1,2], Edin Husić[2,3], Martin Milanič[1,2(✉)], Romeo Rizzi[4], and Alexandru I. Tomescu[5]

[1] UP IAM, University of Primorska, Muzejski trg 2, 6000 Koper, Slovenia
{ademir.hujdurovic,martin.milanic}@upr.si
[2] UP FAMNIT, University of Primorska, Glagoljaška 8, 6000 Koper, Slovenia
edin.husic@student.upr.si
[3] ENS Lyon, Lyon, France
[4] Department of Computer Science, University of Verona, Verona, Italy
romeo.rizzi@univr.it
[5] Department of Computer Science
Helsinki Institute for Information Technology HIIT, University of Helsinki, Helsinki, Finland
tomescu@cs.helsinki.fi

Abstract. Motivated by applications in cancer genomics and following the work of Hajirasouliha and Raphael (WABI 2014), Hujdurović et al. (IEEE TCBB, to appear) introduced the minimum conflict-free row split (MCRS) problem: split each row of a given binary matrix into a bit-wise OR of a set of rows so that the resulting matrix corresponds to a perfect phylogeny and has the minimum number of rows among all matrices with this property. Hajirasouliha and Raphael also proposed the study of a similar problem, referred to as the minimum distinct conflict-free row split (MDCRS) problem, in which the task is to minimize the number of distinct rows of the resulting matrix. Hujdurović et al. proved that both problems are NP-hard, gave a related characterization of transitively orientable graphs, and proposed a polynomial-time heuristic algorithm for the MCRS problem based on coloring cocomparability graphs.

We give new formulations of the two problems, showing that the problems are equivalent to two optimization problems on branchings in a derived directed acyclic graph. Building on these formulations, we obtain new results on the two problems, including: (i) a strengthening of the heuristic by Hujdurović et al. via a new min-max result in digraphs generalizing Dilworth's theorem, (ii) APX-hardness results for both problems, (iii) two approximation algorithms for the MCRS problem, and (iv) a 2-approximation algorithm for the MDCRS problem.

Keywords: The minimum conflict-free row split problem · Branching · Dilworth's theorem · Min-max theorem · Approximation algorithm · APX-hardness

This work is supported in part by the Slovenian Research Agency (I0-0035, research program P1-0285 and research projects N1-0032, J1-5433, J1-6720, J1-6743, and J1-7051) and by the Academy of Finland (grant 274977).

H.L. Bodlaender and G.J. Woeginger (Eds.): WG 2017, LNCS 10520, pp. 303–315, 2017.
https://doi.org/10.1007/978-3-319-68705-6_23

1 Introduction

Motivated by applications in cancer genomics and following the work of Hajirasouliha and Raphael [12], Hujdurović et al. [15] introduced the minimum conflict-free row split problem. Informally, the problem can be stated as follows: given a binary matrix M, split each row of M into a bitwise OR of a set of rows so that the resulting matrix corresponds to a perfect phylogeny and has the minimum number of rows among all matrices with this property. To state the problem formally, we need the following two definitions.

Definition 1. *Given a matrix M, three distinct rows r, r', r'' of M and two distinct columns i and j of M, we denote by $M[(r,r',r''),(i,j)]$ the 3×2 submatrix of M formed by rows r, r', r'' and columns i, j (in this order). Two columns i and j of a binary matrix M are said to be* in conflict *if there exist rows r, r', r'' of M such that*

$$M[(r,r',r''),(i,j)] = \begin{pmatrix} 1\,1 \\ 1\,0 \\ 0\,1 \end{pmatrix}.$$

We say that a binary matrix M is conflict-free *if no two columns of M are in conflict.*

Definition 2. *Let $M \in \{0,1\}^{m\times n}$. Label the rows of M as $r_1, r_2, \ldots, r_m$. A binary matrix $M' \in \{0,1\}^{m'\times n}$ is a* row split *of M if there exists a partition of the set of rows of M' into m sets $R_1, R_2, \ldots R_m$ such that for all $i \in \{1,\ldots,m\}$, r_i is the bitwise OR of the binary vectors in R_i. The set R_i of rows of M' is said to be the set of* split rows *of row r_i (with respect to M').*

For simplicity, we defined a row split as a binary matrix M' for which a suitable partition of rows exists. However, throughout the paper we will make a slight technical abuse of this terminology by considering any row split M' of M as already equipped with an arbitrary (but fixed) partition of its rows $R_1,\ldots,R_m$ satisfying the above condition. For an example of these notions, see Fig. 1.

$$M = \begin{array}{c} \\ r_1 \\ r_2 \\ r_3 \\ r_4 \end{array} \begin{array}{c} \begin{array}{cccccc} c_1 & c_2 & c_3 & c_4 & c_5 & c_6 \end{array} \\ \begin{pmatrix} 1 & 0 & 1 & 1 & 1 & 1 \\ 0 & 0 & 0 & 1 & 1 & 0 \\ 0 & 1 & 1 & 0 & 1 & 1 \\ 0 & 0 & 0 & 0 & 0 & 1 \end{pmatrix} \end{array} \longrightarrow M' = \begin{array}{c} \\ R_1 \\ \\ \\ R_2 \\ R_3 \\ \\ \\ R_4 \end{array} \begin{array}{c} \begin{array}{cccccc} c_1 & c_2 & c_3 & c_4 & c_5 & c_6 \end{array} \\ \left(\begin{array}{cccccc} 1 & 0 & 1 & 0 & 0 & 1 \\ 0 & 0 & 1 & 0 & 0 & 1 \\ 0 & 0 & 0 & 1 & 1 & 0 \\ \hline 0 & 0 & 0 & 1 & 1 & 0 \\ \hline 0 & 0 & 1 & 0 & 0 & 1 \\ 0 & 1 & 0 & 0 & 1 & 0 \\ 0 & 0 & 0 & 0 & 1 & 0 \\ \hline 0 & 0 & 0 & 0 & 0 & 1 \end{array}\right) \end{array}$$

Fig. 1. An example of a binary matrix M and a conflict-free row split M' of M.

We denote by $\gamma(M)$ the minimum number of rows in a conflict-free row split M' of M. Formally, the minimum conflict-free row split problem is defined as follows:

MINIMUMCONFLICT-FREEROWSPLIT (MCRS):

Input: A binary matrix M.
Task: Compute $\gamma(M)$.

We will also consider a variant of the problem, proposed by Hajirasouliha and Raphael [12], in which the task is to compute a row split M' of M such that the number of *distinct* rows in M' is minimized. Let $\eta(M)$ denote the minimum number of distinct rows in a conflict-free row split M' of M. Similarly as above, we consider the corresponding optimization problem.

MINIMUMDISTINCTCONFLICT-FREEROWSPLIT (MDCRS):

Input: A binary matrix M.
Task: Compute $\eta(M)$.

The MCRS and the MDCRS problems are closely related to two well studied families of combinatorial objects: perfect phylogenetic trees and laminar set families. The first connection is well known: the rows of a binary matrix M are the leaves of a perfect phylogenetic tree if and only if M is conflict-free (see [5,11]). Moreover, if this is the case, then the corresponding phylogenetic tree can be retrieved from M in time linear in the size of M [10]. The intuition behind the fact that a conflict-free matrix corresponds to a perfect phylogeny is that one can map each row to a leaf of a tree, and each column to an edge, so that each row has a 1 exactly on those columns that are mapped to the edges on the path from the root to the leaf corresponding to the row. The forbidden 3×2 matrix from Definition 1 as a submatrix leads a contradiction, since then the two distinct edges e_i and e_j to which columns i and j are mapped, respectively, are such that e_i appears both before, and after, e_j on a root-to-leaf path. We refer to [12,15] and to references therein for further details on the biological aspects of the MCRS and the MDCRS problems.

The connection to laminar families follows from the fact that a binary matrix M is conflict-free if and only if the sets of rows indicating the positions of ones in the columns of M form a laminar family. This connection will be exploited in Sect. 4.2. Laminar families of sets play an important role in network design problems [16], in the study of packing and covering problems [3,8,19], and in several other areas of combinatorial optimization, see, e.g., [20].

In [15], Hujdurović et al. proved that the MCRS and the MDCRS problems are NP-hard, gave a related characterization of transitively orientable graphs, and proposed a polynomial-time heuristic algorithm for the problem based on coloring cocomparability graphs.

The aim of this paper is to advance the understanding of structural and computational aspects of the MCRS and the MDCRS problems.

Our Results and Techniques. The first and main result of this paper is a result showing that the MCRS and the MDCRS problems can be equivalently formulated as two optimization problems on branchings in a directed acyclic graph derived from the given binary matrix, the so-called *containment digraph*. (Precise definitions of these notions and the corresponding problems will be given in Sect. 2.) These equivalencies lead to more transparent formulations of the two problems. We will ascertain the applicability and usefulness of these novel formulations by deriving the following results and insights about the MCRS and the MDCRS problems:

– We prove a new min-max result on digraphs strengthening Dilworth's theorem on chain covers and antichains in partially ordered sets. This result, besides being interesting on its own as a generalization of a classical min-max result, connects well to the MCRS problem via the problem's branching formulation. The constructive, algorithmic proof of the result shows that a related problem is polynomially solvable: a problem in which only a subset of all branchings of the containment digraph is examined, namely the so-called linear branchings (branchings corresponding to chain partitions of the poset underlying the containment digraph). This approach leads to a new heuristic for the MCRS problem, improving on a previous heuristic by Hujdurović et al. from [15].
– We strengthen the NP-hardness results for the two problems to APX-hardness results.
– We complement the inapproximability results with three approximation algorithms: a 2-approximation algorithm for the MDCRS problem (implying that the problem is APX-complete) and two approximation algorithms for the MCRS problem, the approximation ratios of which are expressed in terms of two parameters of the containment digraph, corresponding to the height and the width of the underlying partial order, respectively.

Related Work. In [12], Hajirasouliha and Raphael introduced the so-called Minimum-Split-Row problem, in which only a given subset of rows of the input matrix needs to be split and, roughly speaking, the task is to minimize the number of additional rows in the resulting conflict-free row split. All results from [12] actually deal with the variant of the problem in which all rows need to be split (some perhaps trivially by setting $R_i = \{r_i\}$); in this case, the optimal value of the Minimum-Split-Row problem coincides with the difference $\gamma(M) - r(M)$, where $r(M)$ is the number of rows of M. In the same paper, a lower bound on the value of $\gamma(M)$ was derived and, in the concluding remarks of the paper, a study of the MDCRS problem was suggested. In subsequent works by Hujdurović et al. [15], the MCRS problem was introduced and several claims from [12] were proved incorrect, including an NP-hardness proof of the Minimum-Split-Row problem (which would imply NP-hardness of the MCRS problem). However, it was shown in [15] that the MCRS problem is indeed NP-hard, as is the MDCRS problem. Moreover, a polynomially solvable case of the MCRS problem was identified and an efficient heuristic algorithm for the problem on general instances was proposed, based on coloring cocomparability graphs.

The results of this paper improve on the previously known results about the two problems: NP-hardness results are strengthened to APX-hardness results, approximation algorithms for the two problems are proposed, and the heuristic algorithm for the MCRS problem given by Hujdurović et al. from [15] is improved. The key tools leading to most of these results are the newly proposed branching formulations and the new min-max theorem strengthening Dilworth's theorem.

Structure of the Paper. The branching formulations of the two problems are given in Sect. 2. A strengthening of Dilworth's theorem and its connection to the MCRS problem is discussed in Sect. 3. APX-hardness proofs and approximation algorithms are presented in Sect. 4. Due to space constraints, proofs and most figures are omitted; the interested reader is referred to [14] for details.

A Remark on Notation. A *binary matrix* $M \in \{0,1\}^{m \times n}$ is a matrix having m rows and n columns, and all entries 0 or 1. Each row of such a matrix is a vector in $\{0,1\}^n$; each column is a vector in $\{0,1\}^m$. We will usually denote by $R_M = \{r_1, \ldots, r_m\}$ and $C_M = \{c_1, \ldots, c_n\}$ the (multi)sets of rows and columns of M, respectively. The entry of M at row r_i and column c_j will be denoted by $M_{i,j}$ or $M_{r_i,j}$ when appropriate. For brevity, we will often write "the number of distinct rows (resp., columns) of M" to mean "the maximum number of pairwise distinct rows (resp., columns) of M". Two rows (resp., columns) are considered distinct if they differ as binary vectors. All binary matrices in this paper will be assumed to contain no row whose all entries are 0.

2 Formulations in Terms of Branchings in Directed Acyclic Graphs

In this section, we are going to formulate the MCRS and the MDCRS problems in terms of branchings in directed acyclic graphs (DAGs). First, we give the necessary definitions.

Definition 3. *Let $D = (V, A)$ be a DAG. A* branching *of D is a subset B of A such that (V, B) is a digraph in which for each vertex v there is at most one arc leaving v.*

The following construction (see, e.g., [12]) can be performed on any given binary matrix M and results in a directed acyclic graph. Given a column $c_j \in C_M$, the *support of c_j* is the set defined as $\{r_i \in R_M : M_{i,j} = 1\}$ and denoted by $\text{supp}_M(c_j)$. Given a binary matrix $M \in \{0,1\}^{m \times n}$, the *containment digraph D_M* of M is the directed acyclic graph with vertex set $V = \{\text{supp}_M(c) : c \in C_M\}$ and arc set $A = \{(v, v') : v, v' \in V \wedge v \subset v'\}$ where $\subset$ is the relation of proper inclusion of sets.

Let $M \in \{0,1\}^{m \times n}$ be a binary matrix, let $D_M = (V, A)$ be the containment digraph of M, and let B be a branching of D_M. For a vertex $v \in V$, we denote by $N_B^-(v)$ the set of all vertices $v' \in V$ such that $(v', v) \in B$. A *source* of B is a vertex not entered by any arc of B. For a vertex $v \in V$, an element $r \in v$ (that is, a row of M) is said to be *covered* in v with respect to B (or just B-*covered*) if $r \in \cup N_B^-(v)$. (When it is clear to which branching we are referring to, we will say just that "r is covered in v".) Analogously, we say that $r \in v$ is *uncovered* in v with respect to B if r is not covered in v. A B-*uncovered pair* is a pair (r, v) such that r is a row of M, v is a vertex of D_M (that is, the support of a column of M), $r \in v$, and r is uncovered in v with respect to B. For a row r of M, we will denote by $U_B(r)$ the set of all B-uncovered pairs with first coordinate r, and by $U(B)$ the set of all B-uncovered pairs.

For a branching $B \subseteq A$, we say that a vertex $v \in V$ is B-*irreducible* if there exists some element $r \in v$ that is uncovered in v with respect to B (equivalently, if $v \notin \cup N_B^-(v)$). In particular, every source of B is B-irreducible. We denote by $I(B)$ the set of all B-irreducible vertices.

We denote with $\beta(M)$ the minimum number of elements in $U(B)$ over all branchings B of D_M. Similarly, we denote with $\zeta(M)$ the minimum number of elements in $I(B)$ over all branchings B of D_M. The corresponding optimization problems are the following:

MINIMUMUNCOVERINGBRANCHING (MUB):	MINIMUMIRREDUCINGBRANCHING (MIB):
Input: A binary matrix M. *Task:* Compute $\beta(M)$.	*Input:* A binary matrix M. *Task:* Compute $\zeta(M)$.

The announced equivalence between the MCRS and the MUB problems, and between the MDCRS and the MIB problems is captured in the following.

Theorem 1. *For every binary matrix M, the following holds:*

1. *Any branching B of D_M can be transformed in polynomial time to a conflict-free row split of M with exactly $|U(B)|$ rows and with exactly $|I(B)|$ distinct rows.*

2. *Any conflict-free row split M' of M can be transformed in polynomial time to a branching B of D_M such that $|U(B)|$ is at most the number of rows of M' and $|I(B)|$ is at most the number of distinct rows of M'.*

Consequently, for every binary matrix M, we have $\gamma(M) = \beta(M)$ and $\eta(M) = \zeta(M)$.

The proof of the first part of Theorem 1 relies on the notion of a B-*split*, defined as follows.

Definition 4. *Let M be a binary matrix with rows $r_1, \ldots, r_m$ and columns $c_1, \ldots, c_n$. For a branching B of D_M, we define the B-split of M, denoted by M^B, as the matrix with rows indexed by the elements of the set $U(B)$, and columns $c_1', \ldots, c_n'$, as follows. Let $V = V(D_M)$ and for all $j \in \{1, \ldots, n\}$, let $v_j = \mathrm{supp}_M(c_j)$ (so $v_j \in V$). For a vertex $v \in V$, we denote by $B^+(v)$ the set of all vertices in V reachable by a directed path from v in (V, B) (note that $v \in B^+(v)$). For all $(r, v) \in U(B)$ and all $j \in \{1, \ldots, n\}$, set:*

$$M_{(r,v),j}^B = \begin{cases} 1, & if\ v_j \in B^+(v); \\ 0, & otherwise. \end{cases}$$

Note that if $M^B_{(r,v),j} = 1$, then $r \in v_j$.

Lemma 1. *Let M be a binary matrix without duplicated columns, B a branching of D_M, and let M^B be the B-split of M. Then M^B is a conflict-free row split of M with $|U(B)|$ rows, splitting each row r_i of M into rows of M^B indexed by $U_B(r_i)$. Moreover, the number of distinct rows in M^B is $|I(B)|$.*

The following lemma is the key to the converse direction.

Lemma 2. *Given a binary matrix M without duplicated columns and a conflict-free row split M' of M, a branching B of D_M such that M^B can be obtained from M' by removing some rows can be computed in polynomial time.*

3   A Strengthening of Dilworth's Theorem and Its Connection to the MCRS Problem

By Theorem 1, the MCRS problem can be concisely formulated in terms of a problem on branchings in a derived digraph. As shown by Hujdurović et al. in [15], the MCRS problem is NP-hard; consequently, the MUB problem is also NP-hard. In this section we show that a related problem in which we examine only a subset of all the branchings of the containment digraph of the input binary matrix is polynomially solvable. This will be achieved by deriving, in Sect. 3.1, a min-max theorem generalizing the classical Dilworth's theorem on partially ordered sets, which might be of independent interest. The resulting heuristic algorithm will be described in Sect. 3.2.

3.1   A Min-Max Relation Strengthening Dilworth's Theorem

This section can be read independently of the rest of the paper.

Consider a pair (D, π) where $D = (V, A)$ is a DAG and $\pi : V \to \mathbb{Z}_+$ is a *weight function* of D. (We use $\mathbb{Z}_+$ for the set of non-negative integers.) The weight function π is called *monotone* if $\pi_u \leq \pi_v$ for every $(u, v) \in A$.

In D, a *non-trivial path* is a directed path with at least one arc. We denote by D^t the transitive closure of D, that is, the DAG (V, A^t) on the same vertex set as D having an arc $(u, v) \in A^t$ if and only if there exists a non-trivial path in D from u to v. A chain in D is a sequence of vertices $C = (v_1, v_2, \ldots, v_t)$ such that $(v_i, v_{i+1}) \in A^t$ for all $i \in \{1, \ldots, t-1\}$; sometimes we regard C as the set of its vertices $C = \{v_1, v_2, \ldots, v_t\}$. The *price of chain* C is given by $\Pi(C) = \max_{v \in C} \pi_v$. A family of vertex-disjoint chains $P = \{C_1, \ldots, C_p\}$ is called a *chain partition* of D if every vertex of D is contained in precisely one chain of P. The *price of chain partition* P is defined as $\Pi(P) = \sum_{i=1}^{p} \Pi(C_i)$. Consider the following problem.

MINIMUMPRICECHAINPARTITION:

Input: A DAG $D = (V, A)$ and a monotone weight function $\pi : V \to \mathbb{Z}_+$ of D.

Task: Compute a chain partition P of D such that the price $\Pi(P)$ is minimum possible.

In this section we give a polynomial-time algorithm and a min-max characterization for the above problem. As can be expected, the notion of antichain will play a main role in this min-max characterization. An *antichain* of D is a set of vertices $N \subseteq V$ such that N is an independent set (that is, a set of pairwise non-adjacent vertices) in D^t; in other words, no non-trivial path of D has both endpoints in N. Note that $|C \cap N| \leq 1$ for any chain C and any antichain N. The *width* of D, denoted by $wdt(D)$, is the maximum cardinality of an antichain in D. A classical theorem of Dilworth states that the minimum number of chains in a chain partition of D equals $wdt(D)$ [4]. Moreover, a chain partition of D into $wdt(D)$ chains can be computed in time $O(|V(D)|^{5/2})$ [7,13] (see [17, p. 73–74]). Our characterization will be a refinement of Dilworth's theorem and its algorithmic proof makes use of Dilworth's theorem as a subroutine. We must introduce one further notion. A *tower* of antichains of D is a sequence of antichains of D, $T = (N_1, N_2, \ldots, N_{wdt(D)})$, with $|N_i| = i$. The *value of an antichain* N is given by $val(N) = \min_{v \in N} \pi_v$ and the *value* of tower $T = (N_1, N_2, \ldots, N_{wdt(D)})$ is defined as $val(T) = \sum_{i=1}^{wdt(D)} val(N_i)$.

To appreciate the purpose of this notion, we begin with a simple observation.

Lemma 3. *Let D be a DAG, let $P = \{C_1, \ldots, C_p\}$ be a chain partition of D, and let $T = (N_1, N_2, \ldots, N_{wdt(D)})$ be a tower of antichains of D. Then, $\Pi(P) \geq val(T)$ even if the weight function π is not monotone.*

For the case of monotone weight functions, the following min-max strengthening of Dilworth's theorem holds.

Theorem 2. *Let D be a DAG and let π be a monotone weight function of D. Then D admits a chain partition $P = \{C_1, \ldots, C_{wdt(D)}\}$ and a tower of antichains $T = (N_1, N_2, \ldots, N_{wdt(D)})$ such that $\Pi(P) = val(T)$. Such a pair (P, T) can be computed in time $O(|V(D)|^{7/2})$.*

To see that Theorem 2 is a strengthening of Dilworth's theorem, consider an arbitrary DAG $D = (V, A)$ and let π be the weight function of D that is constantly equal to 1. Then, the price of any chain C is $\Pi(C) = \max_{v \in C} \pi_v = 1$ and the price of a chain partition P equals its cardinality. Moreover, the value of any antichain N is $val(N) = \min_{v \in N} \pi_v = 1$, and consequently the value of any tower $T = (N_1, N_2, \ldots, N_{wdt(D)})$ of antichains is $val(T) = \sum_{i=1}^{wdt(D)} val(N_i) = wdt(D)$. Since $wdt(D)$ is a lower bound on the cardinality of any chain partition, applying Theorem 2 to (D, π) gives exactly the statement of Dilworth's theorem for D.

Note also due to the non-linearity of the definitions of the price of a chain and the value of an antichain Theorem 2 is incomparable with the classical weighted generalization of Dilworth's theorem due to Frank [6].

Lemma 3 and Theorem 2 imply the following.

Corollary 1. MINIMUMPRICECHAINPARTITION *can be solved optimally in time $O(|V(D)|^{7/2})$. More specifically, in the stated time a minimum price chain partition P of D can be found with the additional property that $|P| = wdt(D)$ (hence*

P is simultaneously a minimum price chain partition and a minimum size chain partition of D).

A variant of the MINIMUMPRICECHAINPARTITION problem in which the chains used in the partition have to be of bounded size was studied by Moonen and Spieksma in [18]. Contrary to the unrestricted case, which is polynomially solvable (by Corollary 1), Moonen and Spieksma showed that the "chain-bounded" version of the problem is APX-hard.

3.2   Connection with the MCRS Problem

We will now describe a heuristic algorithm for the MCRS problem based on Theorem 2. The basic idea is to search for an optimal solution only among linear branchings, where a branching of D_M is said to be *linear* if it defines a subgraph of maximum in- and out-degree at most one, that is, a disjoint union of directed paths. Note that such branchings correspond bijectively to chain partitions of D_M.

We denote by $\beta_\ell(M)$ the minimum number of elements in $U(B)$ over all linear branchings B of D_M. We now introduce the following problem, referred to as MINIMUMUNCOVEREDLINEARBRANCHING: Given a binary matrix M, compute a linear branching B of D_M such that $|U(B)| = \beta_\ell(M)$.

For a binary matrix M, define a function $\pi : V(D_M) \to \mathbb{Z}_+$ with $\pi(v) = |v|$ (recall that vertices of D_M are pairwise distinct subsets or R_M). By definition of D_M, we have $u \subset v$ whenever (u, v) is an arc in D_M. This implies that π is a monotone weight function of D_M. It is not difficult to see that for a linear branching B and its corresponding chain partition P, we have $\Pi(P) = |U(B)|$. Since linear branchings correspond bijectively to chain partitions, it follows that MINIMUMUNCOVEREDLINEARBRANCHING is a special case of MINIMUMPRICECHAINPARTITION. Using Theorem 2, we obtain that a linear branching B of D_M with $|U(B)| = \beta_\ell(M)$ can be computed in time $O(|V(D_M)|^{7/2})$. This proves the following theorem.

Theorem 3. MINIMUMUNCOVEREDLINEARBRANCHING *can be solved to optimality in time* $O(|V(D_M)|^{7/2})$.

Note that Theorem 3 yields a heuristic polynomial-time algorithm for the MUB problem, and consequently for the MCRS problem. We are now going to explain why this algorithm improves on the heuristic for the latter problem by Hujdurović et al. from [15]. For the sake of simplicity of exposition, suppose that the input matrix M does not have any pairs of identical columns. (It is not difficult to see that this assumption is without loss of generality.) In this case, the algorithm from [15] returns a row split of the input matrix naturally derived from an optimal coloring of the complement of the underlying undirected graph of D_M, which is a cocomparability graph and thus an optimal coloring can be computed efficiently, see, e.g., [9]. Such optimal colorings correspond bijectively to minimum chain partitions of D_M; each color class corresponds to a chain. In the terminology of branchings, the conflict-free row split of the input matrix M returned by the heuristic from [15] is exactly the B-split of M (cf. Definition 4) where B is the linear branching of D_M corresponding to a minimum chain partition of D_M.

In the above approach, any proper coloring could be used instead of an optimal coloring of the derived cocomparability graph. In branching terminology, choosing a proper coloring of the derived cocomparability graph so that the number of rows of the output row split is minimized corresponds exactly to MINIMUMUNCOVEREDLINEARBRANCHING, which can be solved optimally by Theorem 3. Thus, the heuristic algorithm for the MCRS problem that returns the B-split of M where M is an optimal solution to MINIMUMUNCOVERED-LINEARBRANCHING always returns solutions that are at least as good as those computed by the algorithm by Hujdurović et al. from [15]. Moreover, note that by Corollary 1, digraph D_M has a minimum price chain partition that is also minimum with respect to size. This implies the existence of an optimal solution to MINIMUMUNCOVEREDLINEARBRANCHING on M such that the corresponding chain partition is of size $wdt(M)$ and, equivalently, the existence of an optimal coloring of the derived cocomparability graph that minimizes the number of rows in the derived conflict-free row split of M over all proper colorings of the derived graph.

4   (In)approximability Issues

In this section we discuss (in)approximability properties of the four problems studied in this paper, giving both APX-hardness results and approximation algorithms. The approximation ratios of some of our algorithms will be described in terms of the following parameters of the input matrix. Recall that the *width* of a DAG D is the maximum cardinality of an antichain in D. The *height* of a DAG D is the maximum number of vertices in a directed path contained in D. The *width* and the *height* of a binary matrix M are denoted by $wdt(M)$ and by $h(M)$, respectively, and defined as the width, resp. the height, of the containment digraph of M.

4.1   Hardness Results

Our main inapproximability results are summarized in the following theorem, which shows hardness already for very restricted input instances.

Theorem 4. *The MUB and the MIB problems (and consequently the MCRS and the MDCRS problems) are* APX-*hard, even for instances of height* 2.

The above result implies that none of the four problems admits a polynomial-time approximation scheme (PTAS), unless P = NP. Proving that a problem is APX-hard also provides a different proof of NP-hardness. For further background on APX-hardness, we refer to [2]. The APX-hardness for the two branching problems is established by developing L-reductions from the vertex cover problem in cubic graphs, which is known to be APX-hard [1]. The APX-hardness of the other two problems then follows from Theorem 1.

4.2   2-Approximating η and ζ via Laminar Set Families

The result of Theorem 4 raises the question whether the four problems (MCRS, MDRCS, MUB, and MIB) admit constant factor approximations. We now show that this is indeed the case for the MDRCS and MIB problems. This will be achieved by proving a lower and an upper bound for $\eta(M)$, which will together imply a simple 2-approximation algorithm.

The lower bound is based on a connection between conflict-free matrices and laminar families of sets and an upper bound on the size of a laminar family in terms of the size of the ground set. Recall that a *set family* (or a *hypergraph*) is a pair $\mathcal{H} = (V, \mathcal{E})$ where $V = V(\mathcal{H})$ is a set and $\mathcal{E} = E(\mathcal{H})$ is a subset of the power set $\mathcal{P}(V)$. Elements of $V(\mathcal{H})$ are the *vertices* of $\mathcal{H}$; elements of $E(\mathcal{H})$ are its *hyperedges*. A hypergraph $\mathcal{H}$ is said to be *laminar* if every two hyperedges $e, f \in E(\mathcal{H})$ satisfy $e \cap f = \emptyset$, $e \subseteq f$, or $f \subseteq e$. To every binary matrix M we associate a hypergraph, denoted by $\mathcal{H}_M$, which we will refer to as the *column hypergraph* of M, which is the hypergraph having the rows of M as vertices and the support sets of the columns of M as hyperedges. Formally, $\mathcal{H}_M$ has vertex set $V(\mathcal{H}_M) = R_M$ and hyperedge set $E(\mathcal{H}_M) = \{\mathrm{supp}_M(c) : c \in C_M\}$.

The following observation follows immediately from definitions.

Observation 1. *A binary matrix M is conflict-free if and only if its column hypergraph $\mathcal{H}_M$ is laminar.*

The following upper bound on the size of a laminar hypergraph is well known, see, e.g., [20].

Theorem 5. *Every laminar hypergraph $\mathcal{H}$ satisfies $|E(\mathcal{H})| \leq 2|V(\mathcal{H})|$.*

Observation 1 and Theorem 5 imply the following.

Corollary 2. *Every conflict-free binary matrix M with m rows satisfies $k \leq 2m$, where k is the number of distinct columns of M.*

It is not difficult to see that the value of η is invariant under deleting one of a pair of identical columns. Therefore, Corollary 2 together with a simple row splitting strategy imply the following.

Lemma 4. *For every binary matrix M, we have $k/2 \leq \eta(M) \leq k$, where k is the number of distinct columns of M.*

Lemma 4 admits a constructive proof, which can be turned into the announced approximation result.

Theorem 6. *There is a 2-approximation algorithm for the MDCRS (and consequently for the MIB) problem running in time $O(mnk)$ on a given matrix $M \in \{0, 1\}^{m \times n}$ with exactly k distinct columns.*

Note that Theorems 4 and 6 imply that the MDCRS and the MIB problems are APX-complete.

4.3   Two Approximation Algorithms for Computing γ and β

While the question of whether the MCRS (and consequently the MUB) problem admits a constant factor approximation algorithm on general instances remains open, we give in this section two partial results in this direction. We show that the two problems admit constant factor approximation algorithms on instances of bounded height or width.

Theorem 7. *Let M be a binary matrix and let B be an arbitrary branching of D_M. Then, the number of rows in the B-split of M is at most $h(M)\gamma(M)$.*

Remark 1. Since in Theorem 7, there is no restriction on the branching B, an $h(M)$-approximation to $\gamma(M)$ can be obtained simply by taking $B = \emptyset$ and returning the resulting row split.

For instances of bounded width, a constant factor approximation can be obtained by considering any B-split resulting from a linear branching B of D_M consisting of $wdt(M)$ paths.

Theorem 8. *Any algorithm that, given a binary matrix M, computes a linear branching B of D_M consisting of $wdt(M)$ paths and returns the corresponding B-split of M is a $wdt(M)$-approximation algorithm for the MCRS problem.*

5   Conclusion

In this paper, we revised the minimum conflict-free row split problem and a variant of it. We formulated the two problems as optimization problems on branchings in a derived directed acyclic graph and, building on these formulations, obtained several new algorithmic and complexity insights about the two problems, including APX-hardness results and approximation algorithms. Moreover, we proved a min-max result on digraphs strengthening the classical Dilworth's theorem and leading to a new heuristic for the MCRS problem.

The main problem left open by our work is the determination of the exact (in)approximability status of the MCRS problem. In particular, does the problem admit a constant factor approximation? What is the complexity of the MCRS and MDCRS problems on instances of bounded width? Other possibilities for related future research include: (i) the study of the approximability properties of the closely related Minimum-Split-Row problem [12] (our preliminary investigations show that the problem, while being APX-hard, admits a $(2h(M) - 1)$-approximation); (ii) a parameterized complexity study of the considered problems (along with identification of meaningful parameterizations), and (iii) a study of extensions of the model that could be relevant for the biological application, such as the case when the input binary matrix may contain errors or has partially missing data.

References

1. Alimonti, P., Kann, V.: Some APX-completeness results for cubic graphs. Theoret. Comput. Sci. **237**(1–2), 123–134 (2000)
2. Ausiello, G., Crescenzi, P., Gambosi, G., Kann, V., Marchetti-Spaccamela, A., Protasi, M.: Complexity and Approximation. Springer, Berlin (1999). doi:10.1007/978-3-642-58412-1
3. Cheriyan, J., Jordán, T., Ravi, R.: On 2-coverings and 2-packings of laminar families. In: Nešetřil, J. (ed.) ESA 1999. LNCS, vol. 1643, pp. 510–520. Springer, Heidelberg (1999). doi:10.1007/3-540-48481-7_44
4. Dilworth, R.P.: A decomposition theorem for partially ordered sets. Ann. Math. **51**, 161–166 (1950)
5. Estabrook, G.F., et al.: An idealized concept of the true cladistic character. Math. Biosci. **23**(3–4), 263–272 (1975)
6. Frank, a.: Finding minimum weighted generators of a path system. In: Contemporary Trends in Discrete Mathematics (Štiřín Castle, 1997), DIMACS Series in Discrete Mathematics and Theoretical Computur Science, vol. 49, pp. 129–138. American Mathematical Society, Providence (1999)
7. Fulkerson, D.R.: Note on Dilworth's decomposition theorem for partially ordered sets. Proc. Amer. Math. Soc. **7**, 701–702 (1956)
8. Gabow, H.N., Manu, K.S.: Packing algorithms for arborescences (and spanning trees) in capacitated graphs. Math. Program. **82**, 83–109 (1998)
9. Golumbic, M.C.: Algorithmic Graph Theory and Perfect Graphs, 2nd edn. Elsevier Science B.V., Amsterdam (2004)
10. Gusfield, D.: Efficient algorithms for inferring evolutionary trees. Networks **21**(1), 19–28 (1991)
11. Gusfield, D.: Algorithms on Strings, Trees and Sequences: Computer Science and Computational Biology. Cambridge University Press, Cambridge (1997)
12. Hajirasouliha, I., Raphael, B.J.: Reconstructing mutational history in multiply sampled tumors using perfect phylogeny mixtures. In: Brown, D., Morgenstern, B. (eds.) WABI 2014. LNCS, vol. 8701, pp. 354–367. Springer, Heidelberg (2014). doi:10.1007/978-3-662-44753-6_27
13. Hopcroft, J.E., Karp, R.M.: An $n^{5/2}$ algorithm for maximum matchings in bipartite graphs. SIAM J. Comput. **2**, 225–231 (1973)
14. Hujdurović, A., Husić, E., Milanič, M., Rizzi, R., Tomescu, A.I.: Reconstructing perfect phylogenies via branchings in acyclic digraphs and a generalization of Dilworth's theorem. https://arxiv.org/abs/1701.05492
15. Hujdurović, A., Kačar, U., Milanič, M., Ries, B., Tomescu, A.I.: Complexity and algorithms for finding a perfect phylogeny from mixed tumor samples. IEEE/ACM Trans. Comput. Biol. Bioinform. (2017, to appear). For an extended abstract, see Proceedings of WABI 2015, LNCS 9289, pp. 80–92 (2015)
16. Jain, K.: A factor 2 approximation algorithm for the generalized Steiner network problem. Combinatorica **21**(1), 39–60 (2001)
17. Mäkinen, V., Belazzougui, D., Cunial, F., Tomescu, A.I.: Genome-Scale Algorithm Design. Cambridge University Press, Cambridge (2015)
18. Moonen, L.S., Spieksma, F.C.R.: Partitioning a weighted partial order. J. Comb. Optim. **15**(4), 342–356 (2008)
19. Sakashita, M., Makino, K., Fujishige, S.: Minimizing a monotone concave function with laminar covering constraints. Discrete Appl. Math. **156**(11), 2004–2019 (2008)
20. Schrijver, A.: Combinatorial Optimization: Polyhedra and Efficiency. Algorithms and Combinatorics, vol. 24. Springer, Berlin (2003)

Drawing Planar Graphs with Few Geometric Primitives

Gregor Hültenschmidt[1], Philipp Kindermann[1(✉)], Wouter Meulemans[2], and André Schulz[1]

[1] LG Theoretische Informatik, FernUniversität in Hagen, Hagen, Germany
gregorhueltenschmidt@gmx.de,
{philipp.kindermann,andre.schulz}@fernuni-hagen.de
[2] Department of Mathematics and Computer Science, TU Eindhoven, Eindhoven, The Netherlands
w.meulemans@tue.nl

Abstract. We define the *visual complexity* of a plane graph drawing to be the number of basic geometric objects needed to represent all its edges. In particular, one object may represent multiple edges (e.g., one needs only one line segment to draw two collinear edges of the same vertex). Let n denote the number of vertices of a graph. We show that trees can be drawn with $3n/4$ straight-line segments on a polynomial grid, and with $n/2$ straight-line segments on a quasi-polynomial grid. Further, we present an algorithm for drawing planar 3-trees with $(8n - 17)/3$ segments on an $O(n) \times O(n^2)$ grid. This algorithm can also be used with a small modification to draw maximal outerplanar graphs with $3n/2$ edges on an $O(n) \times O(n^2)$ grid. We also study the problem of drawing maximal planar graphs with circular arcs and provide an algorithm to draw such graphs using only $(5n - 11)/3$ arcs. This provides a significant improvement over the lower bound of $2n$ for line segments for a nontrivial graph class.

1 Introduction

The complexity of a graph drawing can be assessed in a variety of ways: area, crossing number, bends, angular resolution, etc. All these measures have their justification, but in general it is challenging to optimize all of them in a single drawing. More recently, the *visual complexity* was suggested as another quality measure for drawings [18]. The visual complexity denotes the number of simple geometric entities used in the drawing.

Typically, we consider as entities either straight-line segments or circular arcs. To distinguish these two types of drawings, we call the former *segment drawings* and the latter *arc drawings*. The idea is that we can use, for example, a single segment to draw a path of collinear edges. The hope is that a drawing that consists of only a few geometric entities is easy to perceive. It is a natural

---

The work of P. Kindermann and A. Schulz was supported by DFG grant SCHU 2458/4-1. The work of W. Meulemans was supported by Marie Skłodowska-Curie Action MSCA-H2020-IF-2014 656741.

H.L. Bodlaender and G.J. Woeginger (Eds.): WG 2017, LNCS 10520, pp. 316–329, 2017.
https://doi.org/10.1007/978-3-319-68705-6_24

question to ask for the best possible visual complexity of a drawing of a given graph. Unfortunately, it is an NP-hard problem to determine the smallest number of segments necessary in a segment drawing [10]. However, we can still expect to prove bounds for certain graph classes.

Related work. For a number of graph classes, upper and lower bounds are known for segment drawings and arc drawings; the upper bounds are summarized in Table 1. However, these bounds (except for cubic 3-connected graphs) do not require the drawings to be on the grid. In his thesis, Mondal [15] gives an algorithm for triangulations with few segments—but more than Durocher and Mondal [9] require—on an exponential grid.

Table 1. Upper bounds on the visual complexity. Here, n is the number of vertices, ϑ the number of odd-degree vertices and e the number of edges. Constant additions or subtractions have been omitted.

Class	Segments		Arcs	
Trees	$\vartheta/2$	[8]	$\vartheta/2$	[8]
Maximal outerplanar	n	[8]	n	[8]
3-trees	$2n$	[8]	$11e/18$	[18]
3-connected planar	$5n/2$	[8]	$2e/3$	[18]
Cubic 3-connected planar	$n/2$	[13,16]	$n/2$	[13,16]
Triangulation	$7n/3$	[9]	**5n/3**	Theorem 5
4-connected triangulation	$9n/4$	[9]	**3n/2**	Theorem 6
4-connected planar	$9n/4$	[9]	**9n/2−e**	Theorem 8
Planar	$16n/3 - e$	[9]	**14n/3−e**	Theorem 7

There are three trivial lower bounds for the number of segments required to draw any graph $G = (V, E)$ with n vertices and e edges: (i) $\vartheta/2$, where ϑ is the number of odd-degree vertices, (ii) $\max_{v \in V} \lceil \deg(v)/2 \rceil$, and (iii) $\lceil e/(n-1) \rceil$. For triangulations and general planar graphs, a lower bound of $2n - 2$ and $5n/2 - 4$, respectively, is known [8]. Note that the trivial lower bounds are the same as for the slope number of graphs [20], that is, the minimum number of slopes required to draw all edges, and that the number of slopes of a drawing is upper bounded by the number of segments. Chaplick et al. [3,4] consider a similar problem where all edges are to be covered by few lines (or planes); the difference to our problem is that collinear segments are only counted once.

Contributions. In this work, we present two types of results. In the first part (Sects. 2, 3 and 4), we discuss algorithms for segment drawings on the grid with low visual complexity. This direction of research was posed as an open problem by Dujmović et al. [8], but only a few results exist; see Table 2. We present an algorithm that draws trees on an $O(n^2) \times O(n^{1.58})$ grid using $3n/4$ straight-line segments. For comparison, the drawings of Schulz [18] need also $3n/4$ arcs on a smaller $O(n) \times O(n^{1.58})$ grid, but use the more complex circular arcs instead. Our segment drawing algorithm for trees can be modified to generate drawings with an optimal number of $\vartheta/2$ segments on a quasi-polynomial grid. We also present algorithms to compute segment drawings of planar 3-trees and maximal outerplanar graphs, both on an $O(n) \times O(n^2)$ grid. In the case of planar 3-trees, the algorithm needs at most $(8n - 17)/3$ segments, and in the case of maximum outerplanar graphs the algorithm needs at most $3n/2$ segments.

Finally, in Sect. 5, we study arc drawings of triangulations and general planar graphs. In particular, we prove that $(5n - 11)/3$ arcs are sufficient to draw any triangulation with n vertices. We highlight that this bound is significantly lower than the $2n - 2$ *lower* bound known for segment drawings [8] and the so far best-known $2e/3 = 2n$ upper bound for circular arc drawings [18]. A straightforward extension shows that $(14n - 3e - 29)/3$ arcs are sufficient for general planar graphs with e edges.

Table 2. Same as in Table 1 but for grid drawings.

Class	Type	Vis.Compl.	Grid	
Trees	Arcs	$3n/4$	$O(n) \times O(n^{1.58})$	[18]
Trees	Segments	$\mathbf{3n/4}$	$\mathbf{O(n^2) \times O(n^{1.58})}$	Theorem 1
Trees	Segments	$\vartheta/2$	**pseudo-polynom.**	Theorem 2
Cubic planar 3-conn.	Segments	$n/2$	$O(n) \times O(n)$	[13,16]
Maximal outerplanar	Segments	$\mathbf{3n/2}$	$\mathbf{O(n) \times O(n^2)}$	Theorem 4
Triangulation	Segments	$8n/3$	$O((3.63n)^{4n/3})$	[15]
Planar 3-tree	Segments	$\mathbf{8n/3}$	$\mathbf{O(n) \times O(n^2)}$	Theorem 3

Preliminaries. Given a triangulated planar graph $G = (V, E)$ on n vertices, a *canonical order* $\sigma = (v_1, \ldots, v_n)$ is an ordering of the vertices in V such that, for $3 \leq k \leq n$, (i) the subgraph G_k of G induced by $v_1, \ldots, v_k$ is biconnected, (ii) the outer face of G_k consists of the edge (v_2, v_1) and a path C_k, called *contour*, that contains v_k, and (iii) the neighbors of v_k in G_{k-1} form a subpath of C_{k-1} [6,7]. A canonical order can be constructed in reverse order by repeatedly removing a vertex without a chord from the outer face.

Most of our algorithms make ample use of *Schnyder realizers* [17]. Assume we selected a face as the outer face with vertices v_1, v_2 and v_n. We decompose the interior edges into three trees: T_1, T_2, and T_n rooted at v_1, v_2, and v_n, respectively. The edges of the trees are oriented to their roots. For $k \in \{1, 2, n\}$, we call each edge in T_k a k-*edge* and the parent of a vertex in T_k its k-*parent*. In the figures of this paper, we will draw 1-edges red, 2-edges blue, and n-edges green. The decomposition is a Schnyder realizer if at every interior vertex the edges are cyclically ordered as: outgoing 1-edge, incoming n-edges, outgoing 2-edge, incoming 1-edges, outgoing n-edge, incoming 2-edges. The trees of a Schnyder realizer are also called *canonical ordering trees*, as each describes a canonical order on the vertices of G by a (counter-)clockwise pre-order traversal [5]. There is a unique *minimal* realizer such that any interior cycle in the union of the three trees is oriented clockwise [2]; this realizer can be computed in linear time [2,17]. The number of such cycles is denoted by Δ_0 and is upper bounded by $\lfloor (n-1)/2 \rfloor$ [21]. Bonichon et al. [1] prove that the total number of leaves in a minimal realizer is at most $2n - 5 - \Delta_0$.

2   Trees with Segments on the Grid

Let $T = (V, E)$ be an undirected tree. Our algorithm follows the basic idea of the circular arc drawing algorithm by Schulz [18]. We make use of the *heavy path decomposition* [19] of trees, which is defined as follows. First, root the tree at some vertex r and direct all edges away from the root. Then, for each non-leaf u, compute the size of each subtree rooted in its children. Let v be the child of u with the largest subtree (one of them in case of a tie). Then, (u, v) is called a *heavy edge* and all other outgoing edges of u are called *light edges*; see Fig. 1a. The maximal connected components of heavy edges form the *heavy paths* of the decomposition.

We call the vertex of a heavy path closest to the root its *top node* and the subtree rooted in the top node a *heavy path subtree*. We define the *depth* of a heavy path (subtree) as follows. We treat each leaf that is not incident to a heavy edge as a heavy path of depth 0. The depth of each other heavy path P is by 1 larger than the maximum depth of all heavy paths that are connected from P by an outgoing light edge. Heavy path subtrees of common depth are disjoint.

Boxes. We order the heavy paths nondecreasingly by their depth and then draw their subtrees in this order. Each heavy path subtree is placed completely inside an L-shaped box (*heavy path box*) with its top node placed at the reflex angle; see Fig. 1b for an illustration of a heavy path box B_i with top node u_i, width $w_i = \ell_i + r_i$, and height $h_i = t_i + b_i$. We require that (i) heavy path boxes of common depth are disjoint, (ii) u_i is the only vertex on the boundary, and (iii) $b_i \geq t_i$.

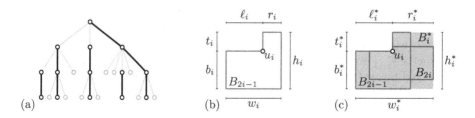

Fig. 1. (a) A heavy path decomposition; (b) the heavy path box B_i with top node u_i and its lengths; (c) the merged box B_i^* for B_{2i-1} and B_{2i} and its lengths.

Note that the boxes will be mirrored horizontally and/or vertically in some steps of the algorithm. We assign to each heavy path subtree of depth 0 a heavy path box B_i with $\ell_i = r_i = t_i = b_i = 1$.

Drawing. Assume that we have already drawn each heavy path subtree of depth d. When drawing the subtree of a heavy path $\langle v_1, \ldots, v_m \rangle$ of depth $d + 1$, we proceed as follows. The last vertex on a heavy path has to be a leaf, so v_m is a leaf. If outdeg(v_{m-1}) is odd, we place the vertices $v_1, \ldots, v_m$ on a vertical line; otherwise, we place only the vertices $v_1, \ldots, v_{m-1}$ on a vertical line and treat v_m as a heavy path subtree of depth 0 that is connected to v_{m-1}. For $1 \le h \le m - 1$, all heavy path boxes adjacent to v_h will be drawn either in a rectangle on the left side of the edge (v_h, v_{h+1}) or in a rectangle on the right side of the edge (v_{h-1}, v_h) (a rectangle that has v_1 as its bottom left corner for $h = 1$); see Fig. 2a for an illustration with even outdeg(v_{m-1}).

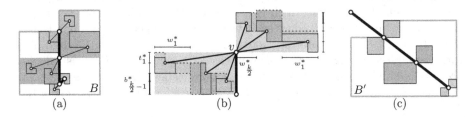

Fig. 2. (a) Placement of a heavy path, its box B, and areas for the adjacent heavy path boxes. (b) Placement of the heavy path boxes adjacent to v. (c) Further improvement on the visual complexity via increasing the size of the boxes.

We now describe how to place the heavy path boxes $B_1, \ldots, B_k$ with top node $u_1, \ldots, u_k$, respectively, incident to some vertex v on a heavy path into the rectangles described above. First, assume that k is even. Then, for $1 \le i \le k/2$, we order the boxes such that $b_{2i} \le b_{2i-1}$. We place the box B_{2i-1} in the lower left rectangle and box B_{2i} in the upper right rectangle in such a way that the edges (v, u_{2i-1}) and (v, u_{2i}) can be drawn with a single segment. To this end, we construct a *merged* box B_i^* as depicted in Fig. 1c with $\ell_i^* = \max\{\ell_{2i-1}, \ell_{2i}\}$, $r_i^* = \max\{r_{2i-1}, r_{2i}\}$, and $w_i^* = \ell_i^* + r_i^*$; the heights are defined analogously. The merged boxes will help us reduce the number of segments. We mirror all merged boxes horizontally and place them in the lower left rectangle (of width $\sum_{l=i}^{k/2} w_i^*$) as follows. We place B_1^* in the top left corner of the rectangle. For $2 \le j \le k/2$, we place B_j^* directly to the right of B_{j-1}^* such that its top border lies exactly t_{j-1}^* rows below the top border of B_{j-1}^*. Symmetrically, we place the merged boxes (vertically mirrored) in the upper right rectangle. Finally, we place each box B_{2i-1} (horizontally mirrored) in the lower left copy of B_i^* such that their inner concave angles coincide, and we place each box B_{2i} (vertically mirrored) in the upper right copy of B_i^* such that their inner concave angles coincide; see Fig. 2b. If k is odd, then we simply add a dummy box $B_{k+1} = B_k$ that we remove afterwards.

Analysis. We will now calculate the width w and the height h of this construction. For the width, we have $w = 2\sum_{i=1}^{k/2} w_i^* = 2\sum_{i=1}^{k/2} \max\{w_{2i-1}, w_{2i}\} \leq 2\sum_{i=1}^{k} w_i$. The height of each rectangle in the construction is at least $2\sum_{i=1}^{k/2} t_i^*$, but we have to add a bit more for the bottom parts of the boxes; in the worst case, this is $\max_{1 \leq j \leq k/2} b_{2j-1}$ in the lower rectangle and $\max_{1 \leq h \leq k/2} b_{2h}$ in the upper rectangle. Since we require $b_i \geq t_i$ for each i, we have

$$h \leq 2\sum_{i=1}^{k/2} t_i^* + \max_{1 \leq j \leq k/2} b_{2j-1} + \max_{1 \leq h \leq k/2} b_{2h} \leq 2\sum_{i=1}^{k} t_i + \sum_{j=1}^{k} b_j \leq \frac{3}{2}\sum_{i=1}^{k} h_i.$$

Since all heavy path trees of common depth are disjoint, the heavy path boxes of common depth are also disjoint. Further, we place only the top vertex of a heavy path on the boundary of its box. Finally, since we order the boxes such that $b_{2i} \leq b_{2i-1}$ for each i, for the constructed box B we have $b \geq t$.

Due to the properties of a heavy path decomposition, the maximum depth is $\lceil \log n \rceil$. Recall that we place the depth-0 heavy paths in a box of width and height 2. Hence, by induction, a heavy path subtree of depth d with n' vertices lies inside a box of width $2 \cdot 2^d \cdot n'$ and height $2 \cdot (3/2)^d \cdot n'$. Thus, the whole tree is drawn in a box of width $2 \cdot 2^{\lceil \log n \rceil} n = O(n^2)$ and height $2 \cdot (3/2)^{\lceil \log n \rceil} n = O(n^{1+\log 3/2}) \subseteq O(n^{1.58})$. Following the analysis of Schulz [18], the drawing uses at most $\lceil 3e/4 \rceil = \lceil 3(n-1)/4 \rceil$ segments.

Theorem 1. *Every tree admits a straight-line drawing that uses at most $\lceil 3e/4 \rceil$ segments on an $O(n^2) \times O(n^{1.58})$ grid.*

We finish this section with an idea of how to get a grid drawing with the best possible number of straight-line segments. Due to the limited space we give only a sketch. Observe that there is only one situation in which the previous algorithm uses more segments than necessary, that is the top node of each heavy path. This suboptimality can be "repaired" by *tilting* the heavy path as sketched in Fig. 2c. Note that the incident subtrees with smaller depth will only be translated. To make this idea work, we have to blow up the size of the heavy path boxes. We are left with scaling in each "round" by a polynomial factor. Since there are only $\log n$ rounds, we obtain a drawing on a quasi-polynomial grid. However, an implementation of the algorithm shows that some simple heuristics can already substantially reduce the drawing area, which gives hope that drawings on a polynomial grid exist for all trees.

Theorem 2. *Every tree admits a straight-line drawing with the smallest number of straight-line segments on a quasi-polynomial grid.*

3 Planar 3-Trees with Few Segments on the Grid

A *3-tree* is a maximal graph of treewidth k, that is, no edges can be added without increasing the treewidth. Each *planar 3-tree* can be produced from the complete graph K_4 by repeatedly adding a vertex into a triangular face and connecting it to all three vertices incident to this face. This operation is also known as *stacking*. Any planar 3-tree admits exactly one Schnyder realizer, and it is cycle-free [2].

Let T be a planar 3-tree. Let T_1, T_2, and T_n rooted at v_1, v_2, and v_n, respectively, be the canonical ordering trees of the unique Schnyder realizer of T. Without loss of generality, let T_1 be the canonical ordering tree having the fewest leaves, and let $\sigma = (v_1, v_2, \ldots, v_n)$ be the canonical order induced by a clockwise pre-order walk of T_2. The following lemma was proven by Durocher and Mondal [9].

Lemma 1 [9]. *Let $a_1, \ldots, a_m$ be a strictly x-monotone polygonal chain C. Let p be a point above C such that the segments a_1p and a_mp do not intersect C except at a_1 and a_m. If the positive slopes of the edges of C are smaller than slope(a_1, p), and the negative slopes of the edges of C are greater than slope(p, a_m), then every a_i is visible from p.* □

Overview and notation. The main idea of the algorithm is to draw the graph such that T_1 is drawn with one segment per leaf. We inductively place the vertices according to the canonical order $\sigma = (v_1, \ldots, v_n)$ and refer to the step in which vertex v_k is placed as *step k*. For the algorithm, we make use of the following additional notation; see Fig. 3. For each vertex v_i, the *1-out-slope* out$_1(i)$ is the slope of its outgoing 1-edge, the *2-out-slope* out$_2(i)$ is the slope of its outgoing 2-edge, and the *in-slope* in(i) is the highest slope of the incoming 1-edges in the current drawing. Further, we denote by par$_1(i)$ the 1-parent of v_i and by par$_2(i)$ the 2-parent of v_i. For two vertices v_i, v_j, we denote by lca(i, j) the lowest common ancestor of v_i and v_j in T_1. For an edge (v_i, v_j) we call the closed region bounded by (v_i, v_j), the path $(v_i, \ldots, \text{lca}(i, j))$, and the path $(v_j, \ldots, \text{lca}(i, j))$ the *domain* dom(i, j) of (v_i, v_j). For each step k, we denote by λ_k the number of leafs in the currently drawn subtree of T_1, by s_k the number of segments that are used to draw T_1, and by η_k the highest slope of the 1-edges in the current drawing. We denote by $C_k^{\rightarrow}$ the part of the contour C_k between v_k and v_2. Note that par$_2(k) \in C_{k-1}^{\rightarrow}$ because the canonical order is induced by T_2 in clockwise order; hence, either v_k and v_{k-1} are connected, or par$_2(k)$ is an ancestor of v_{k-1} on T_2.

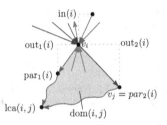

Fig. 3. Definitions for the drawing algorithm for planar 3-trees.

Invariants. After each step k, $3 \leq k \leq n$, we maintain the following invariants:

(I1) The contour C_k is a strictly x-monotone polygonal chain; the x-coordinates along $C_k^{\rightarrow}$ increase by 1 per vertex.

(I2) The 1-edges are drawn with $s_k = \lambda_k$ segments in total with integer slopes between 1 and $\eta_k \leq \lambda_k$.

(I3) For each $(v_i, v_j) \in C_k^{\rightarrow}$, we have $\mathrm{lca}(i,j) = \mathrm{par}_1(j)$ and for each 1-edge $e \neq (\mathrm{par}_1(j), v_j)$ in $\mathrm{dom}(i,j)$ it holds that $\mathrm{slope}(e) > \mathrm{out}_1(j)$.

(I4) For each $v_i \in C_k^{\rightarrow}$, $\mathrm{out}_1(i) > \mathrm{out}_2(i)$.

(I5) The current drawing is crossing-free and for each $(v_i, v_j) \in T_n$, slope $(v_i, v_j) > \mathrm{out}_1(j)$.

(I6) Vertex v_1 is placed at coordinate $(0,0)$, v_2 is placed at coordinate $(k-1, 0)$, and every vertex lies inside the rectangle $(0,0) \times (k-1, (k-1)\lambda_k)$.

The algorithm. The algorithm starts with placing v_1 at $(0,0)$, v_2 at $(2,0)$, and v_3 at $(1,1)$. Obviously, all invariants (I1)–(I6) hold. In step $k > 3$, the algorithm proceeds in two steps. Recall that v_k is a neighbor of all vertices on the contour between $v_l = \mathrm{par}_1(k)$ and $v_r = \mathrm{par}_2(k)$.

In the first step, the *insertion step*, v_k is placed in the same column as v_r. We distinguish between three cases to obtain the y-coordinate of v_k; case (i) is shown in Fig. 4a. (i) If no incoming 1-edge of v_l has been drawn yet, then we draw the edge (v_l, v_k) with slope $\mathrm{out}_1(l)$; (ii) If v_l already has an incoming 1-edge and v_l and v_r are the only neighbors of v_k in the current drawing, then we draw the edge (v_l, v_k) with slope $\mathrm{in}(l) + 1$; (iii) Otherwise, we draw the edge (v_l, v_k) with slope $\eta_{k-1} + 1$. Note that this does not maintain invariant (I1).

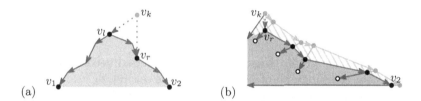

Fig. 4. (a) Inserting vertex v_k in case (i). (b) Shifting $\mathrm{par}_2(k), \ldots, v_2$ along their outgoing 1-edge.

In the second step of the algorithm, the *shifting step*, the vertices between v_r and v_2 on the contour C_k have to be shifted to the right without increasing the number of segments s_k used to draw T_1. To this end, we iteratively extend the outgoing 1-edge of these vertices, starting with v_r, to increase their x-coordinates all by 1; see Fig. 4b. This procedure places the vertices on the grid since the slopes of the extended edges are all integer by invariant (I2).

Theorem 3. *Every planar 3-tree admits a straight-line drawing that uses at most $(8n - 17)/3$ segments on an $O(n) \times O(n^2)$ grid.*

Sketch of Proof. The full proof that the invariants are maintained is given in the full version of the paper [12]. The invariants (I3) and (I4) ensure that we can apply Lemma 1 to insert the new vertices crossing-free and maintain planarity in the shifting step. Recall that we chose T_1 as the canonical ordering tree with the smallest number of leaves and that there are at most $2n - 5$ leaves in total, so T_1

has $\lambda_n \leq (2n - 5)/3$ leaves. Each canonical ordering tree has $n - 2$ edges. By invariant (I2), T_1 is drawn with λ_n segments, so we use $2(n-2)+\lambda_n \leq (8n-17)/3$ segments in total. The area follows immediately from invariant (I6).

4   Maximal Outerplanar Graphs with Segments on the Grid

A graph is *outerplanar* if it can be embedded in the plane with all vertices on one face (called outerface), and it is *maximal outerplanar* if no edge can be added while preserving outerplanarity. This implies that all interior faces of a maximal outerplanar graph are triangles. Outerplanar graphs have degeneracy 2 [14], that is, every induced subgraph of an outerplanar graph has a vertex with degree at most two. Thus, we find in every maximal outerplanar graph a vertex of degree 2 whose removal (taking away one triangle) results in another maximal outerplanar graph. By this, we gain a deconstruction order that stops with a triangle. Let $G = (V, E)$ be a maximal outerplanar graph and let $\sigma = (v_1, \ldots, v_n)$ be the reversed deconstruction order.

Lemma 2. *The edges of G can be partitioned into two trees T_1 and T_2. More-over, we can turn G into a planar 3-tree by adding a vertex and edges in the outerface. The additional edges form a tree T_n. The three trees T_1, T_2 and T_n induce a Schnyder realizer.*

Proof. We build the graph G according to the reversed deconstruction order. Let G_k denote the subgraph of G induced by the set $\{v_1, v_2, \ldots, v_k\}$. Further, let G'_k be the graph obtained by adding the vertex v_n and the edges (v_i, v_n) for all $1 \leq i \leq k$ to G. We prove by induction over k that there exists a Schnyder realizer induced by T_1, T_2, T_n for G'_k such that the trees T_1 and T_2 form the graph G_k and G'_k is a planar 3-tree. Note that Felsner and Trotter [11] already proved this lemma without the statement that G'_k is a planar 3-tree.

For the base case $k = 2$, our hypothesis is certainly true; see G'_2 in Fig. 5. Assume our assumption holds for some k. In order to obtain G_{k+1} from G_k, we have to add the vertex v_{k+1} and two incident edges (v_i, v_{k+1}) and (v_j, v_{k+1}). Assume that v_i is left of v_j on C_k. We add (v_i, v_{k+1}) to T_1 and (v_j, v_{k+1}) to T_2. This is safe since we cannot create a cycle in any of the trees. There is another

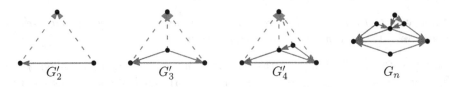

G'_2 $\qquad$ G'_3 $\qquad$ G'_4 $\qquad$ G_n

Fig. 5. Construction of an maximal outerplanar graph embedded in a planar 3-tree as done in the proof of Lemma 2. The tree T_n is dashed.

a new edge (v_{k+1}, v_n) in G'_{k+1} which we add to T_n. Again, no cycle can be created. The three outgoing tree edges at v_i form three wedges (unless v_i is the root of T_1). The new ingoing 1-edge (v_i, v_{k+1}) lies in the wedge bounded by the outgoing 2-edge and the outgoing n-edge. For the edge (v_j, v_{k+1}), we can argue analogously. Hence, the three trees induce a Schnyder realizer; see also Fig. 5. Moreover, the graphs G'_k and G'_{k+1} differ exactly by the vertex v_{k+1} that has been stacked into a triangular face of G'_k; thus, since G'_k is a planar 3-tree, so is G'_{k+1}. We now have proven the induction hypothesis for $k+1$. To obtain the statement of the lemma, we take the Schnyder realizer for the graph G'_{n-1} and move the edge (v_1, v_n) from T_n to T_1 and the edge (v_2, v_n) from T_n to T_2. Now T_1 and T_2 form G, and all three trees induce the Schnyder realizer of a planar 3-tree. $\qquad\square$

Consider a drawing according to Theorem 3 of the planar 3-tree introduced in Lemma 2 that contains the maximal outerplanar graph G as a spanning subgraph. By deleting T_n, we obtain a drawing of G. Note that in this drawing the outer face is realized as an interior face, which can be avoided (if this is undesired) by repositioning v_n accordingly. The Schnyder realizer has $2n-5$ leaves in total, but $n-3$ of them belong to T_n. We can assume that T_1 has the smallest number of leaves, which is at most $n/2 - 1$. We need $n-2$ segments for drawing T_2 (one per edge), and three edges for the triangle v_1, v_2, v_n. In total, we have at most $n/2 - 1 + n - 2 + 3 = 3n/2$ segments. Since the drawing is a subdrawing from our drawing algorithm for planar 3-trees, we get the same area bound as in the planar 3-tree scenario. We summarize our results in the following theorem.

Theorem 4. *Every maximal outerplanar graph admits a straight-line drawing that uses at most $3n/2$ segments on an $O(n) \times O(n^2)$ grid.*

5 Triangulations and Planar Graphs with Circular Arcs

Similar to Schulz [18], a canonical order $v_1, \ldots, v_n$ on the vertices of a triangulation is reversed and used to structure our drawing algorithm. We start by drawing v_1, v_2, and v_n on a circle; see Fig. 6a. We assume that they are placed

Fig. 6. (a) Initial state of the algorithm. (b) State of the algorithm after processing v_n. Hatching indicates undrawn region.

as shown and hence refer to the arc connecting v_1 and v_2 as *the bottom arc*. The interior of the circle is the *undrawn* region $\mathcal{U}$ which we maintain as a strictly convex shape. The vertices incident to $\mathcal{U}$ are referred to as the *horizon* and denoted $h_1, h_2, \ldots, h_{k-1}, h_k$ in order; we maintain that $h_1 = v_1$ and $h_k = v_2$. Initially, we have $k = 3$ and $h_2 = v_n$. We iteratively take a vertex h_i of the horizon (the latest in the canonical order) to process it, that is, we draw its undrawn neighbors and edges between these, thereby removing h_i from the horizon.

Invariant. We maintain as invariant that each vertex v (except v_1, v_2, and v_n) has a segment ℓ_v incident from above such that its downward extension intersects the bottom arc strictly between v_1 and v_2. Observe that, since $\mathcal{U}$ is strictly convex, this and h are the only intersection points for ℓ_h with the undrawn region's boundary for a vertex h on the horizon.

Processing a vertex. To process a vertex h_i, we first consider the triangle $h_{i-1}h_ih_{i+1}$: this triangle (except for its corners) is strictly contained in $\mathcal{U}$. We draw a circular arc A from h_{i-1} to h_{i+1} with maximal curvature, but within this triangle; see Fig. 7a. This ensures a plane drawing, maintaining a strictly convex undrawn region. Moreover, it ensures that h_i can "see" the entire arc A.

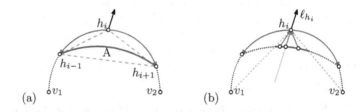

Fig. 7. (a) Arc A lies inside the dashed triangle $h_{i-1}h_ih_{i+1}$. (b) Undrawn neighbors of h_i are placed on A, in the section determined by v_1 and v_2. One neighbor is placed to align with ℓ_{h_i} towards a predecessor of h_i.

Vertex h_i may have a number of neighbors that were not yet drawn. To place these neighbors, we dedicate a fraction of the arc A. In particular, this fraction is determined by the intersections of segments v_1h_i and v_2h_i with A; see Fig. 7b. By convexity of $\mathcal{U}$, these intersections exist. If h_{i-1} is equal to v_1, then the intersection for v_1h_i degenerates to v_1; similarly, the intersection of v_2h_i may degenerate to v_2. We place the neighbors in order along this designated part of A, drawing the relevant edges as line segments. This implies that all these neighbors obtain a line segment that extends to intersect the bottom arc, maintaining the invariant. We position one neighbor to be a continuation of segment ℓ_{h_i}, which by the invariant must extend to intersect the designated part of A as well.

Complexity. We perform our algorithm using the canonical order induced by the canonical ordering tree in the minimal Schnyder realizer having the fewest leaves; without loss of generality, let T_n be this tree. Recall that T_n has at most $(2n - 5 - \Delta_0)/3$ leaves; since $\Delta_0 \geq 0$, we simplify this to $(2n - 5)/3$ for the remainder of the analysis. We start with one circle and subsequently process $v_n, \ldots, v_4$, adding one circular arc per vertex (representing edges in T_1 and T_2) and a number of line segments (representing edges in T_n). Note that processing v_3 has no effect since the edge $v_1 v_2$ is the bottom arc. Counting the circle as one arc, we thus have $n - 2$ arcs in total. At every vertex in T_n, one incoming edge is collinear with the outgoing one towards the root. Hence, we charge each line segment uniquely to a leaf of T_n: there are at most $(2n - 5)/3$ segments.

Thus, the total visual complexity is at most $n - 2 + (2n - 5)/3 = (5n - 11)/3$. In particular, this shows that, with circular arcs, we obtain greater expressive power for a nontrivial class of graphs in comparison to the $2n$ lower bound that is known for drawing triangulations with line segments. Since a triangulation has $e = 3n - 6$ edges, we conclude the following.

Theorem 5. *Every triangulation admits a circular arc drawing that uses at most $5n/3 - 11/3 = 5e/9 - 1/3$ arcs.*

This bound readily improves upon the result for line segments ($7e/9 - 10/3$) by Durocher and Mondal [9]. Schulz [18] proved an upper bound of $2e/3$ arcs. The bound above is an improvement on this result, though only for triangulations.

4-connected triangulations. We may further follow the rationale of Durocher and Mondal [9] by applying a result by Zhang and He [21]. Using regular edge labelings, they proved that a triangulation admits a canonical ordering tree with at most $\lceil (n + 1)/2 \rceil$ leaves [21]. Applying this to our analysis, we find that our algorithm uses at most $n - 2 + \lceil (n + 1)/2 \rceil = \lceil (3n - 3)/2 \rceil \le 3n/2 - 1$ arcs.

Theorem 6. *Every 4-connected triangulation admits a circular arc drawing that uses at most $3n/2 - 1 = e/2 + 2$ arcs.*

General planar graphs with circular arcs. The algorithm for triangulations easily adapts to draw a general planar graph G with $n \ge 3$ vertices and e edges. As connected components can be drawn independently, we assume G is connected. We need to only triangulate G, thereby adding $3n - e - 6$ chords. We then run the algorithm described in Theorem 5. Finally, we remove the chords from the drawing. Each chord may split an arc into two arcs, thereby increasing the total complexity by one. From Theorem 5, it follows that we obtain a drawing of G using $(5n/3 - 11/3) + (3n - e - 6) = 14n/3 - e - 29/3$ arcs.

Theorem 7. *Every planar graph with $n \ge 3$ admits a circular arc drawing with at most $14n/3 - e - 29/3$ arcs.*

Again, this bound improves upon the upper bound for line segments ($16n/3 - e - 28/3$) by Durocher and Mondal [9]. Provided the graph is 3-connected, Schulz's [18] bound of $2e/3 - 1$ is lower than our bound, but only for sparse-enough graphs having $e < 14n/5 - 26/5$. However, there are planar graphs that are not 3-connected with as many as $3n - 7$ edges (one less than a triangulation): there is no sparsity for which planar graphs must be 3-connected and Schulz's bound is lower than our result. In case the original graph G is 4-connected, extending it to a triangulation by adding edges does not violate this property. Repeating the above analysis using the improved bound of Theorem 6 yields us the following result.

Theorem 8. *Every 4-connected planar graph admits a circular arc drawing with at most $9n/2 - e - 7$ arcs.*

In the full version of the paper [12], we investigate a heuristic improvement to obtain a lower visual complexity when a graph has multiple faces of size at least 6. However, for worst-case bounds, this improvement is only noticeable in small graphs.

References

1. Bonichon, N., Le Saëc, B., Mosbah, M.: Wagner's theorem on realizers. In: Widmayer, P., Eidenbenz, S., Triguero, F., Morales, R., Conejo, R., Hennessy, M. (eds.) ICALP 2002. LNCS, vol. 2380, pp. 1043–1053. Springer, Heidelberg (2002). doi:10. 1007/3-540-45465-9_89
2. Brehm, E.: 3-orientations and Schnyder 3-tree-decompositions. In: Master's Thesis, Freie Universität Berlin (2000). http://page.math.tu-berlin.de/~felsner/ Diplomarbeiten/brehm.ps.gz
3. Chaplick, S., Fleszar, K., Lipp, F., Ravsky, A., Verbitsky, O., Wolff, A.: Drawing graphs on few lines and few planes. In: Hu, Y., Nöllenburg, M. (eds.) GD 2016. LNCS, vol. 9801, pp. 166–180. Springer, Cham (2016). doi:10.1007/ 978-3-319-50106-2_14
4. Chaplick, S., Fleszar, K., Lipp, F., Ravsky, A., Verbitsky, O., Wolff, A.: The complexity of drawing graphs on few lines and few planes. In: Ellen, F., Kolokolova, A., Sack, J.R. (eds.) Algorithms and Data Structures. LNCS, vol. 10389, pp. 265–276. Springer, Cham (2017). doi:10.1007/978-3-319-62127-2_23
5. de Fraysseix, H., de Mendez, P.O.: On topological aspects of orientations. Discrete Math. **229**(1–3), 57–72 (2001). doi:10.1016/S0012-365X(00)00201-6
6. de Fraysseix, H., Pach, J., Pollack, R.: Small sets supporting Fary embeddings of planar graphs. In: Simon, J. (ed.) Proceedings of 20th Annual ACM Symposium on Theory of Computing (STOC 1988), pp. 426–433. ACM, 1988. doi:10.1145/62212. 62254
7. de Fraysseix, H., Pach, J., Pollack, R.: How to draw a planar graph on a grid. Combinatorica **10**(1), 41–51 (1990). doi:10.1007/BF02122694
8. Dujmović, V., Eppstein, D., Suderman, M., Wood, D.R.: Drawings of planar graphs with few slopes and segments. Comput. Geom. Theory Appl. **38**(3), 194–212 (2007). doi:10.1016/j.comgeo.2006.09.002
9. Durocher, S., Mondal, D.: Drawing plane triangulations with few segments. In: He, M., Zeh, N. (eds.) Proceedings of 26th Canadian Conference on Computational Geometry (CCCG 2014), Carleton University, pp. 40–45 (2014). http://www.cccg. ca/proceedings/2014/papers/paper06.pdf
10. Durocher, S., Mondal, D., Nishat, R.I., Whitesides, S.: A note on minimum-segment drawings of planar graphs. J. Graph Algorithms Appl. **17**(3), 301–328 (2013). doi:10.7155/jgaa.00295

11. Felsner, S., Trotter, W.T.: Posets and planar graphs. J. Graph Theory **49**(4), 273–284 (2005). doi:10.1002/jgt.20081

12. Hültenschmidt, G., Kindermann, P., Meulemans, W., Schulz, A.: Drawing planar graphs with few geometric primitives (2017). Arxiv report 1703.01691. arXiv:1703.01691

13. Igamberdiev, A., Meulemans, W., Schulz, A.: Drawing planar cubic 3-connected graphs with few segments: algorithms and experiments. In: Di Giacomo, E., Lubiw, A. (eds.) GD 2015. LNCS, vol. 9411, pp. 113–124. Springer, Cham (2015). doi:10.1007/978-3-319-27261-0_10

14. Lick, D.R., White, A.T.: k-degenerate graphs. Can. J. Math. **22**, 1082–1096 (1970). doi:10.4153/CJM-1970-125-1

15. Mondal, D.: Visualizing graphs: optimization and trade-offs. Ph.D. thesis, University of Manitoba (2016). http://hdl.handle.net/1993/31673

16. Mondal, D., Nishat, R.I., Biswas, S., Rahman, M.S.: Minimum-segment convex drawings of 3-connected cubic plane graphs. J. Comb. Optim. **25**(3), 460–480 (2013). doi:10.1007/s10878-011-9390-6

17. Schnyder, W.: Embedding planar graphs on the grid. In: Johnson, D.S. (ed.) Proceedings of 1st Annual ACM-SIAM Symposium on Discrete Algorithms (SODA 1990), pp 138–148. SIAM (1990). http://dl.acm.org/citation.cfm?id=320191

18. Schulz, A.: Drawing graphs with few arcs. J. Graph Algorithms Appl. **19**(1), 393–412 (2015). doi:10.7155/jgaa.00366

19. Tarjan, R.E.: Linking and cutting trees. In: Data Structures and Network Algorithms, pp. 59–70. SIAM (1983). doi:10.1137/1.9781611970265.ch5

20. Wade, G.A., Chu, J.: Drawability of complete graphs using a minimal slope set. Comput. J. **37**(2), 139–142 (1994). doi:10.1093/comjnl/37.2.139

21. Zhang, H., He, X.: Canonical ordering trees and their applications in graph drawing. Discrete Comput. Geom. **33**(2), 321–344 (2005). doi:10.1007/s00454-004-1154-y

Mixed Dominating Set: A Parameterized Perspective

Pallavi Jain[1], M. Jayakrishnan[1], Fahad Panolan[2], and Abhishek Sahu[1($\boxtimes$)]

[1] The Institute of Mathematical Sciences, HBNI, Chennai, India
{pallavij,jayakrishnanm,asahu}@imsc.res.in
[2] Department of Informatics, University of Bergen, Bergen, Norway
fahad.panolan@ii.uib.no

Abstract. In the Mixed Dominating Set (MDS) problem, we are given an n-vertex graph G and a positive integer k, and the objective is to decide whether there exists a set $S \subseteq V(G) \cup E(G)$ of cardinality at most k such that every element $x \in (V(G) \cup E(G)) \setminus S$ is either adjacent to or incident with an element of S. We show that MDS can be solved in time $7.465^k n^{\mathcal{O}(1)}$ on general graphs, and in time $2^{\mathcal{O}(\sqrt{k})} n^{\mathcal{O}(1)}$ on planar graphs. We complement this result by showing that MDS does not admit an algorithm with running time $2^{o(k)} n^{\mathcal{O}(1)}$ unless the Exponential Time Hypothesis (ETH) fails, and that it does not admit a polynomial kernel unless coNP $\subseteq$ NP/poly. In addition, we provide an algorithm which, given a graph G together with a tree decomposition of width tw, solves MDS in time $6^{\text{tw}} n^{\mathcal{O}(1)}$. We finally show that unless the Set Cover Conjecture (SeCoCo) fails, MDS does not admit an algorithm with running time $\mathcal{O}((2 - \epsilon)^{\text{tw}(G)} n^{\mathcal{O}(1)})$ for any $\epsilon > 0$, where $\text{tw}(G)$ is the tree-width of G.

1 Introduction

Dominating (or covering) objects such as vertices and edges in a graph by vertices or edges give rise to several classic problems, such as Vertex Cover, Edge Cover, Dominating Set and Edge Dominating Set (see Table 1). All these problems and their numerous variants have been studied extensively from structural as well as algorithmic points of view. However, all these problems except Edge Cover are known to be NP-complete [11,25], and thus, they have been subjected to intense scrutiny in all the algorithmic paradigms meant for coping with NP-hardness, including approximation algorithms and parameterized complexity. In this paper we consider a well-studied variant of these problems, where the objective is to dominate *vertices and edges* by *vertices and edges*.

In order to define the problems formally, we first define the notion of domination, that is, what a vertex or an edge can dominate. A *vertex dominates* itself, all its neighbors and all the edges incident with it. On the other hand, an

The research leading to these results have received funding from the European Research Council via ERC Advanced Investigator Grant 267959.

© Springer International Publishing AG 2017
H.L. Bodlaender and G.J. Woeginger (Eds.): WG 2017, LNCS 10520, pp. 330–343, 2017.
https://doi.org/10.1007/978-3-319-68705-6_25

Table 1. Different domination problems and their FPT and kernelization status.

Dominating	By	Problem	PC	Poly kernel
Vertices	Vertices	DOMINATING SET	W[2]-hard	No
Vertices	Edges	EDGE COVER	P	$\mathcal{O}(1)$
Edges	Edges	EDGE DOMINATING SET	FPT	Yes
Edges	Vertices	VERTEX COVER	FPT	Yes
Edges+vertices	Vertices	VERTEX COVER	FPT	Yes
Edges+vertices	Edges	EDGE COVER	P	$\mathcal{O}(1)$
Edges+vertices	Edges+vertices	MIXED DOMINATING SET	FPT	No

edge dominates its two endpoints, and all the edges incident with either of its endpoints. We first define the problem of dominating vertices by vertices. A *dominating set* of a graph G is a set $S \subseteq V(G)$ such that every vertex $v \in V(G) \setminus S$ is adjacent to at least one vertex in S. In the DOMINATING SET problem, we are given an input graph G, a positive integer k, and the objective is to check whether there exists a dominating set of size at most k. The edge counterpart of DOMINATING SET is called EDGE DOMINATING SET. The problem we study in this paper is a variant of these domination problems. Towards that we first define the notion of *mixed dominating set (mds)*. Given a graph G, and a set $X \subseteq V(G) \cup E(G)$, X is called a *mds* if every element $x \in (V(G) \cup E(G)) \setminus X$ is either adjacent to or incident with an element of X. More formally, we study the following problem in the parameterized complexity framework.

MIXED DOMINATING SET (MDS)   **Parameter:** k or tw(G)
Input: A graph G on n vertices and m edges and a positive integer k.
Question: Does there exist a mds of size at most k in G?

The notion of mds (also called total cover) was introduced in the 70s by Alavi et al. [1] as a generalization of matching and covering, and after that it has been studied extensively in graph theory [2,9,20,22]. See the chapter in [14] for a survey on mds. The algorithmic complexity of MDS was first considered by Majumdar [18], where he showed that the problem is NP-complete on general graphs and admits a linear-time algorithm on trees. Hedetniemi et al. [15] and Manlove [19] showed that MDS remains NP-complete on bipartite and chordal graphs and on planar bipartite graphs of maximum degree 4, respectively. A decade and half later, Zhao et al. [26] considered MDS and showed that it remains NP-complete on split graphs. Unaware of the older result, they also designed an $\mathcal{O}(n \log n)$ time algorithm on trees. Lan and Chang [17] extended this result and gave a linear time algorithm for MDS on cacti (an undirected graph where any two cycles have at most one vertex in common). Hatami [13] gave a factor 2 approximation algorithm for MDS on general graphs. Recently,

Rajaati et al. [23] studied MDS parameterized by the treewidth of the input graph and designed an algorithm with running time $\mathcal{O}^{\star}(3^{\mathsf{tw}(G)^2})$.[1]

In this paper we undertake a thorough study of MDS with respect to two parameters, the solution size and the treewidth of the input graph, and obtain the following results.

1. MDS admits an algorithm with running time $\mathcal{O}^{\star}(7.465^k)$. We complement the FPT result by observing that (a) MDS does not admit an algorithm with running time $2^{o(k)}n^{\mathcal{O}(1)}$ unless ETH [16] fails; and (b) it does not admit a polynomial kernel unless coNP $\subseteq$ NP/poly. See the last row of Table 1.
2. We design an algorithm with running time $\mathcal{O}^{\star}(6^{\mathsf{tw}(G)})$ for MDS parameterized by $\mathsf{tw}(G)$. This algorithm is a significant improvement over the $\mathcal{O}^{\star}(3^{\mathsf{tw}(G)^2})$ algorithm of Rajaati et al. [23]. We also show that it does not admit an algorithm with running time $\mathcal{O}^{\star}((2 - \epsilon)^{\mathsf{tw}}(G))$, for any $\epsilon > 0$, unless SeCoCo fails [7].

The algorithm for MDS, parameterized by the solution size, is based on a relationship between mds and *vertex cover* (a subset X of vertices such that every edge has at least one endpoint in X) of the input graph. We use this connection to gain structural insights into the problem and give an algorithmically useful characterization of an optimal solution. This characterization leads us to the following algorithm: enumerate all the minimal vertex covers, say C, of size at most $2k$ of the input graph, guess a subset of C and then solve an appropriate auxiliary problem in polynomial time. The algorithm parameterized by treewidth uses standard dynamic programming approach together with subroutines for fast computation of cover-product [3]. Both our hardness results (no polynomial kernel and the lower bound on the running time of MDS parameterized by treewidth) are based on a polynomial time parameter preserving transformation from an appropriate parameterization of RED BLUE DOMINATING SET [8]. For references to algorithms and hardness mentioned in Table 1, and for an introduction to parameterized complexity, we refer to [6].

2   Preliminaries

All graphs in this paper are undirected and simple. For a graph G, $V(G)$ and $E(G)$ denote the set of vertices and edges of G, respectively. An edge between u and v in a graph G is represented by uv. For a set of edges $E' \subseteq E(G)$, we denote by $V(E')$, the set of vertices that are endpoints of edges in E'. For $v \in V(G)$, $N_G(v)$ denotes the set of neighbors of v, and $N_G[v] = N_G(v) \cup \{v\}$. Similarly, for a subset $V' \subseteq V(G)$, $N_G(V') = (\cup_{v \in V'} N_G(v)) \setminus V'$ and $N_G[V'] = N_G(V') \cup V'$. Also, for $V' \subseteq V(G)$, we denote by $G[V]$, the subgraph of G induced on V'. For a graph G and $R \subseteq V(G)$, we use $E(R)$ to denote the set of edges incident with at least one vertex in R. In this paper, V' shall always be a set vertices and E', a

---

[1] $\mathcal{O}^{\star}$ notation suppresses the polynomial factor. That is, $\mathcal{O}(f(k)n^{\mathcal{O}(1)}) = \mathcal{O}^{\star}(f(k))$.

set of edges, unless otherwise specified. We use $\uplus$ to denote the disjoint union of two sets, i.e., for sets S, A and B, $S = A \uplus B$ means $S = A \cup B$ and $A \cap B = \emptyset$.

Treewidth. Let G be a graph. A *tree-decomposition* of a graph G is a pair $(\mathbb{T}, \mathcal{X} = \{X_t\}_{t \in V(\mathbb{T})})$ such that

- $\bigcup_{t \in V(\mathbb{T})} X_t = V(G)$,
- for all $xy \in E(G)$ there is a $t \in V(\mathbb{T})$ such that $\{x, y\} \subseteq X_t$, and
- for all $v \in V(G)$ the subgraph of $\mathbb{T}$ induced by $\{t \mid v \in X_t\}$ is connected.

The *width* of a tree decomposition is $\max_{t \in V(\mathbb{T})} |X_t| - 1$ and the *treewidth* of G is the minimum width over all tree decompositions of G and is denoted by $\mathsf{tw}(G)$.

3   Algorithm for MDS Parameterized by the Solution Size

In this section we design an algorithm for MDS parameterized by the solution size. We start with a simple observation that vertices and endpoints of edges in a mds form a vertex cover.

Lemma 1. *Let G be a graph and $S = V' \cup E'$ be a mds of G. Then $V' \cup V(E')$ is a vertex cover of G, of cardinality at most $2|S|$.*

Proof. Since $S = V' \cup E'$ is a mds of G, where $V' \subseteq V(G)$ and $E' \subseteq E(G)$, every edge in G has at least one of its endpoints in $V' \cup V(E')$. This implies that $V' \cup V(E')$ is a vertex cover of G, of cardinality at most $2|S|$.   □

In order to get a handle on an optimal solution we define what we call a *nice mds*.

> Among all **minimum sized** mixed dominating sets of G, pick the one with **the least number of vertices**. Such a mds is called a **nice mds**. To re-emphasize, a nice mds by definition is minimum sized.

We now prove the following lemma, which forms the crux of our algorithm.

Lemma 2. *Let G be a connected graph and $V' \cup E'$ be a nice mds of G. Then, there is a minimal vertex cover C of G such that $V' \subseteq C \subseteq V' \cup V(E')$.*

Proof. Let $S = V' \cup E'$. Since S is mds, by Lemma 1, $V' \cup V(E')$ is a vertex cover of size at most $2|S|$. Any edge incident on $v \in V'$ dominates v as well as all the edges incident on v. Therefore, if v is such that $S \setminus \{v\}$ dominates $N_G(v)$, then by replacing v in S with some edge incident on v (this is possible since G is connected), we get another minimum sized mds with fewer vertices. This implies every vertex in V' must dominate at least one vertex (other than itself) which no other element in $V' \cup E'$ dominates. More specifically, for every $v \in V'$, there is a vertex $v' \in V(G)$ such that $vv' \in E(G)$ and $v' \notin N_G[(V' \setminus \{v\})] \cup V(E')$. This means, every minimal vertex cover contained in $V' \cup V(E')$ must contain V', because if $C \subseteq V' \cup V(E')$ does not contain $v \in V'$, then edge vv' is not covered by C. Therefore, if the vertex cover $V' \cup V(E')$ is not minimal, keep removing vertices from $V(E') \setminus V'$ until we are left with a minimal vertex cover.   □

Let $V' \cup E'$ be a *nice* mds and C be a minimal vertex cover such that $V' \subseteq C \subseteq V' \cup V(E')$. Let $I = V(G) \setminus C$. Note that I is an independent set and it can partitioned into two sets I_d and I_u, where I_d is the set of vertices dominated by V' and $I_u = I \setminus I_d$. That is, $I_d = N_G(V') \cap I$, and $I_u = I \setminus I_d$. Also, let $Z = C \setminus V'$. We thus have a partition of $V(G)$ into V', Z, I_d and I_u. We call the quadruple (V', Z, I_d, I_u) a *nice partition* of $V(G)$ with respect to the mds $V' \cup E'$ and the minimal vertex cover C (see Fig. 1). Also, we refer to the graph $G' = G[Z \cup I_u]$ as the *companion graph* of G with respect to V' and C.

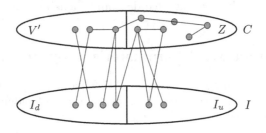

Fig. 1. Partition of $V(G)$ into minimal vertex cover C and independent set I, where $C = V' \uplus Z$ and $I = I_d \uplus I_u$.

Now let us define a new kind of domination called *special domination*. We say a vertex *special dominates* only itself, and an edge *special dominates* its endpoints as well as all the edges incident to at least one of its endpoints. Consequently, we can define a *special dominating set (sds)* as follows. A *sds* of a graph G' is a set $Q' \subseteq V(G') \cup E(G')$ such that every element $x \in (V(G') \cup E(G')) \setminus Q'$ is either adjacent to or incident on an *edge* in Q'. The next lemma shows the relation between mds and sds.

Lemma 3. *Let $V' \cup E'$ be a nice mds of G and C be a minimal vertex cover of G such that $V' \subseteq C \subseteq V' \cup V(E')$. Let (V', Z, I_d, I_u) be a nice partition of $V(G)$ with respect to $V' \cup E'$ and C. Then G has a mds of size at most k if and only if $G' = G[Z \cup I_u]$ has a sds of size at most $k - |V'|$.*

Proof. Assume G has a mds of size at most k. Since $V' \cup E'$ is a nice mds, $|V' \cup E'| \le k$. We can construct a sds Q' for G' as follows. If an edge $e \in E'$ has both its endpoints in $V(G')$, add e to Q'. If an edge $e \in E'$ has exactly one endpoint in $V(G')$, then add that endpoint to Q'.

We now claim that Q' is indeed a sds for G'. Since E' dominates every vertex in $V(E') \supseteq Z \cup I_u = V(G')$, Q' special dominates all vertices of G'. If $e = uv$ is an edge of G' such that there exists an edge $uw \in E'$ (or $vw \in E'$) for some $w \in V(G')$, then $uw \in Q'$ (or $vw \in Q'$) and hence Q' special dominates e.

We claim that Q' special dominates all the edges in G'. By way of contradiction, suppose $e = uv$ is an edge of G' such that there is no edge uw' or vw' in E' for any $w' \in V(G')$. Note that this also means $uv \notin E'$. But $u, v \in V(E') \supseteq Z \cup I_u$. In that case, there must exist $xu, yv \in E'$, where

$x, y \in V' \cup I_d$. Then we claim that $S = V' \cup ((E' \setminus \{xu, yv\}) \cup \{uv\})$ is a mds of G. Notice that $\{xu, yv\}$ dominate the set of vertices $R = \{x, u, y, v\}$ and all the edges $E(R)$ incident to at least one vertex in $R = \{x, u, y, v\}$. Since $V' \cup E'$ is a mds of G, to prove S is a mds of G, it is enough to show that S dominates R and $E(R)$. Since $S \supseteq V' \cup \{uv\}$ and $x, y \in I_d$, we have that $x, y \in N_G[V']$. This implies that S dominates R. Now, what is left to prove is, S dominates $E(R)$. Since $uv \in S$, all the edges incident to at least one of u or v is dominated by S. Finally, we show that S dominates all the edges incident to at least one of x or y. Let e be an edge incident on $z \in \{x, y\}$. If $z \in V'$, then S dominates e, because $z \in V' \subseteq S$. Otherwise $z \in I_d$, because $z \in \{x, y\} \subseteq V' \cup I_d$. Let $e = zw$. Since $z \in I_d$ and $V' \cup Z$ is a vertex cover of G, we have that $w \in V' \cup Z$. If $w \in V'$, then S dominates $e = zw$, because $w \in V' \subseteq S$. If $w \in \{u, v\}$, then S dominates $e = zw$, because $uv \in S$. Otherwise $w \in Z \setminus \{u, v\} \subseteq V(E') \setminus \{u, v\}$. Since $w \in V(E') \setminus \{u, v\}$, there is an edge in $E' \setminus \{xu, yv\} \subseteq S$. This implies that S dominates e. Thus we have shown that S is a mds of cardinality strictly less than that of $V' \cup E'$, a contradiction. Hence we conclude that Q' is a sds of G'.

To prove the other direction, suppose G' has a sds Q' of size atmost $k - |V'|$. We claim that $V' \cup Q'$ is an mds of G. Note that Q' dominates all vertices and edges in graph G' as well as all edges incident on $Z \cup I_u$, and V' dominates all vertices in $V' \cup I_d$ as well as all edges incident on V'. Therefore, $V' \cup Q'$ is a mds of G of cardinality $|V'| + |Q'| \leq k$.   □

Lemma 3 shows that given a graph G, V' and C as defined above, the problem of deciding whether G has a mds of size at most k boils down to deciding whether G' has a sds of size at most $k - |V'|$. This results in solving the following problem.

SPECIAL DOMINATING SET (SDS)
Input: An undirected graph G and a positive integer ℓ.
Question: Does there exist a sds of size at most ℓ in G?

In what follows we first design a polynomial time algorithm for SDS. Towards this, note that an edge has more "special dominating power" than a vertex has, in the sense that an edge special dominates itself, its endpoints and its adjacent edges, whereas a vertex special dominates *only* itself. Therefore, a natural strategy is to first try to special dominate as many vertices and all edges with as *few edges* as possible, and then add to the solution all those vertices that are not special dominated by any of the edges. This intuition leads to following polynomial time algorithm for SDS.

ALGORITHM-SDS (G, ℓ)
Step 1. Find a maximum matching, say M, in G. Let $U = V(G) \setminus V(M)$.
Step 2. If $|M \cup U| \leq \ell$, return Yes; else return No.

The only time consuming step in the above algorithm is Step 1 – finding a maximum matching – and this can be done in time $\mathcal{O}(m\sqrt{n})$ [21]. Thus, ALGORITHM-SDS runs in polynomial time, and the following lemma shows the correctness of the algorithm.

Lemma 4. *Let M be a maximum matching in a graph G and let $U = V(G) \setminus V(M)$. Then $M \cup U$ is a minimum sized sds of G.*

Proof. Since M is a maximum matching, every edge $e \in E \setminus M$ is incident to an edge in M, and thus M special dominates all edges in G. The set M also special dominates all vertices in $V(M)$, and the rest of the vertices in G are special dominated by U. Therefore, $M \cup U$ is indeed a sds of G.

Now we claim that $M \cup U$ is a minimum size sds of G. Since $V(M) \cap U = \emptyset$ and $V(G) = V(M) \cup U$, we have that $|V(G)| = 2|M| + |U|$. Towards proving the minimality of $M \cup U$, we show that any sds $E_1 \cup V_1$ of G, where $E_1 \subseteq E(G)$ and $V_1 \subseteq V(G)$, has cardinality at least $|M \cup U| = |M| + |U|$. Let M_1 be a maximum (w.r.t. E_1) matching contained in E_1. The total number of vertices special dominated by E_1 is at most $2|M_1| + |E_1 \setminus M_1| \le |M_1| + |E_1|$. Since $E_1 \cup V_1$ is a sds of G, we have

$$|M_1| + |E_1| + |V_1| \ge |V(G)|$$
$$|E_1| + |V_1| \ge |V(G)| - |M_1|$$
$$\ge 2|M| + |U| - |M_1| \quad \text{(because } |V(G)| = 2|M| + |U|)$$
$$\ge |M| + |U|. \quad \text{(because } |M| \ge |M_1|)$$

This completes the proof of the lemma.   □

Algorithm-SDS together with Lemma 4 results in the following result.

Lemma 5. *SDS can be solved in time $\mathcal{O}(m\sqrt{n})$.*

We are now fully equipped to give our algorithm for MDS.

Algorithm-MDS (G, k)

Step 1. Enumerate all minimal vertex covers of G of size at most $2k$. Let $\mathcal{C}$ be the collection of such vertex covers.

Step 2. For each $C \in \mathcal{C}$ and each $V' \subseteq C$ such that $|V'| \le k$ and $|C| \le 2k - |V'|$, use Algorithm-SDS to decide if the companion graph G' (w.r.t. C and V') has a sds of size at most $k - |V'|$; if it has, return Yes.

Step 3. Otherwise return No.

The correctness of the algorithm follows from Lemma 3. Now, let us analyze the running time of Algorithm-MDS. Any graph has at most 2^{2k} minimal vertex covers of size at most $2k$. Furthermore, given G and k, all minimal vertex covers of size at most $2k$ can be enumerated in time $2^{2k} n^{\mathcal{O}(1)}$ [10]. This means, Step 1 can be executed in time $2^{2k} n^{\mathcal{O}(1)}$.

For each $C \in \mathcal{C}$, there are at most $2^{|C|} \le 2^{2k}$ choices for V'. For each such choice of C and V', by Lemma 5, a minimum sds in G' can be found in polynomial time. Therefore, the running time of Algorithm-MDS (G, k) can be bounded by $2^{2k} \cdot 2^{2k} \cdot n^{\mathcal{O}(1)} = \mathcal{O}^\star(16^k)$. This, however, is a liberal estimate. A finer analysis shows that the running time can be brought down to $\mathcal{O}^\star(7.465^k)$.

Lemma 6. ALGORITHM-MDS *runs in time* $\mathcal{O}^{\star}(7.465^k)$.

Proof. If (G, k) is a yes-instance of MDS with a nice mds $V' \cup E'$, where $|V'| = j$, then any minimal vertex cover C such that $V' \subseteq C \subseteq V(E')$ can have size at most $|V'| + 2|E'| \leq j + 2(k - j) = 2k - j$. Therefore, in Step 2, we only process those pairs (C, V') such that $|C| \leq 2k - j$, where $|V'| = j$, and there are only 2^{2k-j} such C. Thus Step 2 takes time

$$\sum_{j=0}^{k} 2^{2k-j} \binom{2k-j}{j} = 2^{2k} \sum_{j=0}^{k} 2^{-j} \binom{2k-j}{j}.$$

Since for any $x > 0$, $\binom{n}{i}x^i \leq \sum_{i'=0}^{n} \binom{n}{i'}x^{i'} = (1+x)^n$, we get $\binom{n}{i} \leq (1+x)^n/x^i$. Using this inequality, for any $x > 0$,

$$2^{-j} \binom{2k-j}{j} \leq \frac{(1+x)^{2k-j}}{(2x)^j} = \frac{(1+x)^{2k}}{((1+x) \cdot 2x)^j}.$$

We choose $x = \frac{(\sqrt{3}-1)}{2}$ so that $(1+x) \cdot 2x = 1$. This gives $\frac{(1+x)^{2k}}{((1+x)2x)^j} \leq (1.3661)^{2k}$. Hence, Step 2 can be executed in time $2^{2k} \cdot 1.3661^{2k} \cdot n^{\mathcal{O}(1)} \leq (7.465)^k \cdot n^{\mathcal{O}(1)}$. □

Thus, we get the following theorem.

Theorem 1. MDS *parameterized by* k *can be solved in time* $\mathcal{O}^{\star}(7.465^k)$.

4   Outline of Algorithm for MDS Parameterized by Treewidth

In this section we only give a brief outline of our algorithm for MDS parameterized by treewidth of the input graph. Due to space constraint, we omit the complete description of the algorithm and its analysis. Here, the input is graph G and a tree decomposition of G of width $\text{tw}(G)$.

To design an algorithm we first prove that there is a minimum sized mixed dominating set S of G such that (i) the edges in S form a matching, and (ii) the set of endpoints of the edges in S is disjoint from the vertices in S.

We now give an informal description of our algorithm. Let G be the input graph and $(\mathbb{T}, \mathcal{X} = \{X_t\}_{t \in V(\mathbb{T})})$ be the given tree decomposition of G. Any standard dynamic programming over tree decomposition has three ingredients: for any node $t \in \mathbb{T}$ (i) defining partial solution, (ii) defining equivalence classes among partial solutions (or in other words defining states of DP table according to partial solutions), and (iii) computing a 'best partial solution' for each state from previously computed values. Normally, for any node $t \in \mathbb{T}$, partial solutions are defined according to the properties of the intersection of solutions with the graph G_t. In our case, a partial solution will be a subset of $V(G_t) \cup E(G_t)$. When we define equivalence classes of partial solutions, one of the factors under consideration is the intersection of these partial solutions with the bag X_t. Since partial solutions contain edges, at the first blush, the number of choices for these partial solutions seems to be at least $2^{O(\text{tw}^2)}$. Instead, we prove that it has an equivalent characterization in terms of pairs of vertices which allows us to bound the number equivalence classes. Recall that there is a minimum mixed dominating set $S = V' \cup E'$, where $V' \subseteq V(G)$ and $E' \subseteq E(G)$, with the following properties:

(a) E' is a matching, and
(b) $V' \cap V(E') = \emptyset$.

Let $V' \cup E'$ be a solution satisfying conditions (a) and (b). Consider the pair $(V', V(E'))$. Since $V' \cup E'$ is a solution, we have that (i) $(V', V(E'))$ is a vertex cover of G, (ii) $N_G[V'] \cup V(E') = V(G)$, and (iii) $G[V(E')]$ has a perfect matching. In fact, one can show that any pair of vertex subsets that satisfies these three properties can be turned into a mixed dominating set. That is, these two notions are *equivalent*. As a result, for any node t in the tree decomposition we define partial solutions and equivalence classes among partial solutions as follows. A *partial solution* is a tuple (X, F, Y) satisfying the following conditions, where $X \subseteq V(G_t)$, $F \subseteq E(G_t)$, $Y \subseteq X_t$:

- $X \uplus Y \uplus V(F)$ is a vertex cover of G_t,
- $V(G_t) \setminus X_t \subseteq N_{G_t}[X] \cup V(F)$.

The intuitive meaning of (X, F, Y) is that there will potentially be a solution S such that $X \cup F \subseteq S$ and for each $u \in Y$, there will be an edge $uv \in S \setminus E(G_t)$.

We now define equivalence classes of partial solutions corresponding to a node t in the tree decomposition. We define $\mathcal{P}_t[f]$, where $f : X_t \to \{1, 2, 2', 3, 3'\}$ as the set of partial solutions (X, F, Y), which satisfy the following.

1. $X_t \cap X = f^{-1}(1)$,
2. $X_t \cap V(F) = f^{-1}(2)$,
3. $Y = f^{-1}(2')$, and
4. $(N_{G_t}(X) \cap X_t) \setminus (Y \cup V(F)) \supseteq f^{-1}(3)$.

Informally, each partial solution imposes a partition of X_t, which is defined by f. The set $f^{-1}(1)$ is the set of vertices from X_t which are part of solution. The set $f^{-1}(2)$ is the set of vertices from X_t such that there are edges in the solution which are incident on vertices in $f^{-1}(2)$ and are present in the graph G_t. The set $f^{-1}(2')$ is the set of vertices from X_t such that there are edges in the solution which are incident on vertices in $f^{-1}(2')$ and these edges are not present in the graph G_t. Here, the condition 4 implies that the set $f^{-1}(3)$ is the set of vertices in X_t, which are not part of solution vertices or end points of solution edges in the partial solution, but they are already dominated. The set $f^{-1}(3')$ is the set of vertices in X_t which are not yet dominated and not in $f^{-1}(2')$. The number of equivalence classes is bounded by $5^{\text{tw}(G)}$. Thus, a standard dynamic programming, coupled with fast computation of cover product [3], results in the the following theorem.

Theorem 2. *Given a graph G together with a tree decomposition of width* tw, *MDS can be solved in time $\mathcal{O}^{\star}(6^{\text{tw}})$.*

Theorem 3. *MDS on planar graphs can be solved in time $\mathcal{O}^{\star}(2^{\mathcal{O}(\sqrt{k})})$.*

Proof. By Planar Excluded Grid Theorem [12,24], treewidth of a planar graph that has a vertex cover of size $2k$ is smaller than $(9/2)\sqrt{4k+2}$. So graphs of treewidth larger than $(9/2)\sqrt{4k+2}$ are No-instances of MDS. To exploit it algorithmically, we use the algorithm of Bodlaender et al. [4] which takes as input an n-vertex graph and an integer $\ell > 0$, runs in time $2^{\mathcal{O}(\ell)}n$, and either conclude that treewidth of G is more than ℓ or gives a tree decomposition of with $5\ell + 4$. For convenience, let us call this algorithm ALG-A.

We run ALG-A on input G and $\ell = (9/2)\sqrt{4k+2}$. If it outputs that treewidth of G is greater than ℓ, then we conclude that G is a No-instance of MDS. Otherwise ALG-A will output a tree decomposition of G of width $5\ell + 4 = (45/2)\sqrt{4k+2}+4$. Then, we apply Theorem 2. Both algorithm ALG-A and the algorithm mentioned in Theorem 2 run in time $\mathcal{O}^{\star}(2^{\mathcal{O}(\sqrt{k})})$. This completes the proof of the theorem. $\qquad\square$

5 Lower Bounds

In this section first we prove the following theorem.

Theorem 4. *Unless ETH fails, MDS cannot be solved in time $\mathcal{O}^{\star}(2^{o(\ell)})$, where ℓ is either the solution size or the treewidth of the input graph.*

Proof. Lan and Chang [17] gave a polynomial time reduction from MODIFIED VERTEX COVER (MVC), an NP-complete problem, to MDS on split graphs. There is a reduction from an instance (G, k) of VERTEX COVER to an instance (G', k') of MVC and a reduction from an instance of (G', k') of MVC to an equivalent instance (G'', k'') of MDS in [17]. Here, the input size and the parameter of the instances change as follows. Let $|V(G)| = n$. Then, $|V(G')| = n' \leq n + 2$, $k' = k$, $|V(G'')| = |V(G') \cup E(G')|$, $k'' = (n' + k - 1)/2 \leq (n + k + 1)/2$, where G'' is a split graph with treewidth at most n'.

ETH implies that VERTEX COVER on a graph with n vertices and m edges can not be solved in time $2^{o(n+m)}$ [16]. As a result from the above mentioned reductions, we get that, unless ETH fails, MDS has no $2^{o(\ell)}n^{\mathcal{O}(1)}$ algorithm, where ℓ is the solution size or treewidth of the input graph. $\qquad\square$

Now we prove a kernel lower bound for MDS. That is, we show that unless coNP $\subseteq$ NP/poly, MDS does not admit a polynomial kernel when parameterized by k. We do this by a *polynomial parameter transformation* from an appropriate parameterization of RED BLUE DOMINATING SET (RBDS).

Definition 1 ([5]). *Let P and Q be two parameterized problems. A polynomial parameter transformation (PPT, for short) from P to Q is a polynomial time algorithm, which given an instance, say (x, k) of P, produces an equivalent instance (y, k') of Q such that $k' \leq p(k)$ for some polynomial $p(\cdot)$.*

Proposition 1 ([5]). *If there is a PPT from P to Q and P has no polynomial kernel, then Q has no polynomial kernel.*

In the RBDS problem, the input is a bipartite graph G with bipartition $R \uplus B$ and a positive integer ℓ, and the question is whether there exists a set $X \subseteq R$ of size at most ℓ, which dominates the set B, i.e., $N(X) = B$. (Such a set X is called a red-blue dominating set (rbds, for short) of G). This problem when parameterized by $|R|$ is the same as SMALL UNIVERSE HITTING SET (see [8]) and thus from [8] we get the following result.

Lemma 7 ([8]). RBDS *parameterized by* $|R|$ *and* ℓ *has no polynomial kernel unless* coNP $\subseteq$ NP/poly.

Theorem 5. MDS *parameterized by the solution size has no polynomial kernel, unless* coNP $\subseteq$ NP/poly.

Proof. The proof is by a polynomial parameter transformation from RBDS parameterized by $|R|$ and ℓ. Given an instance $(G = (R \uplus B, E), \ell)$ of RBDS, we construct an equivalent instance $(G', |R| + \ell + 1)$ of MDS. If $B \subseteq V(G)$ contains an isolated vertex, then note that G has no rbds (of any size), so take G' to be a $|R| + \ell + 2$-sized matching. Otherwise, if B has no isolated vertices, then proceed as follows (see Fig. 2).

1. Add all vertices and all edges of G to G', i.e., $V(G') \supseteq V(G)$ and $E(G') \supseteq E(G)$.
2. Corresponding to every vertex $v_i \in R$, add vertices x_i and y_i, and add edges $v_i x_i$ and $x_i y_i$ in G'.
3. Add a vertex z and add edges $z y_i$, for all y_i.
4. Add $|R| + \ell + 2$ additional neighbors to z.

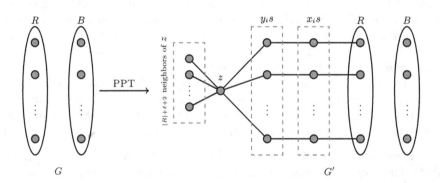

Fig. 2. PPT from RBDS to MDS

We claim that G has a rbds of size at most ℓ if and only if G' has a mds of size at most $|R| + \ell + 1$. Let $X \subseteq R$ be a rbds of G of size at most ℓ. Then $X \cup \{x_i v_i : i = 1, 2, \ldots, |R|\} \cup \{z\}$ is a mds of size at most $|R| + \ell + 1$.

Conversely, assume that G does not have any rbds of size at most ℓ. Let S be a *minimum* sized mds of G'. Let S' be the set of all elements $x \in S$ such that x dominates some element(s) of B. Let $S' = S_1 \uplus S_2 \uplus S_3$, where $S_1 = S' \cap B$, $S_2 = S' \cap E(G')$ and $S_3 = S' \cap R$. Construct $S'' \subseteq R$ as follows: (i) for every $v \in S_1$, add a neighbor of v to S'', (ii) for every edge $ww' \in S_2$, where $w \in R$ and $w' \in B$, add w to S'', and (iii) add all vertices of S_3 to S''. Clearly, $|S''| \leq |S'|$ and S'' is a rbds of G. By assumption, $|S''| > \ell$ which implies that $|S'| > \ell$.

Thus, S' is a subset of the minimum sized mds S and $|S'| > \ell$. Assume that $z \in S$, otherwise $|S| \geq |R| + \ell + 2$. Note that neither the elements of S' nor z can dominate any of the $|R|$ edges $x_i y_i$. And at least $|R|$ elements are required to dominate all of them. Therefore,

$$|S| \geq |\{z\}| + |\{\text{the } |R| \text{ elements that dominate edges } x_i y_i\}| + |S'|$$
$$> 1 + |R| + \ell.$$

That is, G' does not have a mds of size at most $|R| + \ell + 1$. Hence, the theorem follows from the given reduction, Proposition 1 and Lemma 7. □

Now we present an improved lower bound for MDS when parameterized by the treewidth of the input graph. We can reduce an instance of SET COVER problem $(U, \mathcal{F}, \ell)$ to an equivalent instance of RBDS, $(R \uplus B, E, \ell)$, where $R = \mathcal{F}$ and $B = U$. Edge set E consists of edges between $F \in R$ and $x \in B$ if and only if $x \in F$. We now apply the reduction given in the proof of Theorem 5 to an instance of RBDS, $(R \uplus B, E, |R| + \ell)$ to get an equivalent instance of MDS, $(G, |R| + \ell + 1)$. Notice that graph G has treewidth at most $1 + |B| = 1 + |U|$. The Set Cover Conjecture [7] states that SET COVER cannot be solved in $\mathcal{O}^\star((2 - \epsilon)^{|U|})$ time for any $\epsilon > 0$. We thus have the following theorem.

Theorem 6. *Unless the Set Cover Conjecture fails,* MDS *does not admit an algorithm with running time* $\mathcal{O}^\star((2 - \epsilon)^{\mathsf{tw}(G)})$.

6 Conclusion

In this paper we initiated a systematic study of MDS from the viewpoint of parameterized complexity and designed algorithms parameterized by the solution size and the treewidth of the input graph. The algorithm for MDS parameterized by the treewidth significantly improved the known algorithm for the problem. It is curious to note that our algorithm runs in time $\mathcal{O}^\star(5^{\mathsf{pw}})$ on graphs of pathwidth pw, while the same algorithm runs in time $\mathcal{O}^\star(6^{\mathsf{tw}})$ on graphs of treewidth tw. It will be interesting to close this gap as well as prove an optimal lower bound under the Strong Exponential Time Hypothesis (SETH). Another research avenue will be to find families of graph classes on which the problem does admit polynomial kernels. Designing a non-trivial exact exponential time algorithm is another interesting problem.

References

1. Alavi, Y., Behzad, M., Lesniak-Foster, L.M., Nordhaus, E.A.: Total matchings and total coverings of graphs. J. Graph Theor. **1**(2), 135–140 (1977)
2. Alavi, Y., Liu, J., Wang, J., Zhang, Z.: On total covers of graphs. Discrete Math. **100**(1–3), 229–233 (1992)
3. Björklund, A., Husfeldt, T., Kaski, P., Koivisto, M.: Fourier meets möbius: fast subset convolution. In: STOC, pp. 67–74 (2007)
4. Bodlaender, H.L., Drange, P.G., Dregi, M.S., Fomin, F.V., Lokshtanov, D., Pilipczuk, M.: A c^k n 5-approximation algorithm for treewidth. SIAM J. Comput. **45**(2), 317–378 (2016). http://dx.doi.org/10.1137/130947374
5. Bodlaender, H.L., Thomassé, S., Yeo, A.: Kernel bounds for disjoint cycles and disjoint paths. Theor. Comput. Sci. **412**(35), 4570–4578 (2011)
6. Cygan, M., Fomin, F.V., Kowalik, L., Lokshtanov, D., Marx, D., Pilipczuk, M., Pilipczuk, M., Saurabh, S.: Parameterized Algorithms. Springer, Heidelberg (2015). doi:10.1007/978-3-319-21275-3
7. Cygan, M., Dell, H., Lokshtanov, D., Marx, D., Nederlof, J., Okamoto, Y., Paturi, R., Saurabh, S., Wahlström, M.: On problems as hard as CNF-SAT. ACM Trans. Algorithms **12**(3), 41:1–41:24 (2016)
8. Dom, M., Lokshtanov, D., Saurabh, S.: Kernelization lower bounds through colors and IDS. ACM Trans. Algorithms **11**(2), 13:1–13:20 (2014)
9. Erdös, P., Meir, A.: On total matching numbers and total covering numbers of complementary graphs. Discrete Math. **19**(3), 229–233 (1977)
10. Fernau, H.: On parameterized enumeration. In: Ibarra, O.H., Zhang, L. (eds.) COCOON 2002. LNCS, vol. 2387, pp. 564–573. Springer, Heidelberg (2002). doi:10. 1007/3-540-45655-4_60
11. Garey, M.R., Johnson, D.S.: Computers and Intractability: A Guide to the Theory of NP-Completeness. W.H. Freeman, New York (1979)
12. Gu, Q., Tamaki, H.: Improved bounds on the planar branchwidth with respect to the largest grid minor size. Algorithmica **64**(3), 416–453 (2012)
13. Hatami, P.: An approximation algorithm for the total covering problem. Discuss. Math. Graph Theor. **27**(3), 553–558 (2007)
14. Haynes, T.W., Hedetniemi, S., Slater, P.: Fundamentals of Domination in Graphs. CRC Press, Boca Raton (1998)
15. Hedetniemi, S.M., Hedetniemi, S.T., Laskar, R., McRae, A., Majumdar, A.: Domination, independence and irredundance in total graphs: a brief survey. In: Proceedings of the 7th Quadrennial International Conference on the Theory and Applications of Graphs. vol. 2, pp. 671–683 (1995)
16. Impagliazzo, R., Paturi, R., Zane, F.: Which problems have strongly exponential complexity. J. Comput. Syst. Sci. **63**(4), 512–530 (2001)
17. Lan, J.K., Chang, G.J.: On the mixed domination problem in graphs. Theor. Comput. Sci. **476**, 84–93 (2013)
18. Majumdar, A.: Neighborhood hypergraphs: a framework for covering and packing parameters in graphs. Ph.D. thesis, Clemson University (1992)
19. Manlove, D.: On the algorithmic complexity of twelve covering and independence parameters of graphs. Discrete Appl. Math. **91**(1–3), 155–175 (1999)
20. Meir, A.: On total covering and matching of graphs. J. Comb. Theor. Ser. B **24**(2), 164–168 (1978)
21. Micali, S., Vazirani, V.V.: An $\mathcal{O}(\sqrt{|V|}|E|)$ algorithm for finding maximum matching in general graphs. In: FOCS, pp. 17–27 (1980)

22. Peled, U.N., Sun, F.: Total matchings and total coverings of threshold graphs. Discrete Appl. Math. **49**(1–3), 325–330 (1994)
23. Rajaati, M., Hooshmandasl, M.R., Dinneen, M.J., Shakiba, A.: On fixed-parameter tractability of the mixed domination problem for graphs with bounded tree-width. CoRR abs/1612.08234 (2016)
24. Robertson, N., Seymour, P.D., Thomas, R.: Quickly excluding a planar graph. J. Comb. Theor. Ser. B **62**(2), 323–348 (1994)
25. Yannakakis, M., Gavril, F.: Edge dominating sets in graphs. SIAM J. Appl. Math. **38**(3), 364–372 (1980)
26. Zhao, Y., Kang, L., Sohn, M.Y.: The algorithmic complexity of mixed domination in graphs. Theor. Comput. Sci. **412**(22), 2387–2392 (2011)

Simplified Algorithmic Metatheorems Beyond MSO: Treewidth and Neighborhood Diversity

Dušan Knop[2]([⊠]), Martin Koutecký[2], Tomáš Masařík[2], and Tomáš Toufar[1]

[1] Computer Science Institute, Charles University, Prague, Czech Republic
toufi@iuuk.mff.cuni.cz
[2] Department of Applied Mathematics, Charles University, Prague, Czech Republic
{knop,koutecky,masarik}@kam.mff.cuni.cz

Abstract. This paper settles the computational complexity of model checking of several extensions of the monadic second order (MSO) logic on two classes of graphs: graphs of bounded treewidth and graphs of bounded neighborhood diversity.

A classical theorem of Courcelle states that any graph property definable in MSO is decidable in linear time on graphs of bounded treewidth. Algorithmic metatheorems like Courcelle's serve to generalize known positive results on various graph classes. We explore and extend three previously studied MSO extensions: global and local cardinality constraints (CardMSO and MSO-LCC) and optimizing a fair objective function (fairMSO).

We show how these fragments relate to each other in expressive power and highlight their (non)linearity. On the side of neighborhood diversity, we show that combining the linear variants of local and global cardinality constraints is possible while keeping FPT runtime but removing linearity of either makes this impossible, and we provide an XP algorithm for the hard case. Furthemore, we show that even the combination of the two most powerful fragments is solvable in polynomial time on graphs of bounded treewidth.

1 Introduction

It has been known since the '80s that various NP-hard problems are solvable in polynomial time by dynamic programming on trees and "tree-like" graphs. This was famously captured by Courcelle [4] in his theorem stating that any property definable in Monadic Second Order (MSO) logic is decidable in linear time on graphs of bounded treewidth. Subsequently, extensions to stronger logics and optimization versions were devised [2,6] while still keeping linear runtime.

This research was partially supported by project 338216 of GA UK and the grant SVV–2017–260452. M. Koutecký was also supported by the project GA15-11559S of GA ČR. D. Knop and T. Masařík were also supported by the project GA17-09142S of GA ČR.

The full version of this article is available on ArXiv [20].

H.L. Bodlaender and G.J. Woeginger (Eds.): WG 2017, LNCS 10520, pp. 344–357, 2017.
https://doi.org/10.1007/978-3-319-68705-6_26

However, several interesting problems do not admit an MSO description and are unlikely to be solvable in linear time on graphs of bounded treewidth due to hardness results. In the language of parameterized complexity, Courcelle's theorem runs in *fixed-parameter tractable* (FPT) time $f(|\varphi|, \tau)n^{\mathcal{O}(1)}$, where n is the number of vertices of the input graph, τ its treewidth, φ is an MSO formula in prenex form, $|\varphi|$ is the size of the formula, and f is a computable function. On the other hand, the "hard" (specifically, W[1]-hard) problems have algorithms running at best in XP time $n^{g(|\varphi|, \tau)}$, for some computable function $g \in \omega(1)$. This led to examination of extensions of MSO which allow greater expressive power.

Another research direction was to improve the computational complexity of Courcelle's theorem, since the function f grows as an exponential tower in the quantifier depth of the MSO formula. However, Frick and Grohe [12] proved that this is unavoidable unless P = NP which raises a question: is there a (simpler) graph class where MSO model checking can be done in single-exponential (i.e. $2^{k^{\mathcal{O}(1)}}$) time? This was answered in the affirmative by Lampis [24], who introduced graphs of bounded neighborhood diversity. These two classes are incomparable: for example, paths have unbounded neighborhood diversity but bounded treewidth, and vice versa for cliques. Bounded treewidth has become a standard parameter with many practical applications (cf. a survey [3]); bounded neighborhood diversity is of theoretical interest [1,10,13,15,28] because it can be viewed as representing the simplest of dense graphs.

Courcelle's theorem proliferated into many fields. Originating among automata theorists, it has since been reinterpreted in terms of finite model theory [27], database programming [16], game theory [19] and linear programming [22].

1.1 Related Work

For a recent survey of algorithmic metatheorems see Langer et al. [26] and Grohe et al. [17].

Objective functions. A linear optimization version of Courcelle's theorem was given by Arnborg et al. [2]. An extension to further objectives was given by Courcelle and Mosbah [6]. Kolman et al. [23] introduce MSO with a fair objective function (fairMSO) which, for a given MSO formula $\varphi(F)$ with a free edge set variable F, minimizes the maximum degree in the subgraph given by F, and present an XP algorithm. This is justified by the problem being W[1]-hard, as was later shown by Masařík and Toufar [28], who additionally give an FPT algorithm on graphs of bounded neighborhood diversity for MSO_1 and an FPT algorithm on graph of bounded vertex cover for MSO_2.

Extended logics. Along with MSO, Courcelle also considered *counting* MSO (cMSO) where predicates of the form "$|X| \equiv p \mod q$" are allowed, with the largest modulus q constant. Szeider [31] introduced MSO with *local cardinality constraints* (MSO-LCC) and gave an XP algorithm deciding it on graphs

of bounded treewidth. MSO-LCC can express various problems, such as GEN-
ERAL FACTOR, EQUITABLE r-COLORING or MINIMUM MAXIMUM OUTDEGREE,
which are known to be W[1]-hard on graphs of bounded treewidth. Ganian and
Obdržálek [14] study CardMSO, which is incomparable with MSO-LCC in its
expressive power; they give an FPT algorithm on graphs of bounded neighborhood diversity.

1.2 Our Contribution

The contribution of the paper is twofold. First, we survey and enrich the so
far studied extensions of MSO logic – fairMSO, CardMSO, and MSO-LCC. We do
this in Sect. 2.1. Second, we study the parameterized complexity of the associated
model checking problem for various combinations of these MSO extensions. We
completely settle the parameterized complexity landscape for the model checking
problems with respect to the parameters treewidth and neighborhood diversity;
for an overview of the complexity landscape refer to Fig. 1. We postpone formal
definitions of logic extensions and corresponding model checking to Subsect. 2.1.

While both MSO-LCC and CardMSO express certain *cardinality* constraints,
the constraints of CardMSO are inherently *global* and *linear*, yet the constraints
of MSO-LCC are *local* and *non-linear*. This leads us to introduce two more fragments and rename the aforementioned ones: CardMSO becomes MSO^G_{lin}, MSO-
LCC becomes MSO^L and we additionally have MSO^G and MSO^L_{lin}. By this we
give a complete landscape for all possible combinations of global/local and
linear/non-linear.

nd, vc	$\emptyset$	fairMSO	MSO^L_{lin}	MSO^L
MSO	FPT [24]	FPT [28]		W[1]-h, Thm 5
MSO^G_{lin}	FPT [14]		FPT, Thm 3	
MSO^G	W[1]-h, Thm 2			XP, Thm 4

tw	$\emptyset$	fairMSO	MSO^L_{lin}	MSO^L
MSO	FPT [4]	W[1]-hard [28]		XP [31]
MSO^G_{lin}	W[1]-hard [14]			
MSO^G				XP, Thm 1

Fig. 1. Complexity of various logic fragments generalizing MSO on graphs of bounded
vertex cover (vc), neighborhood diversity (nd) and treewidth (tw). Positive results
(FPT, XP) spread to the left and up. W[1]-hardness spreads to the right and down.
Green background (lighter gray in bw print) stands for FPT fragments, while orange
(darker gray) stands for W[1]-hard.

In the following, we do not differentiate between the logics MSO_1 (allowing quantification over vertex sets) and MSO_2 (additionally allowing quantification over edge sets); see a detailed explanation in Subsect. 2.1. For now, it suffices to say that our positive result for graphs of bounded treewidth holds for the appropriate extension of MSO_2, while all remaining results hold for the appropriate extensions of MSO_1.

For graphs of bounded treewidth we give an XP algorithm for the logic MSO^{GL}, which is a composition MSO^G and MSO^L and thus represents the most expressive fragment under our consideration.

Theorem 1. *There is an algorithm that solves the* MSO^{GL} Model Checking *problem in time* $n^{f(|\varphi|,\tau)}$, *where* $\tau = \mathrm{tw}(G)$ *and* f *is a computable function.*

This result is also significant in its proof technique. We connect a recent result of Kolman et al. [22] about the polytope of satisfying assignments with an old result of Freuder [11] about the solvability of constraint satisfaction problem (CSP) of bounded treewidth. This allows us to formulate the proof of Theorem 1 essentially as providing a CSP instance with certain properties, surpassing the typical complexity of a dynamic programming formulation.

This is complemented from the negative side by the following hardness result.

Theorem 2. *The* MSO^G Model Checking *problem is* W[1]-hard *when parameterized by* $vc(G)$.

For graphs of bounded neighborhood diversity we give two positive results. The first is for the logic MSO^{GL}_{lin}, a composition of MSO^L_{lin} and MSO^G_{lin}. We complement them with hardness result for MSO^L.

Theorem 3. *There is an algorithm that solves the* MSO^{GL}_{Lin} Model Checking *problem in time* $f(|\varphi|,\nu) \cdot n^{\mathcal{O}(1)}$, *where* $\nu = nd(G)$ *and* f *is a computable function.*

Theorem 4. *There is an algorithm that solves the* MSO^{GL} Model Checking *problem in time* $n^{f(|\varphi|,\nu)}$, *where* $\nu = nd(G)$ *and* f *is a computable function.*

Theorem 5. *The* MSO^L Model Checking *problem is* W[1]-hard *when parameterized by* $vc(G)$.

2 Preliminaries

Let n be a non-negative integer; by $[n]$ we denote the set $\{1, \ldots, n\}$. For two integers a, b we define a set $[a, b] = \{x \in \mathbb{Z} \mid a \le x \le b\}$. For a vertex $v \in V$ of a graph $G = (V, E)$, we denote by $N_G(v)$ the set of neighbors of v in G, that is, $N_G(v) = \{u \in V \mid \{u, v\} \in E\}$; the subscript G is omitted when clear from the context. For a rooted tree T, $N_T(v)$ denotes the *down-neighborhood* of v, i.e., the set of descendants of v. For a graph $G = (V, E)$ a set $U \subseteq V$ is a *vertex cover* of G if for every edge $e \in E$ it holds that $e \cap U \ne \emptyset$. For more notation in graph theory consult the book [29].

2.1 MSO and Its Extensions

The monadic second order logic MSO extends first order logic using so called monadic variables, which are variables for sets of vertices in MSO_1 and in addition variables for sets of edges in MSO_2.

Regarding MSO_1 and MSO_2. Despite the fact that MSO_2 is strictly stronger than MSO_1 (hamiltonicity is expressible in MSO_2 but not in MSO_1 [27]), it is known [21] that on graphs with bounded treewidth their power is equal. We discuss this in the full version of the paper; it would be great if the full version can be referenced here - the full version is reference number 20. We will show that only a small change in the argument still works even for our extensions of MSO.

On bounded neighborhood diversity, MSO_2 is strictly more powerful than MSO_1; however model checking of an MSO_2 formula is not even in XP unless $E = NE$ [5,25]. Thus, here too we restrict our attention to MSO_1 and use MSO as a shortcut for MSO_1 from now on.

We consider two orthogonal ways to extend MSO logic. In what follows φ is a formula with ℓ free set-variables.

Global cardinality constraints (*global constraints* for short). An MSO formula with c global cardinality constraints contains ℓ-ary predicates $R_1, \ldots, R_c$ where each predicate takes as argument only the free variables of φ. The input to the model checking problem is a graph $G = (V, E)$ on n vertices and a tuple $(R_1^G, \ldots, R_c^G)$, where $R_i^G \subseteq [n]^\ell$.

To define the semantics of the extension, it is enough to define the truth of newly introduced atomic formulae. A formula $R_i(X_1, \ldots, X_\ell)$ is true under an assignment $\mu \colon \{X_1, \ldots, X_\ell\} \to 2^V$ if and only if $(|\mu(X_1)|, \ldots, |\mu(X_\ell)|) \in R_i^G$. We allow the relations to be represented either explicitly as a list of tuples, or implicitly as a linear constraint $a_1|X_1| + \cdots + a_m|X_m| \leq b$, where $(a_1, \ldots, a_m, b) \in \mathbb{R}^{m+1}$.

For example, suppose we want to satisfy a formula $\varphi(X_1, X_2)$ with two sets for which $|X_1| \geq |X_2|^2$ holds. Then, we solve the MSO^G Model Checking problem with a formula $\varphi' := \varphi \wedge [|X_1| \geq |X_2|^2]$, that is, we write the relation as a part of the formula, as this is a more convenient way to think of the problem. However, formally the relation is a part of the input.

Local cardinality constraints. Local cardinality constraints control the size of sets X_i in neighborhood of every vertex. Specifically, we want to control the size of $\mu(X_i) \cap N(v)$ for every v; we define a shorthand $S(v) = S \cap N(v)$ for a subset $S \subseteq V$ and vertex v. *Local cardinality constraints* for a graph $G = (V, E)$ on n vertices and a formula φ with ℓ free variables are mappings $\alpha_1, \ldots, \alpha_\ell$, where each α_i is a mapping from V to $2^{[n]}$.

We say that an assignment μ *obeys local cardinality constraints* $\alpha_i, \ldots, \alpha_\ell$ if for every $i \in [\ell]$ and every $v \in V$ it holds that $|\mu(X_i)(v)| \in \alpha_i(v)$.

The logic that incorporates both of these extensions is denoted as MSO^{GL}. Let φ be an MSO^{GL} formula with c global cardinality constraints. Then the MSO^{GL} Model Checking problem has input:

- graph $G = (V, E)$ on n vertices,
- relations $R_1^G, \ldots, R_c^G \subseteq [n]^\ell$, and,
- mappings $\alpha_1, \ldots, \alpha_\ell$.

The task is to find an assignment μ that obeys local cardinality constraints and such that φ is true under μ by the semantics defined above.

The $\mathsf{MSO}^{\mathsf{GL}}$ logic is very powerful and, as we later show, it does not admit an FPT model checking algorithm neither for the parameterization by neighborhood diversity, nor for the parameterization by treewidth. It is therefore relevant to consider the following weakenings of the $\mathsf{MSO}^{\mathsf{GL}}$ logic:

$\mathsf{MSO}^{\mathsf{G}}$ Only global cardinality constraints are allowed.
$\mathsf{MSO}^{\mathsf{L}}$ (originally MSO-LCC [31]) Only local cardinality constraints are allowed.
$\mathsf{MSO}^{\mathsf{G}}_{\mathsf{lin}}$ (originally CardMSO [13]) The cardinality constraints can only be linear; that is, we allow constraints in the form $[e_1 \geq e_2]$, where e_i is linear expression over $|X_1|, \ldots |X_\ell|$.
$\mathsf{MSO}^{\mathsf{L}}_{\mathsf{lin}}$ Only local cardinality constraints are allowed; furthermore every local cardinality constraint α_i must be of the form $\alpha_i(v) = [l_i^v, u_i^v]$, (i.e., an interval) where $l_i^v, u_i^v \in [n]$. Those constraints are referred to as *linear local cardinality constraints*.
fairMSO Further restriction of $\mathsf{MSO}^{\mathsf{L}}_{\mathsf{lin}}$; now we only allow $\alpha_i(v) = [u_i^v]$.
$\mathsf{MSO}^{\mathsf{GL}}_{\mathsf{lin}}$ A combination of $\mathsf{MSO}^{\mathsf{L}}_{\mathsf{lin}}$ and $\mathsf{MSO}^{\mathsf{G}}_{\mathsf{lin}}$; both local and global constraints are allowed, but only in their linear variants.

The model checking problem for the considered fragments is defined in a natural way analogously to $\mathsf{MSO}^{\mathsf{GL}}$ model checking.

Pre-evaluations. Many techniques used for designing MSO model checking algorithms fail when applied to MSO extensions. A common workaround is first transforming the given $\mathsf{MSO}^{\mathsf{GL}}$ formula into an MSO formula by fixing values of all global constraints to either **true** or **false**. Once we determine which variable assignments satisfy the transformed MSO formula, we can by other means (e.g. integer linear programming or constraint satisfaction) ensure that they obey the constraints imposed by fixing the values to **true** or **false**. This approach was first used for CardMSO by Ganian and Obdržálek [14]. We formally describe this technique as *pre-evaluations*:

Definition 6. *Let φ be an $\mathsf{MSO}^{\mathsf{GL}}$ formula. Denote by $C(\varphi)$ the list of all global constraints. A mapping $\beta \colon C(\varphi) \to \{\mathtt{true}, \mathtt{false}\}$ is called a* pre-evaluation *function on φ. The MSO formula obtained by replacing each global constraint $c_i \in C(\varphi)$ by $\beta(c_i)$ is denoted by $\beta(\varphi)$ and is referred to as a* pre-evaluation *of φ.*

Definition 7. *A variable assignment μ of an $\mathsf{MSO}^{\mathsf{GL}}$ formula φ complies with a pre-evaluation function β if every global constraint $c_i \in C(\varphi)$ evaluates to $\beta(c_i)$ under the assignment μ.*

2.2   Treewidth and Neighborhood Diversity

Treewidth. For notions related to the treewidth of a graph and nice tree decomposition, in most cases we stick to the standard terminology as given by Kloks [18]; the only deviation is in the leaf nodes of the nice tree decomposition where we assume that the bags are empty.

A *tree decomposition* of a graph $G = (V, E)$ is a pair $(T, \mathcal{B})$, where T is a tree and $\mathcal{B}$ is a mapping $\mathcal{B} : V(T) \rightarrow 2^V$. We also use the notion of a *nice* tree decomposition, where T is rooted and the nodes are of four types: empty leaves, introduce and forget nodes which have only one child and their bags differ by exactly one vertex, and join nodes, which have two children with identical bags. An introduce node a with a son b which differs by vertex v is denoted $a = b * (v)$, analogously for a forget node $a = b \dagger (v)$. A join node a with two sons b, b' is denoted $a = \Lambda(b, b')$. Given a graph $G = (V, E)$ and a subset of vertices $V' = \{v_1, \ldots, v_d\} \subseteq V$, we denote by $G[V']$ the subgraph of G induced by V'. Given a tree decomposition $(T, \mathcal{B})$ and a node $a \in V(T)$, we denote by T_a the subtree of T rooted in a, and by G_a the subgraph of G induced by all vertices in bags of T_a, that is, $G_a = G[\bigcup_{b \in V(T_a)} B(b)]$.

Neighborhood diversity [24]. We say that two (distinct) vertices u, v are of the same *neighborhood type* if they share their respective neighborhoods, that is when $N(u) \setminus \{v\} = N(v) \setminus \{u\}$. Let $G = (V, E)$ be a graph. We call a partition of vertices $\mathcal{T} = \{T_1, \ldots, T_\nu\}$ a *neighborhood decomposition* if, for every $i \in [\nu]$, all vertices of T_i are of one neighborhood type. *Neighborhood diversity* (nd(G)) is the size of the unique minimal neighborhood decomposition. Moreover, this decomposition can be computed in linear time.

3   Graphs of Bounded Neighborhood Diversity

For graphs of bounded neighborhood diversity we prove two negative results (Theorems 2 and 5) and two positive results (Theorems 3 and 4).

W[1]-hardness of MSO^L and MSO^G. We begin with a definition of an auxiliary problem:

LCC SUBSET
Input: Graph $G = (V, E)$ with $|V| = n$ and a function $f : V \rightarrow 2^{[n]}$.
Task: Find a set $U \subseteq V$ such that for each vertex $v \in V$ it holds that $|U(v)| \in f(v)$.

Obviously LCC SUBSET is equivalent to MSO^L with an empty formula φ. We call an LCC SUBSET instance *uniform* if, on G with nd(G) = k, the demand function f can be written as $f : [k] \rightarrow 2^{[n]}$, such that vertices of the same type have the same demand set. We show that already uniform LCC SUBSET is W[1]-hard by a reduction from the W[1]-hard k-MULTICOLORED CLIQUE problem [7].

> k-MULTICOLORED CLIQUE   *Parameter:* k
> **Input:** k-partite graph $G = (V_1 \dot\cup \cdots \dot\cup V_k, E)$, where V_a is an independent set for every $a \in [k]$.
> **Task:** Find a clique of size k.

We refer to a set V_a as to a *colorclass* of G. Our proof is actually a simplified proof of W[1]-hardness for the TARGET SET SELECTION problem [8]. Note that Theorem 5 follows easily from Theorem 8.

Theorem 8. *The* LCC SUBSET *problem is* W[1]-*hard when parameterized by the vertex cover number.*

Proof. Denote $G = (V_1 \dot\cup \cdots \dot\cup V_k, E)$ the instance graph for k-MULTICOLORED CLIQUE. We naturally split the set of edges E into sets $E_{\{a,b\}}$ by which we denote the edges between colorclasses V_a and V_b. We may assume that all colorclasses are of the same size which we denote n, and similarly for the number of edges between any two colorclasses which we denote m. Fix $N > n$, say $N = n^2$ and distinct $a, b \in [k]$.

Description of the reduction. We numerate vertices in each color class V_a for $a \in [k]$ using numbers in $[n]$ and denote the numeration of vertex v as n_v^a. We also numerate the edges between color classes a and b by numbers in $[m]$ and denote the numeration of edge e as $m_e^{\{a,b\}}$. Let $I_{ab} = \{n_v^a + Nm_e^{\{a,b\}} \mid v \in e, e \in E_{\{a,b\}}\}$. We build the graph using the following groups of vertices (refer to Fig. 2):

- an independent set S_a of size n for each color class V_a and set $f(v) = \{0\}$ for every $v \in S_a$,
- an independent set $T_{\{a,b\}}$ of size mN for each edge set $E_{\{a,b\}}$, with $f(v) = \{0\}$ for every $v \in T_{\{a,b\}}$,
- a single vertex $\text{Mult}_{\{a,b\}}$ with $f(\text{Mult}_{\{a,b\}}) = \{tN \mid t \in [m]\}$,
- a single vertex Inc_{ab} with $f(\text{Inc}_{ab}) = I_{ab}$.

Finally, we add complete bipartite graphs between S_a and Inc_{ab}, between Inc_{ab} and $T_{\{a,b\}}$, and between $T_{\{a,b\}}$ and $\text{Mult}_{\{a,b\}}$. Denote the resulting graph H. It is straightforward to check that the vertices $\text{Mult}_{\{a,b\}}$ together with vertices Inc_{ab} form a vertex cover of H. It follows that $\text{vc}(H) = \binom{k}{2} + k(k-1)$.

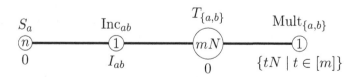

Fig. 2. An overview of the decomposition of a gadget used in the proof of Theorem 8. Numbers inside nodes denote the number of vertices in the independent set represented by the node. Below each node a description of the respective set of admissible numbers is shown.

Correctness of the reduction. Suppose there is a clique of size k in G with set of vertices $\{v_1, \ldots, v_k\}$. We select n_{v_a} vertices in the set S_a and $m_e^{\{v_a, v_b\}}$ vertices in the set $T_{\{a,b\}}$. It is straightforward to check that this is a solution respecting demands in H.

Suppose there is a solution U respecting demands in H. First note that none of vertices $\mathrm{Mult}_{\{a,b\}}$, Inc_{ab} is selected as their neighborhood demands are set to 0. Denote $s_a = |U \cap S_a|$ and $t_{\{a,b\}} = |T_{\{a,b\}} \cap U|$. Now observe that because the demand of vertex $\mathrm{Mult}_{\{a,b\}}$ is fulfilled, then there are $t_{\{a,b\}} = tN$ vertices with $0 \le t < m$. We denote by e_{ab} the edge with numeration $m_{e_{\{a,b\}}} = t$. As the demand of vertex Inc_{ab} is fulfilled the vertex v_a with $n_{v_a} = s_a$ as well as for the vertex Inc_{ba} and vertex v_b. This implies that both v_a and v_b are incident to edge $e_{\{a,b\}}$.   □

Proof idea for Theorem 2. Given an instance of uniform LCC SUBSET on a graph G with $\mathrm{vc}(G) = k$, we construct another graph G' with $\mathrm{vc}(G') = \mathcal{O}(k^2)$ and an $\mathsf{MSO}^{\mathsf{G}}$ formula φ with $\mathcal{O}(k)$ free variables such that $G' \models \varphi$ if and only if G is a "yes" instance of LCC SUBSET. The key insight is that, for each v in the vertex cover, it is possible to express the set $X(v)$ and then use global cardinality constraints to enforce that $|X(v)|$ satisfies the original local cardinality constraints.

Proof idea for Theorem 3. Essentially, we are modifying the algorithm of Ganian and Obdržálek [14] for $\mathsf{MSO}^{\mathsf{G}}_{\mathsf{lin}}$ model checking so that it can deal with the additional constraints introduced by $\mathsf{MSO}^{\mathsf{L}}_{\mathsf{lin}}$. Our crucial lemma states that, given an $\mathsf{MSO}^{\mathsf{GL}}_{\mathsf{lin}}$ instance with arbitrary local linear cardinality constraints α, there is an equivalent instance with constraints α' which are *uniform* with respect to some neighborhood decomposition $\mathcal{T}$ which is of size at most $k4^\ell$. By saying that α' is uniform we mean that $\alpha'(v) = \alpha'(u)$ for all $u, v \in T_j$, for all $T_j \in \mathcal{T}$.

Now it is possible to construct an integer linear program (ILP) in fixed dimension and apply Lenstra's algorithm.

Proof idea for Theorem 4. Let $\varphi \in \mathsf{MSO}^{\mathsf{GL}}$ with just one free variable X; the approach extends to more free variables easily. Let G be a graph with $\nu = \mathrm{nd}(G)$. Fix integers $x_1, \ldots, x_\nu$ such that $x_1 + \cdots + x_\nu \le n$; there are $n^{\mathcal{O}(\nu)}$ choices of such x_i. The goal is to find a set X with $|X \cap T_j| = x_j$. We observe that:

- if X and X' are two sets with $|X \cap T_j| = |X' \cap T_j| = x_j$, their difference on vertices of T_j is irrelevant with respect to the local cardinality constraints of vertices in $V \setminus T_j$, so satisfying the constraints is independent for each type,
- it can be easily checked whether there are x_j vertices of T_j such that the constraints are satisfied.

Then we test if a set X with $|X \cap T_j| = x_j$ satisfies $G \models \varphi(X)$ by a similar argument as before: guess a pre-evaluation β of φ, verify that it satisfies the global cardinality constraints, and then use the FPT model checking algorithm to verify that $G \models \beta(\varphi)$.

4   XP Algorithm for MSO^{GL} on Bounded Treewidth

We believe that the merit of Theorem 1 lies not only in being a very general tractability result, but also in showcasing a simplified way to prove a metatheorem extending MSO. Our main tool is the constraint satisfaction problem (CSP). The key technical result of this section is Theorem 10, which relates MSO and CSP on graphs of bounded treewidth.

In the MSO^{GL} Model Checking problem, we wish to find a satisfying assignment of a formula φ which satisfies further constraints. Simply put, Theorem 10 says that it is possible to restrict the set of satisfying assignments of a formula $\varphi \in \mathsf{MSO}_1$ with CSP constraints under the condition that these additional constraints are structured along the tree decomposition of G. This allows the proof of Theorem 1 to simply be a CSP formulation satisfying this property.

We consider a natural optimization version of MSO^{GL}: the goal is to find a satisfying assignment $X_1, \ldots, X_\ell$ which minimizes $\sum_{j=1}^{\ell} \sum_{v \in X_j} w_v^j$.

4.1   CSP, MSO and Treewidth

An instance $I = (V, \mathcal{D}, \mathcal{H}, \mathcal{C})$ of CSP consists of n *variables* $V = [n]$, their *domains* $\mathcal{D} = D_1, \ldots, D_n$, $D_i \subseteq \mathbb{Z}$ for $i \in [n]$, and relations called *hard constraints* $\mathcal{H} \subseteq \{C_U \mid U \subseteq V\}$. An n-tuple $(z_1, \ldots, z_n) \in \mathbb{Z}^n$ with $z_i \in D_i$ for $i \in [n]$ is a *feasible assignment* if it satisfies all hard constraints. Furthermore, we also consider functions called *weighted soft constraints* $\mathcal{S} \subseteq \{w_U : U \to \mathbb{Z} \mid U \subseteq V\}$ which allow us to solve an optimization variant of CSP.

For a CSP instance $I = (V, \mathcal{D}, \mathcal{H}, \mathcal{S})$, we define its *constraint graph* $G(I)$ as $G = (V, E)$ where two variables $u, v \in V$ share an edge if they appear together in one constraint. The *treewidth of a CSP instance* I (tw(I)) is defined as tw(G(I)). We use D to denote the maximal size of all domains, that is, $D = \max_{u \in V} |D_u|$.

Freuder [11] proved that CSPs of treewidth τ and maximal domain size D can be solved in time $\mathcal{O}(D^\tau + L)$, where L is the length of the CSP instance.

Modeling after the terminology regarding extended formulations of polytopes, we introduce the notion of a *CSP extension*.

Definition 9 (CSP extension). *Let* $I = ([n], \mathcal{D}_I, \mathcal{H}_I, \mathcal{S}_I)$ *be a CSP instance. We say that* $J = ([m], \mathcal{D}_J, \mathcal{H}_J, \mathcal{S}_J)$ *is an* extension *of* I *(or that* J extends I*) if* $\mathrm{Feas}(I) = \{\mathbf{z}^\star|_{[n]} \mid \mathbf{z}^\star \in \mathrm{Feas}(J)\}$.

Using Freuder's algorithm, we can solve CSP instances of bounded treewidth efficiently. Our motivation for introducing CSP extensions is that we are able to formulate a CSP instance I expressing what we need, but having large treewidth and size. However, if an extension J of I existed with bounded treewidth and size, solving J instead suffices.

Let φ be an MSO$_1$ formula with ℓ free variables and let G be a graph on n vertices. We say that a binary vector $\mathbf{y} \in \{0,1\}^{n\ell}$ *satisfies* φ ($G, \mathbf{y} \models \varphi$) if it is a characteristic vector of a satisfying assignment, that is, if $v \in X_i \Leftrightarrow y_v^i = 1$ and $G \models \varphi(X_1, \ldots, X_\ell)$.

Theorem 10. *Let G with $\mathrm{tw}(G) = \tau$ be a σ_2-structure representing a graph, $\varphi \in MSO_1$ with ℓ free variables, $(T, \mathcal{B})$ a nice tree decomposition of G of width τ, and $k \in \mathbb{N}$. Furthermore, let $I = (V, \mathcal{D}, \mathcal{H}, \mathcal{S})$ be a CSP instance with*

$$V = \{y_v^i \mid v \in V(G), i \in [\ell]\} \cup \{x_a^j \mid a \in V(T), j \in [k]\},$$

and $\mathcal{H} = \{\mathbf{y} \mid G, \mathbf{y} \models \varphi\} \cup \mathcal{H}'$, and $\mathcal{H}'$ and $\mathcal{S}$ have the local scope property: $\forall C_U \in \mathcal{H}' \cup \mathcal{S} \; \exists a \in V(T):$

$$U \subseteq \{y_v^i \mid v \in B(a), i \in [\ell]\} \cup \{x_b^j \mid b \in N_T(a), j \in [k]\},$$

i.e., the scope of all constraints is restricted to variables corresponding to the descendants of some node $a \in V(T)$.

Then there exists a CSP instance $J = (V_J, \mathcal{D}_J, \mathcal{H}_J, \mathcal{S}_J)$ which extends I, and,

- *$\mathrm{tw}(J) \leq \mathrm{f}(|\varphi|, \tau) + 2k$,*
- *$\|\mathcal{H}_J\| + \|\mathcal{S}_J\| \leq \mathrm{f}(|\varphi|, \tau) \cdot |V| + (\|\mathcal{H}'\| + \|\mathcal{S}\|)$, and,*
- *$\mathcal{D}_J = \mathcal{D}_I$.*

The proof of Theorem 10 proceeds in three stages:

1. Using a recent result of Kolman et al. [22] we construct a linear program (LP) of bounded treewidth whose integer solutions correspond to feasible assignments of φ.
2. We view this LP as an integer linear program (ILP) and construct an equivalent constraint satisfaction problem (CSP) instance J' of bounded treewidth.
3. We show that if $\mathcal{H}'$ and $\mathcal{S}$ have the local scope property, it is possible to add new constraints derived from $\mathcal{H}'$ and $\mathcal{S}$ to instance J' which results in instance J, such that J is an extension of I.

4.2   CSP Instance Construction

Proof (Theorem 1). As before, we first note that there are at most $2^{|\varphi|}$ different pre-evaluations $\beta(\varphi)$ of φ, so we can try each and choose the best result. Let a pre-evaluation $\beta(\varphi)$ be fixed from now on.

Let $(T, \mathcal{B})$ be a nice tree decomposition of G. We will now construct a CSP instance I satisfying the conditions of Theorem 10, which will give us its extension J with properties suitable for applying Freuder's algorithm.

Let y_v^i be the variables as described above; we use the constraint $G, \mathbf{y} \models \beta(\varphi)$ to enforce that each feasible solution complies with the pre-evaluation $\beta(\varphi)$. Now we will introduce additional CSP variables and constraints in two ways to assure that the local and global cardinality constraints are satisfied. Observe that we introduce the additional CSP variables and constraints in such a way that they have the local scope property of Theorem 10, that is, their scopes will always be limited to the neighborhood of some node $a \in V(T)$.

Global cardinality constraints. In addition to the original $\mathbf{y}$ variables, we introduce, for each node $a \in T$ and each $j \in [\ell]$, a variable s_a^j with domain $[n]$.

The meaning of this variable is $s_a^j = |X_j \cap V(G_a)|$. Thus, in the root node r, s_r^j is exactly $|X_j|$. To enforce the desired meaning of the variables $\mathbf{s}$, we add the following hard constraints:

$$s_a^j = 0 \qquad \text{For all leaves } a \quad \bigwedge \quad s_a^j = s_b^j + y_v^j \qquad\qquad \text{For all } a = b * (v)$$

$$s_a^j = s_b^j \quad \text{For all } a = b \dagger (v) \quad \bigwedge \quad s_a^j = s_b^j + s_{b'}^j - \sum_{v \in B(a)} y_v^j \quad \text{For all } a = \Lambda(b, b')$$

To enforce the cardinality constraints themselves, we add:

$$(s_r^1, \ldots, s_r^\ell) \in R \qquad\qquad \forall R : \beta(R) = \texttt{true} \quad \bigwedge$$

$$(s_r^1, \ldots, s_r^\ell) \in ([n]^\ell \setminus R) \qquad\qquad \forall R : \beta(R) = \texttt{false}$$

Local cardinality constraints. These are dealt with analogously: for every node $a \in I$, every $j \in [\ell]$ and every vertex $v \in B(a)$, we introduce a variable λ_a^{vj} with the meaning $\lambda_a^{vj} = |N_{G_a}(v) \cap X_j|$. Then we set hard constraints to enforce the desired meaning.

Objective function. In order to express the objective function we add soft constraints $\mathcal{S} = \{C_{\{y_v^j\}} \mid v \in V, j \in [\ell]\}$ where $C_{\{y_v^j\}} = w_v^j$ if $y_v^j = 1$ and is 0 otherwise.

In order to apply Theorem 10 to obtain an extension J of our instance I, we determine its parameters. We have introduced ℓ variables s per node, and $\ell\tau$ variables λ per node. Thus, $k = (\tau + 1)\ell$, and $\mathrm{tw}(J) \leq \mathrm{f}(|\varphi|, \tau)$. Clearly, $D_I = n$ and thus $D_J = n$. Let $N = \sum_{j=1}^c |R_j^G| + \sum_{j=1}^\ell \sum_{v \in V(G)} |\alpha_j(v)|$ be the input length of the global and local cardinality constraints. Since $\|\mathcal{H}'\| + \|\mathcal{S}\| \leq N$, we have that $\|\mathcal{H}_J\| + \|\mathcal{S}_J\| \leq f(|\varphi|, \tau) \cdot n + N$. Then, applying Freuder's algorithm to J solves it in time $n^{f(|\varphi|, \tau)} + N$, finishing the proof of Theorem 1. □

5   Conclusions

Limits of MSO extensions, other logics and metatheorems. We have defined extensions of MSO and extended positive and negative results for them. There is still some unexplored space in MSO extensions: Szeider [31] shows that MSO^L where some of the sets of local cardinality constraints are quantified is NP-hard already on graphs of treewidth 2. We are not aware of a comparable result for MSO^G, and no results of this kind are known for graphs of bounded neighborhood diversity. Also, we have not explored other logics, as for example the modal logic considered by Pilipczuk [30]. However, many problems [10,13] are FPT on bounded neighborhood diversity which are not expressible in any of the studied logics. So we ask for a metatheorem generalizing as many such positive results as possible.

Complementary Parameters and Problems. Unlike for treewidth, taking the complement of a graph preserves its neighborhood diversity. Thus our results apply also in the complementary setting, where, given a graph G and a parameter $p(G)$, we are interested in the complexity (with respect to $p(G)$) of deciding a problem P on the *complement* of G. While the complexity stays the same when parameterizing by neighborhood diversity, it is unclear for sparse graph parameters such as treewidth. It was shown very recently [9] that the HAMILTONIAN PATH problem admits an FPT algorithm with respect to the treewidth of the complement of the graph. This suggest that at least sometimes this is the case and some extension of Courcelle's theorem deciding properties of the complement may hold.

References

1. Alves, S.R., Dabrowski, K.K., Faria, L., Klein, S., Sau, I., dos Santos Souza, U.: On the (parameterized) complexity of recognizing well-covered (r, 1) graphs. In: Chan, T.H.H., Li, M., Wang, L. (eds.) COCOA. LNCS, pp. 423–437. Springer, Heidelberg (2016). doi:10.1007/978-3-319-48749-6
2. Arnborg, S., Lagergren, J., Seese, D.: Easy problems for tree-decomposable graphs. J. Algorithms **12**(2), 308–340 (1991)
3. Bodlaender, H.L.: Treewidth: characterizations, applications, and computations. In: Fomin, F.V. (ed.) WG 2006. LNCS, vol. 4271, pp. 1–14. Springer, Heidelberg (2006). doi:10.1007/11917496_1
4. Courcelle, B.: The monadic second-order logic of graphs I: Recognizable sets of finite graphs. Inf. Comput. **85**, 12–75 (1990)
5. Courcelle, B., Makowsky, J.A., Rotics, U.: Linear time solvable optimization problems on graphs of bounded clique-width. Theor. Comput. Syst. **33**(2), 125–150 (2000). http://dx.doi.org/10.1007/s002249910009
6. Courcelle, B., Mosbah, M.: Monadic second-order evaluations on tree-decomposable graphs. In: Schmidt, G., Berghammer, R. (eds.) WG 1991. LNCS, vol. 570, pp. 13–24. Springer, Heidelberg (1992). doi:10.1007/3-540-55121-2_2
7. Cygan, M., Fomin, F.V., Kowalik, Ł., Lokshtanov, D., Marx, D., Pilipczuk, M., Pilipczuk, M., Saurabh, S.: Parameterized Algorithms. Springer, Heidelberg (2015). doi:10.1007/978-3-319-21275-3
8. Dvořák, P., Knop, D., Toufar, T.: Target Set Selection in Dense Graph Classes. CoRR 1610.07530 (October 2016)
9. Dvořák, P., Knop, D., Masařík, T.: Anti-path cover on sparse graph classes. In: Bouda, J., Holík, L., Kofroň, J., Strejček, J., Rambousek, A. (eds.) Proceedings 11th Doctoral Workshop on Mathematical and Engineering Methods in Computer Science, Telč, Czech Republic, 21st–23rd October 2016. Electronic Proceedings in Theoretical Computer Science, vol. 233, pp. 82–86. Open Publishing Association (2016)
10. Fiala, J., Gavenčiak, T., Knop, D., Koutecký, M., Kratochvíl, J.: Fixed parameter complexity of distance constrained labeling and uniform channel assignment problems. In: Dinh, T.N., Thai, M.T. (eds.) COCOON 2016. LNCS, vol. 9797, pp. 67–78. Springer, Cham (2016). doi:10.1007/978-3-319-42634-1_6
11. Freuder, E.C.: Complexity of K-tree structured constraint satisfaction problems. In: Proceedings of the Eighth National Conference on Artificial Intelligence, vol. 1, pp. 49. AAAI 1990, AAAI Press (1990). http://dl.acm.org/citation.cfm?id=1865499.1865500

12. Frick, M., Grohe, M.: The complexity of first-order and monadic second-order logic revisited. Ann. Pure Appl. Logic **130**(1–3), 3–31 (2004). http://dx.doi.org/10.1016/j.apal.2004.01.007
13. Ganian, R.: Using neighborhood diversity to solve hard problems. CoRR abs/1201.3091 (2012). http://arxiv.org/abs/1201.3091
14. Ganian, R., Obdržálek, J.: Expanding the expressive power of monadic second-order logic on restricted graph classes. In: Lecroq, T., Mouchard, L. (eds.) IWOCA 2013. LNCS, vol. 8288, pp. 164–177. Springer, Heidelberg (2013). doi:10.1007/978-3-642-45278-9_15
15. Gargano, L., Rescigno, A.A.: Complexity of conflict-free colorings of graphs. Theor. Comput. Sci. **566**, 39–49 (2015). http://www.sciencedirect.com/science/article/pii/S0304397514009463
16. Gottlob, G., Pichler, R., Wei, F.: Monadic datalog over finite structures with bounded treewidth. In: Proceedings of the 26th ACM SIGACT-SIGMOD-SIGART Symposium on Principles of Database Systems (PODS), pp. 165–174 (2007)
17. Grohe, M., Kreutzer, S.: Methods algorithmic meta theorems. Model Theor. Methods Finite Comb. **558**, 181–206 (2011)
18. Kloks, T. (ed.): Treewidth: Computations and Approximations. LNCS, vol. 842. Springer, Heidelberg (1994). doi:10.1007/BFb0045375
19. Kneis, J., Langer, A., Rossmanith, P.: Courcelle's theorem - a game-theoretic approach. Discret. Optim. **8**(4), 568–594 (2011)
20. Knop, D., Koutecký, M., Masařík, T., Toufar, T.: Simplified algorithmic metatheorems beyond MSO: Treewidth and neighborhood diversity. arXiv preprint. arXiv:1703.00544 (2017)
21. Kolaitis, P.G., Vardi, M.Y.: Conjunctive-query containment and constraint satisfaction. J. Comput. Syst. Sci. **61**(2), 302–332 (2000). http://www.sciencedirect.com/science/article/pii/S0022000000917136
22. Kolman, P., Koutecký, M., Tiwary, H.R.: Extension complexity, MSO logic, and treewidth (v3) (12 July 2016). http://arxiv.org/abs/1507.04907, short version presented at SWAT 2016
23. Kolman, P., Lidický, B., Sereni, J.S.: On Fair Edge Deletion Problems (2009)
24. Lampis, M.: Algorithmic meta-theorems for restrictions of treewidth. Algorithmica **64**(1), 19–37 (2012). http://dx.doi.org/10.1007/s00453-011-9554-x
25. Lampis, M.: Model checking lower bounds for simple graphs. Log. Methods Comput. Sci. **10**(1) (2014). http://dx.doi.org/10.2168/LMCS-10(1:18)2014
26. Langer, A., Reidl, F., Rossmanith, P., Sikdar, S.: Practical algorithms for MSO model-checking on tree-decomposable graphs. Comput. Sci. Rev. **13–14**, 39–74 (2014)
27. Libkin, L.: Elements of Finite Model Theory. Springer-Verlag, Berlin (2004). doi:10.1007/978-3-662-07003-1
28. Masařík, T., Toufar, T.: Parameterized complexity of fair deletion problems. In: Gopal, T.V., Jäger, G., Steila, S. (eds.) TAMC 2017. LNCS, vol. 10185, pp. 628–642. Springer, Heidelberg (2017). doi:10.1007/978-3-319-55911-7_45
29. Matoušek, J., Nešetřil, J.: Invitation to Discrete Mathematics, 2nd edn. Oxford University Press, Oxford (2009)
30. Pilipczuk, M.: Problems parameterized by treewidth tractable in single exponential time: a logical approach. In: Murlak, F., Sankowski, P. (eds.) MFCS 2011. LNCS, vol. 6907, pp. 520–531. Springer, Heidelberg (2011). doi:10.1007/978-3-642-22993-0_47
31. Szeider, S.: Monadic second order logic on graphs with local cardinality constraints. ACM Trans. Comput. Log. **12**(2), 12 (2011). http://doi.acm.org/10.1145/1877714.1877718

Extending Partial Representations of Trapezoid Graphs

Tomasz Krawczyk$^{(\boxtimes)}$ and Bartosz Walczak

Department of Theoretical Computer Science, Faculty of Mathematics
and Computer Science, Jagiellonian University, Kraków, Poland
{krawczyk,walczak}@tcs.uj.edu.pl

Abstract. A trapezoid graph is an intersection graph of trapezoids spanned between two horizontal lines. The partial representation extension problem for trapezoid graphs is a generalization of the recognition problem: given a graph G and an assignment ξ of trapezoids to some vertices of G, can ξ be extended to a trapezoid intersection model of the entire graph G? We show that this can be decided in polynomial time. Thus, we determine the complexity of partial representation extension for one of the two major remaining classes of geometric intersection graphs for which it has been unknown (circular-arc graphs being the other).

Keywords: Graph representations · Trapezoid graphs · Modular decomposition · Partial representation extension

1 Introduction

An *intersection representation* or *model* of a graph G by objects (sets) in a class $\mathcal{S}$ is a mapping ϕ of the vertices of G into $\mathcal{S}$ such that uv is an edge of G if and only if $\phi(u)$ and $\phi(v)$ intersect. Every class of objects $\mathcal{S}$ gives rise to a class of graphs that have a model in $\mathcal{S}$. This paper is devoted to *trapezoid graphs*, that is, graphs that have intersection representations by trapezoids whose bases lie on two common horizontal lines. Trapezoid graphs were introduced by Dagan et al. [8] in connection to channel routing problems in VLSI design. Other significant classes of graphs defined in terms of intersection models (for various choices of $\mathcal{S}$) include interval graphs, unit interval graphs, circular-arc graphs, circle graphs, permutation graphs, and function graphs; see [18,40].

Linear-time recognition algorithms are known for interval graphs [3], unit interval graphs [9], circular-arc graphs [32], and permutation graphs [33], and $O(n^2)$-time recognition algorithms are known for trapezoid graphs [30] and circle graphs [39]. These algorithms also construct a suitable representation if it exists. Function graphs are exactly the complements of comparability graphs, which can be recognized in $O(mn)$ time [17]. By contrast, recognition is NP-complete for

The authors were partially supported by National Science Center of Poland grant 2015/17/B/ST6/01873.

© Springer International Publishing AG 2017
H.L. Bodlaender and G.J. Woeginger (Eds.): WG 2017, LNCS 10520, pp. 358–371, 2017.
https://doi.org/10.1007/978-3-319-68705-6_27

rectangle graphs [28] and string graphs [27,37], and it is ∃ℝ-complete for unit disc graphs [21] and segment graphs [29], where ∃ℝ is the class of problems polynomially equivalent to the existential theory of the reals (NP ⊆ ∃ℝ ⊆ PSPACE).

A generic *partial representation extension* problem asks whether a fixed *partial representation* of a graph G can be extended to a full representation of G. For intersection models in a class of sets $\mathcal{S}$, the problem is formalized as follows.

Partial Representation Extension Problem. Given a graph G with vertex set V, a set $R \subseteq V$, and a partial representation $\xi\colon R \to \mathcal{S}$ that is an intersection model of the subgraph of G induced on R, decide whether there is an intersection model $\phi\colon V \to \mathcal{S}$ of G such that ξ is the restriction of ϕ to R.

Recognition is therefore the special case of partial representation extension where $R = \emptyset$. Being a more general problem, partial representation extension has been considered primarily for classes of graphs recognizable in polynomial time.

Partial representation extension is a relatively recent concept. For intersection representations, it was introduced by Klavík et al. [26] in 2011. It turned into an active area of research providing new insights into classical graph classes. Currently best known running times of partial representation extension algorithms are as follows: $O(m + n)$ for interval graphs [2,25], $O(m + n)$ for proper interval graphs [23], $O(n^2)$ for unit interval graphs [23], $O(m + n)$ for permutation graphs, polynomial for function graphs [22], and polynomial for circle graphs [4]. It has been unknown whether there are polynomial-time partial representation extension algorithms for circular-arc graphs and trapezoid graphs. We solve the problem in the affirmative for trapezoid graphs.

Main Theorem. *There is a polynomial-time partial representation extension algorithm for trapezoid graphs. It runs in $O(n^5)$ time.*

Problems that ask to extend a given partial solution have been also studied in various other contexts, leading to apparent contrasts with the corresponding (classical) problems of deciding whether a solution exists. Chordal graphs can be characterized as intersection graphs of subtrees of a tree. There are linear-time algorithms to recognize chordal graphs and to construct their subtree intersection models [18,36], but all variants of partial subtree representation extension for chordal graphs that were considered are NP-complete [24]. There are linear-time algorithms to extend partial combinatorial embeddings of planar graphs [1] and to construct straight-line drawings of planar graphs from their combinatorial embeddings [38], but extending partial straight-line drawings of planar graphs is NP-hard [35]. Every k-regular bipartite graph has a k-edge-coloring, which can be constructed in $O(m \log k)$ time [6], but extending a partial edge coloring becomes NP-complete even for $k = 3$ [13], and even for planar graphs [31].

An intersection model of a trapezoid graph gives rise to a left-to-right partial order on the trapezoids, which forms a transitive orientation of the complement of the graph. Posets that admit such trapezoid models are *trapezoid posets*, also known as posets with *interval dimension* at most 2. Every trapezoid model of a trapezoid poset P can be transformed into a *normalized model* such that the relative position of any two lateral sides is determined by the structure of P. The lateral sides of trapezoids in a normalized model form a segment model of an appropriately defined 2-dimensional poset called the *split* of P. In general, the interval dimension of P is equal to the dimension of the split of P [12].

Habib et al. [19] proved that interval dimension is a comparability invariant and used this fact to derive the first polynomial-time trapezoid graph recognition algorithm. Namely, they fix an arbitrary transitive orientation of the complement of the graph and test whether the poset thus obtained has interval dimension at most 2 using an $O(n^4)$-time algorithm due to Cogis [5]. Cogis's algorithm as well as later (and faster) algorithms in [12,30] construct the split and test whether it has dimension at most 2, which happens if and only if its incomparability graph is transitively orientable [10]. This two-step approach is the starting point for our partial representation extension algorithm. For another successful approach to recognizing trapezoid graphs, see [34].

The above-mentioned preliminaries are introduced in more detail in Sect. 3. Section 2 is devoted to *modular decomposition*—a structure describing all transitive orientations of a graph, which underlies most technical content of this paper.

In Sect. 4, we focus on the problem of extending a partial representation of a trapezoid poset. Normalized trapezoid models of a poset correspond to transitive orientations of the incomparability graph of the split, which are fully described by modular decomposition. However, the given partial representation may have only non-normalized extensions. To overcome this difficulty, we "normalize" the partial representation, we try to extend it to a normalized model, and finally we try to undo the normalization steps that we have applied to the initial partial representation. We formulate conditions that are necessary and sufficient for successful completion of these steps in terms of a 2-SAT formula, thus reducing the problem of partial representation extension for trapezoid posets to 2-SAT.

Section 5 is devoted to partial representation extension of trapezoid graphs. The choice of a transitive orientation of the complement $\overline{G}$ of the given graph G can influence extendability of a partial representation. We compute a suitable transitive orientation of $\overline{G}$ via dynamic programming, analyzing the modular decomposition of $\overline{G}$ bottom-up and determining, for each module M, at most two *good* transitive orientations of M based on those determined for the submodules of M. To decide whether a transitive orientation of a module M is good, we fix transitive orientations of the submodules of M arbitrarily but respecting some consistency constraints (expressed again in terms of a 2-SAT formula), and we test whether the given partial representation can be extended to a trapezoid model of the induced subgraph $G[M \cup R]$ respecting the fixed transitive orientation of $\overline{G}[M \cup R]$. This way, we reduce the graph problem to the poset problem.

Circular-arc graphs have a lot in common with trapezoid posets; for instance, they admit an analogous notion of a normalized model. Moreover, co-bipartite circular-arc graphs have trapezoid intersection models in which every trapezoid is infinite to the left or to the right. It is easy to derive a polynomial-time partial representation extension algorithm for co-bipartite circular-arc graphs from our algorithm for trapezoid posets. It remains open whether there is a polynomial-time partial representation extension algorithm for general circular-arc graphs.

2 Modular Decomposition and Transitive Orientations

A *graph* is a pair $(V, \sim)$ where $\sim$ is an irreflexive and symmetric *edge* relation on V. The *complement* of $(V, \sim)$ is the graph $(V, \not\sim)$ where $x \not\sim y \iff x \neq y$ and not $x \sim y$. A *poset* is a pair $(V, <)$ where $<$ is a partial order on V. A poset $(V, <)$ is a *transitive orientation* of a graph $(V, \sim)$ when $x \sim y \iff x < y$ or $x > y$. If $\star$ is a binary relation on V and $X, Y \subseteq V$, then $X \star Y$ denotes that $x \star y$ for all $x \in X$ and $y \in Y$. If $(V, \star)$ is a graph or a poset and $X \subseteq V$, then $(X, \star)$ denotes the graph or the poset on X in which $\star$ is restricted to $X \times X$.

A non-empty set $M \subseteq V$ is a *module* of a graph $(V, \sim)$ if $x \sim M$ or $x \not\sim M$ for every $x \in V \setminus M$. A module M of $(V, \sim)$ is a *strong module* if additionally $M \subset N$, $N \subset M$, or $M \cap N = \emptyset$ for every other module N of $(V, \sim)$. In particular, two strong modules are either nested or disjoint. The *modular decomposition* of $(V, \sim)$, denoted by $\mathcal{M}(V, \sim)$, is the family of all strong modules of $(V, \sim)$. The modular decomposition ordered by inclusion forms a tree in which V is the root, the *children* of a strong module M are the maximal proper subsets of M in $\mathcal{M}(V, \sim)$, and the leaves are the singleton modules $\{x\}$ for all $x \in V$. Clearly, $\mathcal{M}(V, \sim) = \mathcal{M}(V, \not\sim)$, and the tree structure implies $|\mathcal{M}(V, \sim)| \leqslant 2|V| - 1$.

Theorem 1 (Gallai [14]). *If $M_1, M_2 \in \mathcal{M}(V, \sim)$ and $M_1 \sim M_2$, then every transitive orientation $(V, <)$ of $(V, \sim)$ satisfies either $M_1 < M_2$ or $M_1 > M_2$.*

The children of a non-singleton module $M \in \mathcal{M}(V, \sim)$ form a partition of M. Such a module is *serial* if $M_1 \sim M_2$ for any two children M_1 and M_2, *parallel* if $M_1 \not\sim M_2$ for any two children M_1 and M_2, and *prime* otherwise. Equivalently, M is serial if the graph $(M, \not\sim)$ is disconnected, parallel if the graph $(M, \sim)$ is disconnected, and prime if both $(M, \sim)$ and $(M, \not\sim)$ are connected. The edge relation $\sim$ restricted to the edges between vertices in different children of M is denoted by $\sim_M$. If $x \sim y$, then $x \sim_M y$ for exactly one module $M \in \mathcal{M}(V, \sim)$.

Theorem 2 (Gallai [14]). *There is a bijection between the transitive orientations $(V, <)$ of $(V, \sim)$ and the families $\{(M, <_M) \colon M \in \mathcal{M}(V, \sim)\}$ such that $(M, <_M)$ is a transitive orientation of $(M, \sim_M)$, given by $x < y \iff x <_M y$, where M is the module in $\mathcal{M}(V, \sim)$ such that $x \sim_M y$.*

When $M \subset \mathcal{M}(V, \sim)$, we call the edges of $(M, \sim_M)$ simply the edges of M, and we call a transitive orientation of $(M, \sim_M)$ simply a transitive orientation of M.

Theorem 3 (Gallai [14]). *A prime module has either exactly two transitive orientations, one the reverse of the other, or no transitive orientations at all.*

A parallel module has exactly one (empty) transitive orientation. The transitive orientations of a serial module correspond to the total orderings of its children.

First polynomial-time algorithms to decide whether a graph has a transitive orientation and to construct it (if it exists) are implicit in [15,16]. Golumbic [17] provided an $O(mn)$-time algorithm. The first polynomial-time algorithm to compute the modular decomposition of a graph is due to James et al. [20]. McConnell and Spinrad [33] provided $O(m+n)$-time algorithms to compute the modular decomposition and a transitive orientation (if it exists). Another linear-time modular decomposition algorithm is presented in [7].

3   Trapezoid Models

We fix two horizontal lines in the plane $L_1 = \mathbb{R} \times \{1\}$ and $L_2 = \mathbb{R} \times \{2\}$. We let $\mathbb{S}$ denote the set of segments with one endpoint on L_1 and the other on L_2. We let $\mathbb{T}$ denote the set of non-degenerate trapezoids with one base on L_1 and the other on L_2 (every trapezoid in $\mathbb{T}$ has four distinct corners). We let $<_1$ and $<_2$ denote the left-to-right orders of points on L_1 and L_2, respectively. When X and Y are distinct segments in $\mathbb{S}$ or trapezoids in $\mathbb{T}$, we let

- $X <_i Y$ denote that $X \cap L_i <_i Y \cap L_i$, for $i \in \{1, 2\}$,
- $X < Y$ denote that $X <_1 Y$ and $X <_2 Y$,
- $X \sim Y$ denote that $X \cap Y \neq \emptyset$.

Such X and Y satisfy exactly one of the relations $X < Y$, $X \sim Y$, and $X > Y$.

A *trapezoid model* or simply a *model* of a graph $(V, \sim)$ or a poset $(V, <)$ is a map $\phi \colon V \to \mathbb{T}$ such that $x \sim y \iff \phi(x) \sim \phi(y)$ or $x < y \iff \phi(x) < \phi(y)$, respectively. The corners of all trapezoids used in a model are assumed to be distinct. A *trapezoid graph/poset* is a graph/poset that has a model. When X is a subset of the domain of a function ϕ, we let $\phi(X) = \{\phi(x) \colon x \in X\}$.

Let $(V, <)$ be a poset with a trapezoid model ϕ. For $x \in V$, let $\phi(|x)$ and $\phi(x|)$ denote the left and the right side, respectively, of the trapezoid $\phi(x)$. We have

$$\phi(x|) < \phi(|y) \iff x < y, \qquad \phi(|x) < \phi(|y) \implies \downarrow x < \uparrow y,$$
$$\phi(|x) < \phi(|y) \implies \downarrow x \subseteq \downarrow y, \qquad \phi(x|) < \phi(y|) \implies \uparrow x \supseteq \uparrow y, \tag{1}$$

for all $x, y \in V$, where $\downarrow x = \{z \in V \colon z < x\}$ and $\uparrow x = \{z \in V \colon z > x\}$. The model ϕ is *normalized* if it also satisfies the following conditions:

$$\downarrow x < \uparrow y \implies \phi(|x) < \phi(y|),$$
$$\downarrow x \subset \downarrow y \implies \phi(|x) < \phi(|y), \qquad \uparrow x \supset \uparrow y \implies \phi(x|) < \phi(y|). \tag{2}$$

They are "almost converse" to (1); there is no condition on the relation between $\phi(|x)$ and $\phi(|y)$ when $\downarrow x = \downarrow y$ or between $\phi(x|)$ and $\phi(y|)$ when $\uparrow x = \uparrow y$.

A usual way of constructing a normalized model is to start with an arbitrary trapezoid model and then apply a series of *normalization steps* of three types:

1. If $\downarrow x < \uparrow y$ and $\phi(|x) >_i \phi(y|)$, where $i \in \{1, 2\}$, then the corner of $\phi(|x)$ on L_i is pulled to the left and the corner of $\phi(y|)$ on L_i is pulled to the right until the two corners pass each other somewhere between $\phi(\downarrow x)$ and $\phi(\uparrow y)$.
2. If $\downarrow x \subset \downarrow y$ and $\phi(|x) >_i \phi(|y)$, where $i \in \{1, 2\}$, then the corner of $\phi(|x)$ on L_i is pulled to the left until it passes the corner of $\phi(|y)$ on L_i.
3. If $\uparrow x \supset \uparrow y$ and $\phi(x|) >_i \phi(y|)$, where $i \in \{1, 2\}$, then the corner of $\phi(y|)$ on L_i is pulled to the right until it passes the corner of $\phi(x|)$ on L_i.

They keep ϕ a model of $(V, <)$, eventually leading to a normalized model. We present a different proof of existence of normalized models further in this section. Nevertheless, normalization steps motivate some of our considerations in Sect. 4.

Let V' be the set obtained from V by splitting each element $x \in V$ into two copies, denoted by $|x$ and $x|$. Thus $V' = \{|x, x| \colon x \in V\}$. We use notation like x' to refer to an element of V' of either form, $|x$ or $x|$, where $x \in V$. We identify functions $\phi\colon V \to \mathbb{T}$ with functions $\phi\colon V' \to \mathbb{S}$ satisfying $\phi(|x) < \phi(x|)$ as follows: $\phi(x)$ is the trapezoid with left side $\phi(|x)$ and right side $\phi(x|)$. This is consistent with the notation $\phi(|x)$ and $\phi(x|)$ introduced before for functions $\phi\colon V \to \mathbb{T}$.

Conditions (1) and (2) motivate the following definitions. Given a partial order $<$ on V, we define a binary relation on V', also denoted by $<$, as follows:

$$
\begin{aligned}
x| < |y &\iff x < y, & |x < y| &\iff \downarrow x < \uparrow y, & (*) \\
|x < |y &\iff \downarrow x \subset \downarrow y, & x| < y| &\iff \uparrow x \supset \uparrow y.
\end{aligned}
\tag{3}
$$

The relation $<$ thus defined is a partial order on V', and $(*)$ implies $|x < x|$. We also define binary relations $\equiv$ and $\approx$ on V' as follows:

$$
\begin{aligned}
|x \equiv |y &\iff \downarrow x = \downarrow y, & x| \equiv y| &\iff \uparrow x = \uparrow y, & |x \not\equiv y|, \\
x' \approx y' &\iff \text{neither of } x' < y',\ x' > y',\ x' \equiv y' \text{ holds, where } x', y' \in V'.
\end{aligned}
$$

Comparing (1), (2), and (3), we conclude that ϕ is a normalized model of $(V, <)$ if and only if the following implications hold for all $x', y' \in V'$:

$$
x' < y' \implies \phi(x') < \phi(y'), \qquad x' \approx y' \implies \phi(x') \sim \phi(y').
\tag{4}
$$

We call the poset $(V', <)$ the *split* of $(V, <)$. The equivalence classes of $\equiv$ are (not necessarily strong) modules of $(V', \approx)$, which we call *irrelevant modules*.

The following fundamental theorem, the proof of which is outlined below, provides a polynomial-time recognition algorithm for trapezoid posets.

Theorem 4 (cf. [5,12,30]). *The following are equivalent for a poset $(V, <)$:*

1. *The poset $(V, <)$ has a trapezoid model (is a trapezoid poset).*
2. *The poset $(V, <)$ has a normalized trapezoid model.*
3. *The graph $(V', \approx)$ is transitively orientable.*

Recognition algorithms for trapezoid posets presented in [5,12,30] are based on the same approach as above except that they use a version of split in which elements equivalent under $\equiv$ become identified (then $\approx$ becomes the incomparability relation of the split). Although such an identification does not matter for recognition, it is not possible when dealing with partial representations.

The implication $2 \Rightarrow 1$ in Theorem 4 is obvious. For $1 \Rightarrow 3$, we need a weaker version of conditions (4) valid for models that are not necessarily normalized.

Lemma 5. *Let ϕ be a trapezoid model of $(V, <)$, and let $x', y', z' \in V'$.*

1. *If $x' < y'$, then $\phi(x') <_1 \phi(y')$ or $\phi(x') <_2 \phi(y')$.*
2. *If $x' \approx y'$, then $\phi(x') \sim \phi(y')$.*
3. *If $x' \approx y' \approx z'$ and $\phi(x') <_k \phi(y') <_k \phi(z')$, then $x' \approx z'$, for $k \in \{1, 2\}$.*

In view of conditions 2 and 3 of Lemma 5, every trapezoid model ϕ of $(V, <)$ gives rise to a transitive orientation $(V', \prec^\phi)$ of the graph $(V', \approx)$, defined as follows:

$$
x' \prec^\phi y' \iff \phi(x') <_1 \phi(y') \iff \phi(x') >_2 \phi(y'), \quad \text{when } x' \approx y'.
\tag{5}
$$

For the implication $3 \Rightarrow 2$ in Theorem 4, let $(V', \prec)$ be a transitive orientation of $(V', \approx)$. There is a normalized trapezoid model ϕ of $(V, <)$ such that

$$\begin{aligned}
x' < y' \text{ or } (x' \approx y' \text{ and } x' \prec y') &\implies \phi(x') <_1 \phi(y'), \\
x' < y' \text{ or } (x' \approx y' \text{ and } x' \succ y') &\implies \phi(x') <_2 \phi(y').
\end{aligned} \tag{6}$$

Indeed, the conditions above determine the relations $<_1$ and $<_2$ on $\phi(V')$ everywhere except within irrelevant modules. Moreover, if I is an irrelevant module and $x' \approx I$, then either $x' \prec I$ or $x' \succ I$ (by transitivity of $\prec$), and (6) imply either $\phi(x') <_i \phi(I)$ or $\phi(x') >_i \phi(I)$ for $i \in \{1, 2\}$. An arbitrary arrangement of $\phi(I)$ within each irrelevant module I yields a function $\phi \colon V' \to \mathbb{S}$ such that $\phi(|x) < \phi(x|)$ for all $x \in V$ and the corresponding function $\phi \colon V \to \mathbb{T}$. It follows from (3) and (6) that ϕ satisfies (1) and (2), so ϕ is a normalized model of $(V, <)$.

In fact, (5) and (6) provide a complete description of all normalized models ϕ of $(V, <)$ in terms of transitive orientations $(V', \prec)$ of $(V', \approx)$, up to an arbitrary rearrangement of $\phi(I)$ within each irrelevant module I.

Theorem 4 and the next result yield a polynomial-time recognition algorithm for trapezoid graphs—we can fix a transitive orientation of the complement of the given graph arbitrarily and test whether the resulting poset is a trapezoid poset.

Theorem 6 (Habib et al. [19]). *If $(V, \sim)$ is a trapezoid graph, then every transitive orientation of the complement of $(V, \sim)$ is a trapezoid poset.*

Theorem 6 can be established by showing that for any model ϕ of $(V, \sim)$ and any prime or parallel module $M \in \mathcal{M}(V, \sim)$, the representation of M in ϕ can be "squeezed" to form contiguous blocks on L_1 and L_2 and then reversed (rearranged if M is parallel) to agree with the requested transitive orientation of M, independently of the rest of ϕ.

4 Extending Partial Representations of Trapezoid Posets

In this section, we provide a polynomial-time algorithm for the partial representation extension problem for trapezoid posets: given a poset $(V, <)$, a set $R \subseteq V$, and a partial representation $\xi \colon R \to \mathbb{T}$ that is a model of $(R, <)$, decide whether ξ can be extended to a model of $(V, <)$. We compute the split $(V', <)$ of $(V, <)$, the graph $(V', \approx)$, and its modular decomposition $\mathcal{M}(V', \approx)$. In view of Theorem 4, we can assume that $(V', \approx)$ is transitively orientable, otherwise we are safe to reject the instance. Theorem 2 provides a correspondence between the transitive orientations $(V', \prec)$ of $(V', \approx)$ and the transitive orientations $(M, \prec_M)$ of the modules $M \in \mathcal{M}(V', \approx)$. We compute the two transitive orientations $(M, \prec_M^0)$ and $(M, \prec_M^1)$ claimed by Theorem 3 for every prime module $M \in \mathcal{M}(V', \approx)$.

Let $R' = \{|x, x| \colon x \in R\}$. We can assume that ξ satisfies conditions 1–3 of Lemma 5 for any $x', y', z' \in R'$, otherwise we are safe to reject the instance. By a formula analogous to (5), ξ yields a transitive orientation $(R', \prec^\xi)$ of $(R', \approx)$.

Under the assumptions above, we reduce the problem of extendability of ξ to a model of $(V, <)$ to satisfiability of a carefully designed set of constraints Φ over boolean variables, where every constraint involves at most two variables, so that Φ is equivalent to a 2-SAT formula. The algorithm just tests whether or not Φ is satisfiable and accepts or rejects the instance accordingly. We design Φ so that every model of $(V, <)$ extending ξ yields an assignment of boolean values 0 and 1 to the variables of Φ that makes Φ satisfied. Therefore, when introducing variables and constraints of Φ, we interpret their meaning in the context of an arbitrarily chosen (but fixed) model ϕ of $(V, <)$ that extends ξ, and we intend to make validity of the constraints self-evident. Finally, we prove that if Φ is satisfiable, then ξ can be extended to a model of $(V, <)$.

Let $Q = V \smallsetminus R$. Through the rest of this section, we implicitly assume that $a, b \in R$, $x \in Q$, and $(u, v) \in (R \times Q) \cup (Q \times R)$ whenever these symbols are used.

Suppose $u \not\succ v$. We use a boolean variable $\beta_1(|u, v|)$ to represent whether or not $\phi(|u) <_1 \phi(|v|)$ and a boolean variable $\beta_2(|u, v|)$ to represent whether or not $\phi(|u) <_2 \phi(|v|)$. At least one of these two conditions must hold when $\phi(u) \sim \phi(v)$ or $\phi(u) < \phi(v)$. We express this fact by adding the following constraint to Φ:

$$\beta_1(|u, v|) \vee \beta_2(|u, v|). \tag{Φ1}$$

Let M be a prime module in $\mathcal{M}(V', \approx)$ with $M \cap R' \neq \emptyset$. Recall that M has exactly two transitive orientations, $(M, \prec_M^0)$ and $(M, \prec_M^1)$, one the reverse of the other. We use a boolean variable $\mu(M)$ whose value represents the choice of one of these transitive orientations: $\prec_M^\phi = \prec_M^{\mu(M)}$. Whenever $a' \approx_M b'$ and $a' \prec^\xi b'$, we must have $a' \prec_M^\phi b'$, which we express by adding the following constraint to Φ:

$$\mu(M) = k, \quad \text{where } k \in \{0, 1\} \text{ is such that } a' \prec_M^k b'. \tag{Φ2}$$

Whenever $|u \approx_M v|$, the relation between $\phi(|u)$ and $\phi(|v|)$ in $<_1$ and $<_2$ is determined by $\prec_M^\phi$ as in (5), which we express by adding the following identities to Φ:

$$\begin{aligned}
\beta_1(|u, v|) = \neg\mu(M) \quad &\text{and} \quad \beta_2(|u, v|) = \mu(M) \quad &&\text{if } |u \prec_M^0 v|, \\
\beta_1(|u, v|) = \mu(M) \quad &\text{and} \quad \beta_2(|u, v|) = \neg\mu(M) \quad &&\text{if } |u \prec_M^1 v|.
\end{aligned} \tag{Φ3}$$

Now, let M be a serial module in $\mathcal{M}(V', \approx)$ with children $M_1, \ldots, M_r$ in an arbitrary order, and let i and j be distinct indices in $\{1, \ldots, r\}$ such that $(M_i \cup M_j) \cap R' \neq \emptyset$. Theorem 1 implies that either $M_i \prec^\phi M_j$ or $M_i \succ^\phi M_j$. We use a boolean variable $\mu(M_i, M_j)$ to represent whether or not $M_i \prec^\phi M_j$, and we add the following obvious identity to Φ:

$$\mu(M_i, M_j) = \neg\mu(M_j, M_i). \tag{Φ4}$$

Whenever $a' \in M_i$, $b' \in M_j$, and $a' \prec^\xi b'$, we have $M_i \prec^\phi M_j$, and by transitivity of $\prec^\phi$, every other child M_k of M satisfies $M_i \prec^\phi M_k$ or $M_k \prec^\phi M_j$ (or both); we express these conditions by adding the following constraints to Φ:

$$\mu(M_i, M_j) \quad \text{and} \quad \mu(M_i, M_k) \vee \mu(M_k, M_j) \quad \text{for } k \in \{1, \ldots, r\} \smallsetminus \{i, j\}. \tag{Φ5}$$

Whenever $|u \in M_i$ and $v| \in M_j$, the relation between $\phi(|u)$ and $\phi(v|)$ in $<_1$ and $<_2$ is determined by the relation between M_i and M_j in $\prec^\phi$ as in (5), which we express by adding the following identities to Φ:

$$\beta_1(|u,v|) = \mu(M_i, M_j) \quad \text{and} \quad \beta_2(|u,v|) = \neg\mu(M_i, M_j). \tag{$\Phi6$}$$

When $|u \approx v|$, the variables $\beta_1(|u,v|)$ and $\beta_2(|u,v|)$ are identified with μ and $\neg\mu$ or vice versa either by ($\Phi3$) for exactly one variable $\mu = \mu(M)$ or by ($\Phi6$) for exactly one variable $\mu = \mu(M_i, M_j)$; then ($\Phi1$) is redundant. By contrast, when $|u < v|$, the variables $\beta_1(|u,v|)$ and $\beta_2(|u,v|)$ are not involved in any condition of type ($\Phi3$) or ($\Phi6$); then ($\Phi1$) cannot be omitted.

Let $A_i = \{(a',b') \in R' \times R' : a' < b' \text{ and } \xi(a') >_i \xi(b')\}$ for $i \in \{1,2\}$. That is, A_1 and A_2 consist of the pairs (a',b') that do not satisfy the normalization condition (4). We add the following constraints to Φ, for $i \in \{1,2\}$:

$$\neg\beta_i(|x,b|) \quad \text{when } (a|,b|) \in A_i \text{ and } a < x \sim b, \tag{$\Phi7$}$$
$$\beta_i(|x,b|) \Rightarrow \beta_i(|x,a|) \quad \text{when } (a|,b|) \in A_i \text{ and } x \not\succ a,b, \tag{$\Phi8$}$$
$$\beta_{3-i}(|x,a|) \Rightarrow \beta_{3-i}(|x,b|) \quad \text{when } (a|,b|) \in A_i \text{ and } x \not\succ a,b, \tag{$\Phi9$}$$
$$\neg\beta_i(|a,x|) \quad \text{when } (|a,|b) \in A_i \text{ and } a \sim x < b, \tag{$\Phi10$}$$
$$\beta_i(|a,x|) \Rightarrow \beta_i(|b,x|) \quad \text{when } (|a,|b) \in A_i \text{ and } a,b \not\succ x, \tag{$\Phi11$}$$
$$\beta_{3-i}(|b,x|) \Rightarrow \beta_{3-i}(|a,x|) \quad \text{when } (|a,|b) \in A_i \text{ and } a,b \not\succ x, \tag{$\Phi12$}$$
$$\neg\beta_i(|a,x|) \vee \neg\beta_i(|y,b|) \quad \text{when } (|a,b|) \in A_i \text{ and } a \not\succ x < y \not\succ b. \tag{$\Phi13$}$$

Here are illustrations of why ϕ must satisfy the conditions represented by ($\Phi7$)–($\Phi9$): ($\Phi10$)–($\Phi12$): ($\Phi13$):

Lemma 7. *There is a model of $(V,<)$ extending ξ if and only if Φ is satisfiable.*

Proof (Sketch). We have constructed Φ ensuring its satisfiability if $(V,<)$ has a model extending ξ. Now, suppose Φ is satisfiable and the variables are assigned values that make Φ satisfied. It suffices to find a model ϕ of $(V,<)$ such that

$$\phi(a') <_i \phi(b') \iff \xi(a') <_i \xi(b'), \quad \text{for } i \in \{1,2\}. \tag{7}$$

By ($\Phi4$) and ($\Phi5$), there is a transitive orientation $(V', \prec)$ of $(V', \approx)$ that agrees with the interpretation of the values assigned to $\mu(\cdot)$. By ($\Phi2$) and ($\Phi5$), $\prec$ agrees with $\prec^\xi$. We modify the assignment on $\beta_1(\cdot)$ and $\beta_2(\cdot)$ to ensure that

$$\begin{aligned} &\text{if } |a<x| \prec |y<b|, \text{ then } \beta_1(|a,x|) = 0 \text{ or } \beta_1(|y,b|) = 0, \\ &\text{if } |a<x| \succ |y<b|, \text{ then } \beta_2(|a,x|) = 0 \text{ or } \beta_2(|y,b|) = 0, \end{aligned} \tag{8}$$

while keeping Φ satisfied. The proof that this is always possible is technical and makes use of all ($\Phi7$)–($\Phi13$). Then, we start with a normalized model ϕ of $(V,<)$ defined from $\prec$ by (6), arranging $\phi(I)$ within each irrelevant module I so as to satisfy (7) for all $a',b' \in I$. By ($\Phi3$) and ($\Phi6$), the model ϕ satisfies the following:

$$\text{if } u \sim v \text{ and } \beta_i(|u,v|) = 1, \text{ then } \phi(|u) <_i \phi(|v), \quad \text{for } i \in \{1,2\}. \tag{9}$$

We transform ϕ gradually into a model of $(V, <)$ satisfying (7) for all $a', b' \in R'$, in each step choosing some a' and b' that violate (7) so that $\phi(a')$ and $\phi(b')$ are consecutive in the order $<_i$ on $\phi(R')$ and moving $\phi(a')$ and $\phi(b')$ (and only them) on L_i to swap them in the order $<_i$ while maintaining (9). These steps can be interpreted as reversed normalization steps 1–3 described in Sect. 3. They always succeed thanks to (Φ7) and (Φ8), (Φ10) and (Φ11), (Φ13) and (8) for reversed step 3, 2, 1, respectively. By (9) and (Φ1), ϕ remains a model of $(V, <)$. □

The constraint set Φ is equivalent to a 2-SAT formula with $O(n^2)$ variables and $O(n^4)$ clauses, so its satisfiability can be decided in $O(n^4)$ time [11], leading to an $O(n^4)$-time partial representation extension algorithm for trapezoid posets.

5   Extending Partial Representations of Trapezoid Graphs

Finally, we provide a polynomial-time algorithm for the main problem of this paper: given a graph $(V, \sim)$, a set $R \subseteq V$, and a partial representation $\xi\colon R \to \mathbb{T}$ that is a model of $(R, \sim)$, decide whether ξ can be extended to a model of $(V, \sim)$. We assume that $(V, \sim)$ has a model, otherwise we reject the instance. Let $(V, \lesssim)$ be the complement of $(V, \sim)$. To decide whether $(V, \sim)$ has a model extending ξ, we look for a transitive orientation $(V, <)$ of $(V, \lesssim)$ that has a model extending ξ.

We assume $R \neq \emptyset$ and let $\mathcal{M}^R = \{M \in \mathcal{M}(V, \lesssim) : M \cap R \neq \emptyset\}$. We call the modules in $\mathcal{M}^R$ *restricted* and the modules in $\mathcal{M}(V, \lesssim) \setminus \mathcal{M}^R$ *unrestricted*.

Lemma 8. *Let M be a serial module in $\mathcal{M}^R$. Let N_M be the union of all unrestricted children of M, except for one (arbitrary) if M has only one restricted child. Then $(V, \sim)$ has a model extending ξ if and only if $(V \setminus N_M, \sim)$ has.*

We apply Lemma 8 repeatedly to all serial modules $M \in \mathcal{M}^R$, removing all sets N_M thus obtained from V. This does not affect extendability of ξ.

Theorem 2 gives a correspondence between the transitive orientations $(V, <)$ of $(V, \lesssim)$ and the transitive orientations $(M, <_M)$ of the modules $M \in \mathcal{M}(V, \lesssim)$. A transitive orientation $(M, <_M)$ of a module $M \in \mathcal{M}^R$ is ξ-*consistent* if $a <_M b$ whenever $a, b \in M \cap R$ and $\xi(a) < \xi(b)$. A map $\xi\colon R \to \mathbb{T}$ is a model of $(R, <)$ if and only if $(M, <_M)$ is ξ-consistent for every $M \in \mathcal{M}^R$. When M is serial, $(M, <_M)$ is ξ-consistent if and only if the restricted children of M can be ordered as $M_1, \ldots, M_r$ so that $\xi(M_1 \cap R) < \cdots < \xi(M_r \cap R)$. As a result of the aforementioned reduction by Lemma 8, if a serial module $M \in \mathcal{M}^R$ has more than two children, then all of them are restricted and thus M has only one ξ-consistent transitive orientation. This and Theorem 3 allow us to derive, for every module $M \in \mathcal{M}^R$, two transitive orientations $(M, <_M^0)$ and $(M, <_M^1)$, one the reverse of the other, that include all (at most two) ξ-consistent transitive orientations of M. If M is a parallel or singleton module, then $<_M^0$ and $<_M^1$ are both empty.

Below, we formulate sets of constraints with one boolean variable $\lambda(M)$ for each module $M \in \mathcal{M}^R$. The assignments of boolean values 0 and 1 to these variables satisfying a constraint set Π correspond to the transitive orientations $(V, <)$ of $(V, \lesssim)$ called Π-*consistent* as follows: $<_M = <_M^{\lambda(M)}$ for $M \in \mathcal{M}^R$. We start by providing a set of constraints Π with the property that every transitive orientation $(V, <)$ of $(V, \lesssim)$ that has a model extending ξ is Π-consistent. When introducing constraints of Π, we interpret their meaning in the context of an arbitrarily chosen (but fixed) transitive orientation $(V, <)$ of $(V, \lesssim)$ that has a model extending ξ, and we intend to make validity of the constraints self-evident.

First, we add the following constraints to Π to express the fact that the transitive orientation $(M, <_M)$ of each module $M \in \mathcal{M}^R$ must be ξ-consistent:

$$\lambda(M) \neq k \quad \text{if } (M, <_M^k) \text{ is not } \xi\text{-consistent, } M \in \mathcal{M}^R,\ k \in \{0,1\}. \tag{Π1}$$

Let $Q = V \setminus R$. For any $a, b \in R$ and $x \in Q$ such that $\xi(b|) < \xi(a|)$ and $a \lesssim x \sim b$, we must have $x < a$, which we express by adding the following constraint to Π:

$$\lambda(M) = k, \quad \text{where } x <_M^k a,\ M \in \mathcal{M}^R,\ k \in \{0,1\}. \tag{Π2}$$

Symmetrically, for any $a, b \in R$ and $x \in Q$ such that $\xi(|a) < \xi(|b)$ and $a \lesssim x \sim b$, we must have $a < x$, which we express by adding the following constraint to Π:

$$\lambda(M) = k, \quad \text{where } a <_M^k x,\ M \in \mathcal{M}^R,\ k \in \{0,1\}. \tag{Π3}$$

For any $a, b \in R$ and $x, y \in Q$ such that $\xi(|b) < \xi(a|)$ and $a \lesssim x \sim y \lesssim b$, we must have $x < a$ or $b < y$, which we express by adding the following constraint to Π:

$$\lambda(M) = k \vee \lambda(N) = \ell, \quad \text{where } \begin{aligned} & x <_M^k a,\ M \in \mathcal{M}^R,\ k \in \{0,1\}, \\ & b <_N^\ell y,\ N \in \mathcal{M}^R,\ \ell \in \{0,1\}. \end{aligned} \tag{Π4}$$

The last set of constraints to be added to Π is technical. For a prime or serial module $M \in \mathcal{M}^R$ and a transitive orientation $(M, <_M)$ of it, let

$$\mathrm{int}^R(M, <_M) = \{|a : a \in M \cap R \text{ and there is } u \in M \text{ such that } u <_M a\} \cup$$
$$\{a| : a \in M \cap R \text{ and there is } u \in M \text{ such that } u >_M a\}.$$

Since $M \cap R \neq \emptyset$ and M is prime or serial, the set $\mathrm{int}^R(M, <_M)$ is non-empty.

Lemma 9. *Let M and N be two disjoint prime or serial modules in $\mathcal{M}^R$ such that $M \sim N$. If a transitive orientation $(V, <)$ of $(V, \lesssim)$ has a model extending ξ, then $\mathrm{int}^R(M, <_M) \prec^\xi \mathrm{int}^R(N, <_N)$ or $\mathrm{int}^R(M, <_M) \succ^\xi \mathrm{int}^R(N, <_N)$, where $X \prec^\xi Y$ denotes that $\xi(X) <_1 \xi(Y)$ and $\xi(X) >_2 \xi(Y)$.*

For any two disjoint prime or serial modules $M, N \in \mathcal{M}^R$ such that $M \sim N$ and any $k, \ell \in \{0,1\}$, if neither $\mathrm{int}^R(M, <_M^k) \prec^\xi \mathrm{int}^R(N, <_N^\ell)$ nor $\mathrm{int}^R(M, <_M^k) \succ^\xi \mathrm{int}^R(N, <_N^\ell)$, then Lemma 9 implies $<_M \neq <_M^k$ or $<_N \neq <_N^\ell$, and we add the following constraint to Π to express this condition:

$$\lambda(M) \neq k \vee \lambda(N) \neq \ell. \tag{Π5}$$

Clearly, the set Π of constraints (Π1)–(Π5) is equivalent to a 2-SAT formula.

For $M \in \mathcal{M}^R$ and $k \in \{0,1\}$, the transitive orientation $(M, <_M^k)$ is *good* if $(V, \lesssim)$ has a Π-consistent transitive orientation $(V, <)$ such that $<_M = <_M^k$ and

$(M \cup R, <)$ has a model extending ξ. Our goal is to decide whether at least one of the two transitive orientations $(V, <_V^0)$ and $(V, <_V^1)$ of the root module V in $\mathcal{M}^R$ is good. We determine which of the two transitive orientations of every module in $\mathcal{M}^R$ are good by dynamic programming with respect to the modular decomposition tree. This is possible thanks to the following key lemma.

Lemma 10. *Let $M \in \mathcal{M}^R$ and $(V, <^\star)$ be a Π-consistent transitive orientation of $(V, \lessgtr)$. If $(M, <_M^\star)$ is good and $(N \cup R, <^\star)$ has a model extending ξ for every restricted child N of M, then $(M \cup R, <^\star)$ has a model extending ξ.*

Proof (Idea). Since $(M, <_M^\star)$ is good, there is a transitive orientation $(V, <)$ of $(V, \lessgtr)$ such that $<_M = <_M^\star$ and $(M \cup R, <)$ has a model ϕ extending ξ. Whenever $<$ and $<^\star$ disagree within an unrestricted child N of M, we "squeeze" N in ϕ to form contiguous blocks on L_1 and L_2 and replace the representation of N within these blocks by a model of $(N, <^\star)$, as in the proof of Theorem 6. For every restricted child N of M, let ϕ_N be a model of $(N \cup R, <^\star)$ extending ξ. The goal is to replace the representation of every such N in ϕ by its representation in ϕ_N. Before replacement, we make appropriate local adjustments to ϕ around N and to ϕ_N. Then, Π-consistency of $(V, <^\star)$ guarantees that simultaneous replacement of the representations of all restricted children of M can be performed in a consistent way, leading to a model of $(M \cup R, <^\star)$ extending ξ. $\qquad\square$

Let $M \in \mathcal{M}^R$ and $k \in \{0, 1\}$. Suppose we have (correctly) computed, for each proper restricted submodule N of M and each $\ell \in \{0, 1\}$, a boolean value $g(N, \ell)$ equal to 1 if $(N, <_N^\ell)$ is good and 0 otherwise. We decide whether $(M, <_M^k)$ is good as follows, saving the result to an analogous boolean value $g(M, k)$. Let Π_M^k be the set of constraints obtained from Π by adding the following constraints:

$$\lambda(M) = k, \tag{$\Pi6$}$$

$$\lambda(N) \neq \ell \quad \text{when } M \supset N \in \mathcal{M}^R, \ \ell \in \{0, 1\}, \text{ and } g(N, \ell) = 0. \tag{$\Pi7$}$$

We set $g(M, k)$ to 0 if Π_M^k is unsatisfiable. Otherwise, we compute an assignment that makes Π_M^k satisfied. It corresponds to a Π_M^k-consistent transitive orientation $(V, <)$ of $(V, \lessgtr)$. We invoke the algorithm from Sect. 4 to test whether $(M \cup R, <)$ has a model extending ξ, and we set $g(M, k)$ to 1 if it has and 0 otherwise. At the end, we accept the instance if $g(V, 0) = 1$ or $g(V, 1) = 1$ and reject it otherwise.

The procedure above sets $g(M, k)$ to 1 only when $(V, <)$ is Π_M^k-consistent and $(M \cup R, <)$ has a model extending ξ. In that case, by ($\Pi6$), we have $<_M = <_M^k$, so indeed $(M, <_M^k)$ is good. Now, suppose $(M, <_M^k)$ is good—there is a Π-consistent transitive orientation $(V, <)$ of $(V, \lessgtr)$ such that $<_M = <_M^k$ and $(M \cup R, <)$ has a model ϕ extending ξ. For each proper restricted submodule N of M, the restriction of ϕ to $N \cup R$ is a model of $(N \cup R, <)$ extending ξ, so $(N, <_N)$ is good. It follows that $(V, <)$ is Π_M^k-consistent. In particular, Π_M^k is satisfiable, so the procedure picks some Π_M^k-consistent transitive orientation $(V, <^\star)$ of $(V, \lessgtr)$. By ($\Pi6$), we have $<_M^\star = <_M^k$, so $(M, <_M^\star)$ is good. Also, $(N, <_N^\star)$ is good for every proper restricted submodule N of M, by ($\Pi7$). We claim that $(N \cup R, <^\star)$ has a model extending ξ for every restricted submodule N of M including $N = M$. This follows from Lemma 10 by straightforward induction. Indeed, by Lemma 10, the claim on N follows from the claim on the children of N and the fact that $(N, <_N^*)$ is good. We conclude that $(M \cup R, <^*)$ has a model extending ξ, so the procedure sets $g(M, k)$ to 1. This proves correctness of the algorithm.

Each of the $O(n)$ many values $g(M, k)$ is computed in $O(n^4)$ time, by solving the corresponding $O(n^2)$-size 2-SAT formula and invoking the $O(n^4)$-time algorithm from Sect. 4. Therefore, the total running time of the algorithm is $O(n^5)$.

References

1. Angelini, P., Di Battista, G., Frati, F., Jelínek, V., Kratochvíl, J., Patrignani, M., Rutter, I.: Testing planarity of partially embedded graphs. ACM Trans. Algorithms **11**(4), Article 32 (2015)
2. Bläsius, T., Rutter, I.: Simultaneous PQ-ordering with applications to constrained embedding problems. ACM Trans. Algorithms **12**(2), Article 16 (2016)
3. Booth, K.S., Lueker, G.S.: Testing for the consecutive ones property, interval graphs, and graph planarity using PQ-tree algorithms. J. Comput. Syst. Sci. **13**(3), 335–379 (1976)
4. Chaplick, S., Fulek, R., Klavík, P.: Extending partial representations of circle graphs. In: Wismath, S., Wolff, A. (eds.) GD 2013. LNCS, vol. 8242, pp. 131–142. Springer, Cham (2013). doi:10.1007/978-3-319-03841-4_12
5. Cogis, O.: On the Ferrers dimension of a digraph. Discrete Math. **38**(1), 47–52 (1982)
6. Cole, R., Ost, K., Schirra, S.: Edge-coloring bipartite multigraphs in $O(E \log D)$ time. Combinatorica **21**(1), 5–12 (2001)
7. Cournier, A., Habib, M.: A new linear algorithm for modular decomposition. In: Tison, S. (ed.) CAAP 1994. LNCS, vol. 787, pp. 68–84. Springer, Heidelberg (1994). doi:10.1007/BFb0017474
8. Dagan, I., Golumbic, M.C., Pinter, R.Y.: Trapezoid graphs and their coloring. Discrete Appl. Math. **21**(1), 35–46 (1988)
9. Deng, X., Hell, P., Huang, J.: Linear-time representation algorithms for proper circular-arc graphs and proper interval graphs. SIAM J. Comput. **25**(2), 390–403 (1996)
10. Dushnik, B., Miller, E.W.: Partially ordered sets. Am. J. Math. **63**(3), 600–610 (1941)
11. Even, S., Itai, A., Shamir, A.: On the complexity of time table and multi-commodity flow problems. SIAM J. Comput. **5**(4), 691–703 (1976)
12. Felsner, S., Habib, M., Möhring, R.H.: On the interplay between interval dimension and dimension. SIAM J. Discrete Math. **7**(1), 22–40 (1994)
13. Fiala, J.: NP-completeness of the edge precoloring extension problem on bipartite graphs. J. Graph Theory **43**(2), 156–160 (2003)
14. Gallai, T.: Transitiv orientierbare Graphen. Acta Math. Acad. Sci. Hung. **18**(1–2), 25–66 (1967)
15. Ghouila-Houri, A.: Caractérisation des graphes non orientés dont on peut orienter les arrêtes de manière à obtenir le graphe d'une relation d'ordre. C. R. Acad. Sci. **254**, 1370–1371 (1962)
16. Gilmore, P.C., Hoffman, A.J.: A characterization of comparability graphs and of interval graphs. Canad. J. Math. **16**, 539–548 (1964)
17. Golumbic, M.C.: The complexity of comparability graph recognition and coloring. Computing **18**(3), 199–208 (1977)
18. Golumbic, M.C.: Algorithmic Graph Theory and Perfect Graphs. Academic Press, New York (1980)

19. Habib, M., Kelly, D., Möhring, R.H.: Interval dimension is a comparability invariant. Discrete Math. **88**(2–3), 211–229 (1991)
20. James, L.O., Stanton, R.G., Cowan, D.D.: Graph decomposition for undirected graphs. In: Hoffman, F., Levow, R.B., Thomas, R.S.D. (eds.) 3rd Southeastern Conference on Combinatorics, Graph Theory, and Computing (CGTC 1972). Congressus Numerantium, vol. 6, pp. 281–290. Utilitas Mathematica, Winnipeg (1972)
21. Kang, R.J., Müller, T.: Sphere and dot product representations of graphs. Discrete Comput. Geom. **47**(3), 548–568 (2012)
22. Klavík, P., Kratochvíl, J., Krawczyk, T., Walczak, B.: Extending partial representations of function graphs and permutation graphs. In: Epstein, L., Ferragina, P. (eds.) ESA 2012. LNCS, vol. 7501, pp. 671–682. Springer, Heidelberg (2012). doi:10.1007/978-3-642-33090-2_58
23. Klavík, P., Kratochvíl, J., Otachi, Y., Rutter, I., Saitoh, T., Saumell, M., Vyskočil, T.: Extending partial representations of proper and unit interval graphs. Algorithmica **77**(4), 1071–1104 (2017)
24. Klavík, P., Kratochvíl, J., Otachi, Y., Saitoh, T.: Extending partial representations of subclasses of chordal graphs. Theor. Comput. Sci. **576**, 85–101 (2015)
25. Klavík, P., Kratochvíl, J., Otachi, Y., Saitoh, T., Vyskočil, T.: Extending partial representations of interval graphs. Algorithmica **78**(3), 945–967 (2017)
26. Klavík, P., Kratochvíl, J., Vyskočil, T.: Extending partial representations of interval graphs. In: Ogihara, M., Tarui, J. (eds.) TAMC 2011. LNCS, vol. 6648, pp. 276–285. Springer, Heidelberg (2011). doi:10.1007/978-3-642-20877-5_28
27. Kratochvíl, J.: String graphs. II. Recognizing string graphs is NP-hard. J. Combin. Theory Ser. B **52**(1), 67–78 (1991)
28. Kratochvíl, J.: A special planar satisfiability problem and a consequence of its NP-completeness. Discrete Appl. Math. **52**(3), 233–252 (1994)
29. Kratochvíl, J., Matoušek, J.: Intersection graphs of segments. J. Combin. Theory Ser. B **62**(2), 289–315 (1994)
30. Ma, T.H., Spinrad, J.P.: On the 2-chain subgraph cover and related problems. J. Algorithms **17**(2), 251–268 (1994)
31. Marx, D.: NP-completeness of list coloring and precoloring extension on the edges of planar graphs. J. Graph Theory **49**(4), 313–324 (2005)
32. McConnell, R.M.: Linear-time recognition of circular-arc graphs. Algorithmica **37**(2), 93–147 (2003)
33. McConnell, R.M., Spinrad, J.P.: Modular decomposition and transitive orientation. Discrete Math. **201**(1–3), 189–241 (1999)
34. Mertzios, G.B., Corneil, D.G.: Vertex splitting and the recognition of trapezoid graphs. Discrete Appl. Math. **159**(11), 1131–1147 (2011)
35. Patrignani, M.: On extending a partial straight-line drawing. Int. J. Found. Comput. Sci. **17**(5), 1061–1070 (2006)
36. Rose, D.J., Tarjan, R.E., Lueker, G.S.: Algorithmic aspects of vertex elimination on graphs. SIAM J. Comput. **5**(2), 266–283 (1974)
37. Schaefer, M., Sedgwick, E., Štefankovič, D.: Recognizing string graphs in NP. J. Comput. Syst. Sci. **67**(2), 365–380 (2003)
38. Schnyder, W.: Embedding planar graphs on the grid. In: Johnson, D. (ed.) 1st Annual ACM-SIAM Symposium Discrete Algorithms (SODA 1990), pp. 138–148. SIAM, Philadelphia (1990)
39. Spinrad, J.P.: Recognition of circle graphs. J. Algorithms **16**(2), 264–282 (1994)
40. Spinrad, J.P.: Efficient Graph Representations, Field Institute Monographs, vol. 19. AMS, Providence (2003)

On Low Rank-Width Colorings

O-joung Kwon[1]([✉]), Michał Pilipczuk[2], and Sebastian Siebertz[2]

[1] Technische Universität Berlin, Berlin, Germany
ojoungkwon@gmail.com
[2] Institute of Informatics, University of Warsaw, Warsaw, Poland
{michal.pilipczuk,siebertz}@mimuw.edu.pl

Abstract. We introduce the concept of *low rank-width colorings*, generalizing the notion of low tree-depth colorings introduced by Nešetřil and Ossona de Mendez in [26]. We say that a class $\mathcal{C}$ of graphs admits *low rank-width colorings* if there exist functions $N \colon \mathbb{N} \to \mathbb{N}$ and $Q \colon \mathbb{N} \to \mathbb{N}$ such that for all $p \in \mathbb{N}$, every graph $G \in \mathcal{C}$ can be vertex colored with at most $N(p)$ colors such that the union of any $i \leq p$ color classes induces a subgraph of rank-width at most $Q(i)$.

Graph classes admitting low rank-width colorings strictly generalize graph classes admitting low tree-depth colorings and graph classes of bounded rank-width. We prove that for every graph class $\mathcal{C}$ of bounded expansion and every positive integer r, the class $\{G^r \colon G \in \mathcal{C}\}$ of rth powers of graphs from $\mathcal{C}$, as well as the classes of unit interval graphs and bipartite permutation graphs admit low rank-width colorings. All of these classes have unbounded rank-width and do not admit low tree-depth colorings. We also show that the classes of interval graphs and permutation graphs do not admit low rank-width colorings. As interesting side properties, we prove that every graph class admitting low rank-width colorings has the Erdős-Hajnal property and is χ-bounded.

1   Introduction and Main Results

We are interested in covering a graph with (overlapping) pieces in such a way that (1) the number of pieces is small, (2) each piece is simple, and (3) every small subgraph is fully contained in at least one piece. Despite the graph theoretic interest in such coverings, it also has nice algorithmic applications. Consider e.g. the subgraph isomorphism problem. Here, we are given two graphs G and

---

The work of O. Kwon is supported by the European Research Council (ERC) under the European Union's Horizon 2020 research and innovation programme (ERC consolidator grant DISTRUCT, agreement No. 648527). The work of M. Pilipczuk and S. Siebertz is supported by the National Science Centre of Poland via POLONEZ grant agreement UMO-2015/19/P/ST6/03998, which has received funding from the European Union's Horizon 2020 research and innovation programme (Marie Skłodowska-Curie grant agreement No. 665778). M. Pilipczuk is supported by the

Foundation for Polish Science (FNP) via the START stipend programme.

© Springer International Publishing AG 2017
H.L. Bodlaender and G.J. Woeginger (Eds.): WG 2017, LNCS 10520, pp. 372–385, 2017.
https://doi.org/10.1007/978-3-319-68705-6_28

H as input, and we are asked to determine whether G contains a subgraph isomorphic to H. In many natural settings the pattern graph H we are looking for is small and in such case a covering as described above is most useful. By the first property, we can then iterate through the small number of pieces, by the third property, one of the pieces will contain our pattern graph. By the second property, we can test each piece for containment of H.

We can formulate the covering problem in an equivalent way from the point of view of graph coloring as follows. *How many colors are required to color the vertices of a graph G such that the union of any p color classes induce a simple subgraph* (understanding any p color classes as a piece in the above formulation)? It remains to specify what we mean by *simple* subgraphs.

From an algorithmic point of view, trees, or more generally, graphs of bounded tree-width are very well behaved graphs. Many NP-complete problems, in fact, all problems that can be formulated in monadic second order logic, are solvable in linear time on graphs of bounded tree-width [5,6]. In particular, the subgraph isomorphism problem for every fixed pattern graph H is solvable in polynomial time on any graph of bounded tree-width.

Taking graphs of small tree-width as our simple building blocks, we can define a *p-tree-width coloring* of a graph G as a vertex coloring of G such that the union of any $i \leq p$ color classes induces a subgraph of tree-width at most $i-1$. Using the structure theorem of Robertson and Seymour [33] for graphs excluding a fixed graph as a minor, DeVos et al. [11] proved that for every graph H and every integer $p \geq 1$, there is an integer $N = N(H, p)$, such that every H-minor-free graph admits a p-tree-width coloring with N colors.

Tree-depth is another important and useful graph invariant. It was introduced under this name in [25], but equivalent notions were known before, including the notion of *rank* [28], *vertex ranking number* and minimum height of an *elimination tree* [1,10,34], etc. In [25], Nešetřil and Ossona de Mendez introduced the notion of p-tree-depth colorings as vertex colorings of a graph such that the union of any $i \leq p$ color classes induces a subgraph of tree-depth at most i. Note that the tree-depth of a graph is always larger (at least by 1) than its tree-width, hence a low tree-depth coloring is a stronger requirement than a low tree-width coloring. Also based on the structure theorem, Nešetřil and Ossona de Mendez [25] proved that proper minor closed classes admit even low tree-depth colorings.

Not much later, Nešetřil and Ossona de Mendez [26] proved that proper minor closed classes are unnecessarily restrictive for the existence of low tree-depth colorings. They introduced the notion of *bounded expansion classes of graphs*, a concept that generalizes the concept of classes with excluded minors and with excluded topological minors. While the original definition of bounded expansion is in terms of density of shallow minors, it turns out low tree-depth colorings give an alternative characterisation: a class $\mathcal{C}$ of graphs has bounded expansion if and only if for all $p \in \mathbb{N}$ there exists a number $N = N(\mathcal{C}, p)$ such that every graph $G \in \mathcal{C}$ admits a p-tree-depth coloring with $N(p)$ colors [26]. For the even more general notion of *nowhere dense classes of graphs* [27], it turns out that a class $\mathcal{C}$ of graphs closed under taking subgraphs is nowhere dense if and only if for all $p \in \mathbb{N}$ and all $\varepsilon > 0$ there exists n_0 such that every n-vertex graph $G \in \mathcal{C}$ with $n \geq n_0$ admits a p-tree-depth coloring with n^ε colors.

Furthermore, there is a simple algorithm to compute such a decomposition in time $\mathcal{O}(n)$ in case $\mathcal{C}$ has bounded expansion and in time $\mathcal{O}(n^{1+\varepsilon})$ for any $\varepsilon > 0$ in case $\mathcal{C}$ is nowhere dense. As a result, the subgraph isomorphism problem for every fixed pattern H can be solved in linear time on any class of bounded expansion and in almost linear time on any nowhere dense class. More generally, it was shown in [13,17] that every fixed first order property can be tested in linear time on graphs of bounded expansion, implicitly using the notion of low tree-depth colorings, and in almost linear time on nowhere dense classes [18].

Note that bounded expansion and nowhere dense classes of graphs are uniformly sparse graphs. In fact, bounded expansion classes of graphs can have at most a linear number of edges and nowhere dense classes can have no more than $\mathcal{O}(n^{1+\varepsilon})$ many edges. This motivates our new definition of *low rank-width colorings* which extends the coloring technique to dense classes of graphs which are closed under taking induced subgraphs.

Rank-width was introduced by Oum and Seymour [32] and aims to extend tree-width by allowing well behaved dense graphs to have small rank-width. Also for graphs of bounded rank-width there are many efficient algorithms based on dynamic programming. Here, we have the important meta-theorem of Courcelle, Makowsky, and Rotics [8], stating that for every monadic second-order formula (with set quantifiers ranging over sets of vertices) and every positive integer k, there is an $\mathcal{O}(n^3)$−time algorithm to determine whether an input graph of rank-width at most k satisfies the formula. There are several parameters which are equivalent to rank-width in the sense that one is bounded if and only if the other is bounded. These include *clique-width* [7], *NLC-width* [36], and *Boolean-width* [3].

Low rank-width colorings. We now introduce our main object of study.

Definition 1. *A class $\mathcal{C}$ of graphs admits low rank-width colorings if there exist functions $N : \mathbb{N} \to \mathbb{N}$ and $Q : \mathbb{N} \to \mathbb{N}$ such that for all $p \in \mathbb{N}$, every graph $G \in \mathcal{C}$ can be vertex colored with at most $N(p)$ colors such that the union of any $i \leq p$ color classes induces a subgraph of rank-width at most $Q(i)$.*

As proved by Oum [29], every graph G with tree-width k has rank-width at most $k + 1$, hence every graph class which admits low tree-depth colorings also admits low rank-width colorings. On the other hand, graphs admitting a low rank-width coloring can be very dense. We also remark that graph classes admitting low rank-width colorings are monotone under taking induced subgraphs, as rank-width does not increase by removing vertices.

Let us remark that due to the model-checking algorithm of Courcelle et al. [8], the (induced) subgraph isomorphism problem is solvable in cubic time for every fixed pattern H whenever the input graph is given together with a low rank-width coloring for $p = |V(H)|$, using $N(p)$ colors. Indeed, it suffices to iterate through all p-tuples of color classes and look for the pattern H in the subgraph induced by these color classes; this can be done efficiently since this subgraph has rank-width at most $Q(p)$. The caveat is that the graph has to be supplied with an appropriate coloring. In this work we do not investigate the algorithmic aspects of low rank-width colorings, and rather concentrate on the combinatorial question of which classes admit such colorings, and which do not.

Our contribution. We prove that for every class $\mathcal{C}$ of bounded expansion and every integer $r \geq 2$, the class $\{G^r \colon G \in \mathcal{C}\}$ of rth powers of graphs from $\mathcal{C}$ admits low rank-width colorings. It is easy to see that there are classes of bounded expansion such that $\{G^r \colon G \in \mathcal{C}\}$ has both unbounded rank-width and does not admit low tree-depth colorings. We furthermore prove that the class of unit interval graphs and the class of bipartite permutation graphs admit low rank-width colorings. On the negative side, we show that the classes of interval graphs and of permutation graphs do not admit low rank-width colorings. Finally, we also prove that every graph class admitting low rank-width colorings has the Erdős-Hajnal property [15] and is χ-bounded [20].

2 Preliminaries

All graphs in this paper are finite, undirected and simple, that is, they do not have loops or parallel edges. Our notation is standard, we refer to [12] for more background on graph theory. We write $V(G)$ for the vertex set of a graph G and $E(G)$ for its edge set. A *vertex coloring* of a graph G with colors from S is a mapping $c \colon V(G) \rightarrow S$. For each $v \in V(G)$, we call $c(v)$ the color of v. The *distance* between vertices u and v in G, denoted $\mathrm{dist}_G(u, v)$, is the length of a shortest path between u and v in G. The *rth power of a graph* G is the graph G^r with vertex set $V(G)$, where there is an edge between two vertices u and v if and only if their distance in G is at most r.

Rank-width was introduced by Oum and Seymour [32]. We refer to the surveys [21,30] for more background. For a graph G, we denote the adjacency matrix of G by A_G, where for $x, y \in V(G)$, $A_G[x, y] = 1$ if and only if x is adjacent to y. Let G be a graph. We define the *cut-rank* function $\mathrm{cutrk}_G \colon 2^V \rightarrow \mathbb{N}$ such that $\mathrm{cutrk}_G(X)$ is the rank of the matrix $A_G[X, V(G) \setminus X]$ over the binary field (if $X = \emptyset$ or $X = V(G)$, then we let $\mathrm{cutrk}_G(X) = 0$).

A *rank-decomposition* of G is a pair (T, L), where T is a subcubic tree (i.e. a tree where every node has degree 1 or 3) with at least 2 nodes and L is a bijection from $V(G)$ to the set of leaves of T. The *width* of e is define as $\mathrm{cutrk}_G(A_1^e)$ where (A_1^e, A_2^e) is the vertex bipartition of G each A_i^e is the set of all vertices in G mapped to leaves contained in one of components of $T - e$. The *width* of (T, L) is the maximum width over all edges in T, and the *rank-width* of G, denoted by $\mathrm{rw}(G)$, is the minimum width over all rank-decompositions of G. If $|V(G)| \leq 1$, then G has no rank-decompositions, and the rank-width of G is defined to be 0.

The exact definitions of tree-decompositions, tree-width, and tree-depth are not needed in our reasoning. We include them in the appendix for completeness.

A graph is an *interval graph* if it is the intersection graph of a family $\mathcal{I}$ of intervals on the real line, an interval graph is a *unit interval graph* if all intervals in $\mathcal{I}$ have the same length. A graph is a *permutation graph* if it is the intersection graph of line segments whose endpoints lie on two parallel lines.

3   Powers of Sparse Graphs

In this section we show that the class of rth powers of graphs from a bounded expansion class admit low rank-width colorings. The original definition of bounded expansion classes by Nešetřil and Ossona de Mendez [26] is in terms of bounds on the density of bounded depth minors. We will work with the characterisation by the existence of low tree-depth colorings as well as by a characterisation in terms of bounds on generalized coloring numbers.

Theorem 2 (Nešetřil and Ossona de Mendez [26]). *A class $\mathcal{C}$ of graphs has bounded expansion if and only if for all $p \in \mathbb{N}$ there exists a number $N = N(\mathcal{C}, p)$ such that every graph $G \in \mathcal{C}$ admits a p-tree-depth coloring with N colors.*

Our main result in this section is the following.

Theorem 3. *Let $\mathcal{C}$ be a class of bounded expansion and $r \geq 2$ be an integer. Then the class $\{G^r : G \in \mathcal{C}\}$ of rth powers of graphs from $\mathcal{C}$ admits low rank-width colorings.*

For a graph G, we denote by $\Pi(G)$ the set of all linear orders of $V(G)$. For $u, v \in V(G)$ and a non-negative integer r, we say that u is *weakly r-reachable* from v with respect to L, if there is a path P of length at most r between u and v such that u is the smallest among the vertices of P with respect to L. We denote by $\text{WReach}_r[G, L, v]$ the set of vertices that are weakly r-reachable from v with respect to L. The *weak r-coloring number* $\text{wcol}_r(G)$ of G is defined as

$$\text{wcol}_r(G) := \min_{L \in \Pi(G)} \max_{v \in V(G)} \left| \text{WReach}_r[G, L, v] \right|.$$

The weak coloring numbers were introduced by Kierstead and Yang [22] in the context of coloring and marking games on graphs. As shown by Zhu [37], classes of bounded expansion can be characterised by the weak coloring numbers.

Theorem 4 (Zhu [37]). *A class $\mathcal{C}$ has bounded expansion if and only if for all $r \geq 1$ there is a number $f(r)$ such that for all $G \in \mathcal{C}$ it holds that $\text{wcol}_r(G) \leq f(r)$.*

In order to prove Theorem 3, we will first compute a low tree-depth coloring. We would like to apply the following theorem, relating the tree-width (and hence in particular the tree-depth) of a graph and the rank-width of its rth power.

Theorem 5 (Gurski and Wanke [19]). *Let $r \geq 2$ be an integer. If a graph H has tree-width at most p, then H^r has rank-width at most $2(r+1)^{p+1} - 2$.*

We remark that Gurski and Wanke [19] proved this bound for clique-width instead of rank-width, but clique-width is never smaller than the rank-width [32].

The natural idea would be just to combine the bound of Theorem 5 with low tree-depth coloring given by Theorem 2. Note however, that when we consider any subgraph H induced by $i \leq p$ color classes, the graph H^r may be completely different from the graph $G^r[V(H)]$, due to paths that are present in G but disappear in H. Hence we cannot directly apply Theorem 5. Instead, we will prove the existence of a refined coloring of G such that for any subgraph H induced by $i \leq p$ color classes, in the refined coloring there is a subgraph H' such that $G^r[V(H)] \subseteq H'^r$ and such that H' gets only $g(i)$ colors in the original coloring, for some fixed function g. We can now apply Theorem 5 to H' and use fact that rank-width is monotone under taking induced subgraphs.

In the following, we will say that a vertex subset X *receives* a color i under a coloring c if $i \in c^{-1}(X)$. We first need the following definitions.

Definition 6. *Let G be a graph, $X \subseteq V(G)$ and $r \geq 2$. A superset $X' \supseteq X$ is called* an r-shortest path hitter *for X if for all $u, v \in X$ with $1 < \text{dist}_G(u, v) \leq r$, X' contains an internal vertex of some shortest path between u and v.*

Definition 7. *Let G be a graph, let c be a coloring of G, and $r \geq 2$ and $d \geq 1$. A coloring c' is a (d, r)-good refinement of c if for every vertex set X that receives at most p colors under c', there exists an r-shortest path hitter X' of X that receives at most $d \cdot p$ colors under c.*

We use the weak coloring numbers to prove the existence of a good refinement.

Lemma 8. *Let G be a graph and $r \geq 2$ be an integer. Then every coloring c of G using k colors has a $(2\text{wcol}_r(G), r)$-good refinement using $k^{2\text{wcol}_r(G)}$ colors.*

Proof. Let Γ be the set of colors used by c, and let $d := 2\text{wcol}_r(G)$. The (d, r)-good refinement c' that we are going to construct will use subsets of Γ of size at most d as the color set; the number of such subsets is at most $k^{2\text{wcol}_r(G)}$. Let L be a linear order of $V(G)$ with $\max_{v \in V(G)} |\text{WReach}_r[G, L, v]| = \text{wcol}_r(G)$. We construct a new coloring c' as follows:

(1) Start by setting $c'(v) := \emptyset$ for each $v \in V(G)$.
(2) For each pair of vertices u and v such that $u \in \text{WReach}_r[G, L, v]$, we add the color $c(u)$ to $c'(v)$.
(3) For each pair u and v of non-adjacent vertices such that $u <_L v$ and $u \in \text{WReach}_r[G, L, v]$, we do the following. Check whether there is a path P of length at most r connecting u and v such that all the internal vertices of P are larger than both u and v in L. If there is no such path, we do nothing for the pair u, v. Otherwise, fix one such path P, chosen to be the shortest possible, and let z be the vertex traversed by P that is the largest in L. Then we add the color $c(z)$ to $c'(v)$.

Thus, every vertex v receives in total at most $2\mathrm{wcol}_r(G)$ colors of Γ to its final color $c'(v)$: at most $\mathrm{wcol}_r(G)$ in step (2), and at most $\mathrm{wcol}_r(G)$ in step (3), because we add at most one color per each $u \in \mathrm{WReach}_r[G, L, v]$. It follows that each final color $c'(v)$ is a subset of Γ of size at most $2\mathrm{wcol}_r(G)$.

We claim that c' is a (d, r)-good refinement of c. Let $X \subseteq V(G)$ be a set that receives at most p colors under c', say colors $A_1, \ldots, A_p \subseteq \Gamma$. Let X' be the set of vertices of G that are colored by colors in $A_1 \cup \cdots \cup A_p$ under c. Since $|A_i| \leq d$ for each $i \in \{1, \ldots, p\}$, we have that X' receives at most $d \cdot p$ colors under c.

To show that X' is an r-shortest path hitter of X, let us choose any two vertices u and v in X with $u <_L v$ and $1 < \mathrm{dist}_G(u, v) \leq r$. If there is a shortest path from u to v whose all internal vertices are larger than u and v in L, by step (3), X' contains a vertex that is contained in one such path. Otherwise, a shortest path from u to v contains a vertex z with $L(z) < L(v)$ other than u and v. This implies that there exists $z' \in \mathrm{WReach}_r[G, L, v] \setminus \{u\}$ on the path such that $c(z') \in c'(v)$, and hence $z' \in X'$ by step (2). Therefore, X' is an r-shortest path hitter of X, as required.   □

Definition 9. *Let G be a graph, let $X \subseteq V(G)$, and let $r \geq 1$ be an integer. A superset $X' \supseteq X$ is called an r-shortest path closure of X if for each $u, v \in X$ with $\mathrm{dist}_G(u, v) = \ell \leq r$, $G[X']$ contains a path of length ℓ between u and v.*

Definition 10. *Let G be a graph, let c be a coloring of G, and let $r \geq 2$ and $d \geq 1$. A coloring c' is a (d, r)-excellent refinement of c if for every vertex set $X \subseteq V(G)$ there exists an r-shortest path closure X' of X such that if X receives p colors in c', then X' receives at most $d \cdot p$ colors in c.*

We inductively define excellent refinements from good refinements.

Lemma 11. *Let G be a graph, $r \geq 2$ an integer, and let $d_r := \prod_{2 \leq \ell \leq r} 2\mathrm{wcol}_\ell(G)$. Then every coloring c of G using at most k colors has a (d_r, r)-excellent refinement using at most k^{d_r} colors.*

Proof. We prove the lemma by induction on r. For $r = 2$, an r-shortest path hitter of a set X is an r-shortest path closure, and vice versa. Therefore, the statement immediately follows from Lemma 8. Now assume $r \geq 3$. By induction hypothesis, there is a $(d_{r-1}, r-1)$-excellent refinement c_1 of c with at most $k^{d_{r-1}}$ colors. By applying Lemma 8 to c_1, we obtain a $(2\mathrm{wcol}_r(G), r)$-good refinement c' of c_1 with at most $(k^{d_{r-1}})^{2\mathrm{wcol}_r(G)} = k^{d_r}$ colors. We claim that c' is a (d_r, r)-excellent refinement of c. Any set X which gets at most p colors from c' can be first extended to an r-shortest path hitter X' for X which receives at most $2\mathrm{wcol}_r(G) \cdot p$ colors. Then X' can be extended by induction hypothesis to an $(r-1)$-shortest path closure X'' of X' which receives at most $d_{r-1} \cdot 2\mathrm{wcol}_r(G) \cdot p = d_r \cdot p$ colors.

It remains to show that X'' is an r-shortest path closure of X. Take any $u, v \in X$ with $\mathrm{dist}_G(u, v) = \ell \leq r$. If $\ell \leq 1$, then u, v are already adjacent in $G[X]$. Otherwise, since X' is an r-shortest path hitter for X, there is a vertex $z \in X'$ that lies on some shortest path connecting u and v in G. In particular, $\mathrm{dist}_G(u, z) = \ell_1$

and $\text{dist}_G(z, v) = \ell_2$ for ℓ_1, ℓ_2 satisfying $\ell_1, \ell_2 < \ell$ and $\ell_1 + \ell_2 = \ell$. Since X'' is an $(r-1)$-shortest path closure of X', we infer that $\text{dist}_{G[X'']}(u, z) = \ell_1$ and $\text{dist}_{G[X'']}(z, v) = \ell_2$. Hence $\text{dist}_{G[X'']}(u, v) = \ell$ by the triangle inequality.   □

Proof (of Theorem 3). Let G be a graph in $\mathcal{C}$ and let $d_r := \prod_{2 \leq \ell \leq r} 2\text{wcol}_\ell(G)$. Since $\mathcal{C}$ has bounded expansion, by Theorem 4, for each r, $\text{wcol}_r(G)$ is bounded by a constant only depending on $\mathcal{C}$. We start by taking c to be a $(d_r \cdot p)$-tree-depth coloring with $N(d_r \cdot p)$ colors, where N is the function from Theorem 2. Then its (d_r, r)-excellent refinement c' has the property that c' uses at most $N(d_r \cdot p)^{d_r}$ colors, and every subset X which receives at most p colors in c' has an r-shortest path closure X' that receives at most $d_r \cdot p$ colors in c. Thus, the graph induced on X in the rth power G^r is the same at the graph induced on X in the rth power $G[X']^r$. Since $G[X']$ has tree-depth at most $d_r \cdot p$, by Theorem 5, $G[X']^r$ has rank-width at most $2(r+1)^{d_r \cdot p+1} - 2$. Therefore, $G^r[X]$ has rank-width at most $2(r+1)^{d_r \cdot p+1} - 2$ as well.   □

We now give two example applications of Theorem 3. A *map graph* is a graph that can be obtained from a plane graph by making a vertex for each face, and adding an edge between two vertices, if the corresponding faces share a vertex. One can observe $(\star)$[1] that every map graph is an induced subgraph of the second power of another planar graph, namely the *radial graph* of the original graph. Thus, map graphs have low rank-width colorings. A similar reasoning can be performed for line graphs of graphs from any bounded expansion graph class $(\star)$. Thus, both map graphs and line graphs of graphs from any fixed bounded expansion graph class admit low rank-width colorings.

4 Other Positive Results

We now prove that unit interval graphs and bipartite permutation graphs admit low rank-width colorings.

Theorem 12. *The class of unit interval graphs and the class of bipartite permutation graphs admit low rank-width colorings.*

Our results follow from characterizations of these classes obtained by Lozin [23]. Let $n, m \geq 1$. We denote by $H_{n,m}$ the graph with $n \cdot m$ vertices which can be partitioned into n independents sets $V_1 = \{v_{1,1}, \ldots, v_{1,m}\}, \ldots, V_n = \{v_{n,1}, \ldots, v_{n,m}\}$ so that for each $i \in \{1, \ldots, n-1\}$ and for each $j, j' \in \{1, \ldots, m\}$, vertex $v_{i,j}$ is adjacent to $v_{i+1,j'}$ if and only if $j' \in \{1, \ldots, j\}$, and there are no edges between V_i and V_j if $|i - j| \geq 2$. The graph $\widetilde{H}_{n,m}$ is the graph obtained from $H_{n,m}$ by replacing each independent set V_i by a clique.

Lemma 13. *The following statements hold:*

1. (Lozin [23]) The rank-width of $H_{n,m}$ and of $\widetilde{H}_{n,m}$ is at most $3n$.

---

[1] The proofs of claims marked with $(\star)$ appear in the appendix.

2. *(Lozin [23]) Every bipartite permutation graph on n vertices is isomorphic to an induced subgraph of $H_{n,n}$.*
3. *(Lozin [24]) Every unit interval graph on n vertices is isomorphic to an induced subgraph of $\widetilde{H}_{n,n}$.*

Hence, in order to prove Theorem 12, it suffices to prove that the graphs $H_{n,m}$ and $\widetilde{H}_{n,m}$ admit low rank-width colorings.

Proof (of Theorem 12). For every positive integer p, let $N(p) := p + 1$ and $Q(i) := 3i$ for each $i \in \{1, \ldots, p\}$. We prove that for all $n, m \geq 1$, the graphs $H_{n,m}$ and $\widetilde{H}_{n,m}$ can be vertex colored using $N(p)$ colors so that each of the connected components of the subgraph induced by any $i \leq p$ color classes has rank-width at most $R(i)$. As rank-width and rank-width colorings are monotone under taking induced subgraphs, the statement of the theorem follows from Lemma 13.

Assume that the vertices of $H_{n,m}$ (and $\widetilde{H}_{n,m}$, respectively) are named $v_{1,1}, \ldots, v_{1,m}, \ldots, v_{n,1}, \ldots, v_{n,m}$, as in the definition. We color the vertices in the ith row, $v_{i,1}, \ldots, v_{i,m}$, with color $j + 1$ where $j \in \{0, 1, \ldots, p\}$ and $i \equiv j \pmod{p+1}$. Then any connected component H of a subgraph induced by $i \leq p$ colors is isomorphic to $H_{i',m}$ ($\widetilde{H}_{i',m}$, respectively) for some $i' \leq i$. Hence, according to Lemma 13, H has rank-width at most $3i = Q(i)$, as claimed.   □

5   Negative Results

In contrast to the result in Sect. 4, we prove that interval graphs and permutation graphs do not admit low rank-width colorings. For this, we introduce twisted chain graphs. Briefly, a twisted chain graph G consists of three vertex sets A, B, C where each of $G[A \cup C]$ and $G[B \cup C]$ is a chain graph, but the ordering of C with respect to the chain graphs $G[A \cup C]$ and $G[B \cup C]$ are distinct.

Definition 14. *For a positive integer n, a graph on the set of $3n^2$ vertices $A \cup B \cup C$, where $A = \{v_1, \ldots, v_{n^2}\}$, $B = \{w_1, \ldots, w_{n^2}\}$, and $C = \{z_{(i,j)} : 1 \leq i, j \leq n\}$, is called a* twisted chain graph *of order n if*

- *for integers $x, y, i, j \in \{1, \ldots, n\}$ and $k = n(x-1) + y$, v_k is adjacent to $z_{(i,j)}$ if and only if either $(x < i)$ or $(x = i$ and $y \leq j)$;*
- *for integers $x, y, i, j \in \{1, \ldots, n\}$ and $k = n(x-1) + y$, w_k is adjacent to $z_{(i,j)}$ if and only if either $(x < j)$ or $(x = j$ and $y \leq i)$;*
- *the edge relation within $A \cup B$ and within C is arbitrary.*

We first show that a large twisted chain graph has large rank-width. We remark that a similar construction based on merging two chain graphs in a mixed order can be found in Brandstädt et al. [2]. Also, a slightly general construction was given by Dabrowski and Paulusma [9]. Obtaining any lower bound seems to follow from a careful examination and modification of the constructions given in [2] or [9]; however, we prefer to give our own direct proof for the sake of completeness. Also, in those papers, authors provided a lower bound of clique-width, and its direct application to rank-width does not provide a linear lower bound.

Lemma 15 ($\star$). *For every positive integer n, a twisted chain graph of order $12n$ has rank-width at least n.*

Proof (sketch). Let $m := 12n$ and let G be a twisted chain graph of order m. Adopt the notation from the Definition 14 for G, and assume for the sake of contradiction that the rank-width of G is less than n. By a well-known fact about graphs of bounded rank-width, there exists a vertex bipartition (S, T) of G such that $\mathrm{cutrk}_G(S) < n$ and at least one third of vertices of C belong to S, and at least one third belong to T.

Suppose now there are vertices $v_{a_1}, \ldots, v_{a_k} \in A \cap S$ and $z_{(b_1, c_1)}, \ldots, z_{(b_k, c_k)} \in C \cap T$ with the following property satisfied:

$$a_1 \leq (b_1 - 1)m + c_1 < a_2 \leq (b_2 - 1)m + c_2 < \cdots < a_k \leq (b_k - 1)m + c_k.$$

Then it can be easily seen that the submatrix of $A_G[S, T]$ induced by rows corresponding to vertices v_{a_i} and columns corresponding to vertices $z_{(b_i, c_i)}$ has ones in the upper triangle and on the diagonal, and zeroes in the lower triangle. The rank of this submatrix is k, so finding such a structure, called *ordered (S, T)-matching*, for $k = n$ would contradict the assumption that $\mathrm{cutrk}_G(S) < n$. Similarly if all vertices of v_{a_i} were contained in T instead of S, and all vertices $z_{(b_i, c_i)}$ were contained in S instead of T. Also, a similar notion can be defined for B and C, but observe that there the vertices $z_{(b_i, c_i)}$ need to be ordered lexicographically with the second coordinate being the leading one, instead of the first. This difference is the key to the proof.

Consider now all elements $(b, c) \in \{1, \ldots, m\} \times \{1, \ldots, m\}$, ordered lexicographically with the first coordinate leading. For each such (b, c), record whether $z_{(b,c)}$ belongs to S or to T, and examine the obtained sequence of length m^2, consisting of symbols S and T. If this sequence had *alternation* at least $4n$, that is, we could see at least $4n$ times a T after an S, then it is not hard to convince oneself that there would be an ordered (S, T)-matching between A and C of order n, a contradiction. The same analysis can be performed between B and C, but now we order pairs from $\{1, \ldots, m\} \times \{1, \ldots, m\}$ lexicographically with the second coordinate leading. It can be now easily seen that since at least a third of vertices of C belong to S and at least a third belong to T, one of these sequences has alternation at least $\frac{m}{3} = 4n$, which gives the desired contradiction. □

We now observe that if a graph class contains arbitrarily large twisted chain graphs, then it does not admit low rank-width colorings.

Theorem 16. *Let $\mathcal{C}$ be a hereditary graph class, and suppose for each positive integer n some twisted chain graph of order n belongs to $\mathcal{C}$. Then $\mathcal{C}$ does not admit low rank-width colorings.*

Proof. We show that for every pair of integers $m \geq 3$ and $n \geq 1$, there is an graph $G \in \mathcal{C}$ such that for every coloring of G with m colors, there is an induced subgraph H that receives at most 3 colors and has rank-width at least n. This implies that $\mathcal{C}$ does not admit low rank-width colorings. We will need the following simple Ramsey-type argument.

Claim 1 (⋆). For all positive integers k, d, there exists an integer $M = M(k, d)$ such that for all sets X, Y with $|X| = |Y| = M$ and all functions $f \colon X \times Y \to \{1, \ldots, d\}$, there exist subsets $X' \subseteq X$ and $Y' \subseteq Y$ with $|X'| = |Y'| = k$ such that f sends all elements of $X' \times Y'$ to the same value.

Claim 1 follows, e.g., from [35, Theorem 11.5], but in the appendix we give a simple proof for the sake of completeness.

Let $M_1 := M(12n, m)$, $M_2 := M(M_1, m)$, and $M_3 := M(M_2, m)$. Let $G \in \mathcal{C}$ be a twisted chain graph of order M_3; adopt the notation from Definition 14 for G. Suppose G is colored by m colors by a coloring c. By Claim 1, there exist $X_1, Y_1 \subseteq \{1, \ldots, M_3\}$ with $|X_1| = |Y_1| = M_2$ such that $\{z_{(x,y)} \colon (x, y) \in X_1 \times Y_1\}$ is monochromatic under c.

Now, for an index $k \in \{1, \ldots, M_3^2\}$, let $(i_1(k), j_1(k)) \in \{1, \ldots, m\} \times \{1, \ldots, m\}$ be the unique pair such that $k = (i_1(k) - 1)M_3 + j_1(k)$, and let $(i_2(k), j_2(k)) \in \{1, \ldots, m\} \times \{1, \ldots, m\}$ be the unique pair such that $k = (j_2(k) - 1)M_3 + i_2(k)$. By reindexing vertices A and C using pairs $(i_1(k), j_1(k))$ and $(i_2(k), j_2(k))$, we may view coloring c on A and C as a coloring on $\{1, \ldots, M_3\} \times \{1, \ldots, M_3\}$. By applying Claim 1 to the vertices from A indexed by $X_1 \times Y_1$, we obtain subsets $X_2 \subseteq X_1$ and $Y_2 \subseteq Y_1$ such that $|X_2| = |Y_2| = M_1$ and the set $\{v_{(x-1)M_3+y} \colon x \in X_2, y \in Y_2\}$ is monochromatic. Finally, by applying Claim 1 to the vertices from B indexed by $X_2 \times Y_2$, we obtain subsets $X_3 \subseteq X_2$ and $Y_3 \subseteq Y_2$ such that $|X_3| = |Y_3| = 12n$ and the set $\{w_{(y-1)M_3+x} \colon (x, y) \in X_3 \times Y_3\}$ is monochromatic. Now observe that the subgraph $G[\{v_{(x-1)M_3+y}, w_{(y-1)M_3+x}, z_{(x,y)} \colon (x, y) \in X_3 \times Y_3\}]$ receives at most 3 colors, and is a twisted chain graph of order $12n$. By Lemma 15 it has rank-width at least n, so this proves the claim.   □

We now observe (⋆) that a twisted chain graph of order n is an interval graph, provided each of A, B, and C is a clique, and there are no edges between A and B. Similarly, for each n there is a twisted chain graph of order n that is a permutation graph (⋆). See Figs. 1 and 2 for examples of intersection models. By Theorem 16, we obtain the following.

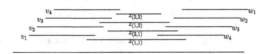

Fig. 1. An interval intersection model of a twisted chain graph of order 2.

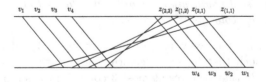

Fig. 2. A permutation intersection model of a twisted chain graph of order 2.

Theorem 17. *The classes of interval graphs and permutation graphs do not admit low rank-width colorings.*

6 Erdős-Hajnal property and χ-boundedness

A graph class $\mathcal{C}$ has the *Erdős-Hajnal property* if there is $\varepsilon > 0$, depending only on $\mathcal{C}$, such that every n-vertex graph in $\mathcal{C}$ has either an independent set or a clique of size n^ε. The conjecture of Erdős and Hajnal [15] states that for every fixed graph H, the class of graphs not having H as an induced subgraph has the Erdős-Hajnal property; cf. [4]. We prove that every class admitting low rank-width colorings has the Erdős-Hajnal property. For this we use the fact that graphs of bounded rank-width have the property, shown by Oum and Seymour [31].

Proposition 18 ($\star$). *Let $\mathcal{C}$ be a class of graphs admitting low rank-width colorings. Then $\mathcal{C}$ has the Erdős-Hajnal property.*

A class $\mathcal{C}$ of graphs is χ-*bounded* if there exists a function $f\colon \mathbb{N} \to \mathbb{N}$ such that for every $G \in \mathcal{C}$ and an induced subgraph H of G, we have $\chi(H) \leq f(\omega(H))$, where $\chi(H)$ is the chromatic number of H and $\omega(H)$ is the size of a maximum clique in H. It was proved by Dvořák and Král' [14] that for every p, the class of graphs of rank-width at most p is χ-bounded. We observe that this fact directly generalizes to classes admitting low rank-width colorings.

Proposition 19 ($\star$). *Let $\mathcal{C}$ be a class of graphs admitting low rank-width colorings. Then $\mathcal{C}$ is χ-bounded.*

7 Conclusions

We introduced the concept of low rank-width colorings, and showed that such colorings exist on rth powers of graphs from any bounded expansion class, for any fixed r, as well as on unit interval and bipartite permutation graphs. These classes are non-sparse and have unbounded rank-width. On the negative side, the classes of interval and permutation graphs do not admit low rank-width colorings.

The obvious open problem is to characterise hereditary graph classes which admit low rank-width colorings in the spirit of the characterisation theorem for graph classes admitting low tree-depth colorings. We believe that Theorem 16 may provide some insight into this question, as it shows that containing arbitrarily large twisted chain graphs is an obstacle for admitting low rank-width colorings. Is it true that every hereditary graph class that does not admit low rank-width colorings has to contain arbitrarily large twisted chain graphs?

In this work we did not investigate the question of computing low rank-width colorings, and this question is of course crucial for any algorithmic applications. Our proof for the powers of sparse graphs can be turned into a polynomial-time algorithm that, given a graph G from a graph class of bounded expansion

$\mathcal{C}$, first computes a low tree-depth coloring, and then turns it into a low rank-width coloring of G^r, for a fixed constant r. However, we do not know how to efficiently compute a low rank-width coloring given the graph G^r alone, without the knowledge of G. The even more general problem of efficiently constructing an approximate low rank-width coloring of any given graph remains wide open.

Finally, we remark that our proof for the existence of low rank-width colorings on powers of graphs from a class of bounded expansion actually yields a slightly stronger result. Precisely, Ganian et al. [16] introduced a parameter *shrub-depth* (or *SC-depth*), which is a depth analogue of rank-width, in the same way as tree-depth is a depth analogue of tree-width. It can be shown that for constant r, the rth power of a graph of constant tree-depth belongs to a class of constant shrub-depth, and hence our colorings for powers of graphs from a class of bounded expansion are actually low shrub-depth colorings. We omit the details.

Acknowledgment. The authors would like to thank Konrad Dabrowski for pointing out the known constructions similar to twisted chain graphs.

References

1. Bodlaender, H.L., Deogun, J.S., Jansen, K., Kloks, T., Kratsch, D., Müller, H., Tuza, Z.: Rankings of graphs. In: Mayr, E.W., Schmidt, G., Tinhofer, G. (eds.) WG 1994. LNCS, vol. 903, pp. 292–304. Springer, Heidelberg (1995). doi:10.1007/3-540-59071-4_56
2. Brandstädt, A., Engelfriet, J., Le, H.O., Lozin, V.V.: Clique-width for 4-vertex forbidden subgraphs. Theory Comput. Syst. **39**(4), 561–590 (2006)
3. Bui-Xuan, B.M., Telle, J.A., Vatshelle, M.: Boolean-width of graphs. Theoret. Comput. Sci. **412**(39), 5187–5204 (2011)
4. Chudnovsky, M.: The Erdös-Hajnal conjecture—a survey. J. Graph Theory **75**(2), 178–190 (2014)
5. Courcelle, B.: Graph rewriting: an algebraic and logic approach. In: Handbook of Theoretical Computer Science, volume B, pp. 193–242 (1990)
6. Courcelle, B.: The monadic second-order logic of graphs. I. Recognizable sets of finite graphs. Inf. Comput. **85**(1), 12–75 (1990)
7. Courcelle, B., Engelfriet, J., Rozenberg, G.: Handle-rewriting hypergraph grammars. J. Comput. Syst. Sci. **46**(2), 218–270 (1993)
8. Courcelle, B., Makowsky, J.A., Rotics, U.: Linear time solvable optimization problems on graphs of bounded clique-width. Theory Comput. Syst. **33**(2), 125–150 (2000)
9. Dabrowski, K., Paulusma, D.: Clique-width of graph classes defined by two forbidden induced subgraphs. Comput. J. **59**(5), 650–666 (2016)
10. Deogun, J.S., Kloks, T., Kratsch, D., Müller, H.: On vertex ranking for permutation and other graphs. In: Enjalbert, P., Mayr, E.W., Wagner, K.W. (eds.) STACS 1994. LNCS, vol. 775, pp. 747–758. Springer, Heidelberg (1994). doi:10.1007/3-540-57785-8_187
11. DeVos, M., Ding, G., Oporowski, B., Sanders, D.P., Reed, B., Seymour, P., Vertigan, D.: Excluding any graph as a minor allows a low tree-width 2-coloring. J. Comb. Theory, Series B **91**(1), 25–41 (2004)
12. Diestel, R.: Graph Theory. Graduate Texts in Mathematics, vol. 173, 4th edn. Springer, Heidelberg (2012)

13. Dvořák, Z., Král', D., Thomas, R.: Testing first-order properties for subclasses of sparse graphs. J. ACM (JACM) **60**(5), 36 (2013)
14. Dvořák, Z., Král', D.: Classes of graphs with small rank decompositions are χ-bounded. Eur. J. Comb. **33**(4), 679–683 (2012)
15. Erdős, P., Hajnal, A.: Ramsey-type theorems. Discret. Appl. Math. **25**(1–2), 37–52 (1989)
16. Ganian, R., Hliněný, P., Nešetřil, J., Obdržálek, J., Ossona de Mendez, P., Ramadurai, R.: When trees grow low: shrubs and fast MSO_1. In: Rovan, B., Sassone, V., Widmayer, P. (eds.) MFCS 2012. LNCS, vol. 7464, pp. 419–430. Springer, Heidelberg (2012). doi:10.1007/978-3-642-32589-2_38
17. Grohe, M., Kreutzer, S.: Methods for algorithmic meta theorems. In: Model Theoretic Methods in Finite Combinatorics, vol. 558, pp. 181–206 (2011)
18. Grohe, M., Kreutzer, S., Siebertz, S.: Deciding first-order properties of nowhere dense graphs. In: STOC 2014, pp. 89–98. ACM (2014)
19. Gurski, F., Wanke, E.: The NLC-width and clique-width for powers of graphs of bounded tree-width. Discret. Appl. Math. **157**(4), 583–595 (2009)
20. Gyárfás, A.: Problems from the world surrounding perfect graphs. Zastosowania Matematyki (Appl. Math.) **19**, 413–441 (1987)
21. Hliněný, P., Oum, S., Seese, D., Gottlob, G.: Width parameters beyond tree-width and their applications. Comput. J. **51**(3), 326–362 (2008)
22. Kierstead, H.A., Yang, D.: Orderings on graphs and game coloring number. Order **20**(3), 255–264 (2003)
23. Lozin, V.V.: Minimal classes of graphs of unbounded clique-width. Ann. Comb. **15**(4), 707–722 (2011)
24. Lozin, V.V., Rudolf, G.: Minimal universal bipartite graphs. Ars Comb. **84**, 345–356 (2007)
25. Nešetřil, J., Ossona de Mendez, P.: Tree-depth, subgraph coloring and homomorphism bounds. Eur. J. Comb. **27**(6), 1022–1041 (2006)
26. Nešetřil, J., Ossona de Mendez, P.: Grad and classes with bounded expansion I. Decompositions. Eur. J. Comb. **29**(3), 760–776 (2008)
27. Nešetřil, J., Ossona de Mendez, P.: On nowhere dense graphs. Eur. J. Comb. **32**(4), 600–617 (2011)
28. Nešetřil, J., Shelah, S.: On the order of countable graphs. Eur. J. Comb. **24**(6), 649–663 (2003)
29. Oum, S.: Rank-width is less than or equal to branch-width. J. Graph Theory **57**(3), 239–244 (2008)
30. Oum, S.: Rank-width: agorithmic and structural results. Discret. Appl. Math. **231**, 15–24 (2017)
31. Oum, S., Seymour, P.: Personal communication
32. Oum, S., Seymour, P.: Approximating clique-width and branch-width. J. Comb. Theory, Series B **96**(4), 514–528 (2006)
33. Robertson, N., Seymour, P.D.: Graph minors. XVI. Excluding a non-planar graph. J. Comb. Theory, Series B **89**(1), 43–76 (2003)
34. Schäffer, A.A.: Optimal node ranking of trees in linear time. Inf. Process. Lett. **33**(2), 91–96 (1989)
35. Trotter, W.T.: Combinatorics and Partially Ordered Sets. Johns Hopkins Series in the Mathematical Sciences. Johns Hopkins University Press, Baltimore (1992)
36. Wanke, E.: k-NLC graphs and polynomial algorithms. Discret. Appl. Math. **54**(2–3), 251–266 (1994)
37. Zhu, X.: Colouring graphs with bounded generalized colouring number. Discret. Math. **309**(18), 5562–5568 (2009)

On Strongly Chordal Graphs
That Are Not Leaf Powers

Manuel Lafond[(✉)]

Department of Mathematics and Statistics, University of Ottawa, Ottawa, Canada
mlafond2@uOttawa.ca

Abstract. A common task in phylogenetics is to find an evolutionary tree representing proximity relationships between species. This motivates the notion of leaf powers: a graph $G = (V, E)$ is a leaf power if there exist a tree T on leafset V and a threshold k such that $uv \in E$ if and only if the distance between u and v in T is at most k. Characterizing leaf powers is a challenging open problem, along with determining the complexity of their recognition. Leaf powers are known to be strongly chordal, but few strongly chordal graphs are known to *not* be leaf powers, as such graphs are difficult to construct. Recently, Nevries and Rosenke asked if leaf powers could be characterized by strong chordality and a finite set of forbidden induced subgraphs.

In this paper, we provide a negative answer to this question, by exhibiting an infinite family $\mathcal{G}$ of (minimal) strongly chordal graphs that are not leaf powers. During the process, we establish a connection between leaf powers, alternating cycles and quartet compatibility. We also show that deciding if a chordal graph is $\mathcal{G}$-free is NP-complete.

1 Introduction

In phylogenetics, a classical method for inferring an evolutionary tree of species is to construct the tree from a distance matrix, which depicts how close or far each species are to one and another. Roughly speaking, similar species should be closer to each other in the tree than more distant species. In some contexts, the actual distances are ignored (e.g. when they cannot be trusted due to errors), and only the notions of "close" and "distant" are preserved. This corresponds to a graph in which the vertices are the species, and two vertices share an edge if and only if they are "close". This motivates the definition of *leaf powers*, which was proposed by Nishimura et al. in [16]: a graph $G = (V, E)$ is a leaf power if there exist a tree T on leafset $V(G)$ and a threshold k such that $uv \in E$ if and only if the distance between u and v in T is at most k. Hence the tree T, which we call a *leaf root*, is a potential evolutionary history for G, as it satisfies the notions of "close" and "distant" depicted by G. It is also worth noting that this type of similarity graph is also encountered in the context of gene *orthology* inference, which is a special type of relationship between genes (see e.g. [12,21]). A similarity graph G is used as a basis for the inference procedure, and being

© Springer International Publishing AG 2017
H.L. Bodlaender and G.J. Woeginger (Eds.): WG 2017, LNCS 10520, pp. 386–398, 2017.
https://doi.org/10.1007/978-3-319-68705-6_29

able to verify that G is a leaf power would provide a basic test as to whether G correctly depicts similarity, as such graphs are known to contain errors [11].

A considerable amount of work has been done on the topic of leaf powers (see [6] for a survey), but two important challenges remain open: to determine the computational complexity of recognizing leaf powers, and to characterize the class of leaf powers from a graph theoretic point of view. Despite some interesting results on graph classes that are leaf powers [4,5,10], both problems are made especially difficult due to our limited knowledge on graphs that are *not* leaf powers. Such knowledge is obviously fundamental for the characterization of leaf powers, but also important from the algorithmic perspective: if recognizing leaf powers is in P, a polynomial time algorithm is likely to make usage of structures to avoid, and if it is NP-hard, a hardness reduction will require knowledge of many non-leaf powers in order to generate "no" instances.

It has been known for many years that leaf powers must be strongly chordal (i.e. chordal and sun-free). Brandstädt et al. exhibited one strongly chordal non-leaf power by establishing an equivalence between leaf powers and NeST graphs [3,5]. Recently [15], Nevries and Rosenke found seven such graphs, all identified by the notion of bad 2-cycles in *clique arrangements*, which are of special use in strongly chordal graphs [14]. These graphs have at most 12 vertices, and in [13], the authors conjecture that they are the only strongly chordal non-leaf powers. This was also posed as an open problem in [6]. A positive answer to this question would imply a polynomial time algorithm for recognizing leaf powers, as strong chordality can be checked in $O(\min\{m \log n, n^2\})$ time [17,19].

In this paper, we unfortunately give a negative answer to this question. We exhibit an infinite family $\mathcal{G}$ of strongly chordal graphs that are not leaf powers, and each graph in this family is minimal for this property (i.e. removing any vertex makes the graph a leaf power). This is done by first establishing a new necessary condition for a graph G to be a leaf power, based on its *alternating cycles* (which are cyclic orderings of vertices that alternate between an edge and a non-edge). Namely, there must be a tree T that can satisfy the edges/non-edges of each alternating cycle C of G after (possibly) subdividing some of its edges (see Sect. 3 for a precise definition). This condition has two interesting properties. First, every graph currently known to not be a leaf power fails to satisfy this condition. And more importantly, this provides new tools for the construction of novel classes of non-leaf powers. In particular, alternating cycles on four vertices enforce the leaf root to contain a specific *quartet*, a binary tree on four leaves. This connection lets us borrow from the theory of *quartet compatibility*, which is well-studied in phylogenetics (see e.g. [1,2,18,20]). More precisely, we use results from [18] to create a family $\mathcal{G}$ of strongly chordal graphs whose 4-alternating cycles enforce a minimal set of incompatible quartets. We then proceed to show that deciding if a chordal graph G contains a member of $\mathcal{G}$ as an induced subgraph is NP-complete. Thus, $\mathcal{G}$-freeness is the first known property of non-leaf powers that we currently ignore how to check in polynomial time. This result also indicates that if the problem admits a polynomial time algorithm, it will have to make use of strong chordality (or some other structural property), since chordality alone is not enough to identify forbidden structures quickly.

The paper is organized as follows: in Sect. 2, we provide some basic notions and facts. In Sect. 3, we establish the connection between leaf powers, alternating cycles and quartets, along with its implications. In Sect. 4, we exhibit the family $\mathcal{G}$ of strongly chordal graphs that are not leaf powers. We then show in Sect. 5 that deciding if a chordal graph is $\mathcal{G}$-free is NP-complete. Due to space constraints, some proofs are omitted from this version. A full version is available at https://arxiv.org/abs/1703.08018.

2   Preliminary Notions

All graphs in this paper are simple and finite. For $k \in \mathbb{N}^+$, we use the notation $[k] = \{1, \ldots, k\}$. We denote the set of vertices of a graph G by $V(G)$, its set of edges by $E(G)$, and its set of non-edges by $\overline{E}(G)$. By $G[X]$ we mean the subgraph induced by $X \subseteq V(G)$. The set of neighbors of $v \in V(G)$ is $N(v)$. The P_4 is the path of length 3 and the $2K_2$ is the graph consisting of two vertex-disjoint edges. A k-sun, denoted S_k, is the graph obtained by starting from a clique of size $k \geq 3$ with vertices $x_1, \ldots, x_k$, then adding vertices $a_1, \ldots, a_k$ such that $N(a_i) = \{x_i, x_{i+1}\}$ for each $i \in [k-1]$ and $N(a_k) = \{x_k, x_1\}$. A graph is a sun if it is a k-sun for some k, and G is sun-free if no induced subgraph of G is a sun.

A graph G is chordal if it has no induced cycle with four vertices or more, and G is strongly chordal if it is chordal and sun-free. A vertex v is simplicial if $N(v)$ is a clique, and v is simple if it is simplicial and, in addition, for every $x, y \in N(v)$, one of $N(x) \subseteq N(y) \setminus \{x\}$ or $N(y) \subseteq N(x) \setminus \{y\}$ holds. An ordering $(x_1, \ldots, x_n)$ of $V(G)$ is a perfect elimination ordering if, for each $i \in [n]$, x_i is simplicial in $G[\{x_i, \ldots, x_n\}]$. The ordering is simple if, for each $i \in [n]$, x_i is simple in $G[\{x_i, \ldots, x_n\}]$. It is well-known that a graph is chordal if and only if it admits a perfect elimination ordering [9], and a graph is strongly chordal if and only if it admits a simple elimination ordering [8].

Denote by $\mathcal{L}(T)$ the set of leaves of a tree T. We say a graph $G = (V, E)$ is a k-leaf power if there exists a tree T with $\mathcal{L}(T) = V$ such that for any two distinct vertices $u, v \in V$, $uv \in E$ if and only if the distance between u and v in T is at most k. Such a tree T is called a k-leaf root of G. A graph G is a leaf power if there exists a positive integer k such that G is a k-leaf power.

A quartet is an unrooted binary tree on four leaves (an unrooted tree T is binary if all its internal vertices have degree exactly 3). For a set of four elements $X = \{a, b, c, d\}$, there exist 3 possible quartets on leafset X which we denote $ab|cd, ac|bd$ and $ad|bc$, depending on how the internal edge separates the leaves. We say that T contains a quartet $ab|cd$ if $\{a, b, c, d\} \subseteq \mathcal{L}(T)$ and the path between a and b does not intersect the path between c and d. We denote $\mathcal{Q}(T) = \{ab|cd : T \text{ contains } ab|cd\}$. We say that a set of quartets Q is compatible if there exists a tree T such that $Q \subseteq \mathcal{Q}(T)$, and otherwise Q is incompatible.

For a tree T and $x, y \in V(T)$, $p_T(x, y)$ denotes the set of edges on the unique path between x and y. We may write $p(x, y)$ when T is clear from the context.

It will be convenient to extend the definition of leaf powers to weighted edges. A *weighted* tree (T, f) is a tree accompanied by a function $f : E(T) \to \mathbb{N}^+$ weighting its edges. If $F \subseteq E(T)$, we denote $f(F) = \sum_{e \in F} f(e)$. The distance $d_{T,f}(x, y)$ between two vertices of T is given by $f(p(x, y))$, i.e. the sum of the weights of the edges lying on the $x - y$ path in T. We may write $d_f(x, y)$ for short. We say that (T, f) is a *leaf root* of a graph G if there exists an integer k such that $xy \in E(G)$ iff $d_f(x, y) \leq k$. We will call k the *threshold* corresponding to (T, f). Note that in the usual setting, the edges of leaf roots are not weighted, though arbitrarily many degree 2 vertices are allowed. It is easy to see that this distinction is merely conceptual, since an edge e with weight $f(e)$ can be made unweighted by subdividing it $f(e) - 1$ times.

A tree T is *unweighted* if it is not equipped with a weighting function. We say an unweighted tree is an *unweighted leaf root* of a graph G if there is a weighting f of $E(T)$ such that (T, f) is a leaf root of G.

A first observation that will be of convenience later on is that, even though the usual definition of leaf powers does not allow edges of weight 0, they do not alter the class of leaf powers.

Lemma 1. *Let G be a graph, and let (T, f) be a weighted tree in which $\mathcal{L}(T) = V(G)$ and $f(e) \geq 0$ for each $e \in E(T)$. If there exists an integer k such that $uv \in E(G) \Leftrightarrow d_f(u, v) \leq k$, then T is an unweighted leaf root of G.*

Proof. If no edge has weight 0, there is nothing to do. Otherwise, we devise a weighting function f' for T. Let $d = \max_{x,y \in V(T)} |p(x, y)|$. Set $f'(e) = (d+1) \cdot f(e)$ for each $e \in E(T)$ having $f(e) > 0$, and $f'(e) = 1$ for each $e \in E(T)$ having $f(e) = 0$. If $d_f(x, y) \leq k$, then $d_{f'}(x, y) \leq (d + 1)k + d$, and if $d_f(x, y) \geq k + 1$, then $d_{f'} \geq (d + 1)k + (d + 1)$. The threshold $(d + 1)k + d$ shows that T is an unweighted leaf root of G. $\qquad\square$

A tree T' is a *refinement* of a tree T if T can be obtained from T' by contraction of edges. A consequence of the above follows.

Lemma 2. *Let T be an unweighted leaf root of a leaf power G. Then any refinement T' of T is also an unweighted leaf root of G.*

Proof. We may take a weighting f such that (T, f) is a leaf root of G, refine it in order to obtain T', weight the newly created edges by 0 and apply Lemma 1. $\qquad\square$

The following was implicitly proved in [4] (see full version for a proof).

Lemma 3. *Suppose that G has a vertex v of degree 1. Then G is a leaf power if and only if $G - v$ is a leaf power.*

3   Alternating Cycles and Quartets in Leaf Powers

In this section, we restrict our attention to alternating cycles in leaf powers, which let us establish a new necessary condition on the topology of unweighted leaf roots. This will serve as a basis for the construction of our family of forbidden induced subgraphs. Although we will not use the full generality of the statements proved here, we believe they may be of interest for future studies.

Let (A, B) be a pair such that $A \subseteq E(G)$ and $B \subseteq \overline{E}(G)$. We say a weighted tree (T, f) *satisfies* (A, B) if there exists a threshold k such that for each edge $\{x, y\} \in A$, $d_f(x, y) \leq k$ and for each non-edge $\{x, y\} \in B$, $d_f(x, y) > k$. Thus (T, f) is a leaf root of G iff it satisfies $(E(G), \overline{E}(G))$. For an unweighted tree T, we say that T *can satisfy* (A, B) if there exists a weighting f of $E(T)$ such that (T, f) satisfies (A, B).

A sequence of $2c$ distinct vertices $C = (x_0, y_0, x_1, y_1, \ldots, x_{c-1}, y_{c-1})$ is an *alternating cycle* of a graph G if for each $i \in \{0, \ldots, c - 1\}$, $x_i y_i \in E(G)$ and $y_i x_{i+1} \notin E(G)$ (indices are modulo c in all notions related to alternating cycles). In other words, the vertices of C alternate between an edge and a non-edge. We write $V(C) = \{x_0, y_0, \ldots, x_{c-1}, y_{c-1}\}$, $E(C) = \{x_i y_i : 0 \leq i \leq c - 1\}$ and $\overline{E}(C) = \{y_i x_{i+1} : 0 \leq i \leq c - 1\}$. A weighted tree *satisfies* C if it satisfies $(E(C), \overline{E}(C))$, and an unweighted tree *can satisfy* C if it can satisfy $(E(C), \overline{E}(C))$. The next necessary condition for leaf powers is quite an obvious one, but will be of importance throughout the paper.

Proposition 1. *If G is a leaf power, then there exists an unweighted tree T that can satisfy every alternating cycle of G.*

As it turns out, every graph that is currently known to not be a leaf power fails to satisfy the above condition (actually, we may even restrict our attention to cycles of length 4 and 6, as we will see). This suggests that it is also sufficient, and we conjecture that if there exists a tree that can satisfy every alternating cycle of G, then G is a leaf power. As a basic sanity check towards this statement, we show that in the absence of alternating cycles, a graph is indeed a leaf power.

Proposition 2. *If a graph G has no alternating cycle, then G is a leaf power.*

Proof. Since a chordless cycle of length at least 4 contains an alternating cycle, G must be chordal. By the same argument, G cannot contain an induced gem (the gem is obtained by taking a P_4, and adding a vertex adjacent to each vertex of the P_4). In [4], it is shown that chordal gem-free graphs are leaf powers. □

We will go a bit more in depth with alternating cycles, by first providing a characterization of the unweighted trees that can satisfy an alternating cycle C. Let T be an unweighted tree with $V(C) \subseteq V(T)$. For each $i \in \{0, \ldots, c-1\}$, we say the path in T between x_i and y_i is *positive*, and the path between y_i and x_{i+1} is *negative* (with respect to C). The proof of the following statement can be found in the full version.

Lemma 4. *An unweighted tree T can satisfy an alternating cycle $C = (x_0, y_0, \ldots, x_{c-1}, y_{c-1})$ if and only if there exists an edge e of T that belongs to strictly more negative paths than positive paths w.r.t. C.*

Lemma 4 lets us relate quartets and 4-alternating cycles easily. If $C = (x_0, y_0, x_1, y_1)$, the edges of the quartets $x_0 x_1 | y_0 y_1$ and $x_0 y_1 | y_0 x_1$ do not meet the condition of Lemma 4, and therefore no unweighted leaf root can contain these quartets. This was already noticed in [15], although this was presented in another form and not stated in the language of quartets.

Corollary 1. *Let* $C = (x_0, y_0, x_1, y_1)$ *be a 4-alternating cycle of a graph* G. *Then a tree* T *can display* C *if and only if* T *contains the* $x_0 y_0 | x_1 y_1$ *quartet.*

We will denote by $RQ'(G)$ the set of *required quartets* of G, that is $RQ'(G) = \{x_0 y_0 | x_1 y_1 : (x_0, y_0, x_1, y_1)$ is an alternating cycle of $G\}$. The only graphs on 4 vertices that contain an alternating cycle are the P_4, the $2K_2$ and the C_4. However, the C_4 contains two distinct alternating cycles: if four vertices $abcd$ in cyclic order form a C_4, then (a, b, d, c) and (d, a, c, b) are two alternating cycles. The first implies the $ab|cd$ quartet, whereas the second implies the $ad|cb$ quartet. This shows that no leaf power can contain a C_4. Thus $RQ'(G)$ can be constructed by enumerating the $O(n^4)$ induced P_4 and $2K_2$ of G. It is worth mentioning that deciding if a given set of quartets is compatible is NP-complete [20]. However, $RQ'(G)$ is not *any* set of quartets since it is generated from P_4's and $2K_2$'s of a strongly chordal graph, and the hardness does not immediately transfer.

Now, denote by $RQ(G)$ the set of quartets that any unweighted leaf root of G must contain, if it exists. Then $RQ'(G) \subseteq RQ(G)$, and equality does not hold in general. Below we show how to find some of the quartets from $RQ(G) \setminus RQ'(G)$ (Lemma 5, which is a generalization of [15, Lemma 2]).

Lemma 5. *Let* $P_1 = x_0 x_1 \ldots x_p$ *and* $P_2 = y_0 y_1 \ldots y_q$ *be disjoint paths of* G *(with possible chords) such that for any* $0 \leq i < p$ *and* $0 \leq j < q$, $\{x_i, x_{i+1}, y_j, y_{j+1}\}$ *are the vertices of an alternating cycle. Then* $x_0 x_p | y_0 y_q \in RQ(G)$.

Proof. First note that in general, if a tree T contains the quartets $ab|c_i c_{i+1}$ for $0 \leq i < l$, then T must contain $ab|c_0 c_l$ (this is easy to see by trying to construct such a T: start with the $ab|c_0 c_1$ quartet, and insert $c_2, \ldots, c_l$ in order - at each insertion, c_i cannot have its neighbor on the $a - b$ path). For any $0 \leq i < p$, we may apply this observation on $\{a, b\} = \{x_i, x_{i+1}\}$. This yields $x_i x_{i+1} | y_0 y_q \in RQ(G)$, since $x_i x_{i+1} | y_j y_{j+1} \in RQ'(G)$ for every j. Since this is true for every $0 \leq i < p$, we can apply this observation again, this time on $\{a, b\} = \{y_0, y_q\}$ (and the c_i's being the x_i's) and deduce that $y_0 y_q | x_0 x_p \in RQ(G)$.   □

In particular, suppose that G has two disjoint pairs of vertices $\{x_0, x_1\}$ and $\{y_0, y_1\}$ such that x_0 and x_1 (resp. y_0 and y_1) share a common neighbor z (resp. z'), and $z \notin N(y_0) \cup N(y_1)$ (resp. $z' \notin N(x_0) \cup N(x_1)$). Then $x_0 x_1 | y_0 y_1 \in RQ(G)$.

In the rest of this section, we briefly explain how all known non-leaf powers fail to satisfy Proposition 2. We have already argued that a leaf power cannot contain a C_4. As for a cycle C_n with $n > 4$ and vertices $x_0, \ldots, x_{n-1}$ in cyclic order, observe that $x_i x_{i+1} | x_{i+2} x_{i+3} \in RQ(C_n)$ since they form a P_4, for each $i \in \{0, \ldots, n-1\}$ (indices are modulo n). In this case it is not difficult to show that $RQ(C_n)$ is incompatible, providing an alternative explanation as to why leaf powers must be chordal.

A similar argument can be used for S_n, the n-sun, when $n \geq 4$. If we let $x_0, \ldots, x_{n-1}$ be the clique vertices of S_n arranged in cyclic order, again $x_i x_{i+1} | x_{i+2} x_{i+3} \in RQ(S_n)$ for $i \in \{0, \ldots, n-1\}$, here because of Lemma 5 and the degree 2 vertices of S_n. Only S_3, the 3-sun, requires an ad-hoc argument, and it is currently the only known non-leaf power for which the set of required quartets are compatible. Figure 1 illustrates how alternating cycles show that S_3 is not a leaf power. There are only two trees that contain $RQ'(S_3) = \{ay|cz, by|cx, bz|ax\}$, and for both, there is an alternating cycle such that each edge is on the same number of positive and negative paths. We do not know if there are other examples for which quartets are not enough to discard the graph as a leaf power. Moreover, an open question is whether for each even integer n, there exists a non-leaf power and a tree that can satisfy every alternating cycle of length $< n$, but not every alternating cycle of length n.

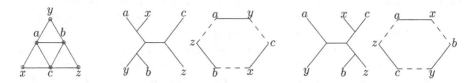

Fig. 1. The 3-sun S_3, and the two trees that contain $RQ'(S_3) = \{ay|cz, by|cx, bz|ax\}$, with each tree accompanied by the alternating cycle of S_3 that it cannot satisfy.

As for the seven strongly chordal graphs presented in [15], they were shown to be non-leaf powers by arguing that $RQ(G)$ was not compatible (although the proof did not use the language of quartet compatibility).

4   Strongly Chordal Graphs That Are Not Leaf Powers

We will use a known set of (minimally) incompatible quartets as a basis for constructing our graph family.

Theorem 1 [18]. *For every integers $r, q \geq 3$, the quartets $Q = \{a_i a_{i+1} | b_j b_{j+1} : i \in [r-1], j \in [q-1]\} \cup \{a_1 b_1 | a_r b_q\}$ are incompatible. Moreover, any proper subset of Q is compatible.*

We now construct the family $\{G_{r,q} : r, q \geq 3\}$ of minimal strongly chordal graphs that are not leaf powers. The idea is to simply enforce that $RQ(G_{r,q})$ contains all the quartets of Q in Theorem 1. Figure 2 illustrate the graph $G_{3,4}$ and a general representation of $G_{r,q}$. For integers $r, q \geq 3$, $G_{r,q}$ is as follows: start with a clique of size $r + q$, partition its vertices into two disjoint sets $A = \{a_1, \ldots a_r\}$ and $B = \{b_1, \ldots, b_q\}$, and remove the edges $a_1 a_r, a_1 b_q, b_1 b_q$ and $b_1 a_r$. Then for each $i \in [r-1]$ insert a node x_i that is a neighbor of a_i and a_{i+1}, and for each $i \in [q-1]$, insert another node y_i that is a neighbor of b_i and b_{i+1}.

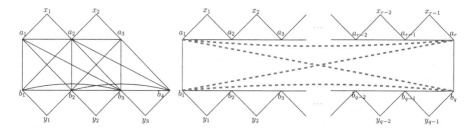

Fig. 2. The graph $G_{3,4}$ on the left, followed by its generalization $G_{r,q}$ on the right. In the latter, all edges between the a_i's and b_i's are present, except the non-edges depicted by red dashed lines. (Color figure online)

We note that in [15], the graph $G_{3,3}$ was one of the seven graphs shown to be a strongly chordal non-leaf power. Hence $G_{r,q}$ can be seen as a generalization of this example. It is possible that the other examples of [15] can also be generalized.

Theorem 2. *For any integers* $r, q \geq 3$, *the graph* $G_{r,q}$ *is strongly chordal, is not a leaf power and for any* $v \in V(G_{r,q})$, $G_{r,q} - v$ *is a leaf power.*

Proof. One can check that $G_{r,q}$ is strongly chordal by the simple elimination ordering: $x_1, x_2, \ldots, x_{r-1}, y_1, \ldots, y_{q-1}, a_1, b_1, a_r, b_q, a_2, \ldots, a_{r-1}, b_2, \ldots, b_{q-1}$.

To see that $G_{r,q}$ is not a leaf power, we note that the incompatible set of quartets of Theorem 1 is a subset of $RQ(G_{r,q})$: $a_i a_{i+1} | b_j b_{j+1} \in R(G_{r,q})$ by Lemma 5 and the paths $a_i x_i a_{i+1}$ and $b_j y_j b_{j+1}$, and $a_1 b_1 | a_r b_q \in RQ(G_{r,q})$ since they induce a $2K_2$.

We now show that for any $v \in V(G_{r,q})$, $G_{r,q} - v$ is a leaf power. First suppose that $v \in A \cup B$, say $v = a_i$ without loss of generality. Then in $G_{r,q} - a_i$, x_i (or take x_{i-1} if $i = r$) has degree one, and so by Lemma 3, $G_{r,q} - a_i$ is a leaf power if and only if $G_{r,q} - a_i - x_i$ is a leaf power. Therefore, it suffices to show that $G_{r,q} - x_i$ is a leaf power. We may thus assume that $v = x_i$ for some i (the $v = y_i$ case is the same by symmetry).

Figure 3 exhibits a leaf root (T, f) for $G_{r,q} - x_i$ (the weighting contains 0 edges, but this can be handled by Lemma 1). In the weighting f, the edges take values depending on variables p, p_1, p_2, p_3 which are defined as follows:

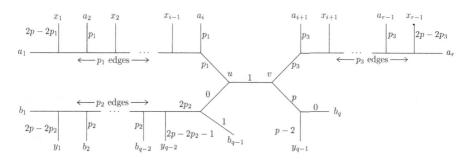

Fig. 3. A leaf root of $G_{r,q} - x_i$.

$$p := 2(2i-1)(2r-2i-1)(2q-3) \quad p_1 := p/(2i-1)$$
$$p_2 := p/(2q-3) \quad p_3 := p/(2r-2i-1)$$

and we set the threshold $k := 2p$. Each edge on the $a_1 - u$, $b_1 - u$ and $a_r - v$ path is weighted by p_1, p_2 and p_3 respectively, with the exception of the last two edges of the $b_1 - u$ path where one edge has weight 0 and the other $2p_2$. One can check that this ensures that $f(p(a_1, u)) = f(p(b_1, u)) = f(p(a_r, v)) = p$, $(p_1, p_2$ and p_3 are chosen so as to distribute a total weight of p across these paths, and p is such that these values are integers). Moreover, $p_1, p_2, p_3 > 2$. Observe that if $i = 1$, then the $a_1 - u$ path is a single edge and $p_1 = p$, and if $i = r - 1$, the $a_r - v$ path is a single edge and $p_3 = p$. It is not hard to verify that (T, f) satisfies the subgraph of $G - x_i$ induced by the a_j's and b_j's (since each pair of vertices has distance at most $2p$, except $a_1 a_r$, $a_1 b_q$, $b_1 a_r$ and $b_1 b_q$).

Now for the x_j's and y_j's. For each $j \in [r-1] \setminus \{i\}$, the edge e incident to x_j has $f(e) = 2p - 2p_1$ if $j < i$ and $f(e) = 2p - 2p_3$ if $j > i$. For $j \in [q-1]$, the edge e incident to y_j has $f(e) = 2p - 2p_2$ if $j \le q - 3$, $f(e) = 2p - 2p_2 - 1$ if $j = q - 2$ and $f(e) = p - 2$ if $j = q - 1$. Each x_j is easily seen to be satisfied, as the only vertices of T within distance $2p$ of x_j are a_j and a_{j+1}. This is equally easy to see for the y_j vertices, with the exception of y_{q-1}. In (T, f), y_{q-1} can reach b_q and b_{q-1} within distance $2p$ as required, but we must argue that it cannot reach a_i nor a_{i+1} (which is enough, since all the other leaves are farther from y_{q-1}. But this follows from that fact that $p_1, p_3 > 2$. This shows that (T, f) is a leaf root of $G_{r,q} - x_i$, and concludes the proof. □

Interestingly, the $G_{r,q}$ graphs might be subject to various alterations in order to obtain different families of strongly chordal non-leaf powers. One example of such an alteration of $G_{r,q}$ is to pick some $j \in \{2, \ldots, r-2\}$ and remove the edges $\{a_i b_q : 2 \le i \le j\}\}$. One can verify that the resulting graph is still strongly chordal, but requires the same set of incompatible quartets as $G_{r,q}$.

5   Hardness of Finding $G_{r,q}$ in Chordal Graphs

We show that deciding if a chordal graph contains an induced subgraph isomorphic to $G_{r,q}$ for some $r, q \ge 3$ is NP-complete. We reduce from the following:

The Restricted Chordless Cycle (RCC) problem:
Input: a bipartite graph $G = (U \cup V, E)$, and two vertices $s, t \in V(G)$ such that $s, t \in U$, both s and t are of degree 2 and they share no common neighbor.
Question: does there exist a chordless cycle in G containing both s and t?

The RCCproblem is shown to be NP-hard in [7, Theorem 2.2][1]. We first need some notation. If P is a path between vertices u and v, we call u and v its

---

[1] Strictly speaking, the problem asks if there exists a chordless cycle with both s and t of size at least k. However, in the graph constructed for the reduction, any chordless cycle containing s and t has size at least k if it exists - therefore the question of existence is hard. Also, s and t are not required to be in the same part of the bipartition, but again, this is allowable by subdividing an edge incident to s or t.

endpoints, and the other vertices are *internal*. Two paths P_1 and P_2 of a graph G are said *independent* if P_1 and P_2 are chordless, do not share any vertex except perhaps their endpoints, and for any internal vertices x in P_1 and y in P_2, $xy \notin E(G)$. Observe that there is a chordless cycle containing s and t if and only if there exist two independent paths P_1 and P_2 between s and t.

From a RCCinstance (G, s, t) we construct a graph H for the problem of deciding if H contains an induced copy of $G_{r,q}$. Figure 4 illustrates the construction.

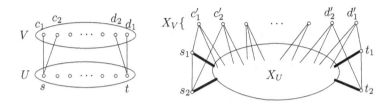

Fig. 4. An illustration of the reduction: G is on the left (only edges incident to s and t are drawn), H is on the right (thick edges mean that every possible edge is present).

Let $V(H) = \{s_1, t_1, s_2, t_2\} \cup X_U \cup X_V$, where $X_U = \{u' : u \in U \setminus \{s, t\}\}$ and $X_V = \{v' : v \in V\}$. Denote $X_U^* = X_U \cup \{s_1, t_1, s_2, t_2\}$. For $E(H)$, add an edge between every two vertices of X_U^* *except* the edges $s_1 t_1, s_1 t_2, s_2 t_1, s_2 t_2$. Moreover, we add an edge between $u' \in X_U$ and $v' \in X_V$ if and only if $uv \in E(G)$. Let $\{c_1, c_2\} = N(s)$ and $\{d_1, d_2\} = N(t)$. Then add edges $s_1 c_1'$ and $s_2 c_2'$, and add the edges $t_1 d_1'$ and $t_2 d_2'$. Notice that X_V forms an independent set.

We claim that H is chordal. Note that each vertex v of X_V is simplicial, since $N(v)$ consists of vertices from X_U and at most one of $\{s_1, t_1, s_2, t_2\}$ (since s and t have no common neighbor). Moreover, $H - X_V$ is easily seen to be chordal, and it follows that H admits a perfect elimination ordering.

Theorem 3. *Deciding if a graph H contains a copy of $G_{r,q}$ for some $r, q \geq 3$ is NP-complete, even if H is restricted to the class of chordal graphs.*

Proof. The problem is in NP, since a subset $I \subseteq V(H)$, along with the labeling of I by the a_i, b_i, x_i and y_i's of a $G_{r,q}$ can serve as a certificate. As for hardness, let G be a graph and H the corresponding graph constructed as above. We claim that G contains two independent paths P_1 and P_2 between s and t if and only if H contains a copy of $G_{r,q}$ for some $r, q \geq 3$. The idea is that s_1, s_2 (resp. t_1, t_2) correspond to the a_1, b_1 (resp. a_r, b_q) vertices of $G_{r,q}$, while the P_1 and P_2 paths give the other vertices. The x_i and y_i's are in X_V, and the a_i and b_i's in X_U^*.

($\Rightarrow$) Let P_1 and P_2 be two independent paths between s and t. Note that both paths alternate between U and V, Let $P_1 = (s = u_1, v_1, u_2, v_2, \ldots, v_{r-1}, u_r = t)$ and $P_2 = (s = w_1, z_1, w_2, z_2, \ldots, z_{q-1}, w_q = t)$. Note that since P_1 and P_2 are independent, every vertex of $G[V(P_1) \cup V(P_2)]$ has degree exactly 2.

We show that the set of vertices $I = \{s_1, t_1, s_2, t_2\} \cup \{x' : x \in V(P_1) \cup V(P_2) \setminus \{s, t\}\}$ forms a $G_{r,q}$. Denote $I_U = I \cap X_U$ and $I_V = I \cap X_V$. First observe that $\{s_1, t_1, s_2, t_2\} \cup I_U$ forms a clique, but minus the edges $\{s_1 t_1, s_1 t_2, s_2 t_1, s_2 t_2\}$. Hence $\{s_1, s_2\}$ will correspond to the vertices $\{a_1, b_1\}$ of $G_{r,q}$, and $\{t_1, t_2\}$ to $\{a_r, b_q\}$, and it remains to find the degree two vertices around this "almost-clique". Observe that $\{v_1, z_1\} = \{c_1, c_2\}$ and $\{v_{r-1}, z_{q-1}\} = \{d_1, d_2\}$. Let $c_{i_1} = v_1, c_{i_2} = z_1$ and $d_{j_1} = v_{r-1}, d_{j_2} = z_{q-1}$, with $\{i_1, i_2\} = \{j_1, j_2\} = \{1, 2\}$. In H, the vertex sequence $(s_{i_1}, u'_2, \ldots, u'_{r-1}, t_{j_1})$ forms a path in $G[I]$ in which every two consecutive vertices share a common neighbor, which lies in I_V. Namely, s_{i_1} and u'_2 share $v'_1 = c'_{i_1}$, u'_i, u'_{i+1} share v'_i, and u'_{r-1}, t_{j_1} share $v'_{r-1} = d'_{j_1}$. The same property holds for the consecutive vertices of the path $(s_{i_2}, w'_2, \ldots, w'_{q-1}, t_{i_2})$. Note that these two paths are disjoint in H and partition I_U. Moreover, by construction each $x' \in I_V$ is a shared vertex for some pair of consecutive vertices, i.e. x' has at least two neighbors in I.

Therefore, it only remains to show that if $x' \in I_V$, then x' has only two neighbors in I. Suppose instead that x' has at least 3 neighbors in I, say y'_1, y'_2, y'_3. Note that all three lie in X_U^*. We must have $|\{s_1, s_2\} \cap \{y'_1, y'_2, y'_3\}| \leq 1$, since s_1 and s_2 share no neighbor in X_V. Likewise, $|\{t_1, t_2\} \cap \{y'_1, y'_2, y'_3\}| \leq 1$. This implies that y'_1, y'_2, y'_3 are vertices corresponding to three distinct vertices of G, say y_1, y_2 and y_3. Then x is a neighbor of y_1, y_2, y_3 and since, by construction, $x, y_1, y_2, y_3 \in V(P_1) \cup V(P_2)$, this contradicts that $G[V(P_1) \cup V(P_2)]$ has maximum degree 2.

($\Leftarrow$) Suppose there is $I \subseteq V(H)$ such that $H[I]$ is isomorphic to $G_{r,q}$ for some $r, q \geq 3$. Add a label to the vertices of I as in Fig. 2 (i.e. we assume that we know where the a_i's, b_i's, x_i's and y_i's are in I). We first show that a_1, b_1, a_r, b_q, which we will call the *corner* vertices, are s_1, s_2, t_1, t_2. If one of a_1 or b_1 is in X_U, then both a_r and b_q must be in X_V, as otherwise there would be an edge between $\{a_1, b_1\}$ and $\{a_r, b_q\}$. But a_r and b_q must share an edge, whereas X_V is an independent set. Thus we may assume $\{a_1, b_1\} \cap X_U = \emptyset$. Suppose that a_1 or b_1 is in X_V, say a_1. Because $b_1 \notin X_U$ as argued above, we must have $b_1 \in \{s_1, s_2, t_1, t_2\}$. Suppose w.l.o.g. that $b_1 = s_1$. Hence $a_1 = c'_1$. Now consider the location of the x_1 vertex of $G_{r,q}$. Then x_1 must be in X_U, in which case x_1 is a neighbor of $s_1 = b_1$, contradicting that I is a copy of $G_{r,q}$. Therefore, we may assume that $\{a_1, b_1\} \cap X_V = \emptyset$. By applying the same argument on a_r and b_q, we deduce that $\{a_1, b_1, a_r, b_q\} = \{s_1, s_2, t_1, t_2\}$. We will suppose, without loss of generality, that $a_1 = s_1, b_1 = s_2$ and $\{a_r, b_q\} = \{t_1, t_2\}$ (otherwise we may relabel the vertices of the $G_{r,q}$ copy, though note that in doing so we cannot make assumptions on which t_i corresponds to which of $\{a_r, b_q\}$).

Now let $(s_1 = a_1, a_2, \ldots, a_r = t_j)$, $j \in \{1, 2\}$ be the path between the "top" corners of the $G_{r,q}$ copy in H, such that $a_i a_{i+1}$ share a common neighbor x_i of degree 2 in $G[I]$, $i \in [r-1]$. Similarly, let $(s_2 = b_1, b_2, \ldots, b_q = t_l)$, ($l \in \{1, 2\}$ and $l \neq j$) be the path between the "bottom" corners of $G_{r,q}$, such that $b_i b_{i+1}$ share a common neighbor y_i of degree 2 for $i \in [q-1]$. We claim that $a_i \in X_U$ for each $2 \leq i \leq r-1$. Suppose instead that some a_i is not in X_U. Since $s_1 = a_1$ is a neighbor of a_i, we must have $a_i = c'_1$ (the only other possibility is $a_i = s_2$, but $s_2 = b_1$). The common neighbor x_{i-1} of a_{i-1} and a_i therefore lies in X_U.

But then, x_{i-1} is a neighbor of $s_2 = b_1$, which is not possible. Therefore, each a_i belongs to X_U. By symmetry, each b_i also belongs to X_U. This implies that every x_i and y_i belong to X_V, with $x_1 = c'_1, x_{r-1} = d'_j, y_1 = c'_2$ and $y_{q-1} = d'_l$.

We can finally find our independent paths P_1 and P_2. It is straightforward to check that $\{s, t, c_1\} \cup \{u : u' \in \{x_i, a_i\}$ for $2 \le i \le r - 1\}$ induces a path P_1 from s to t in G. Similarly, $\{s, t, c_2\} \cup \{u : u' \in \{y_i, b_i\}$ for $2 \le i \le q - 1\}$ also induces a path P_2 from s to t. Moreover, P_1 and P_2 share no internal vertex.

It only remains to show that P_1 and P_2 are independent, i.e. form an induced cycle. We prove that $G[V(P_1) \cup V(P_2)]$ has maximum degree 2. Suppose there is a vertex v of degree at least 3 in $G[V(P_1) \cup V(P_2)]$. Then $v \notin \{s, t\}$ since they have degree 2 in G. Moreover, $v \notin V$, as otherwise, $v' \in I_V$ which implies that v' is an x_i or a y_i and, by construction, v' has at least 3 neighbors in I, a contradiction. Thus $v \in U$, and its 3 neighbors lie in V. Hence, v' is either an a_i or a b_i and has three neighbors in I_V, which is again a contradiction. This concludes the proof.   □

6   Conclusion

In this paper, we have shown that leaf powers cannot be characterized by strong chordality and a finite set of forbidden subgraphs. However, many questions asked here may provide more insight on leaf powers. For one, is the condition of Proposition 1 sufficient? And if so, can it be exploited for some algorithmic or graph theoretic purpose? Also, we do not know if large alternating cycles are important, since so far, every non-leaf power could be explained by checking its alternating cycles of length 4 or 6. A constant bound on the length of "important" alternating cycles would allow enumerating them in polynomial time.

Also, we have exhibited an infinite family of strongly chordal non-leaf powers, but it is likely that there are others. One potential direction is to try to generalize all of the seven graphs found in [15]. The clique arrangement of $G_{r,q}$ may be informative towards this goal. Finally on the hardness of recognizing leaf powers, the hardness of finding $G_{r,q}$ in strongly chordal graphs is of special interest. A NP-hardness proof would now be significant evidence towards the difficulty of deciding leaf power membership. And in the other direction, a polynomial time recognition algorithm may provide important insight on how to find forbidden structures in leaf powers.

References

1. Bandelt, H.-J., Dress, A.: Reconstructing the shape of a tree from observed dissimilarity data. Adv. Appl. Math. **7**(3), 309–343 (1986)
2. Berry, V., Jiang, T., Kearney, P., Li, M., Wareham, T.: Quartet cleaning: improved algorithms and simulations. In: Nešetřil, J. (ed.) ESA 1999. LNCS, vol. 1643, pp. 313–324. Springer, Heidelberg (1999). doi:10.1007/3-540-48481-7_28
3. Bibelnieks, E., Dearing, P.M.: Neighborhood subtree tolerance graphs. Discret. Appl. Math. **43**(1), 13–26 (1993)

4. Brandstädt, A., Hundt, C.: Ptolemaic graphs and interval graphs are leaf powers. In: Laber, E.S., Bornstein, C., Nogueira, L.T., Faria, L. (eds.) LATIN 2008. LNCS, vol. 4957, pp. 479–491. Springer, Heidelberg (2008). doi:10.1007/978-3-540-78773-0_42

5. Brandstädt, A., Hundt, C., Mancini, F., Wagner, P.: Rooted directed path graphs are leaf powers. Discret. Math. **310**(4), 897–910 (2010)

6. Calamoneri, T., Sinaimeri, B.: Pairwise compatibility graphs: a survey. SIAM Rev. **58**(3), 445–460 (2016)

7. Diot, E., Tavenas, S., Trotignon, N.: Detecting wheels. Appl. Anal. Discret. Math. 111–122 (2014)

8. Farber, M.: Characterizations of strongly chordal graphs. Discret. Math. **43**(2–3), 173–189 (1983)

9. Fulkerson, D., Gross, O.: Incidence matrices and interval graphs. Pac. J. Math. **15**(3), 835–855 (1965)

10. Kennedy, W., Lin, G., Yan, G.: Strictly chordal graphs are leaf powers. J. Discret. Algorithms **4**(4), 511–525 (2006)

11. Lafond, M., El-Mabrouk, N.: Orthology and paralogy constraints: satisfiability and consistency. BMC Genomics **15**(6), S12 (2014)

12. Li, L., Stoeckert, C.J., Roos, D.S.: OrthoMCL: identification of ortholog groups for eukaryotic genomes. Genome Res. **13**(9), 2178–2189 (2003)

13. Nevries, R., Rosenke, C.: Towards a characterization of leaf powers by clique arrangements. CoRR, abs/1402.1425 (2014)

14. Nevries, R., Rosenke, C.: Characterizing and computing the structure of clique intersections in strongly chordal graphs. Discret. Appl. Math. **181**, 221–234 (2015)

15. Nevries, R., Rosenke, C.: Towards a characterization of leaf powers by clique arrangements. Graphs Comb. **32**(5), 2053–2077 (2016)

16. Nishimura, N., Ragde, P., Thilikos, D.M.: On graph powers for leaf-labeled trees. J. Algorithms **42**(1), 69–108 (2002)

17. Paige, R., Tarjan, R.E.: Three partition refinement algorithms. SIAM J. Comput. **16**(6), 973–989 (1987)

18. Shutters, B., Vakati, S., Fernández-Baca, D.: Incompatible quartets, triplets, and characters. Algorithms Mol. Biol. **8**(1), 11 (2013)

19. Spinrad, J.P.: Doubly lexical ordering of dense 0–1 matrices. Inf. Process. Lett. **45**(5), 229–235 (1993)

20. Steel, M.: The complexity of reconstructing trees from qualitative characters and subtrees. J. Classif. **9**(1), 91–116 (1992)

21. Tatusov, R.L., Galperin, M.Y., Natale, D.A., Koonin, E.V.: The COG database: a tool for genome-scale analysis of protein functions and evolution. Nucleic Acids Res. **28**(1), 33–36 (2000)

New Results on Weighted Independent Domination

Vadim Lozin[1($\boxtimes$)], Dmitriy Malyshev[2], Raffaele Mosca[3], and Viktor Zamaraev[1]

[1] Mathematics Institute, University of Warwick, Coventry CV4 7AL, UK
V.Lozin@warwick.ac.uk
[2] National Research University Higher School of Economics,
25/12 Bolshaya Pecherskaya Ulitsa, 603155 Nizhny Novgorod, Russia
[3] Dipartimento di Economia, Universitá degli Studi "G. D'Annunzio",
65121 Pescara, Italy

Abstract. Weighted independent domination is an NP-hard graph problem, which remains computationally intractable in many restricted graph classes. Only few examples of classes are available, where the problem admits polynomial-time solutions. In the present paper, we extend the short list of such classes with two new examples.

1 Introduction

INDEPENDENT DOMINATION is the problem of finding in a graph an inclusion-wise maximal independent set of minimum cardinality. This is one of the hardest problems of combinatorial optimization and it remains difficult under substantial restrictions. In particular, it is NP-hard for so-called sat-graphs, where the problem is equivalent to SATISFIABILITY [15]. It is also NP-hard for planar graphs, triangle-free graphs, graphs of vertex degree at most 3 [3], line graphs [14], chordal bipartite graphs [5], etc.

The weighted version of the problem (abbreviated WID) deals with vertex-weighted graphs and asks to find an inclusionwise maximal independent set of minimum total weight. This version is provenly harder, as it remains NP-hard even for chordal graphs [4], where INDEPENDENT DOMINATION can be solved in polynomial time [6].

Not much is known about graph classes allowing an efficient solution of the WID problem. Among rare examples of this type, let us mention cographs and split graphs.

- A *cograph* is a graph in which every induced subgraph with at least two vertices is either disconnected or the complement of a disconnected graph. In the case of cographs, the problem can be solved efficiently by means of modular decomposition.
- A *split graph* is a graph whose vertices can be partitioned into a clique and an independent set. The only available way to solve WID efficiently for a split graph is to examine all its inclusionwise maximal independent sets, of which there are polynomially many.

H.L. Bodlaender and G.J. Woeginger (Eds.): WG 2017, LNCS 10520, pp. 399–411, 2017.
https://doi.org/10.1007/978-3-319-68705-6_30

Let us observe that in both these examples we deal with *hereditary classes*, i.e. with classes of graphs closed under taking induced subgraphs. It is well-known (and not difficult to see) that a class of graphs is hereditary if and only if it can be characterized in terms of minimal forbidden induced subgraphs. For instance, the cographs are precisely P_4-free graphs (i.e. graphs containing no induced P_4), while the split graphs are the graphs which are free of $2K_2, C_4$ and C_5.

The class of sat-graphs (as well as each of the other classes mentioned earlier) also is hereditary. It consists of graphs whose vertices can be partitioned into a clique and a graph of vertex degree at most 1. Therefore, sat-graphs form an extension of split graphs. With this extension the complexity status of the problem jumps from polynomial-time solvability to NP-hardness.

In the present paper, we study two more extensions of split graphs: the class of $(P_5, \overline{P}_5)$-free graphs and the class of $(P_5, \overline{P_3 + P_2})$-free graphs. The first of them also extends the cographs, since both forbidden graphs contain a P_4. From an algorithmic point of view, both extensions are resistant to any available technique. To crack the puzzle for $(P_5, \overline{P}_5)$-free graphs, we develop a new decomposition scheme combining several algorithmic tools. This enables us to show that the WID problem can be solved for $(P_5, \overline{P}_5)$-free graphs in polynomial time. For the second class, we develop a tricky reduction allowing us to reduce the problem to the first class.

Let us emphasize that in both cases the presence of P_5 among the forbidden graphs is necessary, because each of $\overline{P}_5$ and $\overline{P_3 + P_2}$ contains a C_4 and by forbidding C_4 alone we obtain a class where the problem is NP-hard. Whether the presence of P_5 among the forbidden graphs is sufficient for polynomial-time solvability of WID is a big open question. For the related problem of finding a maximum weight independent set (WIS), this question was answered only recently [9] after several decades of attacking the problem on subclasses of P_5-free graphs (see e.g. [2,7,8]). WID is a more stubborn problem, as it remains NP-hard in many classes where WIS can be solved in polynomial time, such as line graphs, chordal graphs, bipartite graphs, etc. Determining the complexity status of WID in P_5-free graphs is a challenging open question. We discuss this and related open questions in the concluding section of the paper. The rest of the paper is organized as follows: Sect. 2 contains preliminary information, in Sect. 3 we solve the problem for $(P_5, \overline{P}_5)$-free graphs, and in Sect. 4 we solve it for $(P_5, \overline{P_3 + P_2})$-free graphs.

2   Preliminaries

All graphs in this paper are finite, undirected, without loops and multiple edges. The vertex set and the edge set of a graph G are denoted by $V(G)$ and $E(G)$, respectively. A subset $S \subseteq V(G)$ is

- *independent* if no two vertices of S are adjacent,
- a *clique* if every two vertices of S are adjacent,
- *dominating* if every vertex not in S is adjacent to a vertex in S.

For a vertex-weighted graph G with a weight function w, by $id_w(G)$ we denote the minimum weight of an independent dominating set in G.

If v is a vertex of G, then $N(v)$ is the *neighbourhood* of v (i.e. the set of vertices adjacent to v) and $V(G) \setminus N(v)$ is the *antineighbourhood* of v. We say that v is *simplicial* if its neighbourhood is a clique, and v is *antisimplicial* if its antineighbourhood is an independent set.

Let S be a subset of $V(G)$. We say that a vertex $v \in V(G) \setminus S$ *dominates* S if $S \subseteq N(v)$. Also, v *distinguishes* S if v has both a neighbour and a non-neighbour in S. By $G[S]$ we denote the subgraph of G induced by S and by $G - S$ the subgraph $G[V \setminus S]$. If S consists of a single element, say $S = \{v\}$, we write $G - v$, omitting the brackets.

If G is a connected graph but $G - S$ is not, then S is a *separator* (also known as a cut-set). A *clique separator* is a separator which is also a clique.

As usual, P_n, C_n and K_n denote a chordless path, a chordless cycle and a complete graph on n vertices, respectively. Given two graphs G and H, we denote by $G + H$ the disjoint union of G and H, and by mG the disjoint union of m copies of G.

We say that a graph G contains a graph H as an induced subgraph if H is isomorphic to an induced subgraph of G. Otherwise, G is H-free.

A class $\mathcal{Z}$ of graphs is hereditary if it is closed under taking induced subgraphs, i.e. if $G \in \mathcal{Z}$ implies that every induced subgraph of G belongs to $\mathcal{Z}$. It is well-known that $\mathcal{Z}$ is hereditary if and only if graphs in G do not contain induced subgraphs from a set M, in which case we say that M is the set of forbidden induced subgraphs for $\mathcal{Z}$.

For an initial segment of natural numbers $\{1, 2, \ldots, n\}$ we will often use the notation $[n]$.

2.1 Modular Decomposition

Let $G = (V, E)$ be a graph. A set $M \subseteq V$ is a *module* in G if no vertex outside of M distinguishes M. Obviously, $V(G)$, $\emptyset$ and any vertex of G are modules and we call them *trivial*. A non-trivial module is also known as a *homogeneous set*. A graph without homogeneous sets is called *prime*. The notion of a prime graph plays a crucial role in *modular decomposition*, which allows to reduce various algorithmic and combinatorial problems in a hereditary class $\mathcal{Z}$ to prime graphs in $\mathcal{Z}$ (see e.g. [12] for more details on modular decomposition and its applications). In particular, it was shown in [3] that the WID problem can be solved in polynomial time in $\mathcal{Z}$ whenever it is polynomially solvable for prime graphs in $\mathcal{Z}$.

In our solution, we will use homogeneous sets in order to reduce the problem from a graph G to two proper induced subgraphs of G as follows. Let $M \subset V$ be a homogeneous set in G. Denote by H the graph obtained from G by contracting M into a single vertex m (or equivalently, by removing all but one vertex m from M). We define the weight function w' on the vertices of H as follows: $w'(v) = w(v)$ for every $v \neq m$, and $w'(m) = id_w(G[M])$. Then it is not difficult to see that

$$id_w(G) = id_{w'}(H). \tag{1}$$

In other words, to solve the problem for G we first solve the problem for the subgraph $G[M]$, construct a new weighted graph H, and solve the problem for the graph H.

2.2   Antineighborhood Decomposition

One of the simplest branching algorithms for the maximum weight independent set problem is based on the following obvious fact. For any graph $G = (V, E)$ and any vertex $v \in V$,

$$is_w(G) = \max\{is_w(G - N(v)), is_w(G - v)\},$$

where w is a weight function on the vertices of G, and $is_w(G)$ stands for the maximum weight of an independent set in G. We want to use a similar branching rule for the WID problem, i.e.

$$id_w(G) = \min\{id_w(G - N(v)), id_w(G - v)\}. \tag{2}$$

However, formula (2) is not necessarily true, because an independent dominating set in the graph $G - v$ is not necessarily dominating in the whole graph G. To overcome this difficulty, we introduce the following notion.

Definition 1. *A vertex v is* permissible *if formula (2) is valid for v*

An obvious sufficient condition for a vertex to be permissible can be stated as follows: if every independent dominating set in $G - v$ contains at least one neighbour of v, then v is permissible.

Applying (2) to a permissible vertex v of G, we reduce the problem from G to two subgraphs $G - v$ and $G - N(v)$. Such a branching procedure results in a decision tree. In general, this approach does not provide a polynomial-time solution, since the decision tree may have exponentially many nodes (subproblems). However, under some conditions this procedure may lead to a polynomial-time algorithm. In particular, this is true for graphs in hereditary classes possessing the following property.

Definition 2. *A graph class $\mathcal{G}$ has the* antineighborhood property *if there is a subclass $\mathcal{F} \subseteq \mathcal{G}$, and polynomial algorithms P, Q and R, such that*

(i)   Given a graph G the algorithm P decides whether G belongs to $\mathcal{F}$ or not;
(ii) Q finds a permissible vertex v in any input graph $G \in \mathcal{G} \setminus \mathcal{F}$ such that the graph $G - N(v)$ induced by the antineighborhood of v belongs to $\mathcal{F}$; we call v a good vertex;
(iii) R solves the WID problem for (every induced subgraph of) any input graph from $\mathcal{F}$.

Directly from the definition we derive the following conclusion.

Theorem 1. *Let $\mathcal{G}$ be a hereditary class possessing the antineighborhood property. Then WID can be solved in polynomial time for graphs in $\mathcal{G}$.*

3 WID in $(P_5, \overline{P}_5)$-Free Graphs •

To solve the problem for $(P_5, \overline{P}_5)$-free graphs, we first develop a new decomposition scheme in Sect. 3.1 that combines modular decomposition and antineighborhood decomposition. In Sect. 3.2 we apply it to $(P_5, \overline{P}_5)$-free graphs.

3.1 Decomposition Scheme

Let $\mathcal{G}$ be a hereditary class such that the class $\mathcal{G}_p$ of prime graphs in $\mathcal{G}$ has the antineighborhood property. We define the decomposition procedure by describing the corresponding decomposition tree $T(G)$ for a graph $G = (V, E) \in \mathcal{G}$. In the description, we use notions and notations introduced in Definition 2.

1. If G belongs to $\mathcal{F}$, then the node of $T(G)$ corresponding to G is a leaf.
2. If $G \notin \mathcal{F}$ and G has a homogeneous set M, then G is decomposed into subgraphs $G_1 = G[M]$ and $G_2 = G[(V \setminus M) \cup \{m\}]$ for some vertex m in M. The node of $T(G)$ corresponding to G is called a *homogeneous node*, and it has two children corresponding to G_1 and G_2. These children are in turn the roots of subtrees representing possible decompositions of G_1 and G_2.
3. If $G \notin \mathcal{F}$ and G has no homogeneous set, then G is prime and by the antineighborhood property of $\mathcal{G}_p$ there exists a good vertex $v \in V$. Then G is decomposed into subgraphs $G_1 = G - N(v)$ and $G_2 = G - v$. The node of $T(G)$ corresponding to G is called an *antineighborhood node*, and it has two children corresponding to G_1 and G_2. The graph G_1 belongs to $\mathcal{F}$ and the node corresponding to G_1 is a leaf. The node corresponding to G_2 is the root of a subtree representing a possible decomposition of G_2.

Lemma 1. *For an n-vertex graph $G \in \mathcal{G}$, the tree $T(G)$ contains $O(n^2)$ nodes.*

Proof. Since $T(G)$ is a binary tree, it is sufficient to show that the number of internal nodes is $O(n^2)$. To this end, we prove that the internal nodes of $T(G)$ can be labeled by pairwise different pairs (a, b), where $a, b \in V(G)$.

Let $G' = (V', E')$ be an induced subgraph of G that corresponds to an internal node X of $T(G)$. If X is a homogeneous node, then G' is decomposed into subgraphs $G_1 = G'[M]$ and $G_2 = G'[(V' \setminus M) \cup \{m\}]$, where $M \subset V'$ is a homogeneous set of G' and m is a vertex in M. In this case, we label X with (a, b), where $a \in M \setminus \{m\}$ and $b \in V' \setminus M$. If X is an antineighborhood node, then G' is decomposed into subgraphs $G_1 = G' - N(v)$ and $G_2 = G' - v$, where v is a good vertex of G'. In this case, X is labeled with (v, b), where $b \in N(v)$.

Suppose, to the contrary, that there are two internal nodes A and B in $T(G)$ with the same label (a, b). By construction, this means that a, b are vertices of both G_A and G_B, the subgraphs of G corresponding to the nodes A and B, respectively. Assume first that B is a descendant of A. The choice of the labels implies that regardless of the type of node A (homogeneous or antineighborhood), the label of A has at least one vertex that is not a vertex of G_B, a contradiction. Now, assume that neither A is a descendant of B nor B is a descendant of A. Let X be the lowest common ancestor of A and B in $T(G)$. If X is a homogeneous node, then G_A and G_B can have at most one vertex in common, and thus A and B cannot have the same label. If X is an antineighborhood node, then one of its children is a leaf, contradicting to the assumption that both A and B are internal nodes. □

Lemma 2. *Let G be an n-vertex graph in $\mathcal{G}$. If time complexities of the algorithms P and Q are $O(n^p)$ and $O(n^q)$, respectively, then $T(G)$ can be constructed in time $O(n^{2+\max\{2,p,q\}})$.*

Proof. The time needed to construct $T(G)$ is the sum of times required to identify types of nodes of $T(G)$ and to decompose graphs corresponding to internal nodes of $T(G)$. To determine the type of a given node X of $T(G)$, we first use the algorithm P to establish whether the graph G_X corresponding to X belongs to $\mathcal{F}$ or not. In the former case X is a leaf node, in the latter case we further try to find in G_X a homogeneous set, which can be performed in $O(n+m)$ time [11]. If G_X has a homogeneous set, then X is a homogeneous node and we decompose G_X into the graphs induced by the vertices in and outside the homogeneous set, respectively. If G_X does not have a homogeneous set, then X is an antineighborhood node, and the decomposition of G_X is equivalent to finding a good vertex, which can be done by means of the algorithm Q. Since there are $O(n^2)$ nodes in $T(G)$, the total time complexity for constructing $T(G)$ is $O(n^{2+\max\{2,p,q\}})$. □

Theorem 2. *If $\mathcal{G}$ is a hereditary class such that the class $\mathcal{G}_p$ of prime graphs in $\mathcal{G}$ has the antineighborhood property, then the WID problem can be solved in polynomial time for graphs in $\mathcal{G}$.*

Proof. Let G be an n-vertex graph in $\mathcal{G}$. To solve the WID problem for G, we construct $T(G)$ and then traverse it bottom-up, deriving a solution for each node of $T(G)$ from the solutions corresponding to the children of that node. The construction of $T(G)$ requires a polynomial time by Lemma 2. For the instances corresponding to leaf-nodes of $T(G)$, the problem can be solved in polynomial time by the antineighborhood property. According to the discussion in Sects. 2.1 and 2.2, the solution for an instance corresponding to an internal node can be derived from the solutions of its children in polynomial time. Finally, as there are $O(n^2)$ nodes in $T(G)$ (Lemma 1), the total running time to solve the problem for G is polynomial. □

3.2   Application to $(P_5, \overline{P_5})$-Free Graphs

In this section, we show that the WID problem can be solved efficiently for $(P_5, \overline{P_5})$-free graphs by means of the decomposition scheme described in Sect. 3.1. To this end, we will prove that the class of prime $(P_5, \overline{P_5})$-free graphs has the antineighborhood property. We start with several auxiliary results. The first of them is simple and we omit its proof.

Observation 1. *Let $G = (V, E)$ be a graph, and let $W \subset V$ induce a connected subgraph in G. If a vertex $v \in V \setminus W$ distinguishes W, then v distinguishes two adjacent vertices of W.*

Proposition 1. *Let $G = (V, E)$ be a prime graph. If a subset $W \subset V$ has at least two vertices and is not a clique, then there exists a vertex $v \in V \setminus W$ which distinguishes two non-adjacent vertices of W.*

Proof. Suppose, to the contrary, that none of the vertices in $V \setminus W$ distinguishes a pair of non-adjacent vertices in W. If $G[W]$ has more than one connected component, then it is easy to see that no vertex outside of W distinguishes W. Hence, W is a homogeneous set in G, which contradicts the primality of G.

If $G[W]$ is connected, then $\overline{G[W]}$ has a connected component C with at least two vertices, since W is not a clique. Then, by our assumption and Observation 1, no vertex outside of W distinguishes C. Also, by the choice of C, no vertex of W outside of C distinguishes C. Therefore, $V(C)$ is a homogeneous set in G. This contradiction completes the proof of the proposition. □

Lemma 3. *If a $(P_5, \overline{P_5})$-free prime graph contains an induced copy of $2K_2$, then it has a clique separator.*

Proof. Let $G = (V; E)$ be a $(P_5, \overline{P_5})$-free prime graph containing an induced copy of $2K_2$. Let $S \subseteq V$ be a minimal separator with the property that $G - S$ contains at least two non-trivial connected components, i.e. connected components with at least two vertices. Such a separator necessarily exists, since G contains an induced $2K_2$. It follows from the choice of S that

- $G - S$ has $k \geq 2$ connected components $C_1, \ldots, C_k$;
- $r \geq 2$ of these components, say $C_1, \ldots, C_r$, have at least two vertices, and all the other components $C_{r+1}, \ldots, C_k$ are trivial;
- every vertex in S has a neighbour in each of the non-trivial components $C_1, \ldots, C_r$ (since S is minimal);
- for every $i \in \{r+1, \ldots, k\}$, the unique vertex of the trivial component C_i has a neighbour in S (since G is connected).

In the remaining part of the proof, we show that G has a clique separator. Let us denote $U_i = V(C_i)$ for $i = 1, \ldots, k$. We first observe the following.

Claim 1. *Any vertex in S distinguishes at most one of the sets $U_1, \ldots, U_r$.*

Proof. Assume $v \in S$ distinguishes U_i and U_j for distinct $i, j \in [r]$. Then by Observation 1 v distinguishes two adjacent vertices a, b in U_i and two adjacent vertices c, d in U_j. But then a, b, v, c, d induce a forbidden P_5.

According to Claim 1, the set S can be partitioned into subsets $S_0, S_1 \ldots, S_r$, where the vertices of S_0 dominate every member of $\{U_1, \ldots, U_r\}$, and for each $i \in [r]$, the vertices of S_i distinguish U_i and dominate U_j for all j different from i. Moreover, for each $i \in [r]$ the set S_i is non-empty, as the graph G is prime. Now we prove two more auxiliary claims.

Claim 2. *For $0 \leq i < j \leq r$, every vertex in S_i is adjacent to every vertex in S_j.*

Proof. Assume that the claim is false, i.e. there exist two non-adjacent vertices $s_i \in S_i$ and $s_j \in S_j$. By Observation 1 there exist two adjacent vertices $a, b \in U_j$ that are distinguished by s_j. But then s_i, s_j, a, b and any vertex in $N(s_i) \cap U_i$ induce a forbidden $\overline{P_5}$, a contradiction.

Claim 3. *For $i \in [r]$, no vertex in U_i distinguishes two non-adjacent vertices in S_i.*

Proof. Assume that there exists a pair of non-adjacent vertices $x, y \in S_i$ that are distinguished by a vertex $u_i \in U_i$. Let $j \in [r] \setminus \{i\}$, and let $s_j \in S_j$ and $u_j \in U_j \setminus N(s_j)$. Then, since s_j dominates S_i, we have that u_j, x, y, s_j, u_i induce a forbidden $\overline{P_5}$, a contradiction.

We split further analysis into two cases.

Case 1: there is at least one trivial component in $G \setminus S$, i.e. $k > r$. For $i \in \{r+1, \ldots, k\}$ we denote by u_i the unique vertex of U_i. Let $U = \{u_{r+1}, \ldots, u_k\}$ and let u^* be a vertex in U with a minimal (under inclusion) neighbourhood. We will show that $N(u^*)$ is a clique, and hence is a clique separator in G. By Claim 2, it suffices to show that $N(u^*) \cap S_i$ is a clique for each $i \in \{0, 1, \ldots, k\}$. Suppose that for some i the set $N(u^*) \cap S_i$ is not a clique. Then, by Proposition 1, there are two nonadjacent vertices $x, y \in N(u^*) \cap S_i$ distinguished by a vertex $z \in V \setminus (N(u^*) \cap S_i)$. It follows from Claims 2 and 3 that either $z \in S_i \setminus N(u^*)$ or $z \in U$. If $z \in S_i \setminus N(u^*)$, then u^*, x, y, z, and any vertex in $U_j, j \in [r] \setminus \{i\}$ induce a forbidden $\overline{P_5}$, a contradiction. Hence, assume that none of the vertices in $S \setminus (N(u^*) \cap S_i)$ distinguishes two nonadjacent vertices in $N(u^*) \cap S_i$. If $z \in U$, with z being nonadjacent to x and adjacent to y, then by the minimality of $N(u^*)$ there is a vertex $s \in N(z)$ that is not adjacent to u^*. Since $N(z) \subseteq S$, vertex s does not distinguish x and y. But then x, u^*, y, z, s induce either a P_5 (if s is adjacent neither to x nor to y) or a $\overline{P_5}$ (if s is adjacent to both x and y), a contradiction.

Case 2: there are no trivial components in $G \setminus S$, i.e. $k = r$. First, observe that $|S_0| \leq 1$, since G is prime and no vertex outside of S_0 distinguishes S_0 (which follows from the definition of S_0, Claim 2 and the fact that $k = r$). Further, Claims 2 and 3 imply that for each $i \in [r]$ no vertex in $V \setminus S_i$ distinguishes two nonadjacent vertices in S_i. Therefore, applying Proposition 1 we conclude that S_i is a clique. Hence $S = \bigcup_{i=0}^{r} S_i$ is a clique separator in G. $\qquad\square$

Lemma 4. *Let G be a $(P_5, \overline{P_5})$-free prime graph containing an induced copy of $2K_2$. Then G contains a permissible antisimplicial vertex.*

Proof. By Lemma 3, G has a clique separator, and therefore it also has a minimal clique separator S. Let $C_1, \ldots, C_k, k \geq 2$, be connected components of $G - S$, and $U_i = V(C_i), i = 1, \ldots, k$. Since S is a minimal separator, every vertex in S has at least one neighbour in each of the sets $U_1, \ldots, U_k$. By Claim 1 in the proof of Lemma 3, any vertex in S distinguishes at most one of the sets $U_1, \ldots, U_k$, and therefore, the set S partitions into subsets $S_0, S_1 \ldots, S_k$, where the vertices of S_0 dominate every member of $\{U_1, \ldots, U_k\}$, and for each $i \in [k]$ the vertices of S_i distinguish U_i and dominate U_j for all j different from i.

If $S_0 \neq \emptyset$, then any vertex in S_0 is adjacent to all the other vertices in the graph, and therefore it is permissible and antisimplicial. Hence, without loss of generality, assume that $S_0 = \emptyset$ and $S_1 \neq \emptyset$.

Let s be a vertex in S_1 with a maximal (under inclusion) neighbourhood in U_1. We will show that s is antisimplicial and permissible. Suppose that the graph induced by the antineighbourhood of s contains a connected component C with at least two vertices. Since G is prime, by Observation 1 it must contain a vertex p outside of C distinguishing two adjacent vertices q and t in C. Then p does not belong to $N(s) \cap U_1$, since otherwise q, t, p, s together with any vertex in U_2 would induce a P_5. Therefore, p belongs to S_1. Since the set $N(s) \cap U_1$ is maximal, it contains a vertex y nonadjacent to p. But now t, q, p, s, y induce either a P_5 or its complement, as y does not distinguish q and t. This contradiction shows that every component in the graph induced by the antineighbourhood of s is trivial, i.e. s is antisimplicial.

Assume now that s is not permissible, i.e. there exists an independent dominating set I in $G - s$ that does not contain a neighbour of s. Since s dominates $U_2 \cup \ldots \cup U_k$, the set I is a subset of $U_1 \setminus N(s)$. But then I is not dominating, since no vertex of U_2 has a neighbour in I, This contradiction completes the proof of the lemma. $\qquad \square$

Lemma 5. *The class of prime $(P_5, \overline{P_5})$-free graphs has the antineighborhood property.*

Proof. Let $\mathcal{F}$ be the class of $(2K_2, \overline{P_5})$-free graphs (this is a subclass of $(P_5, \overline{P_5})$-free graphs, since $2K_2$ is an induced subgraph of P_5). Clearly, graphs in $\mathcal{F}$ can be recognized in polynomial time. The WID problem can be solved in polynomial time for graphs in $\mathcal{F}$, because the problem is polynomially solvable on $2K_2$-free graphs (according to [1], these graphs have polynomially many maximal independent sets).

If a prime $(P_5, \overline{P_5})$-free graph $G = (V, E)$ does not belong to $\mathcal{F}$, then by Lemma 4 it contains a permissible vertex v whose antineighbourhood is an independent set, and therefore, $G - N(v) \in \mathcal{F}$. It remains to check that a permissible antisimplicial vertex in G can be found in polynomial time. It follows from the proof of Lemma 4 that in a minimal clique separator of G any vertex with a maximal neighbourhood is permissible and antisimplicial. A minimal clique separator in a graph can be found in polynomial time [13], and therefore the desired vertex can also be computed efficiently. $\qquad \square$

Now the main result of the section follows from Theorem 2 and Lemma 5.

Theorem 3. *The WID problem can be solved in polynomial time in the class of $(P_5, \overline{P_5})$-free graphs.*

4   WID in $(P_5, \overline{P_3 + P_2})$-Free Graphs

To solve the problem for $(P_5, \overline{P_3 + P_2})$-free graphs, we introduce the following notation: for an arbitrary graph F, let F^* be the graph obtained from F by adding three new vertices, say b, c, d, such that b is adjacent to each vertex of F, while c is adjacent to b and d only (see Fig. 1 for an illustration in the case $F = \overline{P_5}$). The importance of this notation is due to the following result.

Theorem 4 [10]. *Let F be any connected graph. If the WID problem can be solved in polynomial time for (P_5, F)-free graphs, then this problem can also be solved in polynomial time for (P_5, F^*)-free graphs.*

This result together with Theorem 3 leads to the following conclusion.

Corollary 1. *The WID problem can be solved in polynomial time in the class of $(P_5, \overline{P_5}^*)$-free graphs.*

To solve the problem for $(P_5, \overline{P_3 + P_2})$-free graphs, in what follows we reduce it to $(P_5, \overline{P_3 + P_2}, \overline{P_5}^*)$-free graphs, where the problem is solvable by Corollary 1.

Let G be a $(P_5, \overline{P_3 + P_2})$-free graph containing a copy of $\overline{P_5}^*$ induced by vertices $a_1, a_2, a_3, a_4, a_5, b, c, d$, as shown in Fig. 1.

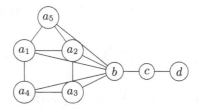

Fig. 1. The graph $\overline{P_5}^*$

Denote by U the set of vertices in G that have at least one neighbour in $\{a_1, a_2, a_3, a_4, a_5\}$, that is, $U = N(a_1) \cup \ldots \cup N(a_5)$. In particular, U includes $\{a_1, a_2, a_3, a_4, a_5, b\}$. We assume that

(**) the copy of $\overline{P_5}^*$ in G is chosen in such a way that $|U|$ is minimum.

Proposition 2. *If a vertex $x \in U$ has a neighbour y outside of U, then x is adjacent to each of the vertices a_1, a_2, a_3, a_4.*

Proposition 2 allows us to partition the set U into three subsets as follows:

U_1 consists of the vertices of U that are adjacent to each of the vertices a_1, a_2, a_3, a_4, and have at least one neighbour outside of U;
U_2 consists of the vertices of U that are adjacent to each of the vertices a_1, a_2, a_3, a_4, but have no neighbours outside of U;
$U_3 = U \setminus (U_1 \cup U_2)$.

Notice that U_1 is non-empty as it contains b. Also $\{a_1, a_2, a_3, a_4, a_5\} \subseteq U_3$, and no vertex in U_3 has a neighbour outside of U.

Proposition 3. U_1 *is a clique in* G.

Proposition 4. *The graph* $G[U_2 \cup U_3]$ *is* $\overline{P_5}^*$-*free.*

Now we describe a reduction from the graph G with a weight function w to a graph G' with a weight function w', where $|V(G')| \leq |V(G)| - 4$, G' is $(P_5, \overline{P_3 + P_2})$-free, and $id_w(G) = id_{w'}(G')$. First, we define G' as the graph obtained from G by

1. removing the vertices of U_3;
2. adding edges between any two non-adjacent vertices in $U_1 \cup U_2$;
3. adding a new vertex u adjacent to every vertex in $U_1 \cup U_2$.

Clearly, $|V(G')| \leq |V(G)| - 4$, as the set U_3 of the removed vertices contains at least 5 elements and we add exactly one new vertex u. In the next proposition, we show that the above reduction does not produce any of the forbidden subgraphs.

Proposition 5. *The graph* G' *is* $(P_5, \overline{P_3 + P_2})$-*free.*

Now we define a weight function w' on the vertex set of G' as follows:

1. $w'(x) = w(x)$, for every $x \in V(G') \setminus (\{u\} \cup U_1 \cup U_2)$;
2. $w'(u) = id_w(G[U_3])$;
3. $w'(x) = w(x) + id_w(G[U \setminus N[x]])$, for every $x \in U_1$;
4. $w'(x) = w(x) + id_w(G[U \setminus (U_1 \cup N[x])])$, for every $x \in U_2$.

Lemma 6. *Given a weighted graph* (G, w), *the weighted graph* (G', w') *can be constructed in polynomial time.*

To show that $id_w(G) = id'_w(G)$, we need two auxiliary propositions.

Proposition 6. *Any independent dominating set in* $G[U_3]$ *dominates* $U_1 \cup U_2$.

Proposition 7. *For every vertex* $x \in U_2$, *any independent dominating set in the graph* $G - U$ *dominates* $U_1 \setminus N(x)$.

Lemma 7. *For any weighted graph* (G, w), *we have* $id_w(G) = id_{w'}(G')$.

Now we are ready to prove the main result of this section.

Theorem 5. *The WID problem is solvable in polynomial time for* $(P_5, \overline{P_3 + P_2})$-*free graphs.*

Proof. Let (G, w) be an n-vertex $(P_5, \overline{P_3 + P_2})$-free weighted graph. If G contains an induced copy of $\overline{P_5}^*$, then by Proposition 5, and Lemmas 6 and 7, the graph (G, w) can be transformed in polynomial time into a $(P_5, \overline{P_3 + P_2})$-free weighted graph (G', w') with at most $n-4$ vertices such that $id_w(G) = id_{w'}(G')$. Repeating this procedure at most $\lfloor n/4 \rfloor$ times we obtain a $(P_5, \overline{P_3 + P_2}, \overline{P_5}^*)$-free weighted graph (H, σ) such that $id_w(G) = id_\sigma(H)$. By Corollary 1 the WID problem for (H, σ) can be solved in polynomial time. To conclude the proof we observe that a polynomial-time procedure computing $id_w(G)$ can be easily transformed into a polynomial-time algorithm finding an independent dominating set of weight $id_w(G)$. $\square$

5   Concluding Remarks and Open Problems

In this paper, we proved that WEIGHTED INDEPENDENT DOMINATION can be solved in polynomial time for $(P_5, \overline{P}_5)$-free graphs and $(P_5, \overline{P_3 + P_2})$-free graphs. A natural question to ask is whether these results can be extended to a class defined by one forbidden induced subgraph.

From the results in [3] it follows that in the case of one forbidden induced subgraph H the problem is solvable in polynomial time *only if* H is a linear forest, i.e. a graph every connected component of which is a path. On the other hand, it is known that this necessary condition is not sufficient, since INDEPENDENT DOMINATION is NP-hard in the class of $2P_3$-free graphs. This follows from the fact that all sat-graphs are $2P_3$-free [15].

In the case of a *disconnected* forbidden graph H, polynomial-time algorithms to solve WEIGHTED INDEPENDENT DOMINATION are known only for mP_2-free graphs for any fixed value of m. This follows from a polynomial bound on the number of maximal independent sets in these graphs [1]. The unweighted version of the problem can also be solved for $P_2 + P_3$-free graphs [10]. However, for weighted graphs in this class the complexity status of the problem is unknown.

Problem 1. Determine the complexity status of WEIGHTED INDEPENDENT DOMINATION in the class of $P_2 + P_3$-free graphs.

In the case of a *connected* forbidden graph H, i.e. in the case when $H = P_k$, the complexity status is known for $k \geq 7$ (as P_7 contains a $2P_3$) and for $k \leq 4$ (as P_4-free graphs are precisely the cographs). Therefore, the only open cases are P_5-free and P_6-free graphs. As we mentioned in the introduction, the related problem of finding a maximum weight independent set (WIS) has been recently solved for P_5-free graphs [9]. This result makes the class of P_5-free graphs of particular interest for WEIGHTED INDEPENDENT DOMINATION and we formally state it as an open problem.

Problem 2. Determine the complexity status of WEIGHTED INDEPENDENT DOMINATION in the class of P_5-free graphs.

We also mentioned earlier that a polynomial-time solution for WIS in a hereditary class $\mathcal{X}$ does not necessarily imply the same conclusion for WID in $\mathcal{X}$. However, in the reverse direction such examples are not known. We believe that such examples do not exist and propose this idea as a conjecture.

Conjecture 1. If WID admits a polynomial-time solution in a hereditary class $\mathcal{X}$, then so does WIS.

Acknowledgements. The results of Sect. 3 were obtained under financial support of the Russian Science Foundation grant No. 17-11-01336. The results of Sect. 4 were obtained under financial support of the Russian Foundation for Basic Research, grant No. 16-31-60008-mol-a-dk, RF President grant MK-4819.2016.1 and LATNA laboratory, National Research University Higher School of Economics.

References

1. Balas, E., Yu, C.S.: On graphs with polynomially solvable maximum-weight clique problem. Networks **19**, 247–253 (1989)
2. Bodlaender, H.L., Brandstädt, A., Kratsch, D., Rao, M., Spinrad, J.: On algorithms for (P_5, gem)-free graphs. Theoret. Comput. Sci. **349**, 2–21 (2005)
3. Boliac, R., Lozin, V.: Independent domination in finitely defined classes of graphs. Theoret. Comput. Sci. **301**, 271–284 (2003)
4. Chang, G.J.: The weighted independent domination problem is NP-complete for chordal graphs. Discret. Appl. Math. **143**, 351–352 (2004)
5. Damaschke, P., Muller, H., Kratsch, D.: Domination in convex and chordal bipartite graphs. Inf. Process. Lett. **36**, 231–236 (1990)
6. Farber, M.: Independent domination in chordal graphs. Oper. Res. Lett. **1**, 134–138 (1982)
7. Giakoumakis, V., Rusu, I.: Weighted parameters in $(P_5, \overline{P}_5)$-free graphs. Discret. Appl. Math. **80**, 255–261 (1997)
8. Karthick, T.: On atomic structure of P_5-free subclasses and maximum weight independent set problem. Theoret. Comput. Sci. **516**, 78–85 (2014)
9. Lokshantov, D., Vatshelle, M., Villanger, Y.: Independent set in P_5-free graphs in polynomial time. In: Proceedings of the Twenty-Fifth Annual ACM-SIAM Symposium on Discrete Algorithms, pp. 570–581 (2014)
10. Lozin, V., Mosca, R., Purcell, C.: Independent domination in finitely defined classes of graphs: polynomial algorithms. Discret. Appl. Math. **182**, 2–14 (2015)
11. McConnell, R.M., Spinrad, J.: Modular decomposition and transitive orientation. Discret. Math. **201**, 189–241 (1999)
12. Möhring, R.H., Radermacher, F.J.: Substitution decomposition for discrete structures and connections with combinatorial optimization. Ann. Discret. Math. **19**, 257–356 (1984)
13. Whitesides, S.H.: An algorithm for finding clique cut-sets. Inf. Process. Lett. **12**, 31–32 (1981)
14. Yannakakis, M., Gavril, F.: Edge dominating sets in graphs. SIAM J. Appl. Math. **38**, 364–372 (1980)
15. Zverovich, I.E.: Satgraphs and independent domination. Part 1. Theoret. Comput. Sci. **352**, 47–56 (2006)

The Parameterized Complexity
of the Equidomination Problem

Oliver Schaudt and Fabian Senger(✉)

Institut für Informatik, Universität zu Köln, Weyertal 80, 50931 Cologne, Germany
schaudt@uni-koeln.de, senger@zpr.uni-koeln.de

Abstract. A graph $G = (V, E)$ is called equidominating if there exists a value $t \in \mathbb{N}$ and a weight function $\omega \colon V \to \mathbb{N}$ such that the total weight of a subset $D \subseteq V$ is equal to t if and only if D is a minimal dominating set. To decide whether or not a given graph is equidominating is referred to as the EQUIDOMINATION problem.

In this paper we show that two parameterized versions of the EQUIDOMINATION problem are fixed-parameter tractable: the first parameterization considers the target value t leading to the TARGET-t EQUIDOMINATION problem. The second parameterization allows only weights up to a value k, which yields the k-EQUIDOMINATION problem.

In addition, we characterize the graphs whose every induced subgraph is equidominating. We give a finite forbidden induced subgraph characterization and derive a fast recognition algorithm.

Keywords: Minimal dominating set · Equidominating graph · Kernelization · Parameterized complexity · Hereditary property

1 Introduction

Let G be a simple, undirected graph. A subset S of the vertices of G is called a **dominating set**, if every vertex of G is an element of S or adjacent to a vertex of S. If a dominating set does not contain another dominating set as a subset, it is called a **minimal dominating set**. Throughout this paper we use the abbreviation **mds** for minimal dominating sets.

While the main stream of the research on dominating sets in graphs focuses on the optimization aspects of the problem, there are several interesting graph classes defined around this concept, for example the class of efficient dominating graphs [2], of well-dominated graphs [4], of domination perfect graphs [13], of upper domination perfect graphs [5] and of strong domination perfect graphs [11].

Another example is the class of **domishold** graphs, introduced by Benzaken and Hammer in [1]. These are the graphs for which there are positive weights associated to the vertices of the graph such that a subset D of vertices is dominating if and only if the sum of the weights of the vertices of D exceeds a certain threshold t. In other words, the characteristic vectors of the dominating sets are exactly the zero-one solutions of a linear inequality, where the coefficients of the inequality correspond to the weights of the vertices.

ⓒ Springer International Publishing AG 2017
H.L. Bodlaender and G.J. Woeginger (Eds.): WG 2017, LNCS 10520, pp. 412–424, 2017.
https://doi.org/10.1007/978-3-319-68705-6_31

This concept motivated Payan to define **equidominating** graphs [10]. Loosely speaking, these are the graphs for which the characteristic vectors of the minimal dominating sets are the zero-one solutions of a linear equality. Formally, equidominating graphs have the following definition.

Definition 1. *A graph* $G = (V, E)$ *is called* ***equidominating*** *if there exists a value* $t \in \mathbb{N} = \{1, 2, 3, \ldots\}$ *and a weight function* $\omega \colon V \to \mathbb{N}$ *such that for all* $D \subseteq V$ *the following equivalence holds:*

$$D \text{ is an mds} \iff \omega(D) := \sum_{v \in D} \omega(v) = t.$$

The pair (ω, t) *is called an* ***equidominating structure***, ω *an* ***equidominating function*** *and* t *a* ***target value***.

Figure 1 shows an equidominating graph. Every mds has a total weight of 23 and further, every subset of weight 23 is an mds. One advantage of having an equidominating structure of the graph at hand is that one can check whether a given vertex subset is an mds in linear time.

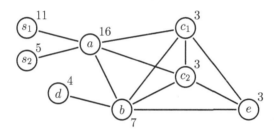

Fig. 1. An equidominating graph on 8 vertices; the weights are drawn next to the vertices and the target value is $t = 23$.

The EQUIDOMINATION problem is to decide whether a given graph is equidominating or not. Unfortunately, the computational complexity of this problem is unknown. It is not even clear whether EQUIDOMINATION is in NP.

It can be seen that the following problem is coNP-complete: given a graph G, a weight function ω and some number $t \in \mathbb{N}$, is (ω, t) an equidominating structure of G? This intractability result remains true in the seemingly simple case when G is just the disjoint union of edges. That can be seen by applying literally the same reduction from the WEAK PARTITION problem given by Milanič et al. [9], who proved coNP-completeness for the the analogous question for maximal stable sets.

We remark that there is no characterization of the class of equidominating graphs in terms of forbidden induced subgraphs: if one attaches a pendant vertex to every vertex of an arbitrary graph (the so-called corona of a graph with K_1), the resulting graph is equidominating. To our knowledge, the only existing result in this direction (see Theorem 2 in [10]) characterizes graphs that are both equidominating and domishold. Further, it is shown in [10] that threshold graphs are equidominating.

To get a grip on the computational complexity of the problem, we introduce the following two parameterized notions of equidomination.

Definition 2. *For a given* $t \in \mathbb{N}$ *a graph* $G = (V, E)$ *is called **target-t equidominating** if there is an equidominating structure of the form* (ω, t) *for G.*

Definition 3. *For a given* $k \in \mathbb{N}$ *a graph* $G = (V, E)$ *is called **k-equidominating** if there exists an equidominating structure* (ω, t) *with* $\omega \colon V \to \{1, \dots, k\}$ *for some* $t \in \mathbb{N}$. *In this case,* ω *is said to be a **k-equidominating function** and the pair* (ω, t) *a **k-equidominating structure**.*

Note that a k-equidominating graph is also k'-equidominating for all $k' \geq k$. It is clear that every target-t equidominating graph is also t-equidominating since every vertex is contained in some mds and thus its weight, with respect to any equidominating function, cannot exceed t. The opposite, however, is not true. Indeed, the edgeless graph on $t + 1$ vertices is k-equidominating for every $k \in \mathbb{N}$ but not target-t equidominating.

In this paper, we study the following two parameterized versions of the EQUIDOMINATION problem:

k-EQUIDOMINATION:

> *Instance:* A graph G and $k \in \mathbb{N}$
> *Parameter:* k
> *Problem:* Decide whether G is k-equidominating

TARGET-t EQUIDOMINATION:

> *Instance:* A graph G and $t \in \mathbb{N}$
> *Parameter:* t
> *Problem:* Decide whether G is target-t equidominating

In the course of this paper we show that both the k-EQUIDOMINATION problem and the TARGET-t EQUIDOMINATION problem are fixed-parameter tractable. We do this by using the so-called kernelization technique: we construct an equivalent instance the size of which is bounded by a function of the parameter. Further, we give FPT algorithms for both problems by applying an XP algorithm to the kernels. For a given graph on n vertices and m edges this leads to a running time of $\mathcal{O}(nm^2 + n^2 + t^{2t^2+3t+1})$ for the TARGET-t EQUIDOMINATION problem and $\mathcal{O}(nm^2 + n^2 + k^{6k^2+7k+1})$ for the k- EQUIDOMINATION problem.

In this extended abstract the proofs and most of the algorithms are omitted due to space constraints. The interested reader is referred to the upcoming full version of the paper[1].

---

[1] http://arxiv.org/abs/1705.05599.

This paper is structured as follows: In Sect. 2 we briefly introduce the twin relation. In Sect. 3 we examine relationships between the twin relation and equidomination and give an algorithm to discover certain structures that can appear. We state the existence of an XP algorithm for the k- EQUIDOMINATION problem (which can also be used for the TARGET-t EQUIDOMINATION problem) in Sect. 4. We make use of this algorithm and of reduction rules described in Sect. 5 to deduce the desired tractability results, stated in Sect. 6. Afterward, we characterize the class of hereditarily equidominating graphs in Sect. 7. In the last section we draw a conclusion and give a brief outlook.

2 Twin Relation

Two vertices v, w of a graph are called **twins** if $N(v) \setminus \{w\} = N(w) \setminus \{v\}$ holds. Here v and w can be either adjacent (**true twins**) or non-adjacent (**false twins**). It is easy to see that the **twin relation** (two vertices are related if they are twins) is an equivalence relation. The equivalence classes are called **twin classes** and the partition of the vertices into twin classes is called the **twin partition**.

All vertices of a twin class are either pairwise adjacent or pairwise non-adjacent. Therefore, twin classes are specified to be **clique classes** in the first and **stable set classes** in the latter case. A twin class can also be a single vertex. Even though a single vertex is strictly speaking a stable set as well as a clique, we use the terms clique class and stable set class only for twin classes with at least two elements. A twin class with one vertex is called a **singleton class**. In Fig. 1, c_1 and c_2 form a clique class, s_1 and s_2 a stable set class, and all other vertices singleton classes.

Let T_1 and T_2 be two twin classes. Then either every vertex of T_1 is adjacent to every vertex of T_2 or every vertex of T_1 is non-adjacent to every vertex of T_2. In the first case we say that T_1 and T_2 **see** each other and that T_1 sees T_2 and vice versa. We also say that a vertex and a twin class see each other, and likewise two vertices. Furthermore, if appropriate we use expressions for twin classes that are usually used for vertices (e.g. a twin class is adjacent to).

The twin partition can be computed in linear time using one of the modular decomposition algorithms of [3,8,12].

3 Properties of Twin Classes Regarding Equidomination

In this section we use the twin relation to obtain structural results regarding equidomination. Initially, we introduce the following definition.

Definition 4. *Let $G = (V, E)$ be a graph. Two vertices $x, y \in V$ are called **mds-exchangeable** if and only if there exists an mds $D \subseteq V$ with $|\{x, y\} \cap D| = 1$ and if for all mds $D \subseteq V$ with $|\{x, y\} \cap D| = 1$ the symmetric difference $(D \setminus \{x, y\}) \cup (\{x, y\} \setminus D)$ is also an mds.*

Loosely speaking, two vertices are mds-exchangeable if they can be exchanged for another in any mds containing exactly one of them. Two mds-exchangeable vertices of an equidominating graph must have the same weight in every equidominating function. Even if two vertices are mds-exchangeable, they can both be elements of one mds (e.g. two non-adjacent vertices of C_4). While c_1, c_2 and e in Fig. 1 are mds-exchangeable, s_1 and s_2 are not since there is no mds containing only one of the two vertices.

As the following observations show, the twin relation is a very helpful instrument with regard to equidomination.

Observation 1. *For every minimal dominating set D and every stable set class S, we have $|D \cap S| \in \{0, 1, |S|\}$.*

Observation 2. *For every minimal dominating set D and every clique class C, we have $|D \cap C| \le 1$.*

Observation 3. *Since the vertices of a clique class are pairwise mds-exchangeable, every equidominating function must be constant on every clique class. Analogously, if there exist an mds containing exactly one vertex of a stable set class, every equidominating function must be constant on that stable set class.*

Observation 4. *Since every maximal stable set is also an mds, every (non-maximal) stable set can be extended to an mds. Therefore, for every stable set S and for every equidominating structure (ω, t) it holds that $\omega(S) \le t$ and hence $|S| \le t$.*

In the following lemmas we examine whether different vertices of an equidominating graph can have equal weights or not. As we will see two vertices can only have the same weight if they lie in the same twin class or are adjacent. That means for one thing, that when trying to construct an equidominating structure one has to consider fewer combinatorial possibilities. And for another thing, that for a given number of weights to be allocated one can bound the diameter of an equidominating graph.

Lemma 1. *Let $G = (V, E)$ be an equidominating graph with equidominating structure (ω, t) and let $x, y \in V$ be two vertices of different twin classes with $dist(x, y) \ge 2$. Then it holds that $\omega(x) \ne \omega(y)$.*

Note that in the previous lemma the two mentioned vertices must be from different twin classes. Two elements of a stable set class of course can have the same weight while always having distance at least two.

In the following we take a closer look at adjacent vertices, where we find a slightly more complicated situation. We begin by showing that vertices of stable set classes and clique classes that see each other cannot have the same weight.

Lemma 2. *Let $G = (V, E)$ be an equidominating graph with equidominating structure (ω, t) and let $S \subseteq V$ be a stable set class and $C \subseteq V$ a clique class that see each other. Then for all $x \in S$ and for all $y \in C$ it holds that $\omega(x) \ne \omega(y)$.*

Further, a vertex of a stable set class and an adjacent singleton class cannot have the same weight in an equidominating structure.

Lemma 3. *Let* $G = (V, E)$ *be an equidominating graph with equidominating structure* (ω, t). *Let* $S \subseteq V$ *be a stable set class and let* $\{y\}$ *be a singleton class with* $y \in N(S)$. *Then for all* $x \in S$ *it holds that* $\omega(x) \neq \omega(y)$.

As the next Lemma shows, adjacent stable set classes can have same weights, but only in a specific situation.

Lemma 4. *Let* $G = (V, E)$ *be an equidominating graph with equidominating structure* (ω, t) *and let* $S_1, S_2 \subseteq V$ *be two adjacent stable set classes with two vertices* $x \in S_1$, $y \in S_2$ *of the same weight. Then the following assertions hold:*

(i) $|S_1| = |S_2| = 2$,
(ii) ω *is constant on* $S_1 \cup S_2$,
(iii) *every twin class seen by* S_1 *is also seen by* S_2 *and vice versa.*

Further, if two adjacent stable set classes of size two have the same closed neighborhood, all vertices of those stable set classes have the same weight in any equidominating structure.

As a consequence of Lemma 4 there can be an arbitrarily large number of stable set classes of size two with vertices of the same weight. Such an occurrence could be a problem when trying to achieve bounded kernels for the parameterized problems. But the good thing is that all of those stable set classes see each other and see the same twin classes in the remainder of the graph. Therefore, as we will see, it is possible to reduce them to a manageable number. For a better handling we introduce the following new term. For that we define $K_{2n} - ne$ ($n \in \mathbb{N}$) to be a complete graph on $2n$ vertices from which n disjoint edges are removed.

Definition 5. *Let* $G = (V, E)$ *be a graph and* $\mathcal{S} \subseteq V$ *be a maximal subset such that* $G[\mathcal{S}] \cong K_{2n} - ne$ *for some* $n \geq 2$ *and such that each vertex of* $\mathcal{S}$ *is adjacent to the same vertices in* $V \setminus \mathcal{S}$. *Then* $\mathcal{S}$ *is called a **stable set bundle**.*

Maximal here means, that no other subset fulfills the two conditions and properly contains $\mathcal{S}$. Every stable set bundle contains several stable set classes of size two. Note that a stable set bundle can be created by adding a false twin to every vertex of a clique class. Following Lemma 4 the vertices of a stable set bundle are pairwise mds-exchangeable and, therefore, every equidominating function is constant on a stable set bundle.

Now, regarding the question whether two vertices can have the same weight in an equidominating structure, the last open question is: can vertices of a clique class or a singleton class have the same weight as its neighboring clique classes or singleton classes? The answer is yes: there can be a clique, that is not a clique class, the vertices of which are pairwise mds-exchangeable.

Definition 6. *Let* G *be a graph and* $\mathcal{C}$ *an inclusion-wise maximal clique of pairwise mds-exchangeable vertices that contains at least two twin classes. Then* $\mathcal{C}$ *is called a **clique bundle**.*

Upon first reading it seems a little bit odd to define clique bundles exactly as what we are looking for: pairwise mds-exchangeable vertices of possibly different twin classes. However, the crucial thing here is that we can identify clique bundles efficiently (see Algorithm 1). We require at least two twin classes to be in a clique bundle in order that a twin class on its own is not a clique bundle and a twin class at the same time.

In a clique bundle there can be both clique classes and singleton classes but no stable set classes. In the graph shown in Fig. 1 the clique class $\{c_1, c_2\}$ and the singleton class $\{e\}$ form a clique bundle. In Fig. 2 you can see an equidominating graph which consists of two clique bundles. In this graph every mds contains exactly one vertex of each clique bundle.

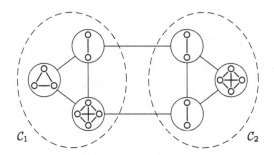

Fig. 2. Example of an equidominating graph consisting of the two clique bundles C_1 and C_2 each containing three clique-classes with two to four vertices. The clique classes are indicated by circles and an edge between the circles of two clique classes represents all edges between the vertices of the corresponding clique classes.

We use the term **bundle** to refer to either a stable set bundle or a clique bundle. Recall that a twin class is either a stable set class, a clique class or a singleton class. Due to the existence of bundles we introduce a sort of generalization of twin classes: a **pseudo class** is either a twin class not contained in a bundle or a stable set bundle or a clique bundle. That is, a pseudo class is exactly one of following: (a) a singleton class, (b) a stable set class, (c) a clique class, (d) a stable set bundle or (e) a clique bundle. By this definition we get the following result.

Corollary 1. *There is a unique partition of the vertices of a graph into pseudo classes.*

The introduction of pseudo classes is motivated by the following corollary, which summarizes this section.

Corollary 2. *Let G be an equidominating graph with equidominating structure (ω, t) and P_1, P_2 be two different pseudo classes. Then for all $x \in P_1$ and $y \in P_2$ it holds that $\omega(x) \neq \omega(y)$.*

To recognize bundles we developed Algorithm 1, which decides for two adjacent vertices whether or not they are mds-exchangeable. More precisely, the algorithm checks if there can be private neighbors of one vertex in any dominating set that are not seen by the other one. After computing the twin partition one can apply Algorithm 1 to adjacent clique classes and singleton classes, and adjacent stable set classes of size two to find all clique bundles and stable set bundles.

Algorithm 1. Checking adjacent vertices for mds-exchangeability

---

Input: Two adjacent vertices $x, y \in V$ of a graph $G = (V, E)$
Output: YES, if x and y are mds-exchangeable, otherwise **NO**
1: **for all** $(v_1, v_2) \in \{(x, y), (y, x)\}$ **do**   ▷ Check both combinations
2:    **for all** $v' \in N(v_1) \setminus N[v_2]$ **do**
3:        **if** $\{v_1\} \cup \Big(V(G) \setminus \big(N[v'] \cup \{v_2\}\big)\Big)$ is a dominating set **then**
4:            return **NO**   ▷ x and y are not mds-exchangeable
5:        **end if**
6:    **end for**
7: **end for**
8: return **YES**   ▷ x and y are mds-exchangeable

Proposition 1. *Algorithm 1 is correct and runs in $\mathcal{O}(nm)$ time.*

To discover all bundles one has to apply Algorithm 1 for every edge, which gives us a total running time of $\mathcal{O}(nm^2)$.

4   An XP Algorithm for the k-Equidomination Problem

In this section we describe an algorithm which decides whether a given graph is k-equidominating for some fixed $k \in \mathbb{N}$ with a running with only k appearing in the exponents. The aim is to apply this algorithm to the constructed kernels of the parameterized problems. The algorithm mainly follows the ideas and the algorithm for the k-Equistability problem of Levit et al. [6,7]. However, it has to be extended due to the existence of clique bundles and stable set bundles.

The basic idea of the algorithm is that by considering the pseudo classes one does not have to examine every possible weight function nor every possible subset of vertices. Since different vertices of the same pseudo class, roughly said, play the same role regarding domination, two weight functions that differ only by switched weights for two vertices of one pseudo class, can be handled as the same. This leads to equivalence classes of weight functions. Further, we reduce the running time from a brute force algorithm by classifying subsets of vertices. It does not matter, for example, which vertex of a clique class is in an mds. Then, we have to check only one subset per class for being a minimal dominating set.

Theorem 1. *For a given $k \in \mathbb{N}$ it is decidable if a graph $G = (V, E)$ is k-equidominating or not in time $\mathcal{O}\left(nm^2 + n^k k^k + n^{2k+2}k^{-k-1} + k^{3k+3}\right)$ with $|V| = n$ and $|E| = m$ and a k-equidominating structure is computed in this time. Further, for $t \in \mathbb{N}$ the same algorithm can be used to decide if G is target-t equidominating.*

The algorithm also provides the domination number, the upper domination number and minimal dominating sets of minimum and maximum cardinality.

5   Reduction Rules

In the following we examine three reduction rules which we use to construct kernels of the TARGET-t EQUIDOMINATION problem as well as the k-EQUIDOMINATION problem. A graph is called **target-t k-equidominating** if there is a k-equidominating structure with target value t. Note that this is stronger than being k-equidominating and target-t equidominating.

The first rule is about reducing a clique class to a certain number $r \in \mathbb{N}$ of vertices.

r-Clique Class Reduction: If a clique class C contains more than r vertices, delete all but r vertices from C.

Lemma 5. *Let G be a graph, $r, k \in \mathbb{N}$ and $C \subseteq V(G)$ a clique class with $|C| > r$. Furthermore, let G' be the graph obtained from G by applying the r-Clique Class Reduction rule with respect to C. Then for all $t \leq r$ the graph G is target-t k-equidominating if and only if G' is target-t k-equidominating.*

The next rule is about the previous defined stable set bundles. As seen before there can be arbitrarily large stable set bundles in an equidominating graph. However, we can reduce them to a suitable size. Again a positive integer $r \in \mathbb{N}$ specifies the reduction rule.

r-Stable Set Bundle Reduction: If a stable set bundle S contains more than r stable set classes, delete all but r stable set classes of S.

The following lemma shows that the r-Stable Set Bundle Reduction rule can be used to construct kernels for the parameterized problems.

Lemma 6. *Let G be a graph, $r, k \in \mathbb{N}$ and $S = \{S_1, \ldots, S_N\}$ a stable set bundle with $N > r$. Further, let G' be the graph obtained from G by applying the r-Stable Set Bundle Reduction rule to S. Then for all $t \leq 2r$ the graph G is target-t k-equidominating if and only if G' is target-t k-equidominating.*

The last reduction rules deals with clique bundles. It is motivated by the fact that if a clique bundle only consists of singleton classes, then more than one vertex of such a clique bundle can be in an mds. Before we can state the reduction rule we need the following definition.

Definition 7. *Let G be a graph with pseudo class partition $\{P_1, \ldots, P_s\}$. For every vertex $v \in V(G)$ we define the vector $\mu_v = (\mu_v(1), \ldots, \mu_v(s)) \in \mathbb{N}_0^s$ as follows: for $i = 1, \ldots, s$, if $v \in P_i$, then we set $\mu_v(i) := 0$. If $v \notin P_i$ and P_i is a clique bundle, then we set*

$$\mu_v(i) := \begin{cases} |P_i \setminus N(v)| + 1, & \text{there is an mds } D \text{ with } N(v) \cap D \subseteq P_i , \\ 0, & \text{otherwise} . \end{cases}$$

If $v \notin P_i$ and P_i is a singleton class, a clique class, a stable set class or a stable set bundle, then we set

$$\mu_v(i) := \begin{cases} 1, & v \text{ is adjacent to the vertices of } P_i , \\ 0, & \text{otherwise} . \end{cases}$$

To decide if there exist an mds D with $N[v] \cap D \subseteq P_i$ we simply check if $V \setminus (N[v] \setminus P_i)$ is a dominating set. The number $\mu_v(i)$ is of particular interest when P_i is a clique bundle. It tells us how many vertices of a pseudo class P_i must be at least in an mds D with $N[v] \cap D \subseteq P_i$. We will only work with vectors of vertices that lie in clique bundles.

The values of the μ_v are bounded by t in every target-t equidominating graph.

Lemma 7. *Let $G = (V, E)$ be a graph with pseudo class partition $\{P_1, \ldots, P_s\}$ and let $r \in \mathbb{N}$. If there is a vertex $v \in V$ with $\mu_v(i) > r$ for some $i \in \{1, \ldots, s\}$, then G is not target-t equidominating for $t \leq r$.*

As before, we reduce the vertices of a subset of a clique bundle to a certain number $r \in \mathbb{N}$.

r-Clique Bundle Reduction: If a subset $M \subseteq \mathcal{C}$ of a clique bundle $\mathcal{C}$ with $\mu_v = \mu_w$ for all $v, w \in M$ contains more than r vertices, delete all but r vertices of M.

Lemma 8. *Let G be a graph, $r, k \in \mathbb{N}$ and $M \subseteq \mathcal{C}$ a subset of a clique bundle $\mathcal{C}$ with $\mu_v = \mu_w$ for all $v, w \in M$ and $|M| > r$. Furthermore, let G' be the graph obtained from G by applying the r-Clique Bundle Reduction rule with respect to M. Then for all $t \leq r$ the graph G is target-t k-equidominating if and only if G' is target-t k-equidominating.*

6 Main Results

In this section we state that both parameterized equidomination problems admit kernels, the sizes of which are bounded by a function of t, resp. k. Thus, both problems are FPT.

Theorem 2. *The TARGET-t EQUIDOMINATION problem admits a $\mathcal{O}(t^{t+1})$-vertex kernel which is computable in polynomial time. Moreover, there is an algorithm to solve the TARGET-t EQUIDOMINATION problem for a graph on n vertices and m edges which runs in time $\mathcal{O}\left(nm^2 + n^2 + t^{2t^2+3t+1}\right)$.*

Theorem 3. *The k-EQUIDOMINATION problem admits a $\mathcal{O}(k^{3k+1})$-vertex kernel which is computable in polynomial time. Furthermore, there is an algorithm to solve the k-EQUIDOMINATION problem for a given graph on n vertices and m edges which runs in time $\mathcal{O}\left(nm^2 + n^2 + k^{6k^2+7k+1}\right)$.*

The proof of this theorem builds upon two more lemmas.

Lemma 9. *A graph $G = (V, E)$ is not k-equidominating ($k \in \mathbb{N}$) if G has two pseudo classes of size at least k^2, where one of those pseudo classes is a stable set class.*

Lemma 10. *Let $G = (V, E)$ be a graph and $k \in \mathbb{N}$. Furthermore, let $\{S, V'\}$ be a partition of V where S is a set of isolated vertices of size at least k^5 and $|V'| \leq k^3$. Then G is k-equidominating if and only if there exists a k-equidominating function that is constant on S.*

7   Hereditarily Equidominating Graphs

A graph G is called **hereditarily equidominating** if every induced subgraph of G is equidominating. In this section, we give a characterization of the class of hereditarily equidominating graphs in terms of the list of forbidden induced subgraphs and a structural decomposition. This decomposition yields an $\mathcal{O}(n(n+m))$ time recognition algorithm.

In order to state our characterization, we need some more notions. A **chain graph** is a bipartite graph where the neighborhoods of the vertices of either side are comparable with respect to inclusion. Let G_1 and G_2 be two disjoint graphs. Let U_i be the (possibly empty) set of universal vertices of G_i, for $i \in \{1, 2\}$. Let B be any chain graph with bipartition U_1, U_2. We call the graph $(G_1 \cup G_2) + E(B)$ a **chain-join** of G_1 and G_2. Note that the disjoint union of any two graphs is a particular chain-join of these two graphs.

Regarding our decomposition theorem below, the class of **basic graphs** equals $\{K_1\} \cup \{K_{2n} - ne : n \geq 2\}$. An equidominating structure of $K_{2n} - ne$, $n \geq 2$, is given by $\omega \equiv 1$ and $t = 2$. Hence, basic graphs are equidominating. It is not hard to prove that basic graphs are in fact hereditarily equidominating. Interestingly, the basic graphs (except for K_1) are those graphs which consist of one stable set bundle.

Let $\mathcal{F} := \{P_5, C_5, bull, banner, house, K_{2,3}, \overline{P_2 \cup P_3}\}$ (see Fig. 3 for an illustration). As the next theorem shows, the set $\mathcal{F}$ is exactly the set of forbidden induced subgraphs of the class of hereditarily equidominating graphs.

Theorem 4. *For any graph G, the following assertions are equivalent.*

(a) G is hereditarily equidominating.
(b) G is $\mathcal{F}$-free.
(c) One of the following assertions holds.
   (i) G is a basic graph.

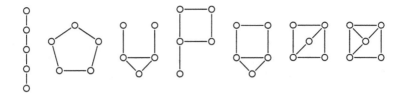

Fig. 3. The forbidden set $\mathcal{F}$; P_5, C_5, *bull, banner, house*, $K_{2,3}$, $\overline{P_2 \cup P_3}$.

(ii) G is obtained from a hereditarily equidominating graph by adding a universal vertex.

(iii) G is the chain-join of two hereditarily equidominating graphs.

Given the fact that hereditarily equidominating graphs admit a finite forbidden subgraph characterization, it is clear that this class can be recognized efficiently. A faster recognition is possible using the decomposition provided by Theorem 4.

Corollary 3. *Let G be a graph on n vertices and m edges. It can be decided in time $\mathcal{O}(n(n+m))$ whether G is a hereditarily equidominating graph.*

8   Conclusion and Outlook

The main result of this paper is that the EQUIDOMINATION problem can be parameterized in two different ways such that the parameterized problems are fixed-parameter tractable. One way is to take the target value t of an equidominating structure as the parameter: this problem is called the TARGET-t EQUIDOMINATION problem. The second way is to allow only vertex weights up to a certain value k, which leads to the k- EQUIDOMINATION problem.

To solve the kernelized instances we developed an XP algorithm that can be used for both problems. Even though there are many analogies between equistability and equidomination – as between the concepts of stability and domination in general – there are new difficulties in the case of equidomination. This fact results in several extensions we needed to discover and formalize. Finally, we characterized hereditarily equidominating graphs in terms of seven forbidden subgraphs, which also leads to a polynomial time recognition algorithm.

There are several open questions that arise in the context of equidomination. Most importantly, we would like to see a hardness proof of the EQUIDOMINATION problem. For this, however, it might be necessary to get a better grip on the combinatorial properties of equidominating graphs. One way in this direction could be a characterization of target-t equidominating and k-equidominating graphs for small $t, k \in \mathbb{N}$.

Further, one could adapt the idea of equidomination – namely characterizing mds by the 0-1-solutions of linear equalities – to other forms of domination like total, multiple or global domination. Thoughts in this direction could concern a characterization of these graph classes as well as the complexity of the according decision problems.

References

1. Benzaken, C., Hammer, P.: Linear separation of dominating sets in graphs. In: Bollobs, B. (ed.) Advances in Graph Theory, Annals of Discrete Mathematics, vol. 3, pp. 1–10. Elsevier (1978), http://www.sciencedirect.com/science/article/pii/S0167506008704928

2. Brandstädt, A., Leitert, A., Rautenbach, D.: Efficient dominating and edge dominating sets for graphs and hypergraphs. In: Chao, K.-M., Hsu, T., Lee, D.-T. (eds.) ISAAC 2012. LNCS, vol. 7676, pp. 267–277. Springer, Heidelberg (2012). doi:10.1007/978-3-642-35261-4_30

3. Cournier, A., Habib, M.: A new linear algorithm for modular decomposition. In: Tison, S. (ed.) CAAP 1994. LNCS, vol. 787, pp. 68–84. Springer, Heidelberg (1994). doi:10.1007/BFb0017474

4. Finbow, A., Hartnell, B., Nowakowski, R.: Well-dominated graphs: a collection of well-covered ones. Ars Combin. **25**, 5–10 (1988)

5. Gutin, G., Zverovich, V.E.: Upper domination and upper irredundance perfect graphs. Discrete Math. **190**(1), 95–105 (1998). http://www.sciencedirect.com/science/article/pii/S0012365X98000363

6. Kim, E.J., Milanič, M., Schaudt, O.: Recognizing k-equistable graphs in FPT time. In: Mayr, E.W. (ed.) WG 2015. LNCS, vol. 9224, pp. 487–498. Springer, Heidelberg (2016). doi:10.1007/978-3-662-53174-7_34

7. Levit, V.E., Milanič, M., Tankus, D.: On the recognition of k-equistable graphs. In: Golumbic, M.C., Stern, M., Levy, A., Morgenstern, G. (eds.) WG 2012. LNCS, vol. 7551, pp. 286–296. Springer, Heidelberg (2012). doi:10.1007/978-3-642-34611-8_29

8. McConnell, R.M., Spinrad, J.P.: Modular decomposition and transitive orientation. Discrete Math. **201**(13), 189–241 (1999). http://www.sciencedirect.com/science/article/pii/S0012365X98003197

9. Milanič, M., Orlin, J., Rudolf, G.: Complexity results for equistable graphs and related classes. Ann. Oper. Res. **188**, 359–370 (2011). doi:10.1007/s10479-010-0720-3

10. Payan, C.: A class of threshold and domishold graphs: equistable and equidominating graphs. Discrete Math. **29**(1), 47–52 (1980). doi:10.1016/0012-365X(90)90286-Q

11. Rautenbach, D., Zverovich, V.: Perfect graphs of strong domination and independent strong domination. Discrete Math. **226**(1), 297–311 (2001). http://www.sciencedirect.com/science/article/pii/S0012365X00001163

12. Tedder, M., Corneil, D., Habib, M., Paul, C.: Simpler linear-time modular decomposition via recursive factorizing permutations. In: Aceto, L., Damgård, I., Goldberg, L.A., Halldórsson, M.M., Ingólfsdóttir, A., Walukiewicz, I. (eds.) ICALP 2008. LNCS, vol. 5125, pp. 634–645. Springer, Heidelberg (2008). doi:10.1007/978-3-540-70575-8_52

13. Zverovich, I.E., Zverovich, V.E.: A characterization of domination perfect graphs. J. Graph Theor. **15**(2), 109–114 (1991). doi:10.1002/jgt.3190150202

Homothetic Triangle Contact Representations

Hendrik Schrezenmaier$^{(\boxtimes)}$

Institut für Mathematik, Technische Universität Berlin, Berlin, Germany
schrezen@math.tu-berlin.de

Abstract. We prove that every 4-connected planar triangulation admits a contact representation by homothetic triangles.

There is a known proof of this result that is based on the Convex Packing Theorem by Schramm, a general result about contact representations of planar triangulations by convex shapes. But our approach makes use of the combinatorial structure of triangle contact representations in terms of Schnyder woods. We start with an arbitrary Schnyder wood and produce a sequence of Schnyder woods via face flips. We show that at some point the sequence has to reach a Schnyder wood describing a representation by homothetic triangles.

Keywords: Contact representation · Schnyder wood · Triangle · Planar triangulation

1 Introduction

A *triangle contact system* $\mathcal{T}$ is a finite system of triangles in the plane such that any two triangles intersect in at most one point. The contact system is *nondegenerate* if every contact involves exactly one corner of a triangle. The graph $G^*(\mathcal{T})$ is the plane graph that has a vertex for every triangle of $\mathcal{T}$ and an edge for every contact of two triangles in $\mathcal{T}$. For a given plane graph G and a triangle contact system $\mathcal{T}$ with $G^*(\mathcal{T}) = G$ we say that $\mathcal{T}$ is a *triangle contact representation* of G.

The main goal of this paper will be to prove the following result.

Theorem 1 [5]**.** *Let G be a 4-connected planar triangulation. Then there is a triangle contact representation of G by homothetic triangles.*

The original proof of Theorem 1 in [5][1] makes use of the following theorem by Schramm.

Theorem 2 (Convex Packing Theorem [10]**).** *Let G be a triangulation with outer face $\{a, b, c\}$. Further let C be a simple closed curve in the plane partitioned into arcs $\mathcal{P}_a, \mathcal{P}_b, \mathcal{P}_c$ and for each interior vertex v of G let $\mathcal{Q}_v$ be a convex set in the plane containing more than one point. Then there is a contact representation of a supergraph of G (on the same vertex set, but possibly with more edges) where each interior vertex v is represented by a homothetic copy of its prototype $\mathcal{Q}_v$ and each outer vertex w by the arc $\mathcal{P}_w$.*

---

[1] The journal version [6] does not contain this proof.

© Springer International Publishing AG 2017
H.L. Bodlaender and G.J. Woeginger (Eds.): WG 2017, LNCS 10520, pp. 425–437, 2017.
https://doi.org/10.1007/978-3-319-68705-6_32

If we want to calculate a homothetic triangle contact representation of G efficiently, this theorem does not help since it is purely existential. On the other hand, Felsner [2] introcuced a combinatorial heuristic that calculates triangle contact representations quite fast in practical experiments [8]. However, to the best of our knowledge, it is not known whether this heuristic terminates for every instance, nor whether it has a good (e.g., polynomial) running time if it terminates.

The heuristic starts by guessing the combinatorial structure of the contact representation in the form of a Schnyder wood. Then a system of linear equations is solved whose variables correspond to the lengths of the segments of the triangles in the contact representation. If the solution is nonnegative, this yields the intended contact representation. Otherwise, the negative variables of the solution can be used as sign-posts indicating how to change the Schnyder wood for another try.

Our new proof of Theorem 1 is based on the theoretical background of this heuristic. Therefore it might help to better understand this heuristic in the future. Felsner and Francis [3] even explicitly ask for a proof of Theorem 1 by this approach. Further, our proof motivates a new heuristic for calculating homothetic triangle contact representations.

A substantial part of this work originates in the author's Masters thesis [12]. In this thesis with a similar approach also the existence of contact representations of 5-connected planar triangulations by homothetic squares has been proved. But in that case there are other known proofs which are not based on the Convex Packing Theorem [7,11]. That is why we will focus on contact representations by homothetic triangles in this paper.

Let us get back to triangle contact representations. In the case that $G^*(T)$ is a planar triangulation, in T the inner (i.e., bounded) faces of $G^*(T)$ are also represented by triangles. We denote these by *dual triangles* and for clear distinction the triangles of T by *primal triangles*.

In Theorem 1 we do not specify what is the shape of the homothetic triangles. The reason is that if we are given a contact representation by homothetic triangles, we can change the shapes of these triangles to homothetic copies of an arbitrary given triangle by a linear transformation of the plane. So we choose to prove the existence of a contact representation by right, isosceles triangles with a horizontal edge at the bottom and a vertical edge at the right hand side. We will even consider a larger class of triangle contact representations. A *right triangle contact representation* is a triangle contact representation by right triangles with a horizontal edge at the bottom and a vertical edge at the right hand side. The *aspect ratio* of such a triangle is the quotient of the lengths of its vertical and its horizontal edge. The *aspect ratio vector* of a right triangle contact representation is the vector of the aspect ratios of its triangles (we assume the vertices of G have a fixed numbering $1, \ldots, n+3$). See Fig. 1 for an example of a right triangle contact representation. Now we can formulate a stronger theorem that implies Theorem 1.

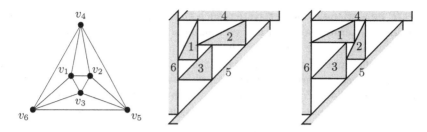

Fig. 1. Two right triangle contact representations of the same graph with aspect ratio vectors $(2, 1/2, 1, 1, 1, 1)$ and $(1/2, 2, 1, 1, 1, 1)$.

Theorem 3. *Let G be a 4-connected triangulation and $\tilde{r} \in \mathbb{R}_{>0}^{n+3}$. Then there is a right triangle contact representation of G with aspect ratio vector $\tilde{r}$.*

The paper is organized as follows: In Sect. 2 we give an introduction to *Schnyder woods* as the combinatorial structure describing triangle contact representations and recall some known results about them. In Sect. 3 we describe a variant of the system of linear equations from the heuristic by Felsner for calculating a right triangle contact representation with given Schnyder wood and given aspect ratio vector. As our main contribution, we prepare the prove of Theorem 3 in Sect. 4 and give the proof in Sect. 5. In Sect. 6 we propose a new heuristic for calculating right triangle contact representation based on this proof.

2   Schnyder Woods

Schnyder woods are a combinatorial structure on triangulations that play a central role in this paper. They were first introduced by Schnyder [9] under the name of *realizers*.

Definition 1. *Let G be a triangulation with outer vertices $v_{n+1}, v_{n+2}, v_{n+3}$ in clockwise order. Then a Schnyder wood of G is an orientation and coloring of the interior edges of G with the colors red, green and blue such that*

- *each edge incident to v_{n+1} is red and incoming, each edge incident to v_{n+2} is green and incoming, and each edge incident to v_{n+3} is blue and incoming,*
- *each inner vertex has in clockwise order exactly one red, green and blue outgoing edge, and in the interval between two outgoing edges there are only incoming edges in the third color (see Fig. 2).*

If we forget about the colors of a Schnyder wood, we obtain a 3-orientation, i.e., each inner vertex has outdegree 3 and the outer vertices have outdegree 0. The converse also holds:

Proposition 1 (de Fraysseix and Ossona de Mendez [13]). *If the graph G is a 3-orientation of a triangulation with outer vertices labeled $v_{n+1}, v_{n+2}, v_{n+3}$ in clockwise order, then there is a unique way of coloring the interior edges of G to receive a Schnyder wood.*

Let $\mathcal{T}$ be a triangle contact system such that $G := G^*(\mathcal{T})$ is a triangulation. If $\mathcal{T}$ is nondegenerate, we can orient each inner edge of G from the triangle whose corner is involved in the contact to the other triangle and obtain an orientation where the outdegree of each inner vertex is at most 3. In the case that $\mathcal{T}$ is degenerate, we can interpret a point where several triangle corners meet as a cyclic sequence of nondegenerate contacts with infinitesimal edge lengths and proceed as in the nondegenerate case. As a consequence of Euler's formula, G has exactly $3n$ inner edges, and therefore the outdegree of each inner vertex has to be exactly 3. Thus $\mathcal{T}$ induces a 3-orientation and hence a Schnyder wood of G. We will call this Schnyder wood the (induced) Schnyder wood of $\mathcal{T}$. Note that the induced Schnyder wood is not unique if $\mathcal{T}$ is degenerate. The following proposition shows that every Schnyder wood is an induced Schnyder wood.

Proposition 2 (de Fraysseix et al. [14]). *Let G be a triangulation and S a Schnyder wood of G. Then there exists a nondegenerate right triangle contact representation of G with induced Schnyder wood S.*

For right triangle contact representations we can obtain the colors of the edges of the associated Schnyder wood also directly, without using Proposition 1. We color an edge red if it corresponds to the upper corner of a triangle, green if it corresponds to the right lower corner of a triangle, and blue if it corresponds to the left lower corner of a triangle. See Fig. 3 for an example.

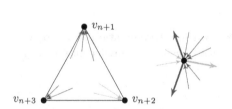

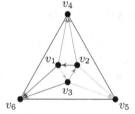

Fig. 2. The local conditions of a Schnyder wood. (Color figure online)

Fig. 3. The Schnyder wood induced by the first example of Fig. 1. (Color figure online)

3   The System of Linear Equations

For the whole section let G be a triangulation with inner vertices $v_1, \ldots, v_n$ and outer vertices $v_{n+1}, v_{n+2}, v_{n+3}$ in clockwise order, let S be a Schnyder wood of G and let $r \in \mathbb{R}_{>0}^{n+3}$ be an aspect ratio vector for G. We will now describe a system of linear equations for calculating the edge lengths of a right triangle contact representation of G with aspect ratio vector r and induced Schnyder wood S. This system has been introduced by Felsner [2] and studied by Rucker [8] for the special case $r = (1, \ldots, 1)$. All results in this section are due to them.

For each inner vertex v of G we have a variable x_v which represents the width of the corresponding primal triangle, and for each inner face f of G a variable x_f which represents the width of the corresponding dual triangle. For a vertex v of G we denote by $\delta_r(v)$ the set of incident faces of v which are located in the interval between the green and blue outgoing edge of v. Analogously we define the sets $\delta_g(v)$ and $\delta_b(v)$. In a right triangle contact representation a primal triangle T hands down his aspect ratio to each dual triangle whose hypotenuse is contained in the hypotenuse of T. If T corresponds to the vertex v of G, these are exactly the dual triangles corresponding to the faces in $\delta_g(v)$. Therefore, if we are given the Schnyder wood and the aspect ratio vector of a right triangle contact representation, we are implicitly also given the aspect ratios of the dual triangles. We denote the aspect ratio of a dual triangle corresponding to the face f of G by r_f. Now we can write down the equation system:

$$\sum_{f \in \delta_r(v_{n+1})} x_f = 1, \tag{1}$$

$$\sum_{f \in \delta_r(v_i)} x_f - x_{v_i} = 0, \quad i = 1, \ldots, n, \tag{2}$$

$$\sum_{f \in \delta_g(v_i)} x_f - x_{v_i} = 0, \quad i = 1, \ldots, n, \tag{3}$$

$$\sum_{f \in \delta_b(v_i)} r_f x_f - r_{v_i} x_{v_i} = 0, \quad i = 1, \ldots, n. \tag{4}$$

Equation (2) says that the length of the horizontal edge of a primal triangle is equal to the sum of the lengths of the adjacent dual triangles. Equations (3) and (4) say the same for the other two edges of a primal triangle. Note that the sum of the lengths of the diagonal edges of the dual triangles corresponding to the faces in $\delta_g(v_i)$ is equal to the length of the diagonal edge of the primal triangle of v_i if and only if the sum of the lengths of their horizontal edges is equal to the length of the horizontal edge of the primal triangle of v_i. The purpose of (1) is to pick one single solution out of the space of solutions of the apart from that homogeneous equation system. We will also use the shorter notation $A_S(r)x = \mathbf{e_1}$ for the equation system.

Proposition 3. *The system $A_S(r)x = \mathbf{e_1}$ is uniquely solvable.*

Because of the way we chose the equations of the system, it is clear that the existence of a right triangle contact representation of G with Schnyder wood S and aspect ratio vector r implies a nonnegative solution. The following proposition shows that also the converse holds.

Proposition 4. *Let $A_S(r)x = \mathbf{e_1}$. There is a right triangle contact representation of G with induced Schnyder wood S and aspect ratio vector r if and only if $x \geq 0$. If $x \geq 0$, the representation is unique inside the three outer triangles.*

Next we will prove a result about the structure of nonnegative solutions with zero entries.

Definition 2. *Let $A_S(r)x = \mathbf{e_1}$ and $x \geq 0$. Then an edge e of G is called a transition edge if it is incident to inner faces f_1 and f_2 of G with $x_{f_1} > 0$ and $x_{f_2} = 0$.*

Lemma 1. *Let $A_S(r)x = \mathbf{e_1}$ and $x \geq 0$. Then the transition edges of G form an edge disjoint union of cycles of length 3. Moreover there is no edge going from a vertex on such a cycle into the interior of this cycle.*

For the sake of completeness we will now briefly describe the heuristic by Felsner [2] for calculating triangle contact representations that is based on the system of linear equations. Felsner only considers homothetic triangle contact representations, but the heuristic can easily be translated to the case of right triangle contact representations with given aspect ratio vector $\tilde{r}$. For the heuristic we need a result similar to Lemma 1 for the case $x \not\geq 0$.

Definition 3. *Let $A_S(r)x = \mathbf{e_1}$. Then an edge e of G is called a sign-separating edge if it is incident to inner faces f_1 and f_2 of G with $x_{f_1} \geq 0$ and $x_{f_2} < 0$.*

Lemma 2. *Let $A_S(r)x = \mathbf{e_1}$. Then the sign-separating edges of G form an edge-disjoint union of directed simple cycles.*

The heuristic starts with an arbitrary Schnyder wood S and solves the system of linear equations $A_S(\tilde{r})x = \mathbf{e_1}$. If the solution is nonnegative, we are done and can compute the contact representation due to Proposition 4. Otherwise, we change in S the orientation of all sign-separating edges. Since the set of sign-separating edges is a disjoint union of simple cycles, we obtain a new 3-orientation of G and thus, due to Propostion 1, a new Schnyder wood S'. Now we proceed with solving the system of linear equations for the new Schnyder wood S' and so on.

In experiments by Rucker [8] this heuristic delivered good results (for the case $r = (1, \ldots, 1)$), i.e., it always terminated after a small number of iterations. But in theory we neither know whether it terminates for every instance, nor know any nontrivial bounds for the number of iterations it takes in the case of termination. Also variants of this heuristic have been studied where the Schnyder wood is changed in some other way, but without success.

4   Preparation of the Proof of Theorem 3

In this section we will introduce some notation and present some lemmas we need for the proof of Theorem 3. Let G be a 4-connected plane triangulation for the whole section.

4.1   Feasible Aspect Ratio Vectors for a Fixed Schnyder Wood

Remember that a triangle contact representation is degenerate if corners of several primal triangles meet in a single point.

Definition 4. *Let S be a Schnyder wood. Then $\mathcal{R}_S$ is defined as the set of aspect ratio vectors of nondegenerate right triangle contact representations of G with induced Schnyder wood S, and $\bar{\mathcal{R}}_S$ as the set of aspect ratio vectors of all (possibly degenerate) right triangle contact representations of G with induced Schnyder wood S.*

In Proposition 4 we have seen that $r \in \mathcal{R}_S$ if and only if $x > 0$ where x is the solution of $A_S(r)x = \mathbf{e}_1$. With Cramer's rule and by bounding the degrees of the occurring polynomials (the determinants) we then get the following:

Lemma 3. *There are polynomials $p_1, \ldots, p_{3n+1}$ in the variables $r_1, \ldots, r_{n+3}$ with $\deg p_j \leq 3n + 1$ for each j such that*

$$\mathcal{R}_S = \{r \in \mathbb{R}_{>0}^{n+3} : p_j(r) > 0 \text{ for } j = 1, \ldots, 3n + 1\},$$
$$\bar{\mathcal{R}}_S = \{r \in \mathbb{R}_{>0}^{n+3} : p_j(r) \geq 0 \text{ for } j = 1, \ldots, 3n + 1\}.$$

In particular $\mathcal{R}_S$ is an open set and $\bar{\mathcal{R}}_S$ its closure.

The following lemma follows from Lemma 3 and shows that the intersection of $\mathcal{R}_S$ with a line segment decomposes into a bounded number of intervals.

Lemma 4. *Let $r_0, r_1 \in \mathbb{R}_{>0}^{n+3}$ be two distinct aspect ratio vectors and for each $0 \leq t \leq 1$ let $r_t := (1 - t)r_0 + tr_1$. Then there are open intervals $I_1, \ldots, I_k$ with $k \leq (3n + 1)\lfloor \frac{3n+1}{2} \rfloor + 1$ such that*

$$I_1 \cup \cdots \cup I_k = \{t \in \mathbb{R} : 0 < t < 1, r_t \in \mathcal{R}_S\},$$
$$\bar{I}_1 \cup \cdots \cup \bar{I}_k \subseteq \{t \in \mathbb{R} : 0 \leq t \leq 1, r_t \in \bar{\mathcal{R}}_S\}.$$

4.2   Neighboring Schnyder Woods

Let G be a planar triangulation. If S and S' are two Schnyder woods of G, then S' can be obtained from S by changing the orientation of the edges in some edge-disjoint directed simple cycles of S. We introduce some notation for Schnyder woods whose difference is small in this sense.

Definition 5. *We call two Schnyder woods S and S' neighboring if the corresponding 3-orientations differ in a single facial cycle C. In this case we call C the* difference cycle *of S and S'.*

The set of Schnyder woods of a fixed graph has been thoroughly studied and it is well known that it has the structure of a distributive lattice with the cover relation being exactly this neighboring relation [1].

Proposition 5. *Let S and S' be two neighboring Schnyder woods and let f be the face of G bounded by the difference cycle of S and S'. Further let r be an aspect ratio vector, and let x and x' be the solutions of $A_S(r)x = \mathbf{e_1}$ and $A_{S'}(r)x' = \mathbf{e_1}$. Then the variables x_f and x'_f corresponding to the face f have different signs, or $x_f = x'_f = 0$.*

Corollary 1. *Let S and S' be two neighboring Schnyder woods. Furthermore let $r \in \mathcal{R}_S$. Then $r \notin \bar{\mathcal{R}}_{S'}$.*

This can be seen as a weak variant of the following conjecture which is motivated by the fact that contact representations of 5-connected triangulations by homothetic squares are unique [7,11].

Conjecture 1. *Let G be a 4-connected triangulation and $\tilde{r} \in \mathbb{R}_{>0}^{n+3}$. Then the right triangle contact representation of G with aspect ratio vector $\tilde{r}$ is unique inside the three outer triangles up to scaling.*

The strategy of the proof of Theorem 3 will be to move along a line segment of aspect ratio vectors keeping the invariant that there exits a right triangle contact representation with the current aspect ratio vector. In this process, the Schnyder wood of this triangle contact representation will stay the same for a whole subsegment of this line segment. The following lemma allows us to switch to a neighboring Schnyder wood if the current one does not work any more.

Lemma 5. *Let $\{s_t = (1-t)s_0 + ts_1 : 0 \le t \le 1\}$ be a line segment of aspect ratio vectors. Let $0 < t_0 < 1$ such that $s_{t_0} \in \bar{\mathcal{R}}_S \setminus \mathcal{R}_S$, in the corresponding solution of the equation system only one face variable x_f is zero, and there is an $\varepsilon > 0$ such that for each $t_0 < t \le t_0 + \varepsilon$ we have $s_t \notin \bar{\mathcal{R}}_S$. Let S' be the neighboring Schnyder wood of S whose difference cycle is the facial cycle of the face f. Then there is an $\varepsilon' > 0$ such that $s_{t_0+\varepsilon'} \in \mathcal{R}_{S'}$.*

Proof. The matrices $A_S(s_{t_0})$ and $A_{S'}(s_{t_0})$ only differ in the column corresponding to the variable x_f. In the solution of $A_S(s_{t_0})x = \mathbf{e_1}$ we have $x_f = 0$. Therefore the solution of $A_S(s_{t_0})x = \mathbf{e_1}$ is also a solution of $A_{S'}(s_{t_0})x = \mathbf{e_1}$, or in other words the solutions of $A_S(s_{t_0})x = \mathbf{e_1}$ and $A_{S'}(s_{t_0})x = \mathbf{e_1}$ are equal.

Now let us view the solutions of $A_{S'}(s_t)x = \mathbf{e_1}$ for t slightly larger than t_0. Since for the aspect ratio vector s_{t_0} all variables except x_f are strictly positive, there is, due to continuity, an $\varepsilon'' > 0$ such that for all aspect ratio vectors s_t with $t_0 \le t \le t_0 + \varepsilon''$ all variables except x_f are strictly positive. Due to Proposition 5 we have $x_f > 0$ for all aspect ratio vectors s_t with $t_0 < t \le t_0 + \min\{\varepsilon, \varepsilon''\}$. Therefore the choice $\varepsilon' := \min\{\varepsilon, \varepsilon''\}$ fulfills $s_{t_0+\varepsilon'} \in \mathcal{R}_{S'}$. □

Since Lemma 5 can only be applied if we run into a degenerate contact representation with only one single degenerate face, we need a more general lemma for the proof of Theorem 3. By $B(m, \rho)$ and $B°(m, \rho)$ we denote the closed and open ball with center m and radius ρ.

Lemma 6. *Let $\{s_t : 0 \le t \le 1\}$ be a line segment of aspect ratio vectors. If there is a $0 < t_0 < 1$ such that $s_{t_0} \in \bar{\mathcal{R}}_S \setminus \mathcal{R}_S$ and an $\varepsilon > 0$ such that for each $t_0 - \varepsilon \le t < t_0$ we have $s_t \in \mathcal{R}_S$, then for each $\varepsilon' > 0$ there is an aspect ratio vector $r \in B(s_{t_0}, \varepsilon')$ and a neighboring Schnyder wood S' of S with $r \in \mathcal{R}_{S'}$.*

5   Proof of Theorem 3

Theorem 3. *Let G be a 4-connected triangulation and $\tilde{r} \in \mathbb{R}_{>0}^{n+3}$. Then there is a right triangle contact representation of G with aspect ratio vector $\tilde{r}$.*

Proof. We assume there is no right triangle contact representation of G with aspect ratio vector $\tilde{r}$. The idea of the proof is to construct under this assumption a line segment contradicting Lemma 4. For that we will construct an infinite sequence $(S_i)_{i \geq 0}$ of Schnyder woods, two sequences $(r_i)_{i \geq 0}$ and $(r'_i)_{i \geq 0}$ of aspect ratio vectors and two sequences $(\varepsilon_i)_{i \geq 0}$ and $(\varepsilon'_i)_{i \geq 0}$ of positive real numbers fulfilling the following invariants:

(I1) For each $r \in B(r_i, \varepsilon_i)$ there is a nondegenerate right triangle contact representation of G with aspect ratio vector r and Schnyder wood S_i.

(I2) For each $r' \in B(r'_i, \varepsilon'_i)$ the line segment $\{(1-t)r' + t\tilde{r} : 0 \leq t \leq 1\}$ intersects the balls $B(r_0, \varepsilon_0), \ldots, B(r_i, \varepsilon_i)$ in this order (with increasing t).

(I3) The points r'_i, r_i and $\tilde{r}$ are collinear and aligned in this order.

(I4) The Schnyder woods S_i and S_{i+1} are neighboring.

It now remains to show how to construct these sequences and why the existence of these sequences contradicts Lemma 4.

Construction of the sequences. Let S_0 be an arbitrary Schnyder wood of G, let $\mathcal{T}_0$ be an arbitrary nondegenerate right triangle contact representation of G with Schnyder wood S_0 (such a contact representation exists due to Proposition 2), and let r_0 be the aspect ratio vector of $\mathcal{T}_0$. Then we know from Lemma 3 that there is a $0 < \varepsilon_0 < 1$ such that for each $r \in B(r_0, \varepsilon_0)$ a nondegenerate right triangle contact representation of G with aspect ratio vector r and Schnyder wood S_0 exists. Furthermore we set $r'_0 := r_0$ and $\varepsilon'_0 := \varepsilon_0$. These initial values obviously fulfill each of the four invariants.

Now we describe how to construct the $(j + 1)$th sequence members from the jth ones. We set $s_t := (1 - t)r'_j + t\tilde{r}$ for $0 \leq t \leq 1$. Then because of (I3) there is a $0 \leq \hat{t} < 1$ with $s_{\hat{t}} = r_j$. Thus Lemma 4 gives us a $\delta > 0$ such that for each $\hat{t} \leq t < \hat{t} + \delta$ there is a nondegenerate right triangle contact representation of G with aspect ratio vector s_t and Schnyder wood S_j, and a degenerate one with aspect ratio vector $s_{\hat{t}+\delta}$ and Schnyder wood S_j. Because of our assumption we have $\hat{t} + \delta < 1$ and because of $s_{\hat{t}+\delta} \notin B(r_j, \varepsilon_j)$ we have $\|r_j - s_{\hat{t}+\delta}\| > \varepsilon_j$. Now we set

$$\delta' := \min \left\{ \left(1 - (\hat{t} + \delta)\right) \varepsilon'_j, \|r_j - s_{\hat{t}+\delta}\| - \varepsilon_j \right\} > 0.$$

Then Lemma 6 gives us an $r_{j+1} \in B^\circ(s_{\hat{t}+\delta}, \delta')$ such that there is a nondegenerate right triangle contact representation of G with aspect ratio vector r_{j+1} and a neighboring Schnyder wood S_{j+1} of S_j. Now we set $r'_{j+1} := r'_j + \frac{1}{1-(\hat{t}+\delta)}(r_{j+1} - s_{\hat{t}+\delta})$. Then

$$\|r'_{j+1} - r'_j\| = \frac{1}{1 - (\hat{t} + \delta)} \|r_{j+1} - s_{\hat{t}+\delta}\| < \frac{1}{1 - (\hat{t} + \delta)} \delta' \leq \varepsilon'_j$$

and therefore we have $r'_{j+1} \in B^\circ(r'_j, \varepsilon'_j)$. Moreover Lemma 3 gives us an $0 < \varepsilon_{j+1} < \delta' - \|r_{j+1} - s_{\hat{t}+\delta}\|$ such that for each $r \in B(r_{j+1}, \varepsilon_{j+1})$ there is a nondegenerate right triangle contact representation of G with aspect ratio vector r and Schnyder wood S_{j+1}. Finally we set $\varepsilon'_{j+1} := \frac{1}{1-(\hat{t}+\delta)}\varepsilon_{j+1}$. See Fig. 4 for an illustration of the construction.

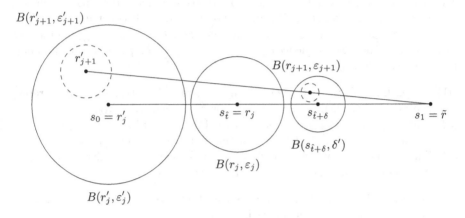

Fig. 4. The construction of the new sequence members.

Clearly the invariants (I1), (I3) and (I4) are fulfilled again. Because of $B(r'_{j+1}, \varepsilon'_{j+1}) \subseteq B(r'_j, \varepsilon'_j)$ for each $r' \in B(r'_{j+1}, \varepsilon'_{j+1})$ the line segment $\{(1-t)r' + t\tilde{r} : 0 \le t \le 1\}$ intersects the balls $B(r_0, \varepsilon_0), \ldots, B(r_j, \varepsilon_j)$ in the right order. From the construction it immediately follows that the ball $B(r_{j+1}, \varepsilon_{j+1})$ is intersected by this line segment, too. Moreover, because of $\delta' \le \|r_j - s_{\hat{t}+\delta}\| - \varepsilon_j$ and $B(r_{j+1}, \varepsilon_{j+1}) \subseteq B^\circ(s_{\hat{t}+\delta}, \delta')$ the intersection point with $B(r_{j+1}, \varepsilon_{j+1})$ is closer to $\tilde{r}$ than the intersection point with $B(r_j, \varepsilon_j)$. Therefore also (I2) is fulfilled again.

Producing a contradiction. Let L be the number of Schnyder woods of G and $c := (3n + 1)\lfloor\frac{3n+1}{2}\rfloor + 1$ (this is the bound from Lemma 4). We set $K := Lc + 1$. Then it follows from the pigeonhole principle that there is a Schnyder wood S such that there are indices $0 \le i_1 < \cdots < i_{c+1} \le K$ with $S = S_{i_1} = \cdots = S_{i_{c+1}}$. For $l = 1, \ldots, c+1$ let $\hat{r}_l$ be an intersection point of the line segment $\{(1-t)r'_K + t\tilde{r} : 0 \le t \le 1\}$ and the ball $B(r_{i_l}, \varepsilon_{i_l})$. Thus for $l = 1, \ldots, c+1$ there is a nondegenerate right triangle contact representation of G with aspect ratio vector $\hat{r}_l$ and Schnyder wood S. From Lemma 4 it follows that the intersection of $\{(1-t)r'_K + t\tilde{r} : 0 \le t \le 1\}$ and $\mathcal{R}_S$ is a disjoint union of at most c open intervals. Therefore there is an l such that $\hat{r}_l$ and $\hat{r}_{l+1}$ belong to the same interval. Particularly for each $0 \le \tau \le 1$ there is a nondegenerate right triangle contact representation of G with aspect ratio vector $(1 - \tau)\hat{r}_l + \tau\hat{r}_{l+1}$ and Schnyder wood S. Because of (I4) the Schnyder woods $S' := S_{i_l+1}$ and S are neighboring. Moreover, because of (I2) there is a $0 \leq \tau' \leq 1$ with $(1 - \tau')\hat{r}_l + \tau'\hat{r}_{l+1} \in B(r_{i_l+1}, \varepsilon_{i_l+1})$. But then because of (I1) there is also a nondegenerate right triangle contact representation of G with aspect ratio vector $(1 - \tau')\hat{r}_l + \tau'\hat{r}_{l+1}$ and Schnyder wood S', contradicting Corollary 1. □

6   A New Heuristic

In this section we will present an new heuristic for computing a right triangle contact representation of a given planar triangulation G with a given aspect ratio vector $\tilde{r}$ that is based on our proof of Theorem 3. The idea of the heuristic is to make progress on a line segment $\{r_t = (1 - t)r_0 + t\tilde{r} : 0 \leq t \leq 1\}$ of aspect ratio vectors. By that we mean that in each iteration the largest t increases for that we know a Schnyder wood S with $s_t \in \mathcal{R}_S$.

We introduce some notation. For a Schnyder wood S and an aspect ratio vector r we denote by $x(S, r)$ the solution of $A_S(r)x = \mathbf{e}_1$. Further, by $S(r)$ we denote the Schnyder wood obtained from S by changing the orientation of the sign-separating edges regarding the solution of $A_S(r)x = \mathbf{e}_1$.

Algorithm 1. Calculation of a right triangle contact representation

---

Input: a 4-connected triangulation G and an aspect ratio vector $\tilde{r} \in \mathbb{R}_{>0}^{n+3}$
Output: a right triangle contact representation of G with aspect ratio vector $\tilde{r}$

$S \leftarrow$ arbitrary Schnyder Wood of G
$T_0 \leftarrow$ arbitrary right triangle contact representation of G with Schnyder wood S
$r_0 \leftarrow$ aspect ratio vector of T_0
while $x(S, \tilde{r}) \not\geq 0$ **do**
    $r_1 \leftarrow \tilde{r}$
    $r_m \leftarrow \frac{r_0 + r_1}{2}$
    while $S(r_m) = S$ **or** $x(S(r_m), r_m) \not\geq 0$ **do**
        if $S(r_m) = S$ **then**
            $r_0 \leftarrow r_m$
        else
            $r_1 \leftarrow r_m$
        end if
        $r_m \leftarrow \frac{r_0 + r_1}{2}$
    end while
    $S \leftarrow S(r_m)$
    $r_0 \leftarrow r_m$
end while
calculate from $x(S, \tilde{r})$ a right triangle contact representation T of G
return T

If we assume that on the line segment we never run into an aspect ratio vector such that in the corresponding contact representation more than one face is degenerate, we can deduce the following from Lemma 5: If we know a

Schnyder wood S with $r_t \in \mathcal{R}_S$, then either $\tilde{r} \in \mathcal{R}_S$ or there is a $t < t' < 1$ such that $S(r_{t'}) \neq S$ and $r_{t'} \in \mathcal{R}_{S(r_{t'})}$ (we could even assume that S and $S(r_{t'})$ are neighboring). This is exactly the idea we realize in Algorithm 1.

We cannot be sure that the inner loop always terminates because we cannot apply Lemma 5 if there are aspect ratio vectors on the line segment such that in the corresponding contact representation more than one face is degenerate. But if the inner loop always terminates, the outer loop terminates after $\mathcal{O}(n^2 L)$ iterations where L is the number of Schnyder woods of G (see Sect. 5). Since the number of Schnyder woods can be exponential in n [4], this yields an exponential running time.

We will conclude by stating some conjectures concerning the computation of triangle contact representations. The strong experimental results we mentioned in the end of Sect. 3, give rise to the following conjecture:

Conjecture 2. The variant of the heuristic by Felsner we described in the end of Sect. 3 terminates for every planar triangulation G, every aspect ratio vector $\tilde{r}$, and every Schnyder wood S of G to start with.

Since the number of iterations has always been small in the experiments, we conjecture that it can be bounded by a polynomial in n. This would yield an algorithm with polynomial running time. Therefore we also conjecture the following:

Conjecture 3. A right triangle contact representation of a planar triangulation G with given aspect ratio vector $\tilde{r}$ can be computed in polynomial time.

References

1. Felsner, S.: Lattice structures from planar graphs. Electron. J. Comb. **11**(1), R15 (2004)
2. Felsner, S.: Triangle contact representations. In: Midsummer Combinatorial Workshop, Praha (2009). http://page.math.tu-berlin.de/~felsner/Paper/prag-report.pdf
3. Felsner, S., Francis, M.C.: Contact representations of planar graphs with cubes. In: Proceedings of SoCG 2011, pp. 315–320. ACM (2011)
4. Felsner, S., Zickfeld, F.: On the number of planar orientations with prescribed degrees. Electron. J. Comb. **15**(1), R77 (2008)
5. Gonçalves, D., Lévêque, B., Pinlou, A.: Triangle contact representations and duality. Graph Drawing, 262–273 (2011)
6. Gonçalves, D., Lévêque, B., Pinlou, A.: Triangle contact representations and duality. Discrete Comput. Geom. **48**(1), 239–254 (2012)
7. Lovász, L.: Geometric representations of graphs (2009). http://www.cs.elte.hu/~lovasz/geomrep.pdf
8. Rucker, J.: Kontaktdarstellungen von planaren Graphen. Diplomarbeit. Technische Universität Berlin (2011). http://page.math.tu-berlin.de/~felsner/Diplomarbeiten/dipl-Rucker.pdf
9. Schnyder, W.: Embedding planar graphs on the grid. In: Proceedings of SODA, pp. 138–148 (1990)

10. Schramm, O.: Combinatorically prescribed packings and applications to conformal and quasiconformal maps. Modified version of Ph.D. thesis from 1990. https://arxiv.org/abs/0709.0710v1

11. Schramm, O.: Square tilings with prescribed combinatorics. Isr. J. Math. **84**(1–2), 97–118 (1993)

12. Schrezenmaier, H.: Zur Berechnung von Kontaktdarstellungen. Masterarbeit. Technische Universität Berlin (2016). http://page.math.tu-berlin.de/~schrezen/Papers/Masterarbeit.pdf

13. de Fraysseix, H., Ossona de Mendez, P.: On topological aspects of orientations. Discrete Math. **229**(1), 57–72 (2001)

14. de Fraysseix, H., Ossona de Mendez, P., Rosenstiehl, P.: On triangle contact graphs. Comb. Probab. Comput. **3**, 233–246 (1994)

Author Index

Printed in the United States
By Bookmasters